Maurice ALLAIS



# The Anisotropy of Space

## The necessary revision of certain postulates of contemporary theories

Original edition:
Maurice ALLAIS. L'Anisotropie de l'Espace. Clément Juglar, 1997.

English translation and page layout by:
Thomas J. GOODEY
René VERREAULT
Arjen DIJKSMAN

This translation was made possible thanks to proofreading by Christine Allais, Jean-Bernard Deloly, Christian Gomez and Christian Marchal.

Funding of the translation was provided by the Maurice Allais Foundation under the aegis of ParisTech Foundation.





**5-7, rue de l'École-Polytechnique – 75005 Paris**

www.editions-harmattan.fr

ISBN : 978-2-343-16870-8
EAN : 9782343168708

Maurice ALLAIS

# The Anisotropy of Space

## The necessary revision of certain postulates of contemporary theories

L'Harmattan

This book is dedicated:

- To all those who do not consider that today's "well-established truths" are untouchable and who do not have blind faith in their immutability;

- And to all those who believe that all real progress in our knowledge can only be based upon data from experiment.

*BOOK PRESENTATION OF THE 1997 FRENCH EDITION*

Maurice Allais, author of a physics book ? This may seem surprising to anyone who is already impressed by the work of the 1988 Nobel Prize in Economics.

And yet Maurice Allais' passion for physics is older than his passion for economics. He declared having devoted a quarter of his time to physics. The results of his work and his experience in this field are entirely original. "*In my experimental and theoretical research (from 1954 to 1996), I demonstrated very significant anomalies, on the one hand in the movement of the paraconical pendulum with anisotropic support and with isotropic support, and on the other hand in optical sightings at marks. I proved their existence, independently of any perverse effect. These anomalies are totally inexplicable in the framework of currently accepted theories*".

This work is based on new experimental data in four fields considered *a priori* as very different, although in close relation with each other: - observations on the paraconical pendulum with anisotropic support and with isotropic support; - observations on the optical deviations of sightings at marks and at collimators; - the regularities characterizing the optical observations of Esclangon and not perceived by him; - the regularities characterizing the interferometric observations of Dayton C. Miller and not perceived by him.

The questioning of modern theories, both of classical theories and of the theory of relativity, has provoked, as could be suspected, fierce opposition. But in physics as in economics, Maurice Allais has an absolute principle: "*All real progress in our knowledge can only be based upon data from experiment*". Facts are the keystone of any theoretical construction.

There is no doubt that this work of Maurice Allais, extraordinarily clear, rigorous and structured, very widely accessible despite the complexity of the subject, will be part of the history of science. Whatever his training, the reader will find in it powerful subjects for reflection.

*"The scientist should be a man willing to listen to every suggestion, but determined to judge for himself. He should not be biased by appearances; have no favorite hypothesis; be of no school; in doctrine have no master. He should not be a respecter of persons but of things. Truth should be his primary object."*

Michael Faraday *

*"Experience teaches us that the opinion of "competent" men is often completely in disagreement with reality, and the history of science is the history of the mistakes of "competent" men."*

Vilfredo Pareto **

*"The physicist who has just given up one of his hypotheses should, on the contrary, rejoice, for he found an unexpected opportunity of discovery. His hypothesis, I imagine, had not been lightly adopted: it took into account all the known factors which seem capable of intervention in the phenomenon. If it is not verified, it is because there is something unexpected and extraordinary about it, because we are on the point of finding something unknown and new."*

Henri Poincaré ***

*"The history of the sciences shows us that the progress of Science has constantly been hampered by the tyrannical influence of certain ideas that have come to be considered as dogmas. For this reason, the principles that we have come to accept without further discussion should periodically be subjected to quite thorough examination."*

Louis de Broglie ****

*"This ostracism of innovators is not at all the exception; very few innovators escape it, and we can, without hesitation, formulate the general rule that every scientist who discovers a principle that departs from classical conformity meets with the impossibility of getting his ideas accepted, whatever may be the rigor of the arguments with which their correctness is formally demonstrated...*

*"The unjust fate of innovators, the ignorance and the oblivion of their work, the negative judgments pronounced against them, even the persecutions that they suffer, are the rule; many scientists and philosophers have mentioned and deplored these facts, but nobody seems yet to have thought how to react against this state of things...*

*"Conformists holding the highest social positions continue, as in the past, to fight against or to suffocate all discoveries that do not accord with their prejudices and with the dogmas applicable in the classical treatises...*

*"We are personally well placed to know this. What is the use of encouraging scientific research, if the fruits of such investigations are destined to be buried and if their authors are condemned in advance to being forgotten or even persecuted?"*

Auguste Lumière *****

---

* Michael Faraday, cited by H.C. Dudley, 1959, *New Principles in Quantum Mechanics*, Exposition Press, New York, p. 6 [113].

** Vilfredo Pareto, 1917, *Traité de Sociologie Générale* (Treatise on General Sociology), Payot, p. 320 [212].

*** Henri Poincaré, 1905, *Science and Hypothesis*, The Walter Scott Publ.Co. Ltd, N.Y., p. 168 [220].

**** Louis de Broglie, 1953, *La Physique Quantique restera-t-elle Indéterministe?* (Will Quantum Mechanics Remain Indeterministic?), Gauthier-Villars, p. 22 [86].

***** Auguste Lumière, 1942, *Les Fossoyeurs du Progrès, Les Mandarins contre les Pionniers de la Science* (The Gravediggers of Progress, The Mandarins against the Pioneers of Science), Léon Sézanne, Lyon, pp. XIII and XVI [188].

Maurice Allais (31 May 1911 – 9 October 2010). Engineer General in the Mines National Corps, professor at the École Nationale des Mines de Paris (1944-1988) and at the Université de Paris X – Nanterre (1970-1985), Maurice Allais was director of the Centre of Economic Analysis (CNRS, École Nationale des Mines de Paris, 1946-2010). Awardee of the CNRS Gold Medal in 1978 and of numerous French and foreign distinctions for an impressive work, Member of the Institute, he was the winner of the Prize in Economic Sciences in Memory of Alfred Nobel in 1988.

# Contents


**INTRODUCTION** .......... 35

A THE MOTIVATION FOR THIS BOOK .......... 37

A.1 New experimental data, contributions to the debate upon the fundamentals of Mechanics and Optics .......... 37

A.2 The objects of this work .......... 41

B MY EXPERIMENTAL AND THEORETICAL RESEARCH .......... 43

B.1 The origin of my experimental research .......... 43

B.2 My experimental researches upon the paraconical pendulum throughout 1954-1960, and upon optical deviations of sightings at marks in 1958 .......... 47

B.3 My theoretical researches, 1950-1996 .......... 54

C THE SUBJECT IN QUESTION .......... 56

C.1 Connections between the observations of the paraconical pendulum and the optical deviations of sightings at marks with the optical experiments of Ernest Esclangon in 1927 1928 and the interferometric experiments of Dayton C. Miller in 1925-1926 .......... 56

C.2 Two fundamental and authentic discoveries .......... 60

D SUPPORTS AND OPPOSITIONS .......... 64

D.1 Exceptional supports .......... 64

D.2 The interest excited by my experiments .......... 67

D.3 Fierce oppositions .......... 69

E TO THE READER .......... 73

E.1 The production of this work .......... 73

E.2 Against the current .......... 76

I **MY EXPERIMENTS ON THE PARACONICAL PENDULUM WITH ANISOTROPIC SUPPORT - 1954-1960** .......... 79
A GENERAL CHARACTERISTICS OF THE EXPERIMENTS ON THE PARACONICAL PENDULUM WITH ANISOTROPIC SUPPORT, AND THEIR RESULTS .......... 81
A.1 Experimental conditions .......... 81
A.2 Enchained experiments - Illustration for the case of the month-long series of June-July 1955 .......... 87
A.3 The experiments performed in 1954-1960 .......... 90
A.4 The Foucault Effect .......... 93
A.5 The diurnal periodic structure of the enchained experiments on the paraconical pendulum with anisotropic support .......... 96
B THREE FUNDAMENTAL QUESTIONS CONCERNING THE PARACONICAL PENDULUM WITH ANISOTROPIC SUPPORT 102
B.1 The reality of the determined periodicities .......... 103
B.2 The observed lunisolar effects and the current theory of gravitation .......... 118
B.3 Periodic components not explicable in terms of known phenomena .......... 130
C THE CRUCIAL EXPERIMENTS OF JUNE-JULY 1958 IN SAINT-GERMAIN AND BOUGIVAL .......... 142
C.1 The two laboratories at Saint-Germain and Bougival .......... 142
C.2 The periodic diurnal structure of the two series of enchained experiments of Bougival and Saint-Germain .......... 144
C.3 Monthly sidereal lunar periodicity .......... 157
C.4 Overall view .......... 160
D DEVIATIONS OBSERVED DURING TWO TOTAL SOLAR ECLIPSES .......... 162
D.1 The total solar eclipse of 30 June 1954 .......... 162
D.2 The total solar eclipse of 2 October 1959 .......... 166
D.3 Comparison of the observed perturbations during the two solar eclipses of 30 June 1954 and 2 October 1959 .......... 168
E ESSENTIAL FACTORS IN THE MOVEMENT OF THE PARACONICAL PENDULUM WITH ANISOTROPIC SUPPORT 171
E.1 An extremely complex movement .......... 171
E.2 Foucault Effect .......... 173
E.3 Anisotropy of the support .......... 176
E.4 Influence of the balls .......... 183
E.5 Periodic astronomical influences .......... 184
E.6 Other factors .......... 188
E.7 Existence of a limit plane, variable with time, in the enchained series of the paraconical pendulum with anisotropic support .......... 193
E.8 Overall view .......... 196

F THE HYPOTHESIS OF ANISOTROPY OF INERTIAL SPACE . . . 197
F.1 Theory and experiment . . . . . . . . . . . . . . . . . . . . . . . . . . . . . . 197
F.2 Accuracy of the verification of the laws of gravitation . . . . . . 201
F.3 The hypothesis of anisotropy of inertial space, and its implications . . . . . . . . . . . . . . . . . . . . . . . . . . . . . . . . . . . . . . . . . 206
G OVERALL VIEW OF MY EXPERIMENTS ON THE PARACONICAL PENDULUM WITH ANISOTROPIC SUPPORT 1954-1960 . . . . . . . . . . . . . . . . . . . . . . . . . . . . . . . . . . . . . . . 213
G.1 Significance and implications of the experiments on the paraconical pendulum with anisotropic support . . . . . . . . . . . 213
G.2 The publication of my work . . . . . . . . . . . . . . . . . . . . . . . . . . 215
G.3 Summary chronology of the experiments performed under the aegis of IRSID. 1954-1960 . . . . . . . . . . . . . . . . . . . . . . . . . 217
G.4 The cost of the experiments . . . . . . . . . . . . . . . . . . . . . . . . . . . 221
G.5 Dogmatic oppositions . . . . . . . . . . . . . . . . . . . . . . . . . . . . . . . 226
G.6 A scientifically incomprehensible decision. The total cessation of my experiments after the full success of the crucial experiments of July 1958 . . . . . . . . . . . . . . . . . . . . . . . 232

**II MY EXPERIMENTS UPON THE PARACONICAL PENDULUM WITH ISOTROPIC SUPPORT 1959-1960** . . . . . . . . . . . . . . . . . . . . . . . 237
A THE DETERMINATION OF THE DIRECTION OF ANISOTROPY OF SPACE . . . . . . . . . . . . . . . . . . . . . . . . . . . . . . . . . . 238
A.1 My experiments upon the paraconical pendulum with the anisotropic support . . . . . . . . . . . . . . . . . . . . . . . . . . . . . . . . . 238
A.2 Determination of the direction of anisotropy of space. A fundamental step . . . . . . . . . . . . . . . . . . . . . . . . . . . . . . . . . . . . 239
B THE IMPLEMENTATION OF AN ISOTROPIC SUPPORT . . . . . . . 241
B.1 Experimental arrangement . . . . . . . . . . . . . . . . . . . . . . . . . . . . 241
B.2 A support having practically no anisotropy . . . . . . . . . . . . . . . 243
C DETERMINATION OF THE AZIMUTH *X* OF ANISOTROPY OF SPACE
METHOD OF ANALYSIS . . . . . . . . . . . . . . . . . . . . . . . . . . . . . . . . 247
C.1 Two possible methods of analysis: enchained series and mobile correlations . . . . . . . . . . . . . . . . . . . . . . . . . . . . . . . . . . 247
C.2 Principle of calculation by the method of mobile correlations 249
C.3 Illustration of the method of mobile correlations . . . . . . . . . . 250
C.4 The method of mobile correlations and its difficulties in practical application . . . . . . . . . . . . . . . . . . . . . . . . . . . . . . . . . 253
D THE VARIATIONS OF THE AZIMUTH OF THE ANISOTROPY OF SPACE NOVEMBER-DECEMBER 1959 AND MARCH-APRIL 1960 . . . . . . . . . . . . . . . . . . . . . . . . . . . . . . . 255
D.1 The series of observations of November-December 1959 and March-April 1960 . . . . . . . . . . . . . . . . . . . . . . . . . . . . . . . 255
D.2 The average anisotropy of space . . . . . . . . . . . . . . . . . . . . . . . . 256
D.3 Determination of the variation over time of the direction of anisotropy of space X during the two periods of November-December 1959 and March-April 1960 . . . . . . . . 259
D.4 Variations over time of $\cos \mathbf{2X}$ corresponding to the direction X of anisotropy of space in November-December 1959 and March-April 1960 . . . . . . . . . . . . . . . . . . . . . . . . . . 263
E THE PERIODIC DIURNAL AND MONTHLY STRUCTURE OF THE DIRECTION OF ANISOTROPY OF SPACE NOVEMBER-DECEMBER 1959 AND MARCH-APRIL 1960 . . . . 269
E.1 Diurnal periodicities of the azimuth X of anisotropy of space 269
E.2 Sidereal monthly lunar periodicity of the azimuth X of anisotropy of space . . . . . . . . . . . . . . . . . . . . . . . . . . . . . . . . . . 273
E.3 Sidereal monthly lunar periodicity of the cumulative values of the differences $(\cos \mathbf{2X} - \overline{\cos \mathbf{2X}})$ . . . . . . . . . . . . . . . 277
F PERIODIC DIURNAL AND MONTHLY STRUCTURE OF THE DISPLACEMENTS OF THE PLANE OF OSCILLATION OF THE PARACONICAL PENDULUM WITH ISOTROPIC SUPPORT FROM THE DIRECTIONS NORTH-SOUTH AND EAST-WEST . . . . . . . . . . . . . . . . . . . . . . . . . . . . . . . . . . . . . . . . . . . 280

F.1 Deviations $\Delta$ of the plane of oscillation of the paraconical pendulum from the meridian . . . 280
F.2 Periodic diurnal components of the deviations of the plane of oscillation of the paraconical pendulum from the meridian 284
F.3 Sidereal monthly lunar periodicity of the deviations $\Delta$ from the meridian of the plane of oscillation of the paraconical pendulum . . . 288
F.4 Structural correspondence between the directions X of anisotropy of space and the deviations $\Delta$ from the meridian of the plane of oscillation of the paraconical pendulum . . . 292
F.5 Cumulative sums of the differences $\Delta - \overline{\Delta}$ between the deviations $\Delta$ and their averages $\overline{\Delta}$ in the azimuths North-South and East-West . . . 295
F.6 Structural similarities between the cumulative values of the $\Delta - \overline{\Delta}$ and the cumulative values of the $\cos \mathbf{2X} - \overline{\cos \mathbf{2X}}$ . 304
G PHASES OF THE SIDEREAL LUNAR MONTHLY PERIODIC COMPONENTS OF 27.322 DAYS OF THE CHARACTERISTICS OF THE MOVEMENT OF THE PARACONICAL PENDULUM AND THE DECLINATION OF THE MOON . . . 306
G.1 Paraconical pendulum with anisotropic support, directions X of anisotropy of space, and deviations $\Delta$ from the directions North-South and East-West of the plane of oscillation of the paraconical pendulum with isotropic support. Phases of the sidereal lunar periodicities of 27.322 days compared together . . . 306
G.2 Deviations $\Delta$ of the plane of oscillation of the paraconical pendulum with isotropic support from the directions North-South and East-West and the cumulative values $\sum(\Delta - \overline{\Delta})$. Phases of the sidereal monthly lunar periodicities of 27.322 days compared . . . 309
G.3 Directions X of anisotropy of space and cumulative values $\sum(\cos \mathbf{2X} - \overline{\cos \mathbf{2X}})$. Phases of sidereal monthly lunar periodicities of 27.322 days compared . . . 312
G.4 Overall view . . . 314
H ENCHAINED EXPERIMENTS FROM 28 SEPTEMBER TO 4 OCTOBER 1959 AND TOTAL SOLAR ECLIPSE OF 2 OCTOBER 1959 . . . 315
H.1 Variations of azimuth $\phi$ of plane of oscillation of paraconical pendulum with isotropic support during the enchained experiments of September-October 1959 . . . 315

H.2 Variations during the enchained experiments of September-October 1959 of $\cos 2\phi$ and $\sin 2\phi$, corresponding to the azimuth $\phi$ of the paraconical pendulum with isotropic support. . . . . . . . . . . . . . . . . . . . . . . 317
I MOVEMENT OF THE PLANE OF OSCILLATION OF THE PARACONICAL PENDULUM WITH ISOTROPIC SUPPORT UNDER THE HYPOTHESIS OF ANISOTROPY OF INERTIAL SPACE . . . . . . . . . . . . . . . . . . . . . . . . . . . . . . . . . . . . . . . . . . . . . . 320
I.1 Equations of movement and calculation of the coefficients of anisotropy . . . . . . . . . . . . . . . . . . . . . . . . . . . . . . . . . . . . . . 320
I.2 The second order differential equation in $\mathbf{u} = \mathbf{X} - \phi$ . . . . . . . 323
J OVERALL VIEW . . . . . . . . . . . . . . . . . . . . . . . . . . . . . . . . . . . . . . . 326
J.1 Determination of the direction $X$ of anisotropy of space (Sections A, B, and C) . . . . . . . . . . . . . . . . . . . . . . . . . . . . . . 326
J.2 Azimuth $X$ of anisotropy of space in November-December 1959 and March-April 1960 (Sections D and E) . . . . . . . . . . . 327
J.3 Displacements $\Delta$ of the plane of oscillation of the paraconical pendulum from the directions North-South and East-West (Section F) . . . . . . . . . . . . . . . . . . . . . . . . . . . . 328
J.4 Remarkable agreements in phase (Section G) . . . . . . . . . . . . . 329
J.5 Enchained series for the paraconical pendulum with isotropic support (Section H) . . . . . . . . . . . . . . . . . . . . . . . . . . 329
J.6 Anisotropy of inertial space (Section I) . . . . . . . . . . . . . . . . . . . 330
J.7 Very significant results . . . . . . . . . . . . . . . . . . . . . . . . . . . . . . . . . 330

**III MY EXPERIMENTS OF JULY 1958 ON OPTICAL DEVIATIONS OF SIGHTINGS AT MARKS, AND THEIR CONTINUATION DURING FEBRUARY-MARCH 1959** . . . . . . . . . . 331

A EXPERIMENTS ON OPTICAL DEVIATIONS OF SIGHTINGS AT MARKS AND AT COLLIMATORS, IRSID, 1958, AND IGN, 1959 . . . . . . . . . . 332

B MY EXPERIMENTS ON OPTICAL DEVIATIONS OF SIGHTINGS AT MARKS, IRSID, JULY 1958 . . . . . . . . . . 334

B.1 Experimental arrangements for the optical observations at IRSID in July 1958 . . . . . . . . . . 334

B.2 Harmonic analysis of the diurnal deviations of the sightings at marks . . . . . . . . . . 335

B.3 Correspondence between the optical deviations of sightings at marks and the azimuths of the paraconical pendulum in July 1958 . . . . . . . . . . 338

B.4 The refusal by the Academy of Sciences to publish my Note of 23 February 1960 on the optical observations at IRSID in July 1958 . . . . . . . . . . 339

B.5 Some supplementary comments on the optical observations of July 1958 at IRSID . . . . . . . . . . 341

B.6 Scientific implications of the experiments performed at IRSID in July 1958 . . . . . . . . . . 345

C EXPERIMENTS ON OPTICAL DEVIATIONS OF SIGHTINGS AT MARKS AND AT COLLIMATORS, INSTITUT GEOGRAPHIQUE NATIONAL, FEBRUARY - MARCH 1959 . . . . 346

C.1 The optical experiments on sightings at marks and at collimators performed during February-March 1959 at IGN . 346

C.2 Diurnal variations of the optical sightings at marks and at collimators at IGN during February-March 1959 . . . . . . . . . . 348

C.3 Monthly sidereal lunar periodicities of sightings at marks and at collimators at IGN in February-March 1959 . . . . . . . . 354

C.4 Scientific implications of the experiments performed at IGN in February-March 1959 . . . . . . . . . . 364

D THE PHASES OF THE PERIODIC COMPONENTS ACCORDING TO THE LUNAR SIDEREAL MONTH OF 27.322 DAYS OF THE DEVIATIONS OF OPTICAL SIGHTINGS AT MARKS AND AT COLLIMATORS OF JULY 1958 AT IRSID AND OF FEBRUARY-MARCH 1959 AT IGN . . . . 366

E OVERALL VIEW OF THE EXPERIMENTS ON OPTICAL SIGHTINGS IN JULY 1958 AT IRSID AND IN FEBRUARY-MARCH 1959 AT IGN . . . . . . . . . . 371

**IV TWO VERY SIGNIFICANT PREVIOUS EXPERIMENTS** ........ 373
A THE OBSERVATIONS OF ESCLANGON AND MILLER ........ 374
B THE OPTICAL OBSERVATIONS OF ERNEST ESCLANGON, 1927-1928 ........ 376
B.1 The researches of Esclangon on the anisotropy of space, 1926-1928 ........ 376
B.2 Analysis of the optical observations of Esclangon in 1927-1928 ........ 378
C THE INTERFEROMETRIC OBSERVATIONS OF DAYTON C. MILLER, 1925-1926 ........ 382
C.1 The experiments and analyses of Dayton C. Miller, 1925-1933 ........ 382
C.2 Miller's experimental procedure ........ 385
C.3 The observations of Miller ........ 386
C.4 Three fundamental questions ........ 391
D THE GREAT UNDERLYING COHERENCE OF MILLER'S OBSERVATIONS ........ 392
D.1 The coherence of the observations in sidereal time ........ 392
D.2 The correspondence between the graphs of the speeds and of the azimuths ........ 394
D.3 The diurnal fittings of the daily values of speeds and azimuths ........ 395
D.4 The hodographs of the velocities ........ 400
D.5 The semi-annual and annual periodic structure of Miller's observations ........ 409
D.6 An undeniable and very great underlying coherence in Miller's observations ........ 410
E CRITICISMS AIMED AT MILLER'S EXPERIMENTS ........ 412
E.1 Miller's results were due to the combined effect of random perturbations and temperature perturbations ........ 412
E.2 The negative results of other interferometric experiments demonstrate that Miller's experiments were invalid ........ 414
E.3 Totally unfounded criticisms ........ 416
F THE INTERPRETATION OF MILLER'S OBSERVATIONS ........ 417
F.1 The interpretation given by Miller to his observations ........ 417
F.2 The interpretation that corresponds to the observational data: an anisotropy of optical space ........ 420
G SIGNIFICANCE AND CONSEQUENCES OF MILLER'S OBSERVATIONS ........ 425
H OVERALL VIEW OF THE OBSERVATIONS OF ESCLANGON AND MILLER ........ 427

V PERIODIC SEMI-ANNUAL AND ANNUAL STRUCTURE OF THE OBSERVATIONS OF THE PARACONICAL PENDULUM, OF THE SIGHTINGS AT MARKS, OF THE OBSERVATIONS OF ESCLANGON AND OF THE INTERFEROMETRIC OBSERVATIONS OF MILLER ..... 429

A THE OBSERVATIONS OF THE PARACONICAL PENDULUM, THE SIGHTINGS AT MARKS, THE OPTICAL OBSERVATIONS OF ESCLANGON, THE INTERFEROMETRIC OBSERVATIONS OF MILLER AND THE MOVEMENT OF THE EARTH ALONG ITS ORBIT ..... 430

B SEMI-ANNUAL AND LONG DURATION PERIODICITIES OF THE OBSERVATIONS OF THE PARACONICAL PENDULUM WITH ANISOTROPIC SUPPORT ..... 432

B.1 Semi-annual periodicity of the average monthly azimuths of the paraconical pendulum with anisotropic support ..... 432

B.2 Semi-annual periodicity of the amplitude of the diurnal component of 24h50m of the paraconical pendulum with anisotropic support ..... 438

B.3 Similarities of the semi-annual and long-term fittings of the average azimuths and the amplitudes of the 25h component ..... 444

B.4 Overall view of the semi-annual and long-term periodicities of the paraconical pendulum with anisotropic support ..... 446

C SEMI-ANNUAL PERIODICITIES OF THE OPTICAL SIGHTINGS AT MARKS AND OF THE OPTICAL SIGHTINGS OF ESCLANGON ..... 447

C.1 Semi-annual periodicity of the optical sightings at marks ... 447

C.2 Semi-annual periodicity of the optical observations of Esclangon ..... 450

D SEMI-ANNUAL AND ANNUAL PERIODICITIES OF THE INTERFEROMETRIC OBSERVATIONS OF MILLER ..... 452

D.1 Determination of the position of the Earth upon its orbit from Miller's observations, independently of any hypothesis 452

D.2 Direct estimation of the characteristic parameters of the eight fundamental graphs of Miller. ..... 453

D.3 Semi-annual and annual periodic structure of the characteristic parameters of the fundamental Graphs of Miller ..... 455

D.4 A double periodicity ..... 462

D.5 Structural interdependences between the speed and azimuth observations of Miller. ..... 465

D.6 Essential aspects of the semi-annual and annual structure of Miller's observations ..... 467

E COMPARED SEMI-ANNUAL AND ANNUAL PERIODIC STRUCTURES FOR THE OBSERVATIONS OF THE PARACONICAL PENDULUM WITH ANISOTROPIC SUPPORT, FOR THE SIGHTINGS AT MARKS, FOR THE OPTICAL OBSERVATIONS OF ESCLANGON, AND FOR THE INTERFEROMETRIC OBSERVATIONS OF MILLER ...... 469
E.1 Phases of the semi-annual periodic components of the four series of observations compared ........................ 469
E.2 A double periodicity centered upon 21 March ............ 471
E.3 A simplified global approach: consideration of only dominant periodicities of the four series of observations, with their maxima or minima on 21 March ............... 480
E.4 A triple conclusion ................................... 483
F OVERALL VIEW OF THE DIURNAL, MONTHLY, SEMI-ANNUAL, ANNUAL, AND LONG DURATION PERIODIC STRUCTURE OF THE OBSERVATIONS OF THE PARACONICAL PENDULUM WITH ANISOTROPIC SUPPORT, OF THE SIGHTINGS AT MARKS, OF THE OPTICAL SIGHTINGS OF ESCLANGON, AND OF THE INTERFEROMETRIC OBSERVATIONS OF MILLER .......... 484
F.1 Similarities between the four series of experiments ........ 485
F.2 Coherence of the periodic structures of the four series of observations ........................................ 487
F.3 Inescapable prolegomena for any synthetic theory of the four series of analyzed observations...................... 489
F.4 The anisotropy of space .............................. 490
F.5 The determination of the position of the Earth upon its orbit by purely terrestrial experiments .................. 491

**VI THE ANISOTROPY OF SPACE** .................................... 493
A VERY SIGNIFICANT CONNECTIONS BETWEEN APPARENTLY VERY DIFFERENT PHENOMENA ............ 494
B THE INTERPRETATION OF THE OBSERVATIONAL DATA .... 496
B.1 Three guidelines .................................... 496
B.2 The anisotropy of space .................................... 499
B.3 Astronomical influences and the anisotropy of space ....... 504
B.4 The incontestable existence of an intermediate medium .... 506
C APPARENT ISOTROPY OR REAL ANISOTROPY - AN ILLUSTRATION .................................... 510
C.1 The evolution of the theory of potentials ................ 510
C.2 The anisotropy of space .................................... 515
D THREE OPPOSING CLAIMS IN CONTEMPORARY THEORIES 518
D.1 The two aspects of light, wave and particle ............... 518
D.2 Continuity and quanta .................................... 520
D.3 Causality and indeterminism .................................... 523
E THE REALITY OF THE ETHER .................................... 536
E.1 Conceptions of the ether .................................... 536
E.2 An experimental evidence .................................... 541
E.3 The foundations of a unitary theory of physics ............ 542
E.4 The ether, the inescapable explanatory factor ............. 545

**VII THE OBSERVATIONAL DATA AND THE NECESSITY FOR REVISION OF CERTAIN POSTULATES OF CONTEMPORARY THEORIES** . . . . . . 547
A THE BIRTH OF THE THEORY OF RELATIVITY - PRECONCEIVED IDEAS . . . . . . 550
A.1 The experimental results in 1900 . . . . . . 552
A.2 The rotation of the Earth and its privileged reference frame . 553
A.3 The movement of translation of the Earth and Poincaré's principle of relativity . . . . . . 556
A.4 Einstein's Theory of Special Relativity . . . . . . 562
A.5 Einstein's General Theory of Relativity . . . . . . 565
A.6 Einstein and his predecessors . . . . . . 569
A.7 The alternative hypothesis of an anisotropy of space . . . . . . 573
B THE THEORY OF RELATIVITY AND EXPERIMENT . . . . . . 577
B.1 A legend: the "negative" result of Michelson's experiment . . 577
B.2 The claimed "negative" results of the experiment of Michelson and the experiments of Miller . . . . . . 581
B.3 The fundamental postulates of the theories of special and general relativity invalidated by experiment . . . . . . 582
B.4 A few commentaries . . . . . . 587
B.5 Rejection of the theories of special and general relativity as being incompatible with the observational data . . . . . . 590
C THE DEVELOPMENT OF THE THEORY OF RELATIVITY: AN EXAGGERATED CONSTRUCTION UPON UNCERTAIN AND FRAGILE FOUNDATIONS . . . . . . 592
C.1 The foundations of the theory of relativity . . . . . . 592
C.2 The Lorentz transformation . . . . . . 594
C.3 The constancy of the speed of light . . . . . . 597
C.4 The principle of the invariance of natural laws . . . . . . 599
C.5 The formulation by Fresnel in 1818 . . . . . . 601
C.6 The experimental data . . . . . . 603
C.7 The mask of mathematics . . . . . . 608
C.8 An acute but one-way critical sense . . . . . . 611
C.9 Obvious exaggerations . . . . . . 614
D THE CLASSICAL THEORIES AND THE ANISOTROPY OF SPACE . . . . . . 617
D.1 From the classical theories to the theory of relativity . . . . . . 617
D.2 Mechanical anomalies of the paraconical pendulum . . . . . . 619
D.3 Optical anomalies of sightings at marks and at collimators . . 622
D.4 Optical anomalies of Esclangon and of Miller . . . . . . 624
D.5 Very remarkable similarities . . . . . . 625
D.6 The theory of universal gravitation . . . . . . 627
E THE FUNDAMENTAL PRINCIPLES OF SCIENTIFIC ANALYSIS . . . . . . 629
E.1 The foundations of the scientific method . . . . . . 629

E.2 Principles of all scientific methodology .................. 633
F A NECESSARY REVISION OF CONTEMPORARY THEORIES . 634
F.1 The facts .............................................. 634
F.2 The interpretation of the facts ........................... 636
F.3 Dogmatism ............................................ 638
F.4 An inevitable revision of the fundamentals of current theories 640
F.5 One criterion only. The absolute supremacy of experimental data over theoretical concepts .............. 643

**VIII A PLAN FOR SIMULTANEOUS EXPERIMENTS** . . . . . . . . . . . . . . . . 645
A ON REPETITION OF THE FIVE SERIES OF EXPERIMENTS ON THE PARACONICAL PENDULUMS WITH ANISOTROPIC AND ISOTROPIC SUPPORT, ON OPTICAL SIGHTINGS AT MARKS AND COLLIMATORS, ON THE OPTICAL SIGHTINGS OF ESCLANGON, AND ON THE INTERFEROMETER OF MILLER . . . . . . . . . . . . . . . . . . . . . . 646
A.1 The observed periodicities . . . . . . . . . . . . . . . . . . . . . . . . . . . . . 646
A.2 An overall project . . . . . . . . . . . . . . . . . . . . . . . . . . . . . . . . . . . 651
B CONDITIONS FOR IMPLEMENTATION OF THESE EXPERIMENTS . . . . . . . . . . . . . . . . . . . . . . . . . . . . . . . . . . . . . . . . 653
B.1 Five separate projects . . . . . . . . . . . . . . . . . . . . . . . . . . . . . . . . 653
B.2 The conditions of success . . . . . . . . . . . . . . . . . . . . . . . . . . . . . 654
B.3 An indispensable progression . . . . . . . . . . . . . . . . . . . . . . . . . . 656
B.4 A set of experiments of exceptional scientific interest . . . . . . 658

**IX FIGHTING AGAINST DOGMA** ..... 659
A THE TYRANNY OF ESTABLISHED TRUTHS ..... 660
A.1 The makers of truth ..... 660
A.2 A significant example ..... 662
A.3 Authority against truth ..... 664
B THE SECULAR OBSCURANTISM OF THE MANDARINS OF PSEUDOSCIENCE ..... 665
B.1 Historical permanence ..... 665
B.2 Two illustrations: Galileo and Kepler ..... 669
B.3 Present-day science ..... 672

**X NEW PERSPECTIVES** .......... 675
A TWO COMPLETELY NEW PHENOMENA .......... 677
A.1 Anomalies of the movement of the paraconical pendulum and anomalies of sightings at marks .......... 677
A.2 New data on two previous experimental series: The optical sightings of Ernest Esclangon and the interferometric observations of Dayton C. Miller .......... 681
A.3 A set of very significant facts .......... 682
B SUPPORTS AND OPPOSITIONS .......... 683
B.1 Invaluable supports .......... 683
B.2 Unrelenting oppositions .......... 685
B.3 The crucial experiments of July 1958 .......... 687
B.4 Cessation of the experiments .......... 688
C MY FAITH IN THE FUTURE .......... 690
C.1 The decree of experiment .......... 690
C.2 Only the future counts .......... 692
C.3 The search for Truth .......... 693

**REFERENCES OF WORKS CITED** . . . . . . . . . . 697

**References** . . . . . . . . . . 697
References . . . . . . . . . . 698

**INDEX OF NAMES** . . . . . . . . . . 709

**Index** . . . . . . . . . . 711

*NOTE BY THE TRANSLATORS*

*Initially, Maurice Allais intended to publish The Anisotropy of Space in two parts:*

- *First volume: The experimental data, which constitutes the present work.*
- *Second volume: Experimental and theoretical supplements, wherein one can find in particular the development of the theoretical elements concerning the pendulum.*

*In this first volume, Maurice Allais frequently refers to the second volume. Unfortunately, as of today, this second volume has never been published. However, its summary and the cross references have been maintained in this translation. Although unpublished, a large part of its content actually exists, as it largely consists of a compilation of well-identified documents, which are in Maurice Allais' archives. Moreover, some of them have been published elsewhere.*


## SECOND VOLUME

## *EXPERIMENTAL AND THEORETICAL SUPPLEMENTS*

### INTRODUCTION

### CHAPTER I - THE PARACONICAL PENDULUM WITH ANISOTROPIC SUPPORT

***A – Experiments on the Paraconical Pendulum with Anisotropic Support***
***B – Theory of the Paraconical Pendulum with Anisotropic Support***

### CHAPTER II - THE PARACONICAL PENDULUM WITH ISOTROPIC SUPPORT

***A – Experiments on the Paraconical Pendulum with Isotropic Support***
***B – Theory of the Paraconical Pendulum with Isotropic Support***
***C – Previous Experiments on the Foucault Pendulum***

**CHAPTER III - EXPERIMENTS ON OPTICAL SIGHTINGS**

*A – Luni-solar and Thermal Effects on the Optical Deviations of the Sightings at Marks*

*B – Attempts at Continuous Automatic Recording of Sightings at Marks performed at Palaiseau and at IGN, April 1992 – January 1993*

*C – Essential Results of the Attempts*

**CHAPTER IV - ANALYSIS AND INTERPRETATION OF MILLER'S INTERFEROMETRIC EXPERIMENTS**

*A – Miller's Eight Fundamental Graphs*

*B – Determination of the Hodograph of the Speed of the Earth*

*C – Critical Analysis of Miller's Theory and of his Interpretation of his Observations*

*D – Two hypotheses. Anisotropy of Space, and the Cosmic Speed of the Earth towards the Hercules Constellation*

**CHAPTER V - A POSSIBLE INTERDEPENDENCE BETWEEN THE AVERAGE AZIMUTHS OF THE PARACONICAL PENDULUM WITH ANISOTROPIC SUPPORT AND THE AVERAGE AZIMUTHS OF MILLER**

*A – Overall view*

*B – Theoretical Analysis*

**CHAPTER VI - HARMONIC ANALYSIS OF THE OBSERVATIONS**

***A – Astronomical Periodicities***
***B – Tests of Periodicity***
***C – Applications of Periodicity Tests***

**CHAPTER VII - REPETITION OF THE EXPERIMENTS**

***A – Repetition of the experiments on the paraconical pendulum***
***B – Repetition of the experiments on optical sightings at marks***

**CHAPTER VIII - DOGMATIC OPPOSITIONS**

***A – Very Strong Opposition of Certain Members of the Academy of Sciences***
***B – Press articles Published about my Experiments – My Letter of 14 February 1958 to the Permanent Secretaries of the Academy of Sciences***
***C – The Refusal of the Academy of Sciences on 5 May 1958 to Publish my Note on the Speed of Light***
***D – The Refusal of the Academy of Sciences to Publish my Note of 23 February 1960 on the Optical Anomalies Demonstrated in the Crucial Experiments of July 1958***

## APPENDICES

*A – Invitation to My Conference of 22 February 1958*
*B – Should the Laws of Gravitation be Reconsidered?* (1958 Memoir)
*C - New Theoretical and Experimental Research Work on Gravity* (Memoir of December 1958)
*D – Test of Periodicity: Generalization of Schuster's Test to the Case of Autocorrelated Time Series, under the Hypothesis of a Process of Random Perturbations of a Stable System* (1961 Memoir)
*E – Frequency, Probability, and Chance, with two Appendices*
*1. Empirical Frequencies and Mathematical Frequencies*
*2. Theorem T. The Simulation of Chance by Almost Periodic Functions* (1982 Memoir)
*F – On the Normal Distribution of the Values of a Sum of Sinusoids at Regularly Spaced Instants* (Note by Maurice Allais and presented by Robert Fortet, CRAS, 30 May 1983)
*G – Considerations relating to Physical Theories* (Extracts from my Memoir of 1983: The Fundamentals of the Theory of Utility and Risk)
*H – Dayton C. Miller's Experiments of 1925-1926 and the Theory of Relativity* (Memoir published in "*La Jaune et la Rouge*", journal of the Ecole Polytechnique, September 1996, pp. 29-37)
*I – Very Significant Regularities in the Interferometric Observations of Dayton C. Miller, 1925-1926* (Proceedings of the Academy of Sciences, March 1997)

## ANNEXES I

*A – Extracts from my Correspondence, 1958-1960*
*B – My mémoir of 1957 on the Speed of Light*
*C – Press Articles*
*D – Scandal at Polytechnique* (Louis Rougier, July 1959)

**ANNEXES II - Notes presented to the Academy of Sciences, but not published in their Proceedings**

***Harmonic Analysis of the Movements of the Paraconical Pendulum***

*1958 - On the Existence of a Periodic Component with Period Near to 24h50m in the Movements of the Paraconical Pendulum with Anisotropic Support in the Observations of November-December 1954 and June-July 1955*, IRSID, 20 November 1958, 11 p.

*1959 - Movement of the Paraconical Pendulum – Research into Hidden Periodicities by Consideration of Frequency Graphs Associated with the Generalization of Schuster's Test*, IRSID, 2 November 1958, 8 p.

*1959 - Movement of the Paraconical Pendulum – Research into Hidden Periodicities by Consideration of Correlograms*, IRSID, 26 November 1959, 9 p.

***Solar Eclipse***

*1959 - Mechanics – Movements of the Paraconical Pendulum and Total Solar Eclipse of 2 October 1959*, IRSID, 10 November 1959, 7 p.

***Characteristics of the Experimental Apparatus***

*1958 - Experimental Determination of the Coefficient of Kinetic Friction of the Pivoting of the Ball of the Paraconical Pendulum*, IRSID, 20 November 1958, 9 p.

***Theory of the Paraconical Pendulum***

*1958 - Application of Bour's Theorem to the most General Case of Terrestrial Movement*, IRSID, 19 March 1958, 9 p.

*1958 - Equations of the Movement of the Paraconical Pendulum with Anisotropic Support for Small Oscillations – First and Second Approximations*, IRSID, 18 August 1958, 7 p.

*1958 - Variations of the Tangential Parameters of the Ellipse Described by the Movement of the Paraconical Pendulum with Anisotropic Support for Small Oscillations – First and Second Approximations*, IRSID, 18 August 1958, 6 p.

*1958 - First-order Corrections to the Paraconical Pendulum with Anisotropic Support for Small Oscillations*, IRSID, 18 August 1958, 4 p.

*1958 - Second-order Corrections to the Paraconical Pendulum with Anisotropic Support for Small Oscillations*, IRSID, 18 August 1958, 5 p.

***Paraconical Pendulum with Isotropic Support – Experimental Researches***

*1959 - Paraconical Pendulum. Implementation of a Support as Isotropic as Possible*, IRSID, 5 November 1959, 10 p.

*1959 - Paraconical Pendulum with Isotropic Support. Determination of the Variations over Time of the Characteristics of the Correlation of the Movement with the Azimuth*, IRSID, 10 November 1959, 7 p.

***Geodesy and Optics – Experimental Researches***

*1958 - Anomalies of the Operations of Triangulation and Levelling – Possible Explanation, and Confrontation with Experiment*, IRSID, 21 May 1958, 5 p.

*1960 - Existence of Periodic Components in the Variations of Readings Corresponding to Sightings Performed with a Fixed Telescope at a Fixed Mark, Correlated with the Movements of the Paraconical Pendulum*, IRSID, 23 February 1960, 6 p.

Photo of the author of this book at the time of the experiments

# INTRODUCTION

*"Theory should give the simplest possible description of the physical world...*

*"It should only appeal to a new physical property or accept a new hypothesis when inescapable necessity compels it to do so."*

*"Thus, when a physicist discovers facts that were previously unknown to him, when his experiments make it possible for him to formulate laws that theory has not anticipated, first of all he should investigate with the greatest possible care whether these laws can be presented, to the required degree of approximation, as consequences of accepted ideas...*

*"It is only after having become certain that the values previously handled by the theory cannot serve as symbols for the qualities observed, and that the established laws cannot be deduced from the accepted hypotheses, that he is authorized to enrich physics with a new physical property and to complicate it with a new hypothesis.*

*"These principles are the very essence of our physical theories."*

Pierre Duhem *

*"Only observable facts have any physical reality."*

Max Born **

*"There is ever a tendency of the most hurtful kind to allow opinions to crystallise into creeds. Especially does this tendency manifest itself when some eminent author, with the power of clear and comprehensive exposition, becomes recognised as an authority...*

*"But any man must err, and the best works should ever be open to criticism. If, instead of welcoming inquiry and criticism, the admirers of a great author accept his writings as authoritative, ... the most serious injury is done to truth.*

*"In matters of philosophy and science authority has ever been the great opponent of truth. A despotic calm is usually the triumph of error...*

*"In science and in philosophy nothing must be held sacred."*

Stanley Jevons ***

# A THE MOTIVATION FOR THIS BOOK

## *A.1 New experimental data, contributions to the debate upon the fundamentals of Mechanics and Optics*

### Debate upon the fundamentals of Mechanics and Optics

1 - This work relates to one of the domains in physics most disputed for three centuries, that of the fundamentals of Mechanics and Optics, and in particular to the questions of the transmission of actions at a distance and the influence of the movement of the Earth upon terrestrial phenomena.[1]

---

* M.P. Duhem, 1899, *Les Théories Electriques de J. Clerk Maxwell. Etude historique et critique* (The Electrical Theories of J. Clerk Maxwell. Historical and Critical Study), Annals of the Scientific Society of Brussels, 24th year, 1899-1900, p. 245 [117].

** Max Born, 1920, *La théorie de la relativité d'Einstein et ses bases physiques: exposé élémentaire* (Einstein's Relativity Theory and its Physical Basis), Gauthier-Villars, 1923, p. 291 [73].

*** Stanley Jevons, 1871, *The Theory of Political Economy*, French translation of the third edition, Giard, 1909, pp. 369-370 [159].

[1] For the pre-relativistic period, see particularly four overviews:

- M. Mascart, 1872-1874, *Sur les modifications qu'éprouve la lumière par suite du mouvement de la source lumineuse et du mouvement de l'observateur* (Upon Modifications of Light due to Movement of the Light Source and Movement of the Observer), Annales de l'Ecole Normale Supérieure, 1872, pp. 157-214; and 1874, pp. 363-420 [191, 192].
- Oliver Lodge, 1893, *Aberration Problems. A Discussion concerning the Motion of the Ether near the Earth, and Concerning the Connexion between Ether and Gross Matter, with some new Experiments*. Philosophical Transactions of the Royal Society of London, 1894, Vol. 184, pp. 727-806 [182].
- Edmund Whittaker, *History of the Theories of Aether and Electricity*, Vol. I, *The Classical Theories*, 1951. See particularly Chapter XIII, *Classical Theory in the age of Lorentz*, pp. 386-428 [279].
- Augustin Sesmat, 1937, *L'Optique des Corps en Mouvement* (The Optics of Moving Bodies), Hermann, Paris [246].

This field has been the scene of very vigorous controversies, in which passion has only too often overwhelmed objectivity, especially after the progressive domination of the theory of relativity in the literature.[2]

*That is why this work will be limited to analysis of the experimental data - the only reliable source of knowledge - and particularly to the analysis of novel experimental data that opens new perspectives.*

**New experimental data in four fields**

2. - The novel experimental data analyzed in this work relates to *four fields* considered *a priori* to be different, although closely related to one another.

- my experiments at the *Institut de Recherche de la Sidérurgie* (Institute of Iron and Steel Research, IRSID) on the paraconical pendulum *with anisotropic support* (*Chapter I*), and my experiments on the paraconical pendulum *with isotropic support* (*Chapter II*);
- my experiments on optical deviations of *sightings at marks*, and the experiments on *sightings at marks and at collimators* that followed, performed at the *Institut Géographique National* (National Geographic Institute, IGN) (*Chapter III*);
- the regularities characterizing the experiments of Esclangon *on optical sightings, which he did not himself notice* (*Chapter IV*);
- the regularities characterizing the *interferometric observations* of Dayton C. Miller, *which he did not himself notice* (*Chapter IV*).

[2] Edmund Whittaker, id., Vol. II, *The Modern Theories*, 1953; see in particular Chapter II, *The Relativity Theory of Poincaré and Lorentz*, pp. 27-77, and Chapter V [280], *Gravitation*, pp. 144-146; and *Chapters IV, VI, and VII* below.

Whittaker planned to publish a third volume covering the period 1926-1950. Due to his death, it was unfortunately not possible for this third volume to be published.

Each of these experiments was distinguished from all previous experiments in its field by an absolutely fundamental characteristic. *They were all based upon very numerous observations, carried out day and night, and spread over long time periods.*

## The danger of preconceived ideas

3 - This work *will avoid suggesting any overall interpretation*, for at least two reasons. First, any overall interpretation would imply analysis requiring too many developments; next and above all, the very nature of the new phenomena demonstrated in this work, for complete definition, would require very numerous preliminary complementary experiments, which would be indispensable for deriving precise laws of these new phenomena.

In the current state of the available information, such an overall theoretical structure *would be utterly premature*. In fact, the new experimental data analyzed in this work - *whose existence is indisputable* - are obviously *so complex* and involve so many phenomena, that *the preliminary experimental development of its structure and regularities* would condition any overall theoretical structure.[3]

We should be on our guard against this trap - *that of preconceived ideas - which has often hamstrung the development of theoretical physics in the last two centuries.*

---

[3] In his memoir of 3 January 1980, "*An Interim Report on a Repeat of the Allais Experiment*", Robert Latham (Imperial College of Science and Technology, 70 p. [170]) wrote [regarding myself] (p. 5):

> "*Indeed in all his work there is a complete absence of any detailed attempt at explanation, coupled however with very great care and thoroughness in the conduct of the experiments*".

In fact this is a principle that I have always followed: not prematurely to present any general theory explaining my results.

All that can be reasonably maintained at the present time, is that *space is anisotropic*.

*The new data deduced from experiment* and presented in this work, when considered overall, seems *to be incompatible* both with the theories of Mechanics and Optics of the pre-relativistic epoch and with the Theory of Relativity, Special or General.

In the unending conflict of doctrines, we must never forget that science is a perpetual process of becoming. *In science there is never any definitive truth.* What fundamentally characterizes scientific research is a constant effort to understand the deep nature of a world that most often remains undecipherable.

I devoutly hope that the new experimental data presented in this work will provoke a new effort of reflection, *totally disengaged from any a priori view, from any prejudgment, from any partisan view, from any passion.*

Little by little, science advances to new heights in its progressive discovery of the profound nature, so hard to decipher, of the world in which we live.

When we perform new experiments, Nature always answers our questions, but she too often appears to answer us like the oracle of Delphi. Her answers in fact do not always appear very clear to us, not so much because they are really ambiguous and incomprehensible, but because we are too often prisoners of fully-formed ideas, of preconceived ideas, and of established truths, *which hamper our understanding*.

## A.2 *The objects of this work*

### Four objectives

1 - The present work fundamentally has four objectives.

• *The first* is to present *an integrated view* of my experiments on the paraconical pendulum with *anisotropic support* performed from 1954 to 1960 (*Chapter I*), of my experiments on the paraconical pendulum with *isotropic support* performed in 1959-1960 (*Chapter II*), of my experiments on optical sightings *at marks* that accompanied the above experiments in 1958 and of the experiments on optical sightings *at marks and at collimators* that followed them in 1959 (*Chapter III*), and of the analyses that I have been making of all these experiments from 1954 to the present.

• *The second objective* is to present *an analysis of two previous series of very significant and absolutely fundamental experiments*: those of Esclangon in 1927-1928, and those of Dayton C. Miller in 1925-1926 (*Chapter IV*).

• *The third objective* is to demonstrate the common characteristics of these five series of experiments: observations that are *inexplicable* in the framework of currently accepted theories, *very marked structural connections*, and *very significant* temporal correlations with astronomical facts, in particular with the position of the Earth upon its orbit (*Chapters V, VI, and VII*).

• This work finally takes on the challenge of *strongly inspiring* anyone who has the possibility of performing, or of causing to be performed, a series of coordinated experiments, for which *every appearance* is that they would yield *major* information upon the very fundamentals of contemporary physical theories (*Chapter VIII*).

## My memoir of 1958

2 - In the journal "*Perspectives X*" of the Ecole Polytechnique in 1958,[1] under the title "*Doit-on Reconsidérer les Lois de la Gravitation?*", I published an overall view, *as of 1 January 1958*, of my work on the paraconical pendulum with anisotropic support, performed from 1954 to 1957. An English version of this memoir was published in 1959 by the American journal *Aero-Space Engineering* of the *Institute of Aeronautical Sciences*, under the title "*Should the Laws of Gravitation be Reconsidered?*"[2]

This memoir was followed by a *Complementary Note* of two pages,[3] in which I gave an account of the *crucial experiments* performed in July 1958 simultaneously with two identical sets of apparatus, on the one hand in my laboratory at the Institute of Steel and Iron Research (IRSID) at Saint-Germain, and on the other hand in a laboratory installed in April 1958 in an subterranean gallery at Bougival, 57 m underground and 6.5 km away.

## Publication of my works in physics

3 - I have naturally always intended to publish, *as soon as possible*, the entirety of my works in physics, *both theoretical and experimental*,[4] but the best is the enemy of the good, and it seems necessary to me, without waiting any further, to present now this *overall view* of the five series of experiments upon which this book is based, which are *very significant* and indeed *absolutely fundamental.*

---

[1] pp. 90-104. This memoir is reproduced in *Appendix B* of the second volume of this work, "*Experimental and theoretical supplements*" (see the *Summary*, p. 31 above).

[2] *Aero-Space Engineering*: September 1959, no. 9, pp. 46-52; October 1959, no. 10, pp. 51-55; November 1959, no. 11, p. 55 [35, 37, 36].

The translation was made in the USA upon the recommendation of Wernher von Braun, director of NASA.

[3] This *Complementary Note*, *subsequently* sent to the editorial offices of *Perspectives X after the crucial experiments of July 1958*, was simply inserted into that publication without pagination after the correction of the proofs of the memoir.

The English version of this *Complementary Note* was integrated into my memoir in *Aero-Space Engineering* (November 1959, p. 55) [36].

[4] The first volume will be published under the title "*Recherches Expérimentales et Théoriques sur les Théories Physiques 1953-1960*" ("Experimental and Theoretical Researches on Physical Theory, 1953-1960"). In particular, it will group together my principal publications and my principal memoirs (see the *References* below at the end of this volume). It will also present the numerical data for all the observations performed from 1954 to 1960, during periods of one month or fifteen days.

This volume will be followed by two others: "*Théorie du Pendule Paraconique*" ("Theory of the Paraconical Pendulum") and "*Analyses Complémentaires des Données de l'Observation*" ("Complementary Analyses of the Observational Data").

As in the case of the two volumes of the present work, these volumes will be published by Clément Juglar Editions, with the help of my friend Guy Berthault.

# B MY EXPERIMENTAL AND THEORETICAL RESEARCH

## *B.1 The origin of my experimental research*

### My passion for exploration of the "Physical Unknown"

1 - First of all, it seems proper for me to explain the origin of all my researches in physics, which I have pursued throughout the last half-century.

I owe my initial passion for physics research, for the exploration of the "*Physical Unknown*", to my course in physics at the Ecole Polytechnique, and *this passion has never stopped motivating me deeply.*

That impelled me, on the one hand to my research in theoretical physics, and on the other hand to my experiments on the movement of the paraconical pendulum. I performed these experiments from 1953 to 1960. After 1960, on multiple occasions, I continued theoretical research on a unitary theory of physics and on the interpretation of my experimental results, particularly in 1967, 1978, 1981, 1985, 1987, 1989, and from 1992 to 1996.[1]

[1] I conducted all these studies in parallel with my activity as an economist, which in 1988 earned me the *Nobel Prize in Economic Sciences.*

In total, since 1950, I have certainly employed *a quarter* of my entire time upon my theoretical and experimental researches on physical theories.

In fact *originally* I wanted to devote myself *entirely* to physics. It was only because of the war that I was led to direct my energies progressively towards economic science (see the Third Edition of my *Traité d'Economie Pure* (Treatise on Pure Economy), 1994, *Third Introduction*, pp. 19 and 26 [53]).

## A conviction

2 - It has always been my conviction that the propagation of gravitation and of electromagnetic action *occurs step by step between adjacent locations, and that this implies* the existence of an intermediate medium, the "*ether*" of Fresnel and the 19th-century physicists; however, without it being possible to consider, *as was generally accepted in the 19th century*, that all the portions of this medium are perfectly stationary with respect to one another, and particularly are stationary with respect to the fixed stars.[2]

This conviction led me, at the start of the 1950s, to consider that a magnetic field corresponds to a local rotation of this intermediate medium. From this I deduced that it should be possible to establish a connection between magnetism and gravitation by observing the action of a magnetic field upon the movement of a pendulum built with a ball of glass suspended by a cord of length around two meters.

## Anomalies in the movement of the Foucault pendulum

3 - In order to detect such an action, I started by observing the movement of a pendulum of the above type in the absence of any magnetic field other than that of the Earth.

*To my very great surprise*, I realized that this movement by no means reduced to the Foucault effect, and that rather it showed *very significant anomalies*, *varying with time* in relation to that effect.[3] The study of these *completely unanticipated* anomalies was the essential object of my experiments from 1954 to 1960.

---

[2] See *Chapter VI* below.

[3] Certainly I was favored by luck. But, although the exploitation of this luck has been very inspiring, it has also proved *extremely difficult from all points of view*.

## The action of a magnetic field on the movement of a pendulum

4 - From the *very limited number* of observations made in 1953,[4] and later in 1954 and 1955 at IRSID, of the movement of a ball of glass oscillating in a magnetic field of the order of a few hundred gauss, *I was not able to draw any definite conclusion at the time.* However, today I consider that *the effects that might be expected are too small to be detected with the magnetic fields that can be produced.*[5]

## The experimental study of the anomalies of the paraconical pendulum

5 - With respect to the anomalies observed in the movement of the pendulum, which were *clearly* demonstrated from February 1953 onward, I devoted myself from 1954 to study of the anomalies of the movement of a short pendulum suspended upon a ball, which I termed the "*paraconical pendulum*".[6]

*Thus, I was not led to perform these experiments by theoretical ideas. They were only a by-product of a completely different line of research which, itself, could not be pursued.*

---

[4] It was thanks to the help of my friend Emmanuel André Martin that I was able to undertake my first experiments in February-June 1953 in an office of the Clemençon company (34 rue Milton, Paris) (see below §D.1).

[5] Thus, in 1989, I stopped suggesting any new experiments upon the action of a magnetic field on a pendulum, as I had formerly intended. In fact, in my experiments from 1953 to 1955, the magnetic field produced at the center of the solenoid where the pendulum was oscillating was only of the order of 400 gauss.

This question will be the subject of a further publication of mine.

[6] From 16 October 1953, thanks to the stalwart support of Pierre Ricard (§D.1 below), I was able to make use of a basement laboratory with two very large rooms one above the other (each six meters by ten) at IRSID in Saint-Germain-en-Laye, with two co-workers, Jacques Bourgeot and Annie Rolland.

The very competent mechanical workshop at IRSID repeatedly gave me most valuable assistance in the construction of the various high-precision apparatuses which I employed from 1954 to 1960.

I carried on my experiments at IRSID from February 1954 to June 1960.

The results I obtained were *completely unexpected* from every point of view, both in their nature and in their magnitude.

*It was experience, and only experience, that inspired me to perform systematic experiments with the paraconical pendulum. Experiment constantly guided me, and it was experiment that finally led me to the conviction that my observations really corresponded to a very real new phenomenon, utterly inexplicable in the framework of currently accepted theories.*

## *B.2 My experimental researches upon the paraconical pendulum throughout 1954-1960, and upon optical deviations of sightings at marks in 1958*

### The existence of a periodic lunar component of 24h50m in the movement of the paraconical pendulum with anisotropic support having an amplitude totally inexplicable in the framework of currently accepted theory

1 - For study of the anomalies observed in the movement of a short pendulum, I principally employed a paraconical pendulum about one meter long, consisting of a vertical bronze disk fixed to a bronze rod and suspended upon a stirrup that rested on a steel ball.

In the absence of any magnetic field other than that of the Earth I observed, during observations *continued* over periods of the order of a month from 1954 to 1960, *very remarkable* occurrences, in particular the existence of a significant diurnal lunar periodicity of 24h50m having an amplitude *considerably greater* than that calculated from currently accepted theories. The *observed* amplitude was around twenty or a hundred million times greater than the *calculated* amplitude, depending upon whether the paraconical pendulum with anisotropic support or the paraconical pendulum with isotropic support was considered.[1] In fact, such a periodic diurnal *lunar* component is *completely inexplicable in the framework of currently admitted theories*.[2]

Moreover, the experiments I performed suggested *the existence, at each instant, of a direction of anisotropy of space.*

[1] See *Chapter I* below, §A.5.3, p. 98 and B.2.1, p. 118, and *Chapter II*, §F.2.2, p. 285.

[2] See *Chapter I* below, §B.2.6 and B.2.7, pp. 123-125, and *Chapter II*, §F.2.2, p. 285.

## The two crucial experiments of July 1958 upon the movement of the paraconical pendulum with anisotropic support

2 - *Identical* results concerning the existence of a diurnal lunar period having significant amplitude were obtained during my two crucial experiments of July 1958 in two laboratories about 6 km apart, one in the basement at Saint-Germain, and the other in a subterranean gallery in Bougival 57 meters underground.[3]

## The associated experiments of July 1958 upon sightings at marks

3 - The existence of anomalies observed in the operations of precision leveling and triangulation,[4] similar to the anomalies observed in the movement of the paraconical pendulum, inspired me to perform, *in parallel with* my experiments upon the paraconical pendulum with anisotropic support in Saint-Germain and Bougival, a series of North-South and South-North optical sightings at fixed marks at Saint-Germain. Due to technical difficulties, these optical sightings were only effectively realized in the second fortnight of July 1958.

In fact, in that period I observed *a remarkable correspondence* between the observations of the azimuths of the paraconical pendulum and the observations corresponding to the azimuths of these reciprocal optical sightings of two azimuthal telescopes at two marks having the same supports as those telescopes.[5]

Nevertheless, the observed amplitudes of the optical deviations, *considered in themselves, are inexplicable in the framework of currently accepted theories.*[6]

---

[3] See *Chapter I* below, Section C.

[4] See my note of 21 May 1958, *Anomalies des opérations de triangulation et de nivellation. Explication possible et confrontation avec l'expérience* (Anomalies of the operations of triangulation and leveling. Possible explanation, and comparison with experiment) [18]. This Note is reproduced in the *Annexes II* of the *Second volume* of this work (see p. 33 above).

[5] My initial intuition was thus *remarkably confirmed*.

[6] *Chapter II*, Section B, below.

## The anomalies observed during the two total solar eclipses of 1954 and 1959

4 - During the total solar eclipse of 30 June 1954, a remarkable deviation of the plane of oscillation of the paraconical pendulum was observed, *quite inexplicable in the framework of currently accepted theories*. A *completely analogous* deviation was again observed during the total solar eclipse of 2 October 1959.[7]

## The existence of a direction of anisotropy of space variable with time, deduced from the observations of the paraconical pendulum with isotropic support

5 - While in all my experiments from February 1954 to July 1958 the support of the paraconical pendulum was anisotropic, finally, in October 1959, I built an *isotropic support*, in order to be able *to determine the direction of anisotropy of space at each instant* by a new method of analysis.

During two month-long series of observations in November-December 1959 and March-April 1960, this approach, which was *completely different* from the previous one, confirmed the existence of a periodic luni-solar structure that was *totally inexplicable* in the framework of current theories, and *demonstrated the existence of a direction of anisotropy of space that varied over time (Chapter II)*.

---

[7] *Chapter I*, Section D, below.

## Two undeniable discoveries

6 - Many experimenters mentioned anomalies in the movement of conical pendulums at the end of the 19th century,[8] *but nothing precise* ever emerged from these experiments, and I think that the *unarguable* proof of the existence of anomalies of the paraconical pendulum and of their periodic structure is *an authentic discovery* of which I can claim the *entire paternity*.

The same holds *a fortiori* for the discovery of the optical deviations of sightings at marks and of their periodic structure which is completely independent of any trivial influence, because *previously nobody even suspected their existence*.

In fact these are two *undeniable* discoveries *of new phenomena* that, in the current state of things, appear only to be explicable by *an anisotropy of space*.[9]

---

[8] See for example Dejean de Fonroque, July 1879, *Du Pendule, Théorie de ses Variations* (On the Pendulum, Theory of its Variations), Chamerot, 32 p. [101], and *Note* of 14 April 1879 to the Academy of Sciences, presented by Cornu, *Sur diverses expériences faites avec un pendule oscillant avec de grandes amplitudes* (On various experiments made with a pendulum oscillating at large amplitudes) [102] (see a detailed analysis of the memoirs of Dejean de Fonroque in the *Second volume* of this work, *Chapter II*, Section C, see p. 28 above).

*The memoirs of Dejean de Fonroque are only qualitative, and cannot be subjected to quantitative analysis.*

[9] See §C.2 below, pp. 60-63.

## My Notes to the Academy of Sciences

7 - From November 1957 to February 1959, my observations of the paraconical pendulum with anisotropic support were the subject of eight Notes to the Academy of Sciences presented by Albert Caquot, and of two Notes on their statistical significance presented by Joseph Kampé de Fériet.[10]

---

[10] These notes are as follows:

Note of 13 May 1957 - *Test de Périodicité - Généralisation du Test de Schuster au cas de séries temporelles autocorrelées* (Test of Periodicity - Generalization of the Schuster Test in the Case of Autocorrelated Time Series) [16]

Note of 13 November 1957 - *Observation des mouvements du pendule paraconique* (Observation of the Movements of the Paraconical Pendulum) [12]

Note of 25 November 1957 - *Analyse harmonique des mouvements du pendule paraconique* (Harmonic Analysis of the Movements of the Paraconical Pendulum) [9]

Note of 4 December 1957 - *Mouvement du pendule paraconique et éclipse totale de soleil du 30 juin 1954* (Movements of the Paraconical Pendulum and the Total Solar Eclipse of 30 June 1954) [11]

Note of 16 December 1957 - *Théorie du pendule paraconique et influence lunisolaire* (Theory of the Paraconical Pendulum and the Luni-Solar Influence) [17]

Note of 23 December 1957 - *Application du Test de Schuster généralisé à l'analyse harmonique des azimuts du pendule paraconique* (Application of the Generalized Schuster Test to Harmonic Analysis of the Azimuths of the Paraconical Pendulum) [10]

Note of 3 November 1958 - *Nouvelles expériences sur le pendule paraconique à support anisotrope* (New Experiments with a Paraconical Pendulum with Anisotropic Support) [25]

Note of 22 December 1958 - *Structure périodique des mouvements du pendule paraconique à support anisotrope à Bougival et Saint-Germain en juillet 1958* (Periodic Structure of the Movements of the Paraconical Pendulum with Anisotropic Support at Bougival and Saint-Germain in July 1958) [27]

Note of 19 January 1959 - *Détermination expérimentale de l'influence de l'inclinaison de la surface portante sur le mouvement du pendule paraconique à support anisotrope* (Experimental Determination of the Influence of Inclination of the Bearing Surface upon the Movement of the Paraconical Pendulum with Anisotropic Support) [31]

Note of 9 February 1959 - *Détermination expérimentale de l'influence de l'anisotropie du support sur le mouvement du pendule paraconique* (Experimental Determination of the Influence of Anisotropy of the Support upon the Movement of the Paraconical Pendulum) [30]

The dates given are those of the publications in the Proceedings (and not the previous dates of their presentations). The dates of presentation were the following: 6 May 1957, 4 November 1957, 18 November 1957, 18 November 1957, 25 November 1957, 4 December 1957, 20 October 1958, 10 November 1958, 1 December 1958, and 26 January 1959.

Apart from their publication in the Proceedings of the Academy of Sciences, these various Notes were also published separately by Gauthier-Villars in two booklets entitled *"Structure périodique des mouvements du pendule paraconique à suspension anisotrope et influence lunisolaire. Résultats expérimentaux et anomalies"* ("Periodic structure of the movements of the paraconical pendulum with anisotropic support, and luni-solar influence. Experimental results and anomalies") [28] (25 p. and 17 p.) The former included the first six of the above Notes, the latter the last four.

**Visits to my laboratories**

8 - From September 1955, I came to the conviction that the movements of the paraconical pendulum with anisotropic support were indeed a new phenomenon that was *totally inexplicable in the framework of currently accepted theory.*

Thus, I circulated the principal results of my researches to various groups and organized visits to my laboratory at IRSID, and then organized visits to my laboratory at Bougival after the crucial experiments of July 1958.[11]

**My Conferences**

9. - My work was the subject of three Conferences organized by the *Cercle Alexandre Dufour*:

- the first, *"Faut-il reconsidérer les lois de la Gravitation ? Sur une nouvelle expérience de Mécanique"* ("Should the Laws of Gravitation be Reconsidered? On a New Experiment in Mechanics"), held on Saturday 22 February 1958 in the Henry Poincaré amphitheater of the Ecole Polytechnique;[12]

- the second, *"Faut-il reconsidérer les lois de la Gravitation ? Nouveaux résultats, bilan et perspectives"* ("Should the Laws of Gravitation be Reconsidered? New Results, Review, and Perspectives"), held on Saturday 7 November 1959 at the French Society of Civil Engineers;[13]

- the third, *"Les périodicités constatées dans le mouvement du pendule paraconique sont-elles réelles ou non? Généralisation du test de Schuster au cas de séries temporelles autocorrélées"* ("Are the Periodicities Observed in the Movement of the Paraconical Pendulum Real or Not? Generalization of the Schuster test to the case of Autocorrelated Time Series"), on Saturday 18 March 1967.[14]

---

[11] In total, 127 persons participated in these visits.

[12] 14 wall posters; 34 projected slides; stenography of the Conference and the Discussion, 80 p. An overall view of this Conference was given in my Memoir of 1958 (§A.2.2 above).

[13] 13 wall posters; 50 projected slides; stenography of the Conference and the Discussion, 69 p.

[14] The text of this Conference was published in Bulletins 120, 121 and 122 of the *Cercle Alexandre Dufour* of April, May, and September 1967, pp. 80-97, 107-124, and 130-132 [41].

In fact, I never had the time to publish my first two Conferences of 22 February 1958 and 7 November 1959, of which I have only preserved the copies and the associated wall posters and slides.

My memoir of 1958 (§A.2.2 above) presented an overall view of my Conference of 22 February 1958.

## Two Prizes

10 - My work on the paraconical pendulum with anisotropic support was marked by two prizes of which I was very appreciative: one French, the 1959 *Galabert Prize* of the *Société Française d'Astronautique* (French Society of Astronautics),[15] and the other American, the Prize of the *Gravity Research Foundation* in 1959.[16]

## The termination of my experiments

11 - Finally, and in spite of the *outstanding success* of the two *crucial* experiments of July 1958, and in spite of the *very promising* results of my experiments on optical sightings at marks performed in July 1958 and of my experiments on the paraconical pendulum with isotropic support performed in November-December 1959 and March-April 1960, I was compelled in June 1960, *due to lack of financial means*, to come to the decision to close my laboratory at IRSID and to separate myself from my two exceptional collaborators, Jacques Bourgeot and Annie Rolland.[17, 18]

---

[15] The memoir presented to the *French Society of Astronautics* had the title "*Recherches théoriques et expérimentales nouvelles sur la Gravitation*" ("New Theoretical and Experimental Researches into Gravitation") (December 1958, 21 p.) [26].

To this memoir were annexed my memoir of 13 May 1958, "*Anomalies du mouvement du pendule paraconique à support anisotrope*" ("Anomalies of the movement of the paraconical pendulum with anisotropic support") (68 p.) [19], my first eight Notes to the Academy of Sciences (note 8 above), and my memoir of 4 November 1957 on the speed of light (see §3.3 below).

[16] The title of the memoir presented to the Gravity Research Foundation of the USA in 1959 was "*New Theoretical and Experimental Research Work on Gravity*" (January 1959, 9 p.) [32].

[17] In 1868 Van der Willigen wrote (*Le Pendule Foucault au Musée Teyler* (The Foucault Pendulum in the Teyler Museum), Arch. Teyler Museum, I, 1868, p. 342) [281]:

> *"In experiments on the Foucault pendulum, at least on the experimental side, people have stopped at the precise point that the real difficulties started."*

However, the responsible French scientific authorities deliberately put an end to my experiments on the paraconical pendulum in 1959, *although paradoxically the essential difficulties had been overcome*.

In any case a refusal, *without real justification*, to examine the anomalies of the paraconical pendulum carefully and in an appropriate manner, could never be conducive to real progress.

In fact, this *completely unjustified* decision to terminate my experiments was *scientifically totally incomprehensible*.

See below §D.3.1, pp. 69-70, and Chapter I, Section G, pp. 213-236.

[18] On my request, and in view of their exceptional qualities, my two collaborators were immediately integrated into the technical services of IRSID.

## *B.3 My theoretical researches, 1950-1996*

### Analysis of the movements of the paraconical pendulum

1 - Since 1953, I never stopped working on the analysis and the meaning of the movements of the paraconical pendulum and of the optical deviations that I considered to be associated with those movements, in particular in regard to their relationship to the search for a unitary theory of physics. In particular, I elaborated a general theory of the movements of the paraconical pendulum and drew up a number of memoirs on physical theories and on the statistical meaning of the observations obtained.[1]

### Two major difficulties

2 - In fact, I had to overcome two *major* difficulties in my theoretical work on the paraconical pendulum.

First, although it may seem incredible in view of the immense literature on the Foucault pendulum, *no author had ever calculated the luni-solar influence upon a Foucault pendulum.* Considering the *fundamental* importance of theoretical estimation of that influence, I therefore was compelled to formulate a complete theory of the movements of the paraconical pendulum.[2]

---

[1] §A2.3.

[2] See my overall memoir "*Théorie du Pendule Paraconique*" ("Theory of the Paraconical Pendulum"), September 1956, 441 p [8].

In 1958, I prepared summaries of some of the essential principles followed in this Memoir in five *Notes* destined for the Academy of Sciences, which however were not published (see above, p. 33).

- *Application du Théorème de Bour au cas des mouvements terrestres dans le cas le plus général* (Application of Bour's Theorem to the case of terrestrial movements in the most general case), 14 March 1958, 9 p. [20]

- *Equations du mouvement du pendule paraconique à support anisotrope à petites oscillations. Première et deuxième approximation* (Equations of movement of the paraconical pendulum with anisotropic support for small oscillations. First and second approximations), 18 August 1958, 7 p. [24]

- *Variations des paramètres osculateurs de l'ellipse décrite dans le mouvement du pendule paraconique à support anisotrope et à petites oscillations. Première et deuxième approximation* (Variations of the osculating parameters of the ellipse described by the movement of the paraconical pendulum with anisotropic support, for small oscillations. First and second approximations), 18 August 1958, 6 p. [29]

- *Corrections de première approximation du pendule paraconique à support anisotrope et à petites oscillations* (Corrections to the first approximation of the paraconical pendulum with anisotropic support, for small oscillations), 18 August 1958, 4 p. [22]

- *Corrections de deuxième approximation du pendule paraconique à support anisotrope et à petites oscillations* (Corrections to the second approximation of the paraconical pendulum with anisotropic support, for small oscillations), 18 August 1958, 5 p. [21]

I asked Henri Villat to present the first Note, but I met with a refusal.

In the second place I realized that, in the immense literature on the search for periodicities, both in physics and in economics, there existed no test of periodicity applicable *to the general case of autocorrelated time series*. It was only in April 1957 that I was able to overcome this difficulty by elaborating a test that generalizes *Schuster's test*, which is only valid for time series consisting of *independent* terms.[3]

### Research on the foundations of a unitary theory of physics

3 - From 1950 to 1960 I pursued my researches on the foundations of a unified theory, and prepared various Notes.[4]

---

[3] My generalization of Schuster's test *in the case of autocorrelated time series* was the subject of my two Notes to the Academy of Sciences of 13 May and 13 November 1957 [16, 10], and of my overall memoir "*Test de Périodicité. Généralisation du test de Schuster au cas de séries temporelles autocorrelées dans l'hypothèse d'un processus de perturbations aléatoires d'un système stable*" ("Test of periodicity. Generalization of the Schuster test to the case of autocorrelated time series, under the hypothesis of a process of random perturbations of a stable system"), presented in 1961 to the *International Statistical Institute* [40]. This text is reproduced in Appendix D to the second volume of this work.

See *Chapter I* below, §B.1.3, note (6).

[4] Several of these Notes were given a limited circulation in mimeographed form, notably:

- *Sur une interprétation possible du champ magnétique terrestre* (On a possible interpretation of the terrestrial magnetic field), 24 October 1957, 7 p. [14]

- *Sur une solution de l'équation aux dérivées partielles*

$$\frac{1}{\sqrt{|g|}}\partial_i\left(\sqrt{|g|}g^{ij}\partial_j\varphi\right)-\frac{1}{c^2}\frac{\partial^2\varphi}{\partial t^2}-2\frac{k_0}{c}\frac{\partial\varphi}{\partial t}-k_0^2\varphi+4\pi K\delta=0$$

*et sur une interprétation de la constance de la vitesse de la lumière* (On a solution of the partial differential equation [...] and on an interpretation of the constancy of the speed of light), 4 November 1957, 12 p. [15]

- *Sur une interprétation possible des anomalies de la gravité et ses applications* (On a possible interpretation of gravitational anomalies, and applications thereof), 5 November 1957, 34 p. [13]

- *Interprétation des anomalies de la pesanteur comme un effet d'écran des actions gravifiques* (Interpretation of anomalies of weight as a screening effect of gravitational actions), March 1960, 29 p. [39]

On my memoir of 4 November 1957 on the speed of light, see *Chapter I* below, §G.5.2, p. 227; *Chapter VI*, §C.1, pp. 511-514; and *Chapter VII*, §C.4, note 4, p. 599.

# C THE SUBJECT IN QUESTION

## *C.1 Connections between the observations of the paraconical pendulum and the optical deviations of sightings at marks with the optical experiments of Ernest Esclangon in 1927-1928 and the interferometric experiments of Dayton C. Miller in 1925-1926*

### The experiments of Esclangon and Miller

1 - The anomalies of the paraconical pendulum and the anomalies of sightings at marks that I have demonstrated present *striking connections* with anomalies encountered during the study of many other phenomena.

In the following I examine in particular two series of anomalies, those corresponding to the optical experiments of Esclangon in 1927-1928 and those corresponding to the interferometric experiments of Dayton C. Miller in 1925-1926. Both of these, *whose existence is alike very real*, appear upon analysis to be endowed with *absolutely exceptional* importance.[1]

### A general correlation with the position of the Earth upon its orbit

2 - What notably characterizes the observations of the movements of the paraconical pendulum with anisotropic support and with isotropic support, the optical observations associated therewith by myself, the optical observations of Esclangon, and the interferometric observations of Miller, *is their correlation with the position of the Earth upon its orbit*, which contradicts a fundamental postulate of the *Theory of Relativity, both Special and General.*[2]

---

[1] *Chapter IV* below.

[2] *Chapters V, VI,* and *VII* below.
*Chapter V* is entirely devoted to the quantitative analysis of this correlation.

## No interpretation

3 - In my *Notes to the Academy of Sciences* in 1957-1959, and in particular in my overall memoir of 1958, "*Doit-on reconsidérer les lois de la gravitation ?*" ("Should the Laws of Gravitation be Reconsidered?") [23], I *systematically* abstained from any interpretation of the anomalies observed in the movement of the paraconical pendulum, for two reasons: the first, that in my view *what was essential was the observed facts*; and the second, that I wanted to hold myself aloof from any useless polemics *on dogmas that were considered* in contemporary theories *to be definitely established* and that were considered by certain members of the Academy of Sciences *to be untouchable*.[3]

## Mutually consistent anomalies

4 - In fact, analysis of the observations of the paraconical pendulum with anisotropic support and with isotropic support, analysis of the observations of sightings at marks that I associated therewith in 1958 and of the observations of sightings at marks and at collimators that followed in 1959, analysis of the optical observations of Esclangon, and analysis of the interferometric observations of Miller, *all* demonstrate the existence *of a very remarkable underlying consistency between all these observations*, and they *all* lead to the same conclusion:

- *It is possible to determine the position of the Earth upon its orbit by purely terrestrial experiments.*

*All of them* also lead to three directing conceptions:

- the existence of an *anisotropy of space*,
- the fact that this anisotropy of space is determined by *astronomical influences*,
- the *existence of an intermediate medium* which is the material support for transmission of these influences.

---

[3] In order to avoid any difficulties with certain members of the Academy of Sciences who were very attached to Einstein's Theory of Relativity, Albert Caquot always *implored* me to abstain from any interpretation (see particularly note 3 of §A 1.3 above, and §D.3 below; similarly, see Section G of *Chapter I* below).

Today it seems preferable to me to free myself *totally* from all restraints of any kind.

The deep analysis of these five series of observations of course does not make it possible to assert the intrinsic validity of these three guiding conceptions, but it does allow us to assert that *everything happens as though* these three guiding conceptions *really* correspond *to an underlying reality*.[4]

In any case, *the validity, the mutual coherence, and the properties of the observations analyzed in the first five Chapters of this work are totally independent of any hypothesis*[5] *and of any theory whatsoever.*

**Only one conclusion is certain**

5 - Any theoretical interpretation of the totality of these five series of observations *would be completely premature at this time*. Obviously, many complementary experiments will be *needed* for enough regularities to appear to lead to precise laws and to a general theory.[6]

*Only one thing seems certain at the moment. New phenomena have been demonstrated that currently accepted theories are incapable of explaining.*

Thus in this work I have limited myself *to the unadorned observational data, and have systematically abstained from presenting any theoretical synthesis whatever.*[7]

---

[4] See *Chapter VI* below.

[5] *In fact, the anisotropy of space and its determination by astronomical influences are not hypotheses. They are given by observation.*

[6] See §A. 1.1 above.

[7] *The only exception* that I have made to this principle is to point out that the observations of the paraconical pendulum *can easily be explained from the hypothesis of an anisotropy of inertial space* (see below, *Chapter I*, §F.3, pp. 206-212, and *Chapter II*, Section I, pp. 320-325).

## The Theory of Relativity

6. - In reality, the reason why so many discussions and passions were provoked, and are still provoked, by the *Theories of Special and General Relativity* is very simple: *the initial error of judgment reached as to the claimed negative character of Michelson's experiment, and the willful discounting of Miller's observations in 1925-1926.* From these roots has sprung a sort of persistent deviation in contemporary physics, accompanied by an intolerant dogmatism.[8]

Nothing illustrates better this type of deviation than this judgment of Fénelon:[9]

> *"Most errors that men make are not because they reason badly from true principles, but because they reason correctly from false principles or inexact judgments."*

## A golden rule

7. - For my part, all my researches and all my works *have been dominated by an absolute conviction: that, to be valuable, any theory whatever must be confirmed by the observational data, both in its hypotheses and in its consequences.*

This conviction explains the maxim that has constantly inspired me through my life in all fields: "*Submission to experimental data is the golden rule which dominates any scientific discipline.*"[10]

---

[8] See *Chapter VII* below, §A.4

[9] *Letter of Fénelon, so-called of Port-Royal*, for the education of the Duke of Chevreuse.

[10] Maurice Allais, 1989, *La Philosophie de ma Vie, Autoportraits* (The Philosophy of my Life, Self-Portraits), Montchrestien, 1989, p. 70 [48].

## *C.2 Two fundamental and authentic discoveries*

### Two new phenomena

1 - As I have already mentioned,[1] in my experiments from 1954 to 1960 I demonstrated *very significant anomalies*, on the one hand in the movement of the paraconical pendulum with anisotropic support and with isotropic support, and on the other hand in optical sightings at marks. *I proved their existence, independently of any perverse effect.* These anomalies are *totally inexplicable* in the framework of currently accepted theories.

Moreover, the optical anomalies I demonstrated in July 1958 presented a striking connection with the anomalies of the paraconical pendulum with anisotropic support. *By this fact alone, a relationship was established between two phenomena which were* a priori *completely foreign to one another, in Optics and in Mechanics.*[2]

*This was undeniably a discovery of two new phenomena, both of which were without precedent in the literature*, and both having implications of *major importance* for the very foundations of contemporary theories.

*These two fundamental and authentic discoveries, which were made in 1958, have now been totally ignored by official science for a period of thirty-eight years.*[3, 4]

---

[1] §B.2.6 above, p. 50.

[2] *Chapter III* below, §B.3, p. 338, and §B.6, p. 345.

[3] Perhaps some readers will accuse me of lacking somewhat in modesty here, but the fact that I can claim the paternity of these two discoveries *does not change their nature in any way*, and I do not see why I should underestimate them, *since their capital importance is obvious*.

*For thirty-eight years* official science has succeeded in maintaining *total silence* about these two discoveries. In the interest of science itself, an end should be put to this silence.

Claiming the paternity of discoveries does not really amount to any lack of ethics. The great Ampère himself wrote:

> "*The importance of these experiments cannot be denied, and neither can there be a refusal to admit that the discovery of the action of the Earth upon conducting wires belongs to me as completely as does that of the mutual action of two bodies.*"

(André-Marie Ampère, 1826, *Théorie mathématique des phénomènes électrodynamiques uniquement déduite de l'expérience* (Mathematical theory of electrodynamic phenomena deduced only from experiment), p. 103 [57]).

[4] Even more so, because the fact that these two fundamental discoveries were made "*by an amateur*" is *utterly incomprehensible, indeed impossible*.

For my detractors, it surpasses the understanding that absolutely essential phenomena could have escaped the cunning of alert experimenters for decades. From their point of view, these claimed discoveries must therefore be based only upon errors and illusions. There is thus no need for them to be taken into consideration.

See §D.3 below.

## The two crucial experiments of July 1958 on the paraconical pendulum with anisotropic support

2. - In fact, *the two crucial experiments* which were carried out in July 1958 *in a continuous manner and under identical conditions* over a month, one in a basement at the Institute of Steel Research (IRSID) at Saint-Germain, and the other in a subterranean gallery at Bougival, six kilometers away and 57 meters underground, *radically swept aside all the objections previously presented*, by yielding *the same results* concerning the existence of a lunar periodicity of 24h50m in the movement of the paraconical pendulum with anisotropic support, *having an amplitude inexplicable in the framework of currently accepted theories.*

It is very remarkable that, during the same month of July 1958 in Saint-Germain, the optical deviations of North-South and South-North sightings at marks by two azimuthal telescopes demonstrated *the same periodic component* of 24h50m, and that this *component was exactly in phase with those of the paraconical pendulums* observed at Saint-Germain and at Bougival.[5]

How can one not remember here what André-Marie Ampère wrote in 1826:[6]

> *"Epochs in which phenomena previously considered as being due to completely different causes are brought under a single principle, have almost always been accompanied by the discovery of a great number of new facts, because the new way of conceiving their causes suggests a multitude of experiments to try, a multitude of explanations to verify."*

[5] The validity and the implications of these two discoveries are today *considerably reinforced* by analysis *based upon the new calculations* presented in this work.

[6] André-Marie Ampère, 1826, *Théorie mathématique des phénomènes électro-dynamiques uniquement déduite de l'expérience* (Mathematical theory of electrodynamic phenomena deduced only from experiment), p. 118 [57].

## Effects of relatively significant magnitude

3. - Considering contemporary research, when the considerable efforts exerted to demonstrate extremely small effects and the *extremely complex and very costly* apparatus deployed are compared with my experiments upon the paraconical pendulum and the associated optical effects, one cannot fail to be struck by the *very great simplicity* of the experimental procedures that I employed and *by the relatively significant quantitative level of the observed effects*, in particular in the case of the paraconical pendulum, for which the observed amplitude of the periodic component of 24h50m is from *twenty to a hundred million times greater* than the amplitude calculated by the currently accepted theory of gravitation.[7, 8]

## New perspectives

4 - Like all new phenomena that, at a given moment, seem to be *inexplicable in the framework of currently admitted theories and that compel questioning of those theories*, the anomalies of the paraconical pendulum and the optical anomalies that I demonstrated open *new perspectives* from very many points of view, notably in terms of the existence *of an anisotropy of space that is variable over time.* [9]

---

[7] According to whether the paraconical pendulum with anisotropic support or with isotropic support is considered.

See below, *Chapter I*, §B.2.1, p. 118, and *Chapter II*, §F.2.2, p. 285.

[8] If we consider, for example, the advance of *42 sexagesimal seconds per century* in the perihelion of Mercury, whose explanation to an accuracy of 5" is considered to be a great success of the theory of relativity (see *Chapter VII* below, §C.6.2), and if we consider the 24h50m lunar influence on the paraconical pendulum which is *of the order of* $10^{-5}$ *radians per second of time* (note 5 above), this corresponds to 6.51 billion sexagesimal seconds per century.* The orders of magnitude are thus *completely different*.

$$^{*}\ 10^{-5}\text{rad/sec} = 10^{-5}\left(\frac{180}{\pi}\times 60\times 60\right)(100\times 365.25\times 24\times 60\times 60)\ \text{seconds per century}$$
$$= 6.51\times 10^{9}\ \text{seconds per century}$$

[9] Some may ask why I have waited so long to publish this work. The reasons are simple. In the 1960s, the hostility that faced me was *so powerful* and the rumors about the non-validity of my experiments were *so numerous and insistent*, some emanating from persons having great reputation and great influence, that it was impossible for me to contend against them. *These rumors still persist today.*

Moreover, in the years following the termination of my experiments, I was completely absorbed by my work in economics on monetary analysis, on the theory of the capitalist optimum, on the theory of surplus, on the theory of random choices, and on the theory of probability, all of which subjects afforded me scope for very numerous publications (see Allais, 1989, *Autoportraits*, pp. 121-144 [48]).

As Max Planck underlined:[10]

> *"When a revision or a transformation of a physical theory takes place, we find that, at the starting point, there are almost always one or more facts that cannot be brought into the framework of the theory in its current form. In fact, the facts are always the keystone upon which the stability of every theory, no matter how important, depends.*
>
> *"For the theoretician truly deserving of the name, there is therefore nothing more interesting than a fact that contradicts a theory previously held to be true, and that is when, for him, the real work begins."*

And as Henri Poincaré wrote:[11]

> *"The physicist who has just given up one of his hypotheses should, on the contrary, rejoice, for he found an unexpected opportunity of discovery. His hypothesis, I imagine, had not been lightly adopted: it took into account all the known factors which seem capable of intervention in the phenomenon. If it is not verified, it is because there is something unexpected and extraordinary: therefore we are on the point of finding something unknown and new."*

---

[10] Max Planck, 1941, *Initiation to Physics*, Flammarion, p. 40 [218].

[11] Henri Poincaré, 1905, *Science and Hypothesis*, The Walter Scott Publ. Co. Ltd, N.Y., p. 168 [220].

# D SUPPORTS AND OPPOSITIONS

## D.1 *Exceptional supports*

• I owe *immense debts of gratitude* to Emmanuel André-Martin, Pierre Ricard, Albert Caquot, and René Dugas,[1] now deceased, who had confidence in me, and thanks to whose support I was able to pursue my experiments from 1954 to 1960 upon the paraconical pendulum and my associated experiments on optical sightings at marks.

It is to *Emmanuel André-Martin* that I owe the availability of my first laboratory in January 1953.[2]

It is to *Pierre Ricard* that I owe the availability in October 1953 of a laboratory at the *Institute of Steel Research* in Saint-Germain, with two collaborators, Jacques Bourgeot and Annie Rolland.[3] It was thanks to them, and particularly to Jacques Bourgeot, whose efficiency, professionalism, intelligence, and devotion were exemplary, that my experiments could be carried out with full success.[4]

---

[1] Emmanuel André-Martin (16 July 1900 - 23 June 1978), Pierre Ricard (3 April 1899 - 4 April 1956), Albert Caquot (1 July 1881 - 27 November 1976), René Dugas (11 August 1897 - 15 June 1957).

[2] In the works of the Clemençon Company, 34 Rue Milton, Paris, whose president was André-Martin, with the very active and effective assistance of M. Coupry, Engineer of the Clemençon Company. See §B.1.4 above.

[3] At that time, Pierre Ricard was the President of the *Union des Industries Métallurgiques et Minières* (Union of Metallurgical and Mining Industries).

In May 1953 I telephoned to Pierre Ricard in order to ask him for help in carrying on, with sufficient financial means, the experiments that I had implemented in the works of the Clemençon Company. His response was astonishing:

> *"I have read your work of 1943 on economics. I consider it to be absolutely fundamental, and comparable, for economics, to Lagrange's "Analytical Mechanics" for mechanics. I therefore have full confidence in you. What do you need?*
>
> *"Unfortunately I am leaving for the United States tomorrow, but you should address yourself on my behalf to the Director of the Institute for Iron and Steel Research. He will do what is necessary."*

Ten days later the decision was definitively taken to put at my disposal, at IRSID, a large laboratory on two floors with two collaborators.

In all my life I have never seen such a capacity for decision based upon confidence in a person.

[4] When in October 1953 I engaged Jacques Bourgeot (who had previously already worked for several months in the heart of the *Institut Géographique National*), he was only 23 years old, but his exceptional capabilities were continuously displayed in the years that followed.

My laboratory at IRSID was operational from 16 October 1953 until 30 June 1960.

To *Albert Caquot* I owe the presentation of eight *Notes* to the Academy of Sciences on the paraconical pendulum, *in spite of numerous and persistent oppositions.*[5] These *Notes* related respectively to the experimental arrangements, to the observational procedure, and to the harmonic analysis of the movements of the paraconical pendulum, to the effects of the total solar eclipse of 30 June 1954, to the theory of the paraconical pendulum and the luni-solar influence, to the new crucial experiments of July 1958 on the paraconical pendulum with anisotropic support at Bougival and at Saint-Germain, to the periodic structure of the azimuths observed during those experiments, to the experimental determination of the effects of the inclination of the bearing surface, and to the anisotropy of the support. My last communication was that of 9 February 1959.[6]

To *René Dugas*[7] I owe constant and effective support vis-a-vis many personalities, and particularly many members of the Academy of Sciences.

The premature deaths in 1956 and 1957 of Pierre Ricard and René Dugas deprived me of two *absolutely essential* supports.

To Joseph Kampé de Fériet I owe the presentation of my *two Notes on harmonic analysis* to the Academy of Sciences, the first on the generalization of the Schuster test to the case of autocorrelated time series, and the second on the application of this test to harmonic analysis of the azimuths of the paraconical pendulum.[8]

---

[5] CRAS, 13 and 25 November 1957, 4 and 16 December 1957, 3 November and 22 December 1958, 19 January and 9 February 1959 (see above §B.2.7, p. 51).

[6] After this date, a powerful cabal opposed itself both to my work in physics and also to my liberal positions in economy.

See *Chapter I* below, §G.5, pp. 226-231, and *Chapter X*, §B.2, pp. 685-686.

See also Louis Rougier, July 1959, *Scandale à Polytechnique* (Scandal at Polytechnique) [236]. (This text is reproduced in *Annex I.D* of the *Second Volume* of this work; see p. 31 above).

[7] René Dugas was director at the SNCF, but his hobby was analysis of the fundamentals of mechanics, and he had published two fundamental works: *History of Mechanics* (1950) [114], and *Mechanics in the 17th Century* (1954) [116], both with prefaces by Louis de Broglie.

René Dugas made me a prediction at the time: *"The day will arrive when your paraconical pendulum is installed in all the observatories of the world."* His prediction has however not yet come to pass.

[8] CRAS, 13 May and 23 December 1957 (see §B.2.7 above, p. 51).

• It was thanks to the constant support of several members of the Academy of Sciences, notably Albert Caquot, Donatien Cot, Georges Darrieus, Joseph Kampé de Fériet, André Léauté, Albert Pérard, Maurice Roy, Pierre Tardi, and René Thiry, and of Generals Paul Bergeron and Jean Guérin, successive presidents of the *Comité d'Action Scientifique de la Défense Nationale* (Committee for Scientific Action of National Defense), and to the impact of my Conference of 22 February 1958, that finances were made available to me by the *Comité d'Action Scientifique de la Défense Nationale* and by the *Centre National de la Recherche Scientifique* (National Center for Scientific Research), and that it was thus possible, in July 1958, to perform *the two crucial experiments*, implemented in a continuous manner over one month and in identical conditions, one in the basement of the *Steel Research Institute* at Saint-Germain, and the other in a subterranean gallery at Bougival six kilometers away and 57 meters underground.[9]

*The full success of these experiments* was marked by the *1959 Galabert Prize of the French Society of Astronautics*, and by a prize from the *Gravity Research Foundation* in 1959.[10]

These experiments of July 1958 were accompanied at IRSID by experiments on sightings at marks, which yielded *completely decisive results*.[11]

---

[9] See §C.2.2 above.

[10] See §B.2.10 above.

[11] *Chapter III*, Section B, pp. 334-344 below.

It was thanks to my friend Guy Berthault that *tests of continuous registration* of sightings at marks could be implemented in 1992 and 1993. An analysis of these is given in the *Second Volume* of this work, *Chapter III*, Section B (see p. 29 above).

It is also thanks to Guy Berthault that it will be possible to publish the present volume and the greater part of my works on physics.

## D.2 *The interest excited by my experiments*

*From 1953 to 1956 I abstained from any publication*, because I wanted to be absolutely sure of the reality of the new phenomenon that I had demonstrated: luni-solar effects whose amplitudes were *completely inexplicable* in the framework of currently accepted theories.

In 1956 I considered that I was ready to publish the essentials of my results. They were the subject of five communications by Albert Caquot and Joseph Kampé de Fériet to the Academy of Sciences.[1]

On 22 February 1958, I presented a summary of my results in my Conference, "*Faut-il reconsidérer les lois de la gravitation ? Sur une nouvelle expérience de mécanique*" ("Should the laws of gravitation be reconsidered? On a new experiment in mechanics") in the Henri Poincaré amphitheater at the Ecole Polytechnique, with Albert Caquot as chairman.[2]

From 1956 to 1958, a hundred and twenty seven figures from the scientific world, of whom more than fifty were specialists in mechanics and geophysics,[3] came to visit my laboratory at Saint-Germain, and then that at Bougival. *Not one of them was able to present a valid explanation of the observed effects in the framework of accepted theories.*

In 1958 *Perspectives X*, the journal of the Polytechnique, published my overall memoir "*Doit-on reconsidérer les lois de la gravitation ?*" ("Should the Laws of Gravitation be Reconsidered?") [23]. The English translation of this memoir was published in 1959 by the American journal *Aero-Space Engineering*, under the title "*Should the Laws of Gravitation be Reconsidered?*" [35, 37, 36].[4]

---

[1] See §B.2.7 above, p. 51, and §D.3.1 below, note 1, p. 69.

[2] §B.2.9 above, p. 52.

The venue was jam-packed. More than six hundred people were present at my Conference.

Very wide publicity was given in the Press. See particularly: Pierre de Latil, *Le pendule fatal aux lois de la mécanique* (The pendulum fatal for the laws of mechanics), Le Figaro Littéraire, 18 January 1968 [171]; René Sudre, *L'énigme de la gravitation* (The enigma of gravitation), Revue des Deux Mondes, 1 February 1958 [255]; Henri François, *Faut-il reconsidérer les lois de la gravitation ?* (Should the laws of gravitation be reconsidered?), Le Monde, 22 February 1958 [146]; Pierre Devaux, *Manifeste scientifique à Polytechnique* (Scientific Manifest at the Polytechnique), Le Figaro, 25 February 1958 [111].

The principal analyses are reproduced in Annex C to the *Second Volume* of this work.

[3] Of whom fourteen were members of the Academy of Sciences: Albert Caquot, Pierre Chévenard, Donatien Cot, Jean Coulomb, André Danjon, Georges Darmois, Joseph Kampé de Fériet, André Léauté, Albert Pérard, Joseph Pérès, René Perrin, Maurice Roy, Pierre Tardi, and René Thiry.

[4] See §A.2.2 above, p. 42.

*Overall, my experiments everywhere excited considerable interest, both in France and abroad.*[5]

In fact, everybody who analyzed my work seriously was impressed by the coherence of my analysis and of my results,[6] and by their scientific value.[7]

---

[5] In particular, I received an immense mailbag from many countries.

Thus for example, quite recently on 1 August 1996, I received a letter from a Chinese physicist, S.W. Zhou of the Department of Physics, *Huazhong University of Science and Technology*, telling me of his researches and publications on anomalies observed in mechanics, optics, and atomic physics during solar eclipses in 1987, 1992, and 1995 *pursuant to my observations of the anomalies of the paraconical pendulum during the eclipse of 30 June 1954*, and underlining my role as a pioneer in research upon the anomalies of gravitation.

[6] See notably the judgment of the English physicist Robert Latham (note 3 of §A.1.3 above, p. 39). See also the appreciation in May 1959 by General Paul Bergeron, president of the Committee for Scientific Action of National Defense, in his letter to Wernher von Braun, *Chapter I* below, §G.6, note 2, p. 232.

[7] After the publication in September 1996 of my article "*Les expériences de Dayton C. Miller et la Théorie de la relativité*" ("The experiments of Dayton C. Miller and the Theory of Relativity") in the Polytechnique journal *La Jaune et la Rouge* [55], one of my correspondents in 1958, Paul Ernest de Montaigne, an old student of the Ecole Polytechnique, did not hesitate to write to me:

> *"I have followed your conferences on the paraconical pendulum with lively interest...*
>
> *"I must say, I am a bit frustrated to see you receive the Nobel Prize in Economics. I rather awaited your receipt of the Nobel Prize in Physics. Now you have returned to Physics, and how! I hope you will not abandon her again."*

When I received the 1988 Nobel Prize in Economic Sciences, Robert Latham of *Imperial College of Science and Technology* wrote to me:

> *"Please accept my most hearty congratulations ...*
>
> *"It is a pity, science being what it is, that you can't get a similar recognition for the pendulum work. I know I am in a minority but my personal view is that it is just as important, and will be acknowledged as such in due course".*

## D.3 *Fierce oppositions*

### A steady build-up of strong opposition

1 - In fact, beginning with the publication of the results of my experiments in 1956, I never ceased to collide with "*established truths*" and with the dogmatism of all sorts of "*establishments*" that ensured their domination.[1]

Multiple objections continued to be put forward against my work, of which the greater part were ill-founded or even totally unfounded, and of which the most formidable were based only *upon nebulous rumors* spread in corridors, too often by very prominent people, to which it was totally impossible for me to answer. I requested to be heard by a Commission of the Academy of Sciences, but in vain. It was to no avail.

*A leaden silence entombed the full success of the crucial experiments of 1958 and their significance.*[2]

---

[1] My first two Notes to the Academy of Sciences, "*Mouvements périodiques du pendule paraconique*" ("Periodic movements of the paraconical pendulum") and "*Analyse harmonique des mouvements du pendule paraconique*" ("Harmonic analysis of the movements of the paraconical pendulum"), presented in 1956 by Albert Caquot, were initially refused by the two permanent Secretaries, Robert Courrier and Louis de Broglie (see their letter of 20 November 1956 to Albert Caquot, reproduced in *Annex I.A. of the Second Volume* of this work; and see p. 31 above). It was only possible to publish these two Notes one year later, on 13 and 25 November 1957 [12, 9].

*In fact constant opposition manifested itself at the Academy of Sciences to the publication of all my Notes, more and more strongly.*

[2] In a recent work, "*L'univers de la relativité générale*" ("The universe of general relativity"), Editions Voies nouvelles, May 1996 [189], Marcel Macaire, an old student of the Ecole Polytechnique, wrote (p. 11 and pp. 66-67):

> *"(Maurice Allais's analysis) ought to have excited passionate interest; curiously, there were only a few echos and some generally hostile criticism. However his arguments were solid...*
>
> *"The criticism that was made in 1958 of Maurice Allais's results was lacking in objectivity.*
>
> *"What strikes one in Maurice Allais's work is the profusion of results that contradict Newton's laws. All his experiments showed anomalies. If there had been only one, we might be doubtful; but their number and their repetition should have inspired scientific circles to analyze the results and to take them into account. However, their publication only excited indifference and hostility. Truthfully speaking, the arguments invoked by the adversaries of Maurice Allais's contentions were contradictory. Some - for example - denied the periodicity of the curves that represented the observed results, while others admitted them, but claimed that Newton's laws were sufficient to explain them.*
>
> *"But there is something worse than indifference: silence. It should have been possible to find, in the specialized literature, a reasoned criticism of Maurice Allais's results, and, in the wider press, editorials admitting the novelty of the results obtained and reporting the arguments of his adversaries. However, nothing of the sort appeared; in 1958, the lack of interest in his experiments that questioned theories accepted for three centuries was total."*

Finally, for lack of funds, and *in spite of the dazzling success of the crucial experiments of July 1958*, I was compelled in June 1960 to close my laboratories at IRSID and Bougival, and to terminate all my experimental research.[3]

## Instant protest

2 - Truth to tell, this resistance to new ideas, as virulent as it is ignorant and incompetent, derives from a constantly underlying postulate: *any theory, any model, any experiment, any study, that departs from the established truths or contradicts them can only be erroneous.*

---

[3] See *Chapter I* below, §G.2, G.5 and G.6, pp. 215-216 and pp. 226-236; and *Chapter X*, §B.2, B.3, and B.4, pp. 685-689.

See also, in the second volume of this work, *Chapter VIII, Dogmatic Oppositions*, and *Annexes I.A and I.D* (see the Summary above, pp. 30-31).

This resistance to new ideas, too often blind and stubborn, certainly constitutes one of the greatest obstacles to the progress of science in all fields. *In all epochs*, discoveries *have met with the fanatical opposition of the mandarins of science.*[4]

However, whatever oppositions I may have encountered, and whatever obstacles of all sorts may have reared themselves up in my path, *they have never succeeded in stopping me defending what I think is the truth, and they never will.*[5]

[4] The most effective and most wicked tactic against new ideas is the conspiracy of silence, against which no defense is possible.

If in the end an idea finally triumphs in spite of all obstacles, the first person who defended it is often robbed of credit, and William James so justly wrote (Allais, 1966, *L'Economique en tant que Science* (Economics as a Science) [42]):

> *"Every doctrine passes through three stages. First it is attacked as absurd. Later it is granted to be true, evident but trivial. At last its true importance is recognised, and its opponents lay claim to the honour of having discovered it."*

[5] I still felt deeply *as a great injustice* the refusal of the Academy of Sciences to publish my *Note of 23 February 1960* on the *purely experimental* results of my optical observations of July 1958 (see *Chapter III* below, §B.4), and of the development of a real cabal against me, which finally forced me to close my laboratory at IRSID in June 1960 and, for lack of financial means, to cease all experimental research (see *Chapter I* below, Section G, pp. 213-236, and *Chapter X*, §B.2 and §B.3, pp. 685-689, and *Chapter VIII* of the *Second Part* of this work, see p. 30 above).

*That the mandarins of official science were able to join in this cabal did not, unhappily, change its nature at all. It only made it worse. A cabal remains a cabal.*

This cabal was not limited to the Academy of Sciences. Louis Armand, president in 1958-1959 of the Promotion Committee of the Ecole Polytechnique, took advantage of the rumors of the invalidity of my work on the paraconical pendulum to contribute to defeating my candidature in 1959 for the chair in economics of the Ecole Polytechnique.

See the *Third Introduction* to my *Traité d'Economie Pure* (Treatise on Pure Economics), §34, pp. 124-126 [53], and Louis Rougier, July 1959, "*Scandale à Polytechnique*" ("Scandal at Polytechnique") [236]. The memoir of Louis Rougier is reproduced in *Annex I.D* in the *Second Volume* of this work (see p. 31 above).

I am perfectly aware of the risks that I am taking by persisting in fighting on ground which, according to all official criteria, I do not hold. But is that a good reason for me to hold my tongue?[6]

*In fact, this book aims at being an immediate protest against the rooted prejudices and blind fanaticism of all those who oppose the progress of science with every power they possess.* As Rabelais wrote so long ago: *"Ignorance is the mother of all evil."*

---

[6] In my *Conference* of 28 February 1958 at the Ecole Polytechnique, and *before* the crucial experiments of July 1958, I had already said:

> *"Without doubt, as an economist, I am taking very great risks in performing physics research, and in running the risk of making mistakes. A professional can be forgiven if he falls into error, but nobody will ever pardon someone who is not in the profession if he goes wrong."*

As to whether I am wrong, or my adversaries are wrong, *only the facts can decide that question.*

# E TO THE READER

## *E.1 The production of this work*

### The presentation in two volumes

1. - Upon reflection, it has seemed best to me to present this work "*The anisotropy of space. The necessity for revision of some postulates of modern theories*" in two volumes, respectively entitled "*The experimental data*" and "*Experimental and theoretical supplements*", and to publish the first volume at this time.[1]

### The first volume

2 - In the following I examine successively:[2]

- My experiments on the paraconical pendulum with anisotropic support (*Chapter I*),
- My experiments on the paraconical pendulum with isotropic support (*Chapter II*),
- My experiments on optical deviations of sightings at marks of 1958, and their continuation in 1959 (*Chapter III*),
- Two very significant series of previous experiments, those of Esclangon and those of Miller (*Chapter IV*),
- The semi-annual and annual periodic structure of the observations analyzed, and their interdependences (*Chapter V*),
- The anisotropy of space (*Chapter VI*),
- The interpretation and implications of the observations analyzed (*Chapter VII*),
- A plan for implementation of simultaneous experiments (*Chapter VIII*),
- The unceasing opposition to new ideas down through history (*Chapter IX*),
- Finally, new perspectives that are opening up today (*Chapter X*).

[1] It will only be possible to publish the second volume in some months.

[2] The present work owes an *enormous* debt to my wife Jacqueline, whose suggestions and criticisms have always been very constructive and extremely useful.

The reader will find all the references relating to the developments of this work[3] and a list of names at the end of this *First Volume*.

**The second volume**

3 - *In the second volume* of this work, I shall examine certain *absolutely essential* developments of the questions discussed in the various Chapters of this *First Volume*, which have not been analyzed here for lack of space.

Moreover I shall attach, as *Appendices* to that second volume, various *Memoirs* that are directly connected with the developments in the ten *Chapters* of this *First Volume*, as well as texts relating to the oppositions that were raised against me and which I have had to face (*Annexes I*) and the Notes prepared for the Academy of Sciences that I was not able to publish in the Proceedings (*Annexes II*).[4]

**Inevitable difficulties**

4 - Certain passages in the present work are *very technical*. Certainly, as much as possible, I have avoided all mathematical formalism, and have in principle banished all discussions requiring mathematical developments to the *Second Volume* of this work.

However certain questions, such as for example the presentation of the principles of the theoretical calculation of the luni-solar influence upon the movement of the paraconical pendulum according to the current theory of gravitation, are *so important* that it has seemed necessary to me to include them in my explanation.[5]

The text also includes very numerous notes and very numerous references, which may give some difficulty to the reader; but, above all, I wished all my developments to be rigorous and completely unambiguous.

---

[3] I particularly recommend the reader always to refer to *the original works, and not to second-hand commentaries*.

[4] See the *Summary* above, pp. 28-33.

[5] See notably *Chapter I*, §B.2 and F.3.2, pp. 118-129 and 206-212; and *Chapter II*, Section I, pp. 320-325.

Finally, the following text includes some repetitions. In fact they have been *inevitable*, because the explanation of each Chapter constitutes, in itself, an organic whole in connection with the explanations in the other Chapters, and because all of the questions studied are mutually interlinked by numerous and relatively complex relationships.

### Quantitative analyses

5 - My analyses of 1954-1960 have been completed by very numerous calculations performed *from January 1995*. All these have *entirely confirmed* the very great underlying coherence of all the observations performed from 1954 to 1960, and have clarified their significance and implications. In fact, these observations are *completely inexplicable* in the framework of currently accepted theories.

In any case, *the quantitative analyses* presented in the first five Chapters of this *First Volume* are *totally independent of any hypothesis and any theory whatever.*

### A composition entirely dedicated to rigor and clarity

6 - Such as it is, and although it has passed through very numerous successive versions during its preparation from July 1995 to February 1997, this work is certainly still *very imperfect*. Its composition has indeed involved very many difficulties resulting from the complexity of the questions analyzed and from the inevitable limitations upon explanation.[6] These problems have only been partially overcome.

One sole principle has guided me: "*To sacrifice everything to rigor and clarity.*"[7]

---

[6] I have also had to dig up again all the necessary materials for the preparation of this work. After more than thirty-five years, that was not always easy.

[7] Here I must give heartfelt thanks to Anne-Marie and Alain Villemur, who have been exceptional collaborators. Anne-Marie Villemur has been able to prepare the successive versions of this work with remarkable efficiency. Alain Villemur has executed the very numerous calculations and prepared the graphics corresponding to the quantitative analyses with great intelligence.

## E.2 Against the current

• Undoubtedly the analyses in this work, which run *totally against the current of the "established truths" of today*, will excite violent opposition. It will be hard for anyone to admit that an economist such as myself will have been capable, by his experiments and his analyses, of interposing a check to physical theories that are taught everywhere as definitive truths.

I remain convinced, as I was long ago in 1959, that there is nothing to be done against the fanatics. Blind and deaf, stubborn in their certainty, they will deny everything *en bloc*.[1] But today, as in 1958,[2] there are some honest men, quite ready to examine the facts, even if at first sight they appear to be opposed to their own convictions.

I have been told that the claim on my part to two fundamental discoveries can only exasperate certain readers. Certainly such a warning does carry some weight; it is unquestionable. But, yet again, underestimating the significance of the new phenomena that I have demonstrated is not really the way to alert the scientific world, or at least the part of the scientific world that is not blinded by its prejudices, by its entrenched positions, and by a blind faith in established truths.

It is undeniable that the implications of the *incontrovertible* existence of the anomalies of the paraconical pendulum and of the optical anomalies that I have demonstrated, and of the optical observations of Esclangon and the interferometric observations of Miller,[3] are such as to compel profound revision to the very fundamentals of current theories.

---

[1] More likely, the opposition will try simply to maintain silence. In *Chapter X*, I shall give a particularly significant recent example in connection with the quantitative analysis of the observations of Miller presented in *Chapter IV* below.

[2] In 1960 at least nine members of the Academy of Sciences (Caquot, Cot, Darrieus, Kampé de Fériet, Léauté, Pérard, Roy, Tardi, and Thiry) were convinced of the necessity for continuation of my experiments, whose results seemed to them absolutely essential, and they put forth the greatest possible efforts for me to receive the necessary means.

[3] A large part of these implications remained *unnoticed* by these two authors.

This is a certainty, but, as long ago in the 1950s, it will collide with the opposition, as blind as it is fierce, of all those who only base their thoughts upon established truths.

• At first glance some of my judgments in this Introduction and the following Chapters may perhaps seem somewhat excessive. But what is really excessive, and in fact inadmissible, is the type of indifference with which the observations of the movement of the paraconical pendulum and the optical observations that I associated with them, and the interferometric observations of Miller have been ignored and buried.[4] As Bouasse once said:

> *"One would not be obliged to use such harsh language if one was speaking to pure spirits; but naked and fleshless truths make little impression and only leave weak traces in the brain that are easily effaced... The great advantage of a powerful attitude is that it obliges people to reflect."*

I here paraphrase what Alexis de Tocqueville said long ago in another context:

> *"I hope to have written the present book without prejudice, but I do not claim to have written it without passion.*
>
> *"Whenever I have met with obvious errors in accepted theories or in recognized facts, I have taken care to turn the light upon them, so that, seeing the obstacles placed in the way of scientific progress, one can better understand their nature.*
>
> *"In order to attain this goal, I confess that I have not feared to wound anyone, neither individuals nor opinions... however respectable they may have been. I have often done this with regret, but never with remorse. I can only hope that those I may have thus displeased can pardon me, in consideration of the disinterested and honest goals I have pursued."*

*The fundamental motivation for my efforts has been to explain what I conceive to be the truth.*

Saint-Cloud, 15 February 1997

---

[4] See *Chapter I* below, Section G, pp. 213-236; *Chapter X*, §B.2, §B.3, and §B.4, pp. 685-689; and, in the second volume of this work, *Chapter VIII* and Annexes IA to ID (see the *Summary* above, pp. 30-31).

## Message to my readers

*I would be very much obliged to the readers of this work if they could share their observations with me.*

*I thank them gratefully in advance.*

# Chapter I
# MY EXPERIMENTS ON THE PARACONICAL PENDULUM WITH ANISOTROPIC SUPPORT - 1954-1960

*The important facts are the crucial facts - in other words, those that can confirm or invalidate a theory. After that, if the results do not agree with expectations, true scientists will not experience a sentiment of embarrassment that they will hastily dispel with some magical helping hand; on the contrary, their curiosity will become highly over-excited; they know that their efforts, their momentary setback, will be repaid a hundredfold, because truth is there, somewhere very near, still hidden and decorated with the attraction of mystery, so to speak, but on the point of being unveiled.*

Henri Poincaré *

My experiments on the paraconical pendulum with anisotropic support[1] were carried out from 1954 to 1960. They gave rise to ten Notes to the Academy of Sciences in 1957, 1958, and 1959,[2] and to an overall account in 1958, "*Doit-on Reconsidérer les Lois de la Gravitation?*" ("Should the Laws of Gravitation be Reconsidered?") [23].[3]

Moreover, I was led to prepare a very great number of memoirs, the references to which are given at the end of this work, and which will be published in a future volume.[4]

My work was the subject of three Conferences organized by the *Alexandre Dufour Circle*: "*Should the Laws of Gravitation be Reconsidered? On a New Experiment in Mechanics*" of 22 February 1958; "*Should the Laws of Gravitation be Reconsidered? New Results, Review and Perspectives*" of 7 November 1959; and "*Are the Periodicities Observed in the Movement of the Paraconical Pendulum Real or Not? A Generalization of Schuster's Test to the Case of Autocorrelated Time Series*" of 18 March 1957.[5]

In view of the very concept of the present work, the following description of the anomalies of the paraconical pendulum with anisotropic support *will necessarily limit itself to essential matters.*[6]

---

* Henri Poincaré, 1913, *Dernières Pensées* (Last Thoughts), Flammarion, p. 336 [226].

[1] My experiments in 1959-1960 on the paraconical pendulum with *isotropic support* are examined in *Chapter II* below.

[2] See above, *Introduction*, §B.2.7, p. 51.

[3] See above, *Introduction*, §A.2.2, p. 42.

[4] See above, *Introduction*, §A.2.3, p. 42.

[5] See above, *Introduction*, §B.2.9., p. 52.

[6] All the useful adjuncts are given in the *Second Volume* of this work (see *Introduction*, §E.1.3 above, p. 74).

# A GENERAL CHARACTERISTICS OF THE EXPERIMENTS ON THE PARACONICAL PENDULUM WITH ANISOTROPIC SUPPORT, AND THEIR RESULTS

## *A.1 Experimental conditions*

### The pendulum utilized

1 - Although I successively used various types of pendulum, I shall limit myself here to a very summary description of the arrangements most generally used.[1]

The photographs opposite represent the overall arrangement of the pendulum and its support.[2]

The pendulum was an *asymmetric* pendulum consisting of a vertical bronze disk of 7.5 kg, 21.8 cm in diameter, fixed upon a bronze rod suspended from a stirrup E made from bronze, which itself rested upon a steel ball 6.5 mm in diameter which was able to roll in any direction upon a plane horizontal surface S.

This surface itself rested upon a cutaway circular support S' made from aluminum having an appendix A, and 4.5 cm in thickness. The cutaway allowed the swinging pendulum to rotate through a total angle of 210 grads. This support S' was supported by three micrometric screws V attached to a support S" which was bolted to a beam, itself clamped against the ceiling by a system of transverse beams.[3]

---

[1] This pendulum was used, in particular, in the month-long series of observations of November-December 1954, June-July 1955, July 1958, November-December 1959, and March-April 1960.

[2] These four photographs are reproductions of *Annexes I through IV* of my *Note* of 13 November 1957 to the Academy of Sciences "*Observations of the movements of the paraconical pendulum*" [12].

Moreover, I reproduce a photo (p. 83) of my laboratory chief performing an observation (photo taken in 1958 by Georges Lacoste).

[3] The direction of these transverse beams is shown in *Annex I* by the vector $\overrightarrow{PQ}$. This vector is perpendicular to the beam which supports the pendulum.

*PHOTOGRAPHS I TO IV*

**PARACONICAL PENDULUM AND SUPPORT**

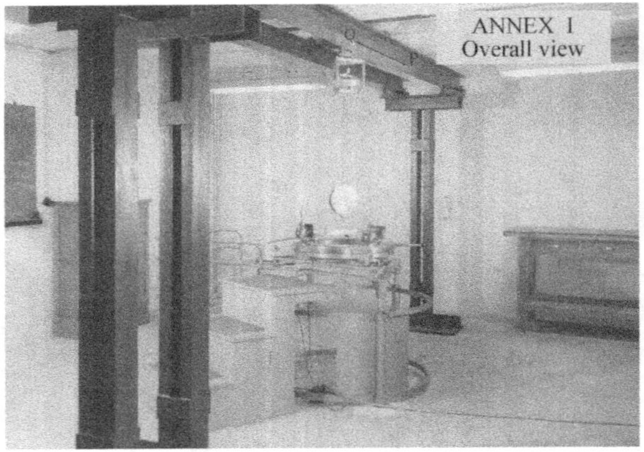
ANNEX I
Overall view

ANNEX II Measuring circle

ANNEX III
Suspension system

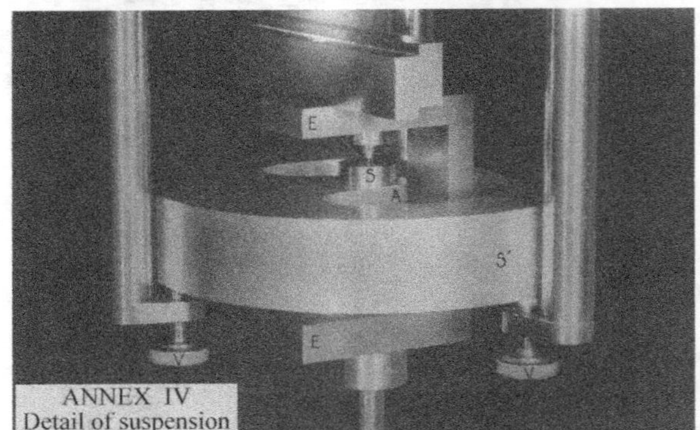
ANNEX IV
Detail of suspension

*Sources: Note* of 13 November 1957 to the Academy of Sciences "*Observation of the movements of the paraconical pendulum*"

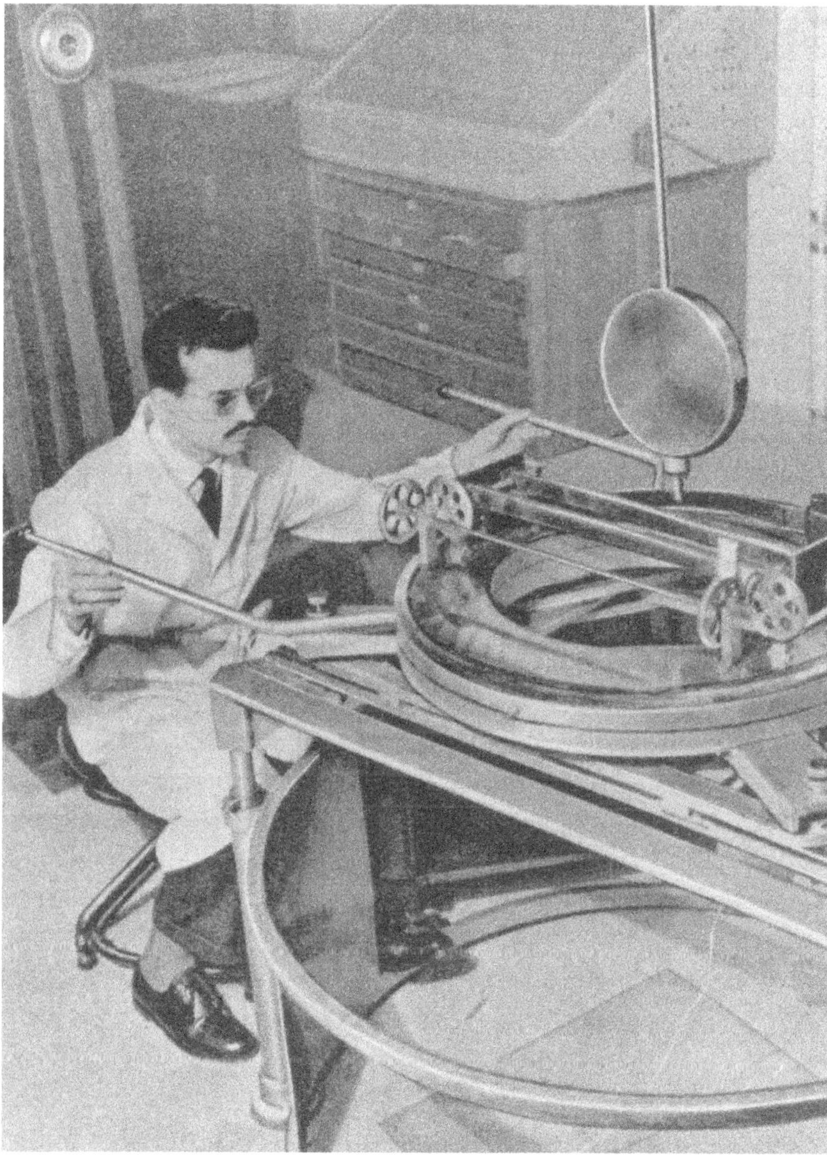

Photo of my laboratory chief, Jacques Bourgeot

Georges Lacoste, *Les progrès dans nos connaissances physiques des champs ouvrent la voie à de passionnantes recherches*, (Progress in our physical knowledge of fields, opening the door to fascinating researches), Sciences et Avenir, no. 135, May 1958, p. 272 [167].

With the rod of the pendulum and its stirrup weighing 4.5 kg, the total weight of the pendulum was 12 kg, and the length of the equivalent simple pendulum was about 83 cm.[4]

The steel balls were high precision S.K.F. balls, and the bearing surfaces were made of tungsten and cobalt carbide. I called this pendulum *paraconical* because of its suspension upon a ball.

### Experimental procedure

2 - The experiments were performed in the basement where my laboratory was located, and the center of gravity of the pendulum was about 1.5 m below natural ground level.

The pendulum was released from a stationary position every 20 minutes with an initial amplitude of about 0.11 radian, by burning a thread.[5] The movement of the pendulum was observed for about 14 minutes, by sighting at the point of a needle which projected from its lower end to 105 cm from the center of the ball.[6]

Generally, the point described a curve which could be considered as a flattened ellipse, of which the plane of the major axis was observed with a system of sights placed upon a circle C, centered upon the axis of the pendulum at rest, and bearing a division into grads and a vernier. This system made it possible to determine the azimuth of the plane of oscillation of the pendulum *to an accuracy of the order of a tenth of a grad.*

---

[4] The corresponding period of oscillation of the pendulum $T = 2\pi\sqrt{l/g}$ was 1.826 seconds.

In CGS units, the moment of inertia $B$ of the pendulum about an axis passing through the center of the ball and perpendicular to the disk was $83.11 \times 10^6$, and the moment of inertia $A$ of the pendulum about an axis passing through the center of the ball and parallel to the disk was $82.89 \times 10^6$. The *coefficient of asymmetry* $\delta = 2(B-A)/(B+A)$ was accordingly $0.269 \times 10^{-2}$. The moment of inertia $C$ of the pendulum about its vertical axis was $270 \times 10^3$. The *gyrostatic coefficient* $\gamma = 2C/(A+B)$ was accordingly $\gamma = 0.325 \times 10^{-2}$.

[5] The amplitude I selected was based upon the maximum value for which the ball did not skid upon the surface S.

During a 14 minute experiment the amplitude decreased from 11 cm to about 9 cm.

[6] This distance $l' = 105$ cm is of course different from the length $l = 83$ cm of the equivalent simple pendulum.

Furthermore, a system of two movable parallel bars B which could be shifted with respect to the measurement circle made it possible to measure the two axes of the pendulum in centimeters, and to determine the azimuth of the plane of the disk, i.e. of the central trihedral of inertia of the pendulum.

The pendulum was stopped after 14 minutes. Six minutes later it was released again *in the plane of the last observed azimuth.* The series of azimuth observations were accordingly *enchained*, successive releases being performed every twenty minutes, day and night. Each period of 24 hours accordingly comprised 72 enchained series of observations.[7]

In order to avoid any systematic influence, the ball supporting the pendulum was changed *for each experiment*, every 20 minutes, and the surface S was changed at the start of each week of observation.

## Anisotropy of the support

3 - The support S" was characterized by a very *slight* difference of elasticity in two perpendicular planes. Due to this anisotropy of the support, the plane of oscillation tended toward a plane in a direction Σ perpendicular to the beam, whose azimuth was approximately 171 grads, counting the azimuths from the North in the anticlockwise sense.[8]

Another result, therefore, was an *average tendency to form ellipses* if the pendulum was released in some other plane.[9]

---

[7] Thus, a month-long series of observations included 2160 twenty-minute experiments.

[8] This direction Σ was parallel to the vector $\overrightarrow{PQ}$ (note 3 above).

[9] This influence of the support was determined in a precise manner by experiments in which the pendulum was released in different azimuths, while eliminating the influence of time by a random choice of the starting azimuths (see §E.3 below).

## Continuous enchained experiments

4 - During each continuous series of enchained experiments, the observers worked average shifts of 3 hours each, day and night.[10]

As far as I know, *this is the only example in the literature of observations carried out in a continuous manner over durations of the order of a month.*[11, 12]

[10] The number of observers was seven on average. During the simultaneous experiments of 1958 at Bougival and Saint-Germain (Section C below) the total number of observers was accordingly fourteen.

These observers were IRSID technicians working overtime. Their professional conscientiousness was remarkable.

[11] The experiments of Esclangon (*Chapter IV*, §B.2, below) were performed over about a year, but they only included, on average, about 15 observations per month.

Miller's experiments were indeed performed in a continuous manner, but only for durations of 6 to 8 days at four different epochs of the year (*Chapter IV*, §C.3, below).

[12] It was possible to perform certain experiments in 1954 with a long pendulum suspended on a wire, thanks to a circular opening about a meter in diameter which was pierced between the two rooms of the laboratory, of which one was above the other (note 6 of §B.1.5 of *Introduction*, above).

## *A.2 Enchained experiments - Illustration for the case of the month-long series of June-July 1955*

During each *continuous* series of enchained experiments, the tendency of the plane of oscillation was not to fix itself near the direction $\Sigma$ of anisotropy of the support; rather, the variation of its azimuth as a function of time appeared as an oscillation that was very irregular, at least at first inspection.

For example, during the series of continuous experiments in 1955 from 12h UT on 7 June to 12h UT on 7 July, quite large variations were observed.[1] During a single period of 24h, the variation of the azimuth sometimes reached and even exceeded 100 grads. The average azimuth $\overline{\phi}$ was 150 grads, *i.e.* 21 grads less than the anisotropy azimuth $\Sigma$ which was 171 grads, all the azimuths being reckoned *in the anticlockwise sense from the North.*

*Graph I* represents the azimuths of enchained experiments from 12h on 7 June until 14h on 12 June. The times of meridian passage of the Moon are shown by the legend "12h LT".[2] And *Graph II* represents the azimuth variations over the entire period, 12h on 7 June until 12h on 7 July.[3] Quite analogous azimuth variations were observed during the other month-long series of experiments.[4]

[1] In all my experiments, the time was reckoned in *universal time* (UT).

[2] The legend "12h LT" means 12 o'clock by lunar time.

[3] All the graphic illustrations from 1954 to 1960 are here photographically reproduced *without any alteration.*

[4] As an illustration, see *Graph XXII* which corresponds to the month-long series of enchained experiments of July 1958 at Bougival (§C.2.4 below).

*GRAPH I*

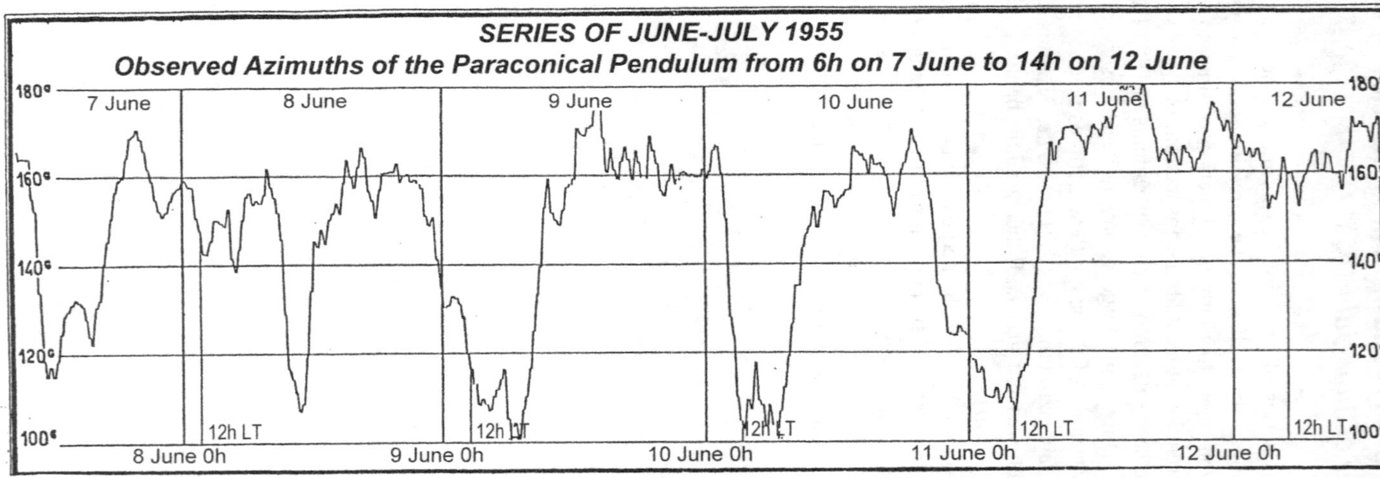


***Legend :*** The angles are reckoned in grads from North in the anticlockwise sense. The azimuth of 100 grads corresponds to the direction perpendicular to the meridian. The azimuth of 200 grads corresponds to the meridian.
12h LT: time of meridian passage of the Moon

***Sources :*** *Note* of 18 November 1959 to the Academy of Sciences, *Analyse harmonique des mouvements du pendule paraconique* (Harmonic analysis of the movements of the paraconical pendulum), and *Graph III A* of my Conference of 22 February 1958.

*GRAPH II*

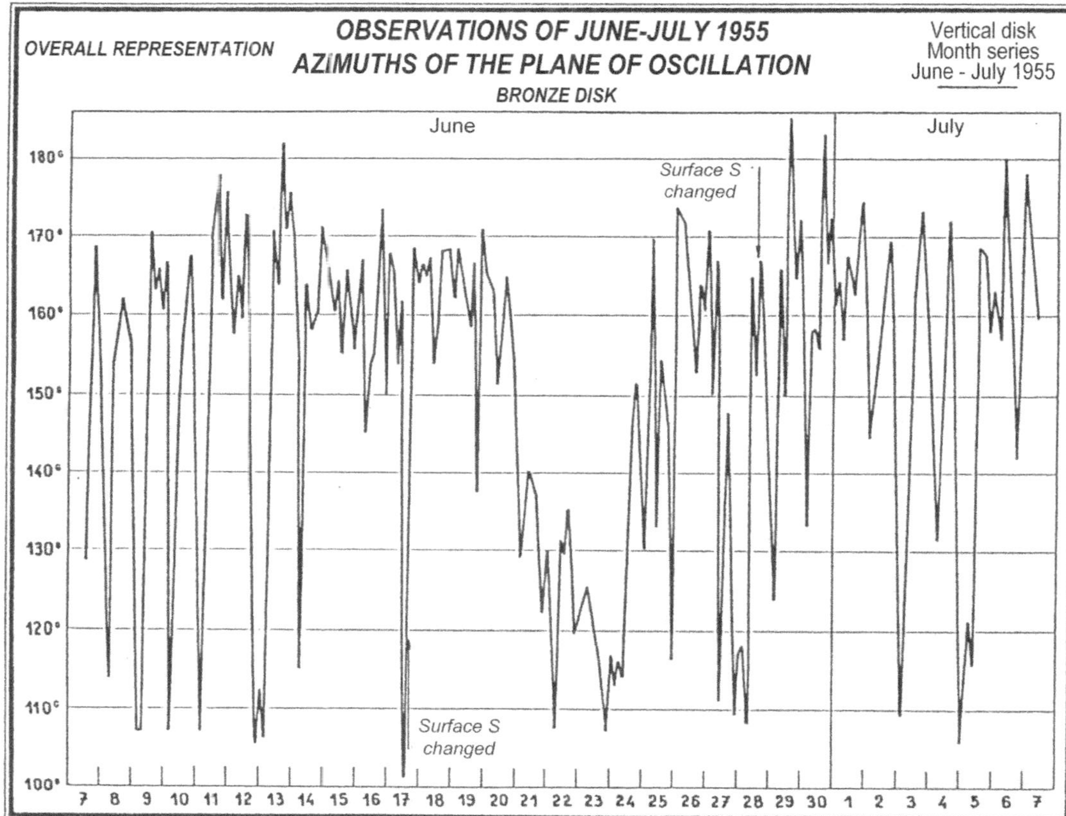


***Legend :*** The angles are reckoned in grads from North in the anticlockwise sense. The azimuth of 100 grads corresponds to the direction perpendicular to the meridian. The azimuth of 200 grads corresponds to the meridian.

***Source :*** *Graph II B 3* of my Conference of 22 February 1958.

## A.3 *The experiments performed in 1954-1960*

From 1954 to 1960, a great number of experiments were performed with pendulums of different types, either over one-month periods or over shorter periods. Below, I shall limit myself to the one-month series and the two-week series.

### General characteristics of the seven one-month series of enchained experiments

1 - From 1954 to 1960, *seven continuous one-month series of enchained observations* of the asymmetric paraconical pendulum with anisotropic support were performed.

*Table I* below shows, for each one-month series, the duration in days, the average azimuth $\overline{\phi}$, the minimum azimuth $\phi_m$, the maximum azimuth $\phi_M$, their average $(\phi_m+\phi_M)/2$, their ratio $(\phi_m+\phi_M)/2\overline{\phi}$, the total azimuth variation $D=\phi_M-\phi_m$, the amplitude $2R_{24}$ of the 24h wave, the amplitude $2R_{25}$ of the 25h wave, the ratio $R_{25}/R_{24}$, and the ratios $R_{24}/D$ and $R_{25}/D$.[1] These values are shown *in both grads and in degrees*.[2, 3]

The total variation of the azimuth always stays less than 166 grads, *due to the returning tendency of the support* that results from its anisotropy. In fact, the average azimuth of 164 grads is relatively close to the direction of anisotropy of the support, which is 171 grads.

The pendulum used during these monthly series of enchained experiments was always identical to that of the series of June-July 1955,[4] except for the series of June-July 1954, in which the pendulum consisted of a vertical disk and two horizontal disks of bronze.[5]

---

[1] For the diurnal waves of 24h and 25h, see §A.5 below, pp. 96-101.

[2] The Graphs of *Chapter V*, Section B are presented in degrees, in order to make it easier to compare them with the Graphs representing Miller's experiments.

[3] In view of the *very restricted* means for calculation available at the time, most of the calculations were performed for a period of 25h instead of 24h50m, which avoided interpolating the observations every ten minutes for application of the Buys-Ballot filter (for the Buys-Ballot filter, see §5 below, note 1, p. 96).

[4] §A.1 above.

[5] Its total weight was 19.8 kg. After the series of experiments of June-July 1954, I lightened the pendulum in order to reduce the disturbing influence of the balls (see §E.4 below).

As results from *Table I*, the amplitudes of the periodic components of 24h and 25h are *relatively* much greater for the two series of November-December 1954 and June-July 1955, than for all the other series. It is only *recently*, in 1995, that I have been able to give a plausible explanation.[6]

**The two-week series of enchained experiments**

2 - Two other 15-day series of enchained experiments with the bronze disk were performed from 18 March to 2 April 1955, and from 14 June to 30 June 1958 at Saint-Germain and at Bougival.

Moreover, a continuous series of enchained experiments was performed from 21 September to 6 October 1955, using a symmetrical pendulum ($A = B$) consisting of a sphere of lead weighing 12.2 kg. The variations of azimuth were completely comparable to those corresponding to the asymmetric pendulum consisting of the bronze disk.[7]

**Presentation of the observations**

3 - The azimuths observed at the end of each 14-minute experiment were presented in large Tables, each column corresponding to a given day.[8]

---

[6] See below *Chapter V*, §B.2.

[7] §A.1 above.

[8] These Tables will be published in the *Second Volume* of this work, *Chapter I*, Section A (see p. 28 above).

*TABLE I*

## *ASYMMETRIC PARACONICAL PENDULUM WITH ANISOTROPIC SUPPORT MONTH-LONG ENCHAINED EXPERIMENTS, 1954-1960*

***Azimuths and Periodic Components of 24h and 25h in grads and in degrees***

***$\Sigma$ = Azimuth of anisotropy of the support = 171.16 grads = 154.04 degrees***

| Periods | Length in days (1) | Average date (2) | $\overline{\phi}$ (3) | $\phi_m$ (4) | $\phi_M$ (5) | $\frac{\phi_m + \phi_M}{2}$ (6) | $\frac{\phi_m + \phi_M}{2\overline{\phi}}$ (7) | D $= \phi_M - \phi_m$ (8) | $2R_{24}$ (9) | $2R_{25}$ (10) | $R_{25}/R_{24}$ (11) | $R_{24}/D$ (12) | $R_{25}/D$ (13) |
|---|---|---|---|---|---|---|---|---|---|---|---|---|---|
| 1 1954 9 June - 9 July | 30 | 174.5 | 164 (148) | 102 (92) | 268 (241) | 185 (166) | 1.13 | 166 (149) | 2.0 (1.8) | 3.2 (2.9) | 1.58 | 0.012 | 0.019 |
| 2 1954 16 Nov - 22 Dec | 36 | 337.5 | 161 (145) | 93 (84) | 253 (228) | 173 (156) | 1.08 | 160 (144) | 10.3 (9.3) | 12.9 (11.6) | 1.25 | 0.064 | 0.080 |
| 3 1955 7 June - 7 July | 30 | 537.8 | 150 (135) | 99 (89) | 180 (162) | 140 (126) | 0.93 | 81 (73) | 11.7 (10.5) | 14.0 (12.6) | 1.20 | 0.129 | 0.155 |
| 4 1958 B 2 July - 1 August | 30 | 1658.5 | 161 (145) | 145 (130) | 177 (159) | 161 (145) | 1.00 | 32 (29) | 1.4 (1.3) | 2.2 (2.0) | 1.60 | 0.044 | 0.068 |
| 5 1958 2 July - 1 August | 30 | 1658.5 | 164 (148) | 141 (127) | 187 (168) | 164 (148) | 1.00 | 46 (41) | 0.8 (0.7) | 2.1 (1.9) | 2.71 | 0.017 | 0.045 |
| 6 1959 20 Nov - 15 Dec | 25 | 2161.75 | 171 (154) | 142 (128) | 200 (180) | 171 (154) | 1.00 | 58 (52) | 2.5 (2.3) | 1.3 (1.2) | 0.54 | 0.043 | 0.023 |
| 7 1960 16 March - 16 April | 31 | 2282 | 174 (157) | 150 (135) | 206 (185) | 178 (160) | 1.02 | 56 (50) | 1.8 (1.6) | 1.5 (1.4) | 0.84 | 0.032 | 0.027 |
| Averages | | | 164 (148) | 125 (112) | 210 (189) | 167 (150) | 1.02 | 86 (77) | 4.4 (4.0) | 5.3 (4.8) | 1.39 | 0.049 | 0.060 |

*Notes :* 1. - All the experiments were performed at IRSID in Saint-Germain, except for experiment 4 which was performed at Bougival.

2. - All the measurements are given in grads. The angles are reckoned from North in the anticlockwise sense. The measurements in degrees are given in parentheses.

3. - The average date of each month-long series is reckoned in days from 1 January 1954.

4. - $\phi_m$ and $\phi_M$ represent the minimum and maximum values of the azimuth of the plane of oscillation. And $\overline{\phi}$ represents the average values of the azimuths $\phi$.

## *A.4 The Foucault Effect*

In fact it is *particularly significant* that, in the enchained series, the tangent at the start of the *average* curve of the various azimuth curves corresponding to the 14-minute series of elementary observations *corresponded exactly to the Foucault effect.*[1]

*Graph III* represents the displacements of the azimuths of the plane of oscillation and of the inertial trihedral of the lead sphere (the symmetrical pendulum) during the series of enchained experiments from 7 to 13 December 1955, and for the bronze disk (asymmetrical pendulum) during the enchained experiments of 4 January 1956.[2]

*In both these cases*, the average of the azimuth of the plane of oscillation, at the start, corresponds *exactly* to the Foucault movement. This only changes when the minor axis of the ellipse has a significant value.

For the lead sphere, *Graph IV* represents the averages during the period 21 September to 5 October 1955 of the azimuths of the oscillation plane and of the central trihedral of inertia from 0h to 12h, from 12h to 24h, and from 0h to 24h. Here also at the start, on average, we see the Foucault effect. It disappears when ellipses appear.[3]

In general, the azimuthal motion of the central trihedral of inertia is different from that of the plane of oscillation. In the case of the asymmetrical pendulum *we show, and experiment confirms*, that the plane of the disk tends to coincide with the plane of oscillation of the pendulum.[4]

---

[1] At Saint-Germain, the angular rotation due to the Foucault effect $-\omega \sin L$ is $-0.55 \times 10^{-4}$rad/s, which, *over 14 minutes*, corresponds to an angular displacement of

$$-0.55 \times 10^{-4} \frac{200}{\pi} 60 \times 14 = -2.94 \text{ grads in 14 minutes}$$

[2] The choice of a lead sphere, which corresponds to a symmetrical pendulum ($B = A$), had the advantage of demonstrating the movement of the trihedral of inertia better than in the case of the asymmetrical pendulum ($B \neq A$) consisting of a disk.

[3] *Graphs III and IV* show that, during a 14 minute experiment, the amplitude reduces from 11 cm to around 9 cm, *i.e.*, with $l' = 105$ cm, from around 0.105 to around 0.086 radians, with an average value of around 0.10 radians.

[4] A detailed analysis of the movement of the asymmetrical paraconical pendulum with anisotropic support is presented in the *Second Volume* of this work, *Chapter I*, Section B (see above, p. 28).

*GRAPH III*

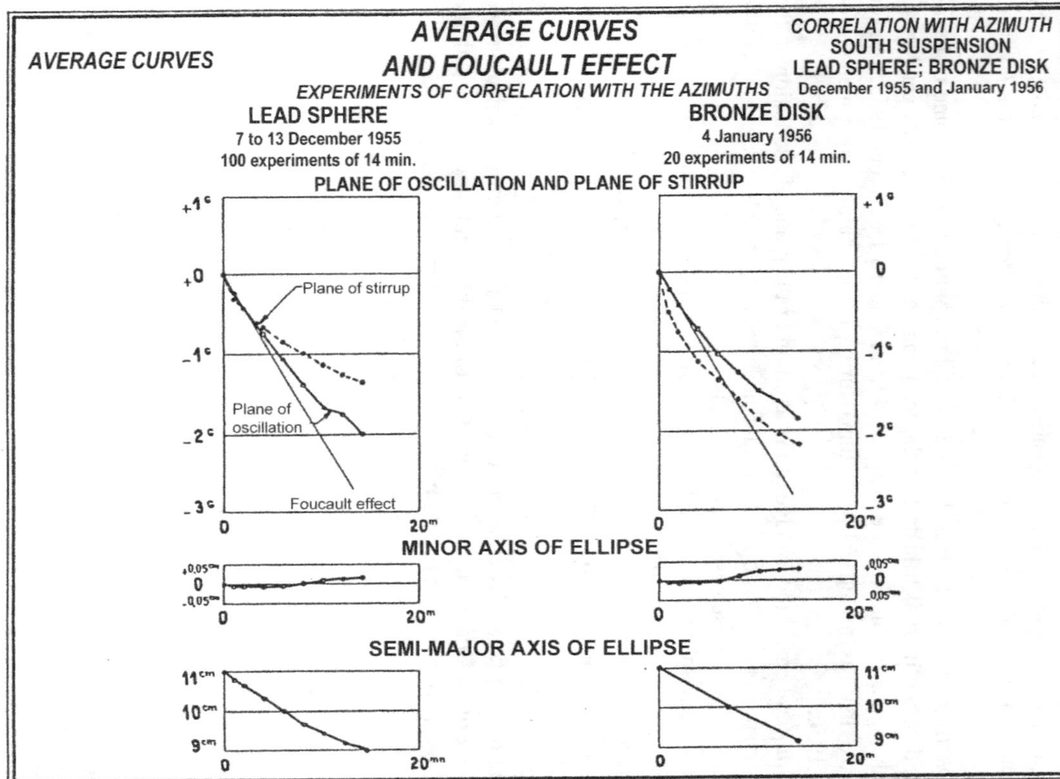


**Legend :** azimuth of the plane of oscillation - - - azimuth of the principal trihedral of inertia

**Source :** *Graph III A 2* of my Conference of 22 February 1958.

*GRAPH IV*

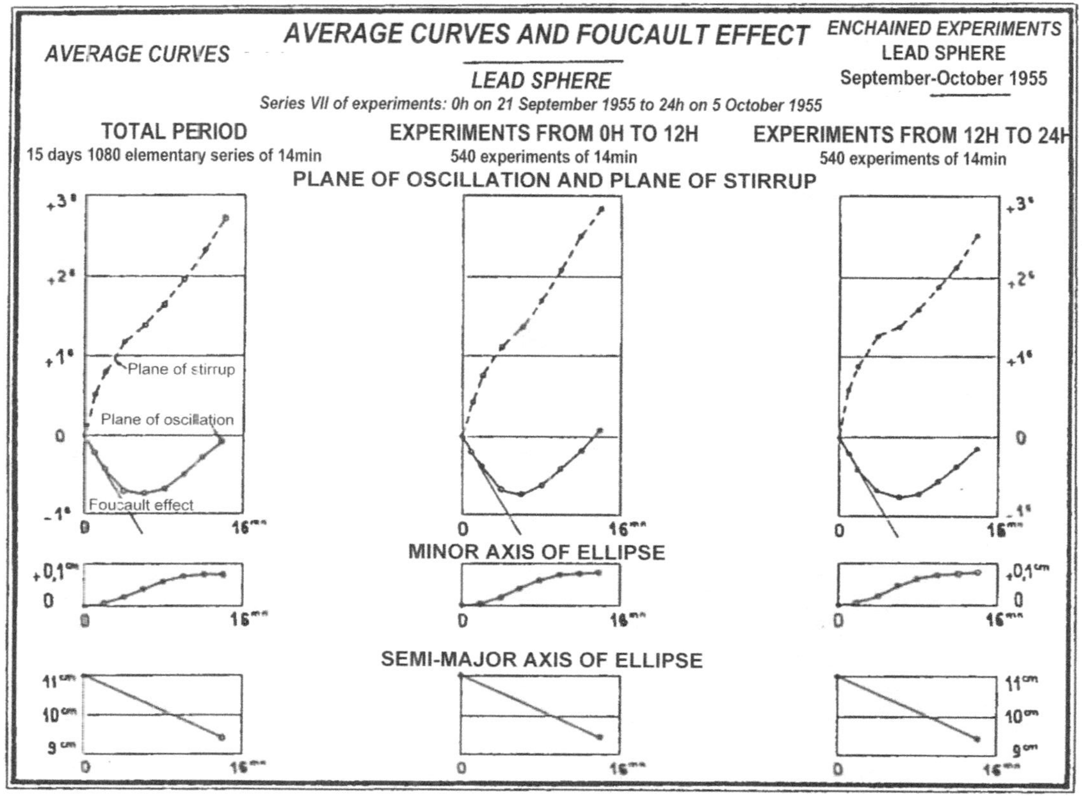


***Legend :*** ●—●—● azimuth of the plane of oscillation – – – ● – – – ● azimuth of the principal trihedral of inertia

***Source :*** *Graph III A 1* of my Conference of 22 February 1958.

## A.5 *The diurnal periodic structure of the enchained experiments on the paraconical pendulum with anisotropic support*

### Harmonic analysis of the enchained experiments

1 - The sequence of azimuths of the plane of oscillation of the paraconical pendulum with anisotropic support observed during a series of enchained experiments forms a time series that can be analyzed in different ways: graphic representation, harmonic analysis (Buys-Ballot filter, fitting to a group of waves by Darwin's method or the method of least squares, periodogram, and correlogram), or representation by auto-regressive schemes.[1]

The results obtained can be considered according to three criteria:

- the probability of obtaining, by chance, an amplitude greater than a certain value for a certain harmonic component;[2]

---

[1] Due to its convenience, and in view of the very limited means for calculation available at the time, we made great use of the Buys-Ballot method.

The principle of this method is essentially as follows:

Suppose that we have a series of $N = pq$ values $x_i$, where $q$ is the period to be considered. We arrange the observations in a table having $p$ lines each of which contains $q$ successive values, and we calculate the averages by column,

$$\begin{array}{rcccc} & x_1 & x_2 & \cdots\cdots\cdots & x_q \\ & x_{q+1} & x_{q+2} & \cdots\cdots\cdots & x_{2q} \\ & \vdots & \vdots & \cdots\cdots\cdots & \vdots \\ & x_{(p-1)q+1} & x_{(p-1)q+2} & \cdots\cdots\cdots & x_{pq} \\ \hline \text{Averages :} & \overline{x_1} & \overline{x_2} & \cdots\cdots\cdots & \overline{x_q} \end{array}$$

Finally we eliminate the trend by considering a $(q+1)^{\text{th}}$ column whose average is $\overline{x_{q+1}}$. The trend is defined by the ratio $(\overline{x_{q+1}} - \overline{x_q})/q$.

[2] See §B.1.3. below, and *Second Volume* of this work, *Chapter VI.*

- the agreement of phases for the original series and for the sum of its periodic components, for each of the two series of 15 days into which a series over 30 days can be decomposed;[3]
- the quality of the fittings, as characterized by the tightness of the dispersion of the points around the fitting sinusoids.

From the totality of all these analyses applied to the various series of experiments, I have become *certain* that the series of experiments we performed presented a remarkable periodic structure, and in particular possessed a periodic component of 24h50m.[4]

### The series of enchained experiments of June-July 1955

2 - For illustration, I limit myself here to description of the results obtained from an *overall harmonic analysis* into the thirteen waves of the theory of tides, applied to the month-long series of 2163 enchained experiments of June-July 1955[5] *(Table II).*

For comparison, I also give the results corresponding to the atmospheric pressure series observed at Le Bourget during the same period *(Table II).*

The components $K_1$ ($T = 23.93$h) and $M_1$ ($T = 24.84\text{h} = 24\text{h}50\text{m}$) of the series of azimuths appear to be particularly *significant.*

It should be remarked that, for the set of thirteen waves, the total of the percentages in relation to the atmospheric pressures is about four times smaller than for the azimuths, even though the atmospheric pressure is not an entirely random value, and even though it includes well-known lunisolar periodicities.[6]

---

[3] See §B.1.2. below.

[4] We have

$$\frac{1}{24} - \frac{1}{29.5305 \times 24} = \frac{1}{24.8412}$$

$$24.8412\text{h} = 24\text{h}50\text{min}28" \qquad (24.8412/24 = 1.03505)$$

where 29.5305 days represents the synodic period of the Moon.

[5] The analysis of this series and of the atmospheric pressure series was performed by the *National Hydrographic Service of Paris* and by the *Hydrographic Institute of Hamburg.*

The thirteen waves considered are those generally used by hydrographic institutes.

[6] For the coefficients corresponding to the theory of tides, see §E.5 below, *Table XI*, p. 187.

For illustration, *Graph V* represents the fitting for the wave of 25h, obtained by direct application of the Buys-Ballot method to the paraconical pendulum series of June-July 1955.[7] The amplitude of this periodic component is 14 grads.

## Orders of magnitude

3 - The minor elliptic wave $M_1$ corresponding to the period of 24h50m (24.84h) of amplitude equal to 10.46 grads (*Table II*) corresponds to a speed of angular displacement of $0.37 \times 10^{-5}$rad/s, *i.e.* about a fifteenth of the Foucault effect which is equal to $0.55 \times 10^{-4}$rad/s.[8]

Moreover, we see that *the total* of the amplitudes of the thirteen components above for the series of azimuths of June-July 1955 is of the order of half of the Foucault effect.[9] *The forces in play are accordingly of the order of magnitude of the force that generates the Foucault effect, which corresponds to the Coriolis acceleration.*[10]

---

[7] I again point out that, in view of the *very limited* means of calculation that we had at the time (calculation by hand with an electrical calculating machine) (see §A.3 above), as a first approximation, most of the calculations were performed by substituting a period of 25h for the period of 24h50m.

In fact calculation shows that, if a wave of period $T$=24h50m is analyzed using a period of $T_1$=25h, the amplitude of the wave is reduced by 6%, and the phase is shifted by 2.25 h.

(See *Chapter VI* of the *Second Volume* of this work, p. 30 above).

[8] In fact, for the period of 24.84 hours and an amplitude of 10.46 grads, we have an average variation equal to

$$\dot{\phi} = 10.46\frac{\pi}{200}\frac{1}{12.42 \times 3600} = 0.367 \times 10^{-5}\text{rad/s} \quad 0.55 \times 10^{-4}/0.367 \times 10^{-5} = 15.0$$

If, instead of taking 2R=10.46 grads (*Table II*), we take 2R=5.3 grads (the average corresponding to *Table I*), we have

$$\dot{\phi} = \frac{5.3}{10.46}0.367 \times 10^{-5} = 0.186 \times 10^{-5}\text{rad/s}$$

which corresponds to about a thirtieth of the Foucault effect ($0.186 \times 10^{-5}/0.55 \times 10^{-4} = 1/29.6$).

[9] 67.04 /10.46 = 6.41 6.41 /15.0 = 0.427

[10] The overall results corresponding to this §A.5 were presented in my Note of 25 November 1957 to the Academy of Sciences, "*Analyse harmonique des mouvements du pendule paraconique*" ("Harmonic Analysis of the Movements of the Paraconical Pendulum") [9].

*TABLE II*

***AZIMUTHS OF THE PARACONICAL PENDULUM AND ATMOSPHERIC PRESSURE***

***Month-long series of June-July 1955***

***Fittings to the 13 periods of the theory of tides***

*Hydrographic Service of Paris, and Hydrographic Institute of Hamburg*

| | | Amplitude of the wave 2R | | | | | | | | | | | | | |
|---|---|---|---|---|---|---|---|---|---|---|---|---|---|---|---|
| Series | Total observed variation D | $K_1$ $23^h$ 93 | $M_1$ $24^h$ 84 | $O_1$ $25^h$ 82 | $Q_1$ $26^h$ 87 | $M_2$ $12^h$ 42 | $S_2$ $12^h$ | $M_3$ $8^h$ 28 | $S_3$ $8^h$ | $\mu_2$ $12^h$ 87 | $L_2$ $12^h$ 19 | $N_2$ $12^h$ 66 | $M_4$ $6^h$ 21 | $M_{S4}$ $6^h$ 10 | TOTAL |
| Azimuth values in grads | 83.10 | 13.00 | 10.46 | 4.78 | 7.78 | 1.40 | 3.94 | 2.54 | 4.88 | 3.70 | 5.30 | 5.30 | 1.64 | 2.32 | 67.04 |
| Pressure in 1/10 millibar | 281.00 | 11.20 | 4.24 | 1.20 | 3.00 | 4.40 | 8.80 | 1.46 | 1.96 | 5.20 | 1.40 | 3.40 | 3.80 | 3.60 | 53.66 |
| | | Amplitude of the wave in % of the total observed variation 2R/D | | | | | | | | | | | | | TOTAL OF % |
| Azimuths | | 15.64 | 12.59 | 5.75 | 9.36 | 1.69 | 4.74 | 3.06 | 5.87 | 4.45 | 6.38 | 6.38 | 1.97 | 2.79 | 80.67 |
| Pressure | | 3.99 | 1.50 | 0.43 | 1.07 | 1.57 | 3.13 | 0.52 | 0.70 | 1.85 | 0.50 | 1.21 | 1.35 | 1.28 | 19.10 |

***Sources :*** *Note* of 25 November 1957 to the Academy of Sciences, *Analyse harmonique des mouvements du pendule paraconique* (Harmonic Analysis of the Movements of the Paraconical Pendulum), and *Table III A* of my Conference of 22 February 1958.

*GRAPH V*

**AZIMUTH OF THE PARACONICAL PENDULUM**
***Month-long series of June-July 1955***
***Fitting to a wave of 25h***
***by the Buys-Ballot method***

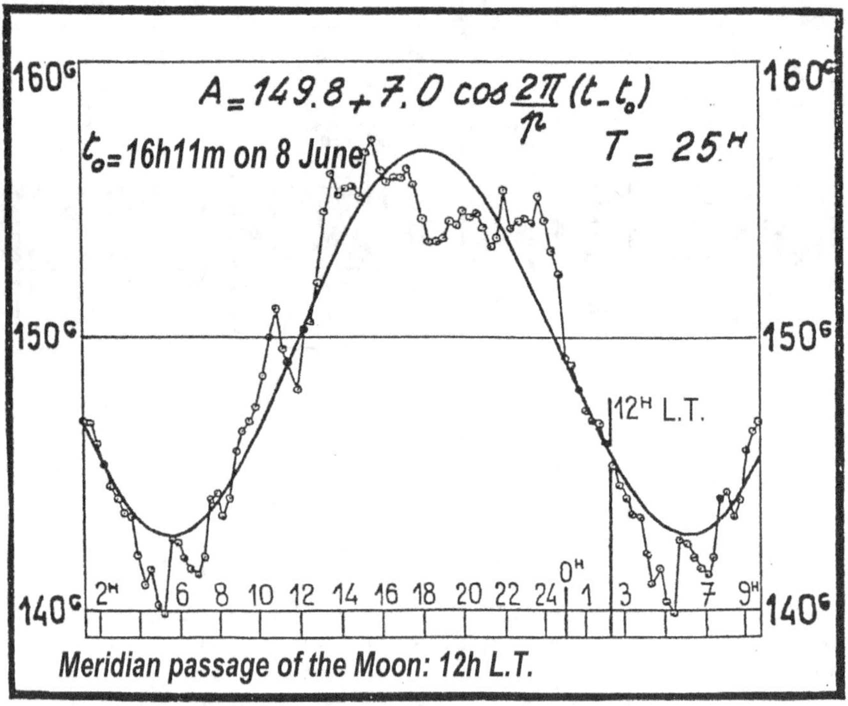


***Sources :*** My *Note* of 25 November 1957 to the Academy of Sciences, *Analyse harmonique des mouvements du pendule paraconique* (Harmonic Analysis of the Movements of the Paraconical Pendulum), and *Graph III A* of my Conference of 22 February 1958.

## Almost periodical structure

4 - In a general way, the series of values corresponding to the enchained series of azimuths of oscillation of the paraconical pendulum present all the characteristics of *almost periodic functions.*[11]

In fact, they present many symmetries or double symmetries with respect to certain dates, many translational similarities, and local periodicities.[12]

## Fittings by least squares

5 - In the above and in the following, continual use is made of fittings by the method of least squares, based upon the general theory of simple and multiple linear correlations.[13]

---

[11] An *almost periodic* function is a sum of sinusoidal components having incommensurable periods (see *Appendix E* of the *Second Volume* of this work, Allais 1983, "*Frequency, Probability, and Chance, Appendix II*" [44], p. 31 above).

[12] I gave very numerous examples of them as pertains to symmetries and to translations in my Conference of 22 February 1958.

On this property of almost periodic functions, see Allais, 1983, id., *Appendix II*, §P.9, *Local regularities of almost periodic functions* [44].

[13] Here it seems useful to remind the reader very briefly of the principle of these calculations in the case of a simple correlation.

Let us consider two functions $z(x)$ and $y(x)$, where $z(x)$ is supposed to depend linearly upon $y(x)$ to the first approximation, and for which we have $n$ pairs of observations $(z_n, y_n)$ which are functions of $x_n$. Estimation of the correlation between $z$ and $y$ reduces to determination of the function

(1) $z^*(x) = ay(x) + b$

where $a$ and $b$ are constants such that we have

(2) $z(x) = z^*(x) + \varepsilon(x)$

and such that, for the $n$ available pairs of observations, the sum $\sum \varepsilon_n^2$ of the squares of the residues $\varepsilon_n$ is minimal.

The coefficient of correlation $R$, which is a measure of the dependence in question, is such that we have

(2) $1 - R^2 = \sigma^2/\Sigma^2$

where $\Sigma$ and $\sigma$ respectively represent the standard deviations of $z(x)$ and $\varepsilon(x)$.

In principle, for all the correlations in this volume, I give the values of $R$, $\Sigma$, and $\sigma$.

On correlation calculations and their meaning, see particularly the remarkable work of Harald Cramer, *Mathematical Methods of Statistics*, Princeton University Press, 1946 [96].

## B THREE FUNDAMENTAL QUESTIONS CONCERNING THE PARACONICAL PENDULUM WITH ANISOTROPIC SUPPORT

The interpretation of the experimental results for the paraconical pendulum with anisotropic support leads us to ask three fundamental questions:

*First question*: Do the enchained observational series contain *statistically significant* periodic terms having periods near 24h or 24h50m, or not?

*Second question*: If yes, *is it possible or not to identify* the periodic effects thus observed with periodic effects resulting from the *current* theory of gravitation (as resulting from the double principle of inertia and universal gravitation, considered valid with respect to any Galilean reference frame), either completed by the corrections of the theory of relativity or not, and in the form in which it is applied in the framework of the *current* theory of relative movements?

*Third question*: If the above periodic effects cannot be thus identified, can the existence of significant periodic terms in the obtained series of enchained observations *be attributed, or not, to the indirect influence of a known periodic phenomenon*?

## *B.1 The reality of the determined periodicities*

Experiment allows us to give a *completely affirmative answer to the first fundamental question*, for three absolutely convincing reasons: observation of triply enchained series, comparison of the structures of the observed series and those reconstituted from their periodic components as determined by harmonic analysis, and application of the *Generalization of Schuster's Test to the case of autocorrelated series*.

### Triply enchained experiments

1 - If the observed variations were purely fortuitous, it would be necessary to admit that they were essentially determined by the random influence of the balls. If they were so determined, then three triply-enchained series - the experiment (3n+3) enchained with the experiment 3n, the experiment (3n+4) enchained with the experiment (3n+1), and the experiment (3n+5) enchained with the experiment (3n+2) - ought to have behaved *independently* of one another, because the ball was changed for each experiment.

But in fact, three triply-enchained series of 14-minute experiments, performed over 87 hours from 5 to 10 May 1957, showed that the movements of the plane of oscillation in the three series *were similar* (*Graph VI*).

These experiments made it possible to evaluate the standard deviation of the random influence of the balls for each elementary experiment of 14 minutes as being approximately $\varepsilon$=2.5 grads. Taking account of the influence of the returning tendency of the support,[1] this implies that the 95% confidence interval for the deviation that could exist between two series of independent observations is around ±12.5 grads. *The influence of the balls is accordingly very important, but it cannot explain the observed similarity of the azimuth variations* (*Graph VI*).[2]

It is *very remarkable* that, starting from the same azimuth of 220 grads, *all these three* series converged *very rapidly* to the same azimuth of around 160 grads.

---

[1] §A.1.3 above, p. 85.

[2] My *Note* of 25 November 1957 to the Academy of Sciences, *Analyse harmonique des mouvements du pendule paraconique* (Harmonic Analysis of the Movements of the Paraconical Pendulum) [9].

GRAPH VI

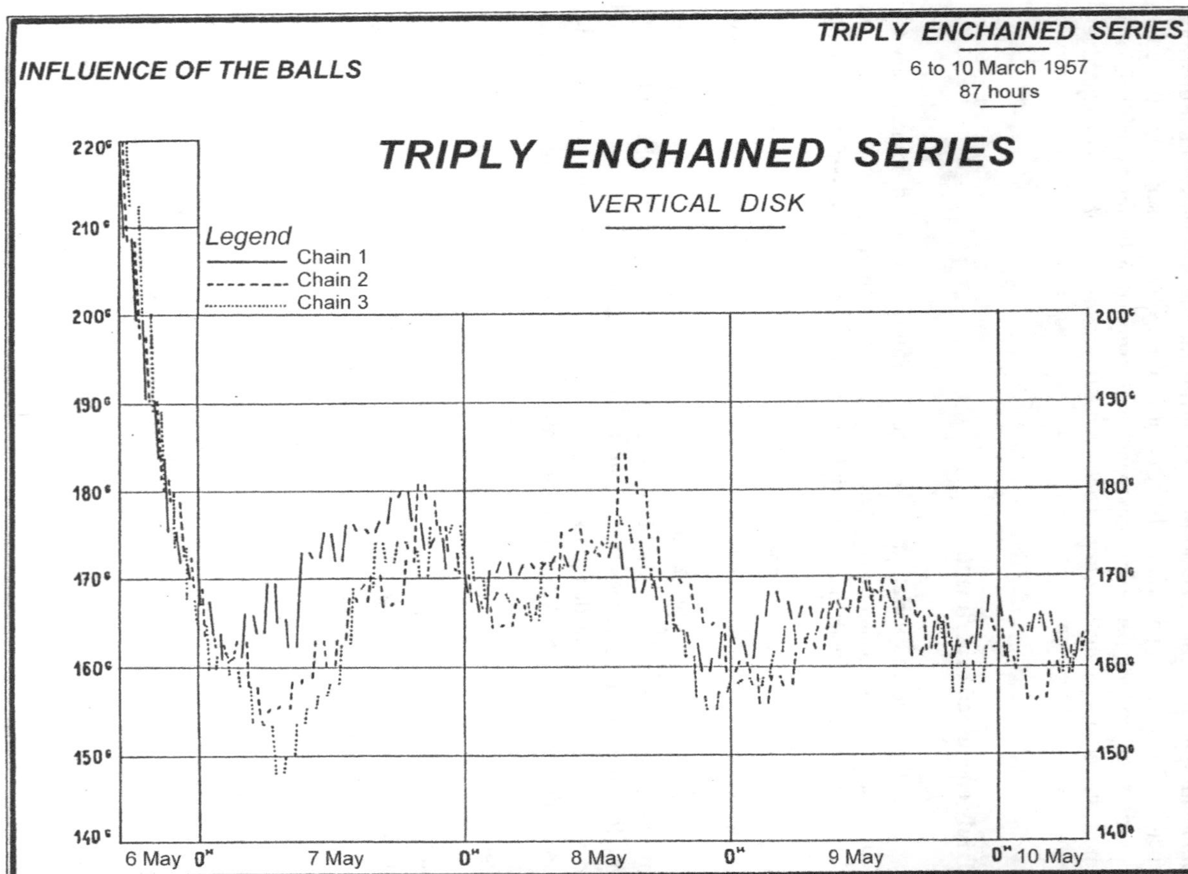


*Source* : *Graph IV A 2* of my Conference of 22 February 1958.

## Observed series and reconstituted series

2 - Both elementary analysis with the Buys-Ballot filter for different periods, on the one hand, and also overall harmonic analysis (by the least squares method) applied *simultaneously* for the thirteen waves of the theory of tides, on the other hand, show the existence of significant components having periods of 24h and 24h50m (the waves $K_1$ and $M_1$ of the theory of tides).[3]

Since the analysis covers a month, we may in fact ask ourselves if these waves are actually present *at every moment of time*. Now, any discrete series of 2n+1 numbers can be represented as a sum of a constant and n sinusoids. Thus, obtaining a sinusoid of a certain period by some method of harmonic analysis *can only have real meaning* if, not only its *relative* amplitude is sufficiently great, but also the periodic structure observed *for the entire* series can be *effectively* found in the different elementary periods into which the period of observation in question can be decomposed.

In fact, it is easy to verify that the periodic structure of the two month-long series of enchained observations of November-December 1954 and June-July 1955 can be considered as having been maintained for the two periods of fifteen days and even for the four periods of a week into which each of these two one-month series can be decomposed.

• For an observed one-month series, the *simultaneous* estimation by the method of least squares of the amplitudes of the thirteen waves usually considered in the theory of tides effectively provides thirteen sinusoids, whose sum can be obtained with the aid of the *Tide Predictor* of Lord Kelvin.[4] The calculated series obtained in this way can be analyzed for periods of 24h and 25h by the same Buys-Ballot filter method as for the observed series.

---

[3] §A.5 above.

[4] *Table II* of §A.5.3 above, p. 99.

If the sinusoids that have been obtained *really* are present in the observed series, then the Buys-Ballot method should give, for each of its components (fifteen-day periods or weeks) sinusoids of comparable phases $H$ and $H'$ for the observed series and for the calculated series.

• *Table III* shows the differences $H - H'$ of these phases in hours and minutes for the two month-long series of November-December 1954 and June-July 1955, for their entire months, their two half-months, and their four single weeks. *The differences of phases $H - H'$ remain relatively very small.*[5]

In view of the fact that each week can be considered as an independent experiment in view of the experimental procedure, *such remarkable agreements of phases must be considered as proving the existence of real periodicities having periods near to 24h and 24h50m.*

By way of illustration, *Graphs VII, VIII, and IX* represent the results of this analysis *of the month-long series of June-July 1955* for the periods of 24h and 25h.[6]

[5] The differences $H - H'$ remain relatively very small, except for the period of 24h and the first week of November-December 1954, and for the period of 25h and the second week of June-July 1955, for which we have $H - H'$ = -4h02m and +2h20m respectively.

*Table III* gives the algebraic averages of $H - H'$ and the averages of their absolute values for the months, for the half-months, and for the weeks.

It should be underlined that the observed series probably contained other periodic components than the thirteen waves considered, and also that there are random perturbations, such as those due to the balls. Both of these were capable of displacing the peaks of the sinusoids. The two relatively large differences of $H - H'$ are not surprising. Actually, what is surprising is that the differences remain so relatively small in all the other cases.

[6] In *Graph VII* and for the wave of 24h, we have $H - H' = 4\text{h}54\text{m} - 4\text{h}48\text{m} = 6\text{m}$ , and in *Graph VIII* and for the wave of 25h, we have $H - H' = 18\text{h}11\text{m} - 18\text{h}12\text{m} = -1\text{m}$.

I remind the reader that at the time, in order to simplify the calculations, the wave of 25h was considered instead of the wave of 24h50m (see above §A.3.1, note 3, p. 90).

We must accept the *identity in practice* of the results of the two methods of analysis, not only for the month, but also for the two half-months.[7]

This analysis shows that the periodic components of 24h and 24h50m brought out by the overall harmonic analysis *were really present in each of the elementary periods into which the month could be divided.*

• We also realize that while, for the two fifteen-day periods, the 25h filter gives from the raw series two sinusoids whose phases differ by 3h23m, this difference is due to the influence of the wave of 24h which is not completely eliminated by an analysis that covers only fifteen days, because *the same difference* is present in connection with the reconstituted series, which we know to be a *sum of sinusoids* (*Graph X*).[8]

• In total, analysis with the Buys-Ballot filter for the periods of 24h and 25h gives comparable results for the *observed* series of November-December 1954 and of June-July 1955, and for the series *reconstituted* from the results of overall harmonic analysis of these two series according to the thirteen waves of the theory of tides, whatever be the elementary period considered in the context of each month. *We can thus conclude that these periodic components actually do exist.*[9]

---

[7] In general, such an identity would not appear if the observed series did not really contain the waves considered in each of its elementary periods, because the least squares fitting that gave the amplitude of the thirteen waves considered was only obtained by an operation that covered the *entire* month.

[8] In fact we have (*Graph X*)

19h59m – 16h36m = 3h23m

and (*Graph IX*)

19h59m – 16h36m = 3h23m

19h41m – 16h16m = 3h15m

[9] The previous analysis, *Table III*, and *Graphs VII, VIII, IX, and X* were presented in my *Note* of 20 November 1958, which was not published but which was very widely circulated: *Sur l'existence d'une composante périodique de période de 24 h. 50 mn. dans les mouvements du pendule paraconique à support anisotrope* (On the existence of a periodic component having period 24h50m in the movements of the paraconical pendulum with anisotropic support).

*TABLE III*

## PARACONICAL PENDULUM WITH ANISOTROPIC SUPPORT OBSERVED AND CALCULATED MONTH-LONG SERIES OF NOVEMBER-DECEMBER 1954 AND OF JUNE-JULY 1955

***Comparison of the phases H and H' corresponding to analysis of the series observed and calculated by the Buys-Ballot method for the periods of 24 hours and 25 hours***

| | Values of H - H' | | | |
|---|---|---|---|---|
| | Nov-Dec 1954 | | June-July 1955 | |
| | $24^h$ | $25^h$ | $24^h$ | $25^h$ |
| 1 + 2 + 3 + 4 | $0^h\ 30^{mn}$ | $-0^h\ 23^{mn}$ | $0^h\ 06^{mn}$ | $-0^h\ 01^{mn}$ |
| 1 + 2<br>3 + 4 | $-1^h\ 02^{mn}$<br>$-0^h\ 56^{mn}$ | $-0^h\ 20^{mn}$<br>$0^h\ 06^{mn}$ | $-1^h\ 09^{mn}$<br>$0^h\ 43^{mn}$ | $0^h\ 20^{mn}$<br>$0^h\ 15^{mn}$ |
| 1<br>2<br>3<br>4 | $0^h\ 16^{mn}$<br>$-4^h\ 02^{mn}$<br>$-1^h\ 41^{mn}$<br>$1^h\ 43^{mn}$ | $1^h\ 31^{mn}$<br>$-1^h\ 27^{mn}$<br>$-1^h\ 42^{mn}$<br>$0^h\ 36^{mn}$ | $-1^h\ 08^{mn}$<br>$0^h\ 59^{mn}$<br>$-1^h\ 23^{mn}$<br>$0^h\ 59^{mn}$ | $-0^h\ 33^{mn}$<br>$2^h\ 20^{mn}$<br>$-0^h\ 56^{mn}$<br>$0^h\ 57^{mn}$ |
| Averages | $\overline{H - H'}$ | | $\overline{\lvert H - H' \rvert}$ | |
| month<br>fortnights<br>weeks | $3^{mn}$<br>$-15^{mn}$<br>$-15^{mn}$ | | $15^{mn}$<br>$37^{mn}$<br>$83^{mn}$ | |

***Legend :*** The symbols 1, 2, 3, and 4 represent the first, the second, the third, and the fourth week; the symbols 1+2 and 3+4 represent the first and the second fortnight; and the notation 1+2+3+4 represents the entire month.

Thus for the first fortnight of June-July 1955 and for the period of 25h (*Graph IX*), for example, we have: 16h 36min - 16h 16min = 20min.

$\overline{H - H'}$ = algebraic average of **H - H'**

$\overline{\lvert H - H' \rvert}$ = average of the absolute values of **H - H'**

***Source :*** My *Note* of 20 November 1958, *Sur l'existence d'une composante périodique voisine de 24h50mn dans les mouvements du pendule paraconique à support anisotrope* (On the existence of a periodic component having period near 24h50m in the movements of the paraconical pendulum with anisotropic support)

*GRAPH VII*

*HARMONIC ANALYSIS*

24h
AZIMUTHS
June- July 1955
1+2+3+4
Actual and reconstituted series

## *HARMONIC ANALYSIS OF SERIES VI OF JUNE-JULY 1955*

BUYS-BALLOT FILTER **24h PERIOD** - Complete actual and reconstituted series (1+2+3+4)

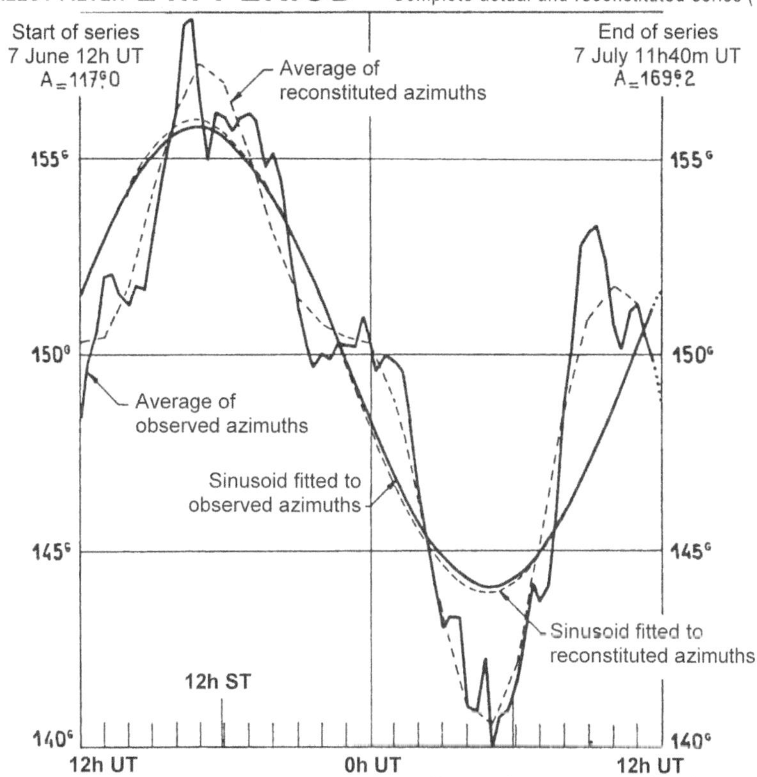


*Fitting sinusoid*

$$y = \bar{y} + R \cos 2\pi \frac{t-\theta}{T}$$

T = 24 θ = time of peak with respect to start of series

θ′ = UT time of peak

| | q | $\bar{y}$ | R | θ | θ′ |
|---|---|---|---|---|---|
| Observed azimuths | 30 | 149.96 | 5.832 | $4^h54$ | $16^h54$ |
| Reconstituted azimuths | 30 | 149.93 | 5.957 | $4^h48$ | $16^h48$ |

q = number of cycles

*N.B.*: 1) The reconstituted azimuths are those corresponding to the sum of the 13 elementary waves of the tides, as determined by the method of least squares.

2) The sidereal time shown is that corresponding to the average date, i.e. 22 June 1955.

3) The scales of the reconstituted azimuths have been offset so that the two fitting sinusoids have the same axis.

***Source*** : My *Note* of 20 November 1958 (see *Table III*)

*GRAPH VIII*

*HARMONIC ANALYSIS*

25h
AZIMUTHS
June- July 1955
1+2+3+4
Actual and reconstituted series

## *HARMONIC ANALYSIS OF SERIES VI OF JUNE-JULY 1955*

BUYS-BALLOT FILTER **25h PERIOD** - Complete actual and reconstituted series (1+2+3+4)

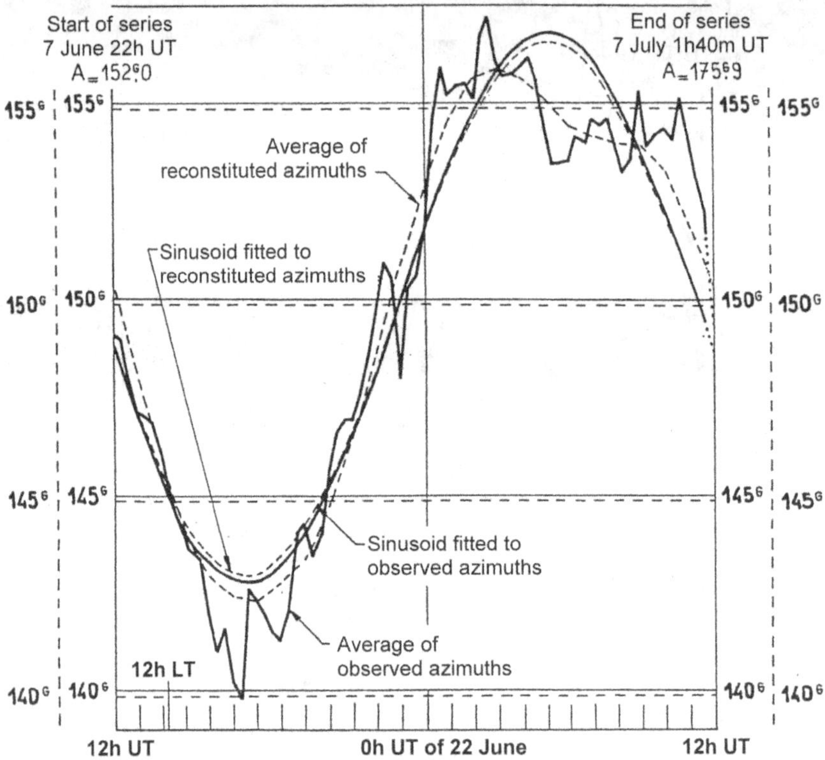


*Fitting sinusoid*

$$y=\bar{y}+R\cos 2\pi\frac{t-\theta}{T}$$

| | q | $\bar{y}$ | R | θ | θ′ |
|---|---|---|---|---|---|
| Observed azimuths | 28 | $149^{G}80$ | $7^{G}007$ | $18^{H}11$ | $15^{H}58$ |
| Reconstituted azimuths | 28 | $149^{G}92$ | $6^{G}766$ | $18^{H}12$ | $15^{H}59$ |

T= 25 θ = time of peak with respect to start of series

θ′= time of peak with respect to time of meridian passage of Moon on average day (considered in average time)

q = number of cycles

*N.B.*: 1) The reconstituted azimuths are those corresponding to the sum of the 13 elementary waves of the tides, as determined by the method of least squares.

2) The shown lunar time is that corresponding to the average date, i.e. 22 June 1955 (12h LT corresponds to the meridian passage of the Moon).

*Source* : My *Note* of 20 November 1958 (see *Table III*)

*GRAPH IX*

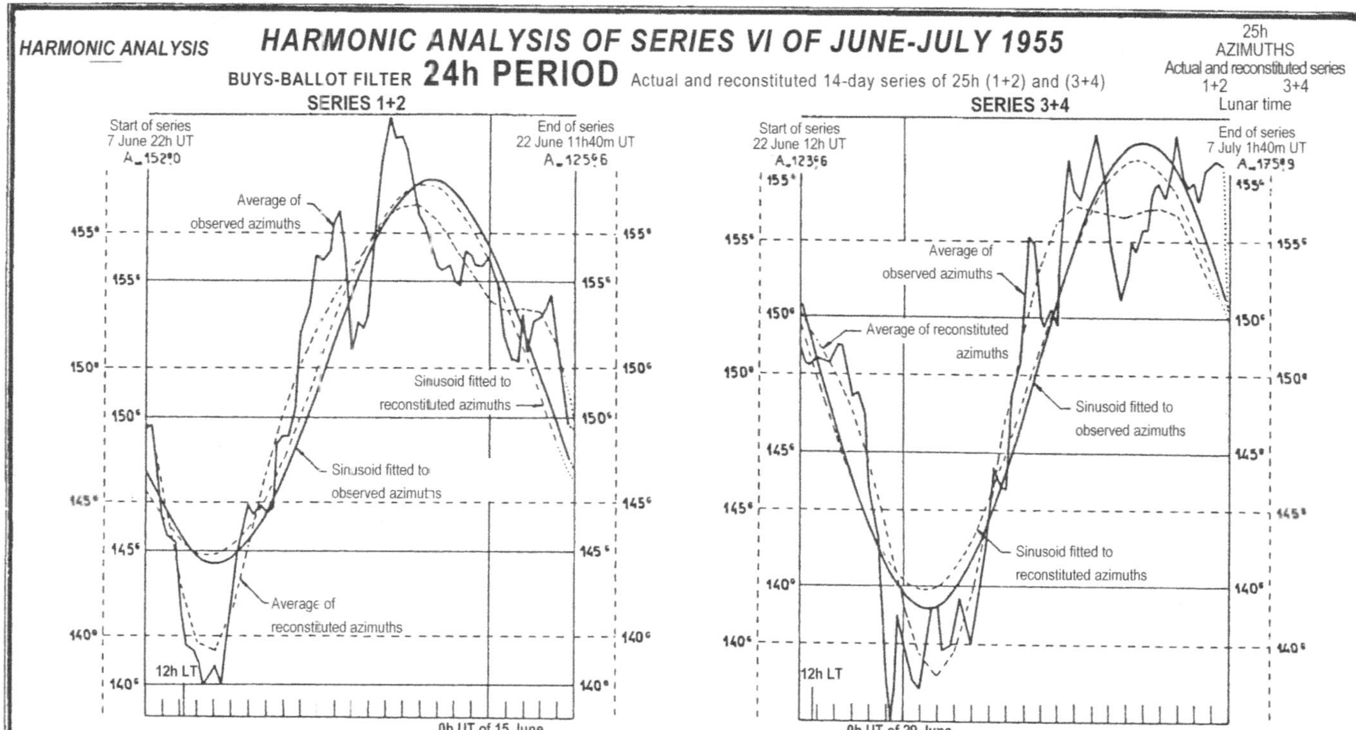


*Fitting sinusoid*

$y = \bar{y} + R \cos 2\pi \frac{t-\theta}{T}$

T = 25

| Period 1+2 | q | $\bar{y}$ | R | θ | θ' |
|---|---|---|---|---|---|
| Observed azimuths | 14 | 151°71 | 7.171 | 16"36 | 14"25 |
| Reconstituted azimuths | 14 | 149°85 | 6.887 | 16"16 | 14"05 |

| Period 3+4 | q | $\bar{y}$ | R | θ | θ' |
|---|---|---|---|---|---|
| Observed azimuths | 14 | 147°89 | 8.693 | 19"59 | -5"42 |
| Reconstituted azimuths | 14 | 149°99 | 8.033 | 19"41 | -5"57 |

θ = time of peak with respect to start of series
θ' = time of peak with respect to time of meridian passage of Moon on average date (considered in average time)
q = number of cycles

*N.B.: 1) The reconstituted azimuths are those corresponding to the sum of the 13 elementary waves of the tides, as determined by the method of least squares.*
*2) The shown lunar time is that corresponding to the average date, i.e. 15 June 1955 (series 1, 2) and 29 June 1955 (series 3, 4). 12h LT corresponds to the meridian passage of the Moon.*

***Source :*** My *Note* of 20 November 1958 (see *Table III*)

*GRAPH X*

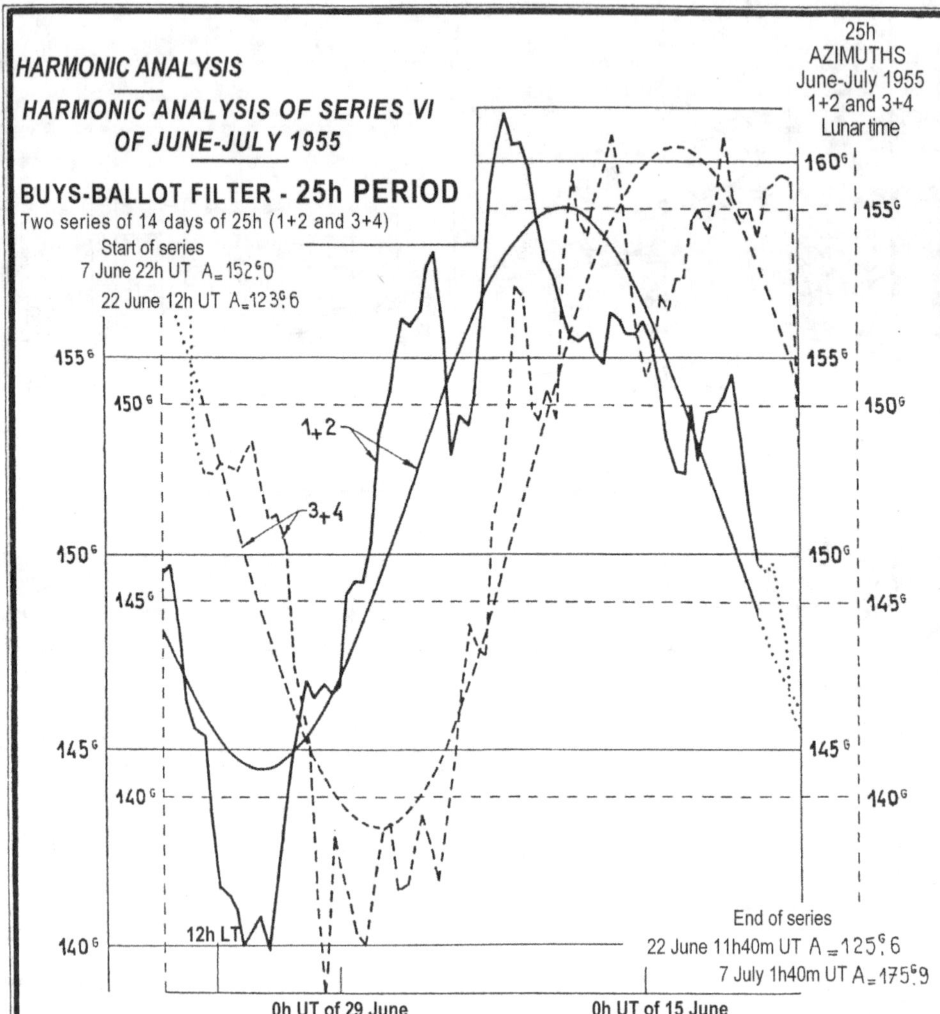


*Fitting sinusoid*

$$y = \bar{y} + R\cos 2\pi \frac{t-\theta}{T}$$

| Period | q | $\bar{y}$ | R | θ | θ′ |
|---|---|---|---|---|---|
| 1 + 2 | 14 | $151^G71$ | 7.171 | $16^H36$ | $14^H25$ |
| 3 + 4 | 14 | $147^G89$ | 8.693 | $19^H59$ | $-5^H42$ |

T = 25 θ = time of peak with respect to start of series

θ′ = time of peak with respect to time of meridian passage of Moon on average date (considered in average time)

q = number of cycles

*N.B.: 1) The shown lunar time is that corresponding to the average date, i.e. 15 June 1955 (series 1+2) and 29 June 1955 (series 3+4). (12h LT corresponds to the meridian passage of the Moon)*

*Source:* My *Note* of 20 November 1958 (see *Table III*)

## Application of the generalization of Schuster's Test to the case of autocorrelated time series

3 - Any discrete series of $N = 2n+1$ numbers $x_i$ can be represented as the sum of a constant and $n$ sinusoids. Obtaining a sinusoid of a given period by any method of harmonic analysis can therefore only have real significance if its relative amplitude is sufficiently great.

Schuster established a significance test that has become classic for the case in which the $x_i$ are *independent*, but *this test is not applicable to a case in which the $x_i$ are autocorrelated.* In 1957 I showed how this test of Schuster can be generalized, and I specified the conditions for a test of periodicity that is applicable to autocorrelated series.[10]

• Let us suppose, for example, that we consider a series of $N$ observations $x_i$, and suppose that application of the Buys-Ballot filter for a period $T$ determines a sinusoid

(1) $X = R\sin\omega(t-t_0) \qquad \omega = 2\pi/T$

and let $P_\omega$ be the probability for $R$ to have a value greater than or equal to a given value $R_\omega$, *under the hypothesis that the series considered does not possess any periodicity*. The formulation that I determined makes it easy to calculate the probability $P_\omega$.[11]

---

[10] See my *Notes* of 13 May and 23 December 1957 to the Academy of Sciences: *Test de périodicité. Généralisation du test de Shuster aux cas de séries temporelles autocorrélées*

(Test of periodicity. Generalization of Schuster's test to the case of autocorrelated time series) [16], and *Application du test de Schuster généralisé à l'analyse harmonique des azimuts du pendule paraconique* (Application of the generalized Schuster test to harmonic analysis of the azimuths of the paraconical pendulum) [10]. These *Notes* were presented by Jean-Marie Kampé de Fériet, a member of the Academy of Sciences.

In 1961, at the 33rd Session of the International Statistical Institute, I presented a detailed demonstration of this test, with application to the case of the observations of the paraconical pendulum at Bougival in 1958 (see Section C below), in my Communication *Test de périodicité. Généralisation du test de Schuster au cas de séries temporelles autocorrélées dans l'hypothèse d'un processus de perturbations aléatoires d'un système stable* (Test of periodicity. Generalization of the Schuster test to the case of autocorrelated time series, under the hypothesis of a process of random perturbations of a stable system) (Bulletin of the International Statistical Institute, 1962, Vol. 39, 2nd. printing, pp 143-194) [40]. This memoir is reproduced in *Appendix D* in the *Second Volume* of this work (see the *Summary* above, p. 31).

[11] We have

(1) $P_\omega = \text{Prob}\,(\mu_\omega \geq \mu) = e^{-\mu}$

with

(2) $\mu_\omega = \frac{I_\omega}{E[I_\omega]} = \frac{NI_\omega}{4\sigma^2 k_\omega} \qquad \sigma^2 = \left(1-\frac{1}{N}\right)s^2 \qquad \omega = 2\pi/p$

(3) $I_\omega = A_\omega^2 + B_\omega^2 = R_\omega^2 \qquad A_\omega = \frac{2}{N}\sum_{s=0}^{N-1} x_{s+1}\cos s\omega \qquad B_\omega = \frac{2}{N}\sum_{s=0}^{N-1} x_{s+1}\sin s\omega$

(4) $k_\omega = 1 + 2\sum_{\nu=1}^{\nu=h}\left(1-\frac{\nu}{N}\right) r_\nu \cos\nu\omega$

$p$ is the period. The frequency $m$ is equal to $N/p$. $s^2$ is the variance of the $x_i$, and $h$ is the value of $\nu$ after which the coefficient of autocorrelation $r_\nu$ is no longer significantly different from zero.

Thus, for example, the Buys-Ballot filter applied to the series of enchained experiments of *June-July 1955* and to the wave of 25h yields $R = 7.0$ grads.[12] For this value, we find $P = 0.48 \times 10^{-5}$.[13] *This means that, under the hypothesis of the absence of any periodicity, there is less than one chance in 100,000 of obtaining a value of R greater than or equal to 7.0 grads by analysis using a Buys-Ballot filter.* From the point of view of statistical analysis, the significance level is accordingly 0.00048%. We may therefore conclude that *the existence of the 25h wave that was found is a practical certainty.*

---

[12] §A.5.2 above, *Graph V*, p. 100.

[13] My *Note* of 23 December 1957 to the Academy of Sciences (note 8 above).

In the case of the series of June-July 1955, with the notation of note (11) above, we have: $N = 2161$, $\sigma = 20.24$ grads; and for $T = 25h$, we have $p = 25 \times 3 = 75$, $k_{75} = 5.292$. We have $R_{75} = 7.01$ grads (*Graph V* of §A.5.2 above). Thus we have $I_{75} = 7.01^2$, and

$$\mu_{75} = (2161 \times 7.01^2) / (4 \times 20.24^2 \times 5.292) = 12.25 \qquad P_{75} = e^{-\mu_{75}} = 0.481 \times 10^{-5}$$

• By way of illustration, *Graph XI* is a frequency graph of the 721 hourly values for the series of *November-December 1954*, with isopleths of statistical significance.[14]

We see that the wave of 24h50m (*i.e.* 24.84h) whose amplitude is approximately $2R = 11$ grads has a significance level of $P$=0.062%.[15] Thus there is *less than one chance in a thousand* for such an amplitude to be observed in a series that has no real periodicity.

---

[14] This Graph is a photographic reproduction of *Graph VI C 1* of my Conference of 7 November 1959, *Faut-il reconsidérer les lois de la gravitation? Nouveaux résultats, Bilan et Perspectives* (Should the Laws of Gravitation be Reconsidered? New Results, Review and Perspectives).

It is useful to compare this frequency graph with the frequency graph for the month-long series of Bougival of July 1958 (*Graph XXVI* of §C.2.4 below, p. 156).

Significant periods appear on these two Graphs, corresponding to harmonics characterized by the values $m$=29, 30, 31, 32, 33, and 34, which correspond to the periods ($p = 721/m$): 24.86h, 24.03h, 23.26h, 22.53h, 21.84h, and 21.21h.

I remind the reader that periods of 23.09h, 22.31h and 21.67h and neighboring periods appear in the analysis of the theory of tides (Schureman, 1941, *Manual of Harmonic Analysis and Prediction of Tides*, pp. 164-165) [244].

*For simplification*, the analysis in the present work will *principally* concentrate upon the periods of 24h50m (24.84h), 24h, 12h25m (12.42h), and 12h, and *in particular* upon the period of 24h50m.

[15] The frequency graph considered corresponds to $N = 721$ hourly values. For the harmonic corresponding to $m = 29$, we have the period $p = 721/29 = 24.86$h $=$ 24h52m, and an amplitude of $2R = 11.08$ grads.

For this period, we have (*Graph XI*): $\mu = 7.39$ and $P = e^{-\mu} = 0.062\%$.

For $m = 30$ we have the period $p = 721/30 = 24.03$h $=$ 24h2m, with $R = 4.66$ grads, $\mu = 5.5$ and $P = e^{-\mu} = 0.4\%$.

The choice of the value $N = 721$ hourly values corresponds to the condition that the number of values should make it possible to separate the amplitudes of the two periods of 24h and 24h50m.

In fact we have $721/24 = 30.04$ and $721/24.84 = 29.03$. These values thus differ very little from integers.

*GRAPH XI*

## APPLICATION OF THE GENERALIZED SCHUSTER TEST

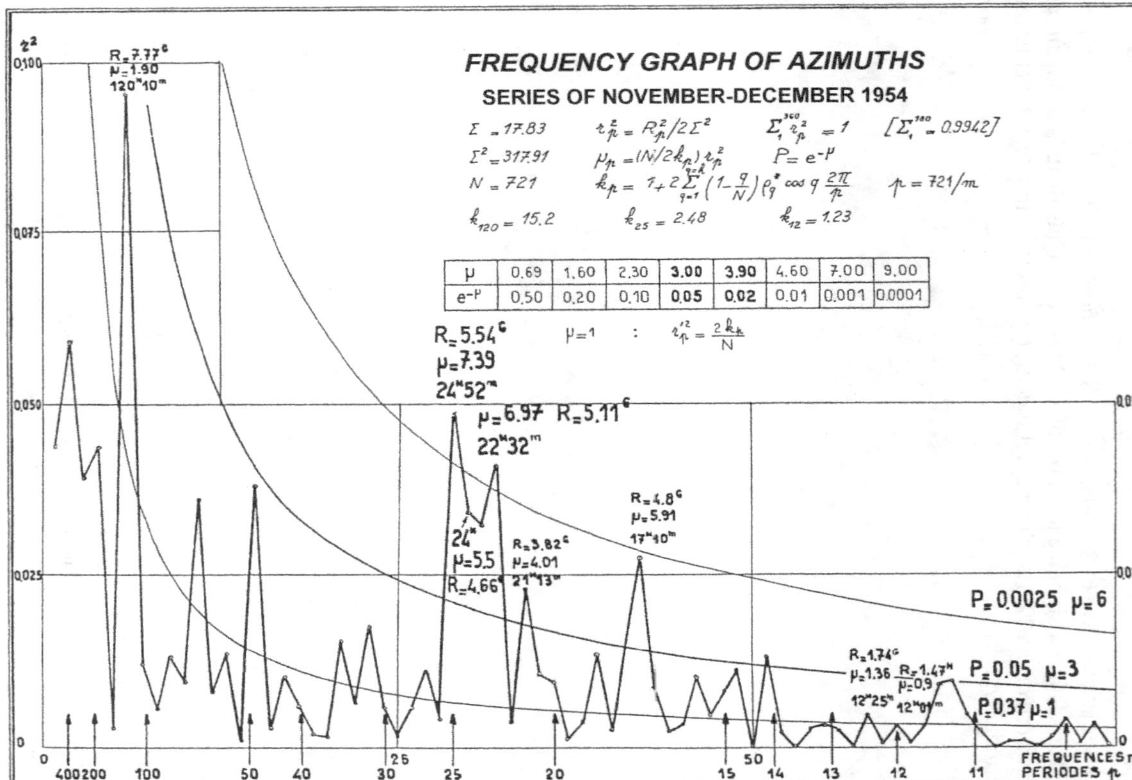


**Legend :** The formulation of this test is deduced from my Note of 4 December 1957 to the Academy of Sciences. See also my memoir of 1961, *Test de périodicité. Généralisation du Test de Schuster au cas de séries temporelles autocorrélées dans l'hypothèse d'un processus de perturbations aléatoires d'un système stable* (Test of periodicity. Generalization of Schuster's Test to the case of autocorrelated time series, under the hypothesis of a process of random perturbations of a stable system).

**Source :** *Graph VI C 1* of my Conference of 7 November 1959.

## Absolute certainty as to the existence of the detected lunisolar effects

4 - From the three analyses above, *i.e.* from analysis of the enchained experiments, from comparison of the periodic structures of the observed series and the calculated series, and from application of the *generalized Schuster test*, we can be *absolutely certain*[16] *of the real existence of the detected diurnal lunisolar effects, and in particular of the existence of the lunar periodicity of 24h50m.*

*Thus we must answer the first fundamental question posed at the beginning of this Section in the affirmative.* The month-long series in question of enchained observations of the azimuths of the paraconical pendulum does *indeed contain statistically significant periodic lunisolar terms, absolutely certainly.*

[16] Of course *the certainty in this case is certainty in practice, not metaphysical certainty.*

## B.2 *The observed lunisolar effects and the current theory of gravitation*

With regard to the observed amplitudes of the lunisolar effects in the case of the anisotropic support, it is easy to verify that the detected effects are *totally inexplicable* in the framework of any current theory of gravitation, be it the Newtonian theory or the theory of relativity.

### Observed and calculated values of the lunar influence upon the azimuth of the paraconical pendulum with anisotropic support

1 - If, for example, we consider the action of the Moon upon the pendulum during a 14-minute experiment, we can estimate that according to currently accepted theory, in the first approximation, it leads to an average variation of the azimuth of less than $10^{-13}$ radians per second.[1]

In fact, the average amplitude of 5.3 grads found for the wave of 25h *over the seven month-long series of observations that were performed* corresponds to an average variation of approximately $0.19 \times 10^{-5}$ radians per second.[2] *The observed effect is thus at least 18 million times greater than the calculated effect.*[3]

### The extreme minuteness of the lunisolar influence on the movement of the paraconical pendulum with anisotropic support

2 - In fact, the theoretical effects of the lunisolar influence upon the movement of the pendulum are so small that *not one of the 19th century authors who worked upon the theory of the pendulum attempted to calculate them, even though some of them were excellent mathematicians.*

The extraordinary minuteness of the calculated effects can be easily explained if we consider that, in order to obtain the effective gradient of the attraction of the Sun or of the Moon at a point on the surface of the Earth, it is necessary to take the difference of the gradients at this point and at the center of the Earth.

---

[1] *Table VII*, and equation (8), p. 129 below.

[2] §A.5.3, note 8, p. 98 above.

[3] $0.186 \times 10^{-5}/10^{-13} = 18.6 \times 10^{6}$

Moreover, the plane of oscillation of the pendulum can only rotate as a result of the influence of the lunisolar attraction due to variations of this gradient around the point in question. It is therefore necessary to consider, for this gradient, the difference between its value at the average position of the pendulum and its value at a neighboring point.[4]

The confrontation between the observed values and the calculated values of the influence of the Moon and the Sun upon the movement of the pendulum is so important from the point of view of this work that it seems necessary to me to present, in the four *Tables IV to VII*, the principles of the calculations that enable this confrontation.[5, 6]

## Table IV - Classical theory of the movement of the Foucault pendulum in the case of a rectilinear or elliptical trajectory

3 - *Table IV* briefly summarizes the results of theoretical analysis of the movement of the pendulum under the action of terrestrial rotation, *as presented in all the standard works.*

---

[4] This is the difference $\nabla_G U_i - \nabla_S U_i$ in equation (5) of *Table V* below, p. 127.

[5] In order to simplify the discussion, the following calculations *do not take account* of the centrifugal force caused by the Earth's rotation and of its variations in the space swept by the pendulum. *The following discussion attempts to calculate, to the first approximation, only the effects of the lunisolar attraction upon the movement of the paraconical pendulum, but not the effects of terrestrial rotation.*

In order to take account of the centrifugal force caused by the Earth's rotation in the following formulas, to the first approximation, it is sufficient to replace the acceleration $g$ of gravity due to the attraction of the Earth by the apparent weight $g^* = g - \omega^2 r_T \cos^2 L$. All works on the subject take this approach.

Thus we have $\omega^2 r_T \cos^2 L = \left(0.729 \times 10^{-4} \times \cos 48.9°\right)^2 \times 6.3712 \times 10^8 = 1.46$, *i.e.* a fraction of about $1.5 \times 10^{-3}$ of the weight.

For a complete calculation, see Allais, 1956, *Théorie du pendule paraconique* (Theory of the Paraconical Pendulum), Part VA, *Théorie des mouvements relatifs* (Theory of the Relative Movements) (pp. V2-V28), and *Mouvement du Pendule conique sous l'influence de la rotation terrestre* (Movement of the Conical Pendulum under the Influence of the Earth's Rotation) (pp. V29-V46) [8]. See also Allais, 13 March 1958, *Application du Théorème de Bour au cas des mouvements terrestres dans le cas le plus général* (Application of Bour's Theorem to the case of terrestrial movements in the most general case) [20].

[6] The *detailed* study of the influence of the celestial bodies upon the movement of the paraconical pendulum with anisotropic support and with isotropic support is given in the *Second Volume* of this work (*Chapters I and II*, Sections B, p. 28 above).

Taking into account the ellipticity $\beta$ of the trajectory, the variation $\dot{\phi}$ of the azimuth of the plane of oscillation is

(1) $\dot{\phi} = -\omega \sin L + (3/8)\, p\alpha\beta \qquad p = \sqrt{g/l}$

$\omega$ is the speed of rotation of the Earth, $L$ the latitude of the location in question, and $\alpha$ and $\beta$ are the major and minor axes of the ellipse described by the pendulum in radians. The second component of $\dot{\phi}$ corresponds to the Airy precession. The plane of oscillation rotates *in the same sense* as that in which the pendulum traverses its elliptical trajectory.[7]

**Table V - Forces acting upon the Foucault pendulum relative to axes linked to the Earth, under the actions of gravity, of terrestrial rotation, and of the attraction of the celestial bodies, according to the current theory of gravitation.**

4 - *Table V* shows how, with respect to axes $Sxyz$ linked to the Earth, where $S$ corresponds to the point of suspension of the pendulum, the acceleration of the center of gravity of the pendulum is determined from the fundamental equation

(2) $\vec{F} = M\vec{\gamma}$

which, according to the theory of gravitation, *is only applicable* (up to uniform translation) *with respect to a Copernican trihedral* $S'x'y'z'$ whose origin is the center of the Sun and whose axes pass through three fixed stars.

Equation (3) of *Table V* gives the expression for the absolute acceleration $\vec{\gamma}$ of the center of gravity $G$ as a function of the acceleration of $G$ with respect to the Earth, of the Coriolis acceleration, and of the entrainment acceleration of the Earth's centre.

An *essential* condition is that we have[8]

(3) $\nabla_G U_i - \nabla_T U_i = (\nabla_S U_i - \nabla_T U_i) + (\nabla_G U_i - \nabla_S U_i)$

[7] The ellipse is traversed in the anticlockwise sense, or in the clockwise sense, according as to whether $\beta$ is positive or negative.

[8] The notation $\nabla_G U_i$ means that the gradient of the gravitational potential of the celestial body $i$ is considered at the point $G$.

*The first term, which corresponds to the deviation of the vertical due to the celestial body i and which is independent of the center of gravity G of the pendulum, has no influence upon the movement of the pendulum.*

*The only cause influencing that movement is the difference between the actions of the gravitation of the celestial body i at the point G and at the point S which is the support point of the pendulum.* This difference is naturally extremely small.[9]

## Table VI - Influence of the Sun and the Moon upon the movement of the paraconical pendulum according to the current theory of gravitation

5 - *Table VI* shows how the right hand sides of the differential equations in $m = x/l$ and $n = y/l$ which depend upon the action of the celestial body $i$ are determined. The influence of the body $i$ on the movement of the paraconical pendulum is determined by the equations[10]

(4) $\ddot{m} + (g/l)\,m = K_i\,(m\cos 2A_i + n\sin 2A_i)$

(5) $\ddot{n} + (g/l)\,m = K_i\,(m\sin 2A_i - n\cos 2A_i)$

(6) $K_i = \frac{3}{2} C_i \sin^2 z_i \qquad C_i = \frac{M_i}{M_T}\frac{r_T^2}{d_i^3}\,g$

where $z_i$ and $A_i$ represent the zenith distance and the azimuth of the celestial body $i$, $M_i$ and $M_T$ are the masses of the body $i$ and of the Earth, $r_T$ is the radius of the Earth, and $d_i$ is the distance from the celestial body $i$ to the center of the Earth.

---

[9] Its order of magnitude is given by equation (10) of *Table V*.

In fact all the authors who have most deeply understood the application of the theory of relative movements, such as Bour and Gilbert, have supposed the force of gravitation to be constant over the entire space swept by the pendulum. *They thus suppose that the second term of the right hand side of equation (3) above (p. 120) is null.*

See Allais, September 1956, *Théorie du pendule paraconique* (Theory of the Paraconical Pendulum), Part 5, The influence of terrestrial rotation, pp. V.1-V.28 [8]. See also Allais, 13 March 1958, *Application du Théorème de Bour au cas des mouvements terrestres dans le cas le plus général* (Application of Bour's theorem to the case of terrestrial movements in the most general case), 32 p [20]. This memoir extends Bour's results to the most general case, when account is taken of the variation of the gravitational field in the space swept by the moving system under consideration.

For relative movements, see E. Bour, *Mémoire sur les mouvements relatifs* (Memoir on relative movements), Journal of Pure and Applied Mathematics, Vol. VIII, 1863, pp. 1-51 [78]. See also Gilbert, *Mémoire sur l'application de la méthode de Lagrange à divers problèmes du mouvement relatif* (Memoir on the application of Lagrange's method to various problems of relative movement), Gauthiers-Villars, Paris 1889, 197 p [151]. Gilbert relies upon Bour's memoir of 1863.

[10] These equations are obtained from the series expansion of the expression (2) of *Table VI*. The second members in the equations (4) and (5) above are considered as being perturbations.

From equations (4) and (5), *use of the method of variation of constants* allows us to deduce that, *in the first approximation*, the influence of the celestial body $i$ can be represented by the two equations[11]

(7) $\dot{\phi} = -\omega \sin L + \frac{3}{8} p\alpha\beta$

(8) $\dot{\beta} = \frac{\alpha}{2p} K_i \sin 2(A_i - \phi)$

Of course if there are several such bodies, we have[12]

(9) $\dot{\beta} = \sum_i \dot{\beta}_i$

---

[11] In fact, it is easy to show that, for the values of $\beta$ observed during each 14-minute experiment, *i.e.* 840 s ($|\beta| < 0.001$) , the *direct* effect (equation 6 of *Table VI* below)

(1) $\dot{\phi} = \frac{1}{p} \frac{\alpha\beta}{\alpha^2 - \beta^2} K_i \cos 2(A_i - \phi)$

is much smaller than the *indirect* effect corresponding to the equation (equation 7 of *Table VI* below)

(2) $\dot{\beta} = \frac{\alpha}{2p} K_i \sin 2(A_i - \phi)$

In fact the order of magnitude $e_1$ of the direct effect is (equation 6 of *Table VI* below)

(3) $e_1 = \frac{1}{p} \frac{\beta}{\alpha} K_i$

while the order of magnitude of the indirect effect is (equation 6 of *Table VII* below)

(4) $e_2 = \left(\frac{3}{8} p\alpha\right)\left(\frac{\alpha}{2p} K_i \frac{\Delta t}{2}\right) = \frac{3}{16} \alpha^2 \frac{\Delta t}{2} K_i$

whence

(5) $\frac{e_2}{e_1} = \left(\frac{3}{16} \alpha^2 \frac{\Delta t}{2} K_i\right) / \frac{1}{p} \frac{\beta}{\alpha} K_i = \frac{3}{16} p \frac{\alpha^3}{\beta} \frac{840}{2}$

*i.e.*, for $|\beta| < 0.001$, $\overline{\alpha} = 1/10$, $p = 3.44$

(6) $\frac{e_2}{e_1} \geq \frac{3}{16} 3.44 \frac{840}{2} = 270.9$

This is a *very general* circumstance for *all* perturbations, that the indirect effect overcomes the direct effect.

From *Graphs III and IV* of §A.4 above (pp. 94-95) and *Table X* of §E.3 below (p. 180), we can certainly take $b$(in cm) $< 0.1$, whence, for $l' = 105$ cm ((§A.1.2, note 6, above), $\beta < 0.001$.

On this question, see my *Note* to the Academy of Sciences of 16 December 1957, *Théorie du pendule paraconique et influence lunisolaire* (Theory of the Paraconical Pendulum and the Lunisolar Influence), note 3 [17]. In this *Note* I gave the minimal value of 130 for the ratio $e_2/e_1$. *This new estimate is better.*

[12] In fact, and *to the first approximation, the small effects add together.*

If we limit ourselves to the influence of the Sun and the Moon, we have

(10) $\dot{\beta} = \frac{\alpha}{2p}\left[K_s \sin 2\left(A_s - \phi\right) + K_l \sin 2\left(A_l - \phi\right)\right]$

where $\phi$ represents the azimuth of the plane of oscillation of the pendulum, and $A_s$ and $A_l$ represent the azimuths of the Sun and the Moon.

We have[13]

(11) $K_s = \frac{3}{2} C_s \sin^2 z_s$ $\qquad C_s \frac{M_s}{M_T} \frac{r_T^2}{d_s^3} g$

(12) $K_l = \frac{3}{2} C_l \sin^2 z_l$ $\qquad C_l \frac{M_l}{M_T} \frac{r_T^2}{d_m^3} g$

(13) $C_s = 0.396 \times 10^{-13}$ $\qquad C_l = 0.862 \times 10^{-13}$

It is *very remarkable* that $C_s$ and $C_l$ are of the same order of magnitude. We have $C_l / C_s = 2.177$.

**Table VII - Observed and calculated values of the influence of the Moon upon the movement of the paraconical pendulum for the periodic component of 24h50m**

6 - Equations (7) and (8) above enable us to determine the theoretical influence of the Moon upon the movement of the pendulum. *Table VII* shows that, for the *average approached value*

(14) $2R = 5.3\text{grads} = 0.0833\text{radians}$

of the wave of 24h50m, the average observed variation $\dot{\phi}$ corresponds to a variation $\Delta\phi = 0.0833$ radians in $24.84/2$ hours, whence the average *observed* value (Equation 1 of *Table VII* below) is

(15) $\dot{\phi}_0 = 0.186 \times 10^{-5}$ radians per second.

[13] The expressions for $K_s$ and $K_m$ are effectively identical to those given in note (3) of my *Note* to the Academy of Sciences of 16 December 1957 (*Théorie du Pendule Paraconique et Influence Lunisolaire* (Theory of the Paraconical Pendulum, and Lunisolar Influence), note 3 [17]), *except* that, by mistake, I included the multiplicative factor $1 + k - h$ corresponding to the deviation of the vertical in my expressions for $K_s$ and $K_m$. Actually the distance *SG* is *totally independent* of the lunisolar deformation of the ground.

In fact the factor $1 + k - h$ (which can be considered as approximately equal to 2/3) is only applicable to the deviation of the vertical, the expression for which is given by equation (6) of *Table V*, p. 127 below (see also §F.1.3 below, and note 6 thereof, p. 200).

The *theoretical value* of the variation $\Delta\dot{\phi}_t$ due to the Moon calculated according to the current theory of gravitation is deduced from the two equations

(16) $\Delta\dot{\phi}_t = \frac{3}{8} p\alpha\Delta\beta$

where $\Delta\beta$ represents the average growth $\beta$ of the semi-minor axis of the ellipse described by the pendulum, corresponding to the equation

(17) $\Delta\beta = \dot{\beta}_t \Delta t = \dot{\beta}(t)\frac{\Delta t}{2} < \frac{\alpha}{2p} K_l \frac{\Delta t}{2}$

which is deduced from equation (8) above.

From this, for a duration $\Delta t = 14$ minutes =840 seconds, we deduce (*Table VII*, equation 8):

(18) $\left|\dot{\phi}_t\right| < 10^{-13}$ radians per second

from which, for the ratio between the observed value and the theoretical value, we obtain (relation 9 of *Table VII* below)

(19) $\left|\dot{\phi}_0/\dot{\phi}_t\right| > 18.3 \times 10^6$

*In the case in question in which the periodicity is 24h50m, the observed value is accordingly at least 18 millions of times greater than the calculated value.*[14, 15]

[14] This figure is smaller than the figure of 50 millions ($10^{-5}/2 \times 10^{-13}$) given in my *Note* of 16 December 1957 to the Academy of Sciences, *Théorie du Pendule Paraconique et Influence Lunisolaire* (Theory of the Paraconical Pendulum and Lunisolar Influence) (§3) [17]; *but it remains of the same order of magnitude*.

The difference essentially is due to consideration, for the wave of 24h50m, of the amplitude $2R = 10.46$ grads of June-July 1955 in my Note of 16 December 1957 (*Table II* du §A.5.3 above, p. 99), instead of the *average* amplitude of 5.3 grads considered here (*Table I* of §A.3.1, p. 92).

[15] The astronomical estimates considered in the calculations of *Tables VI and VII* are the following, in CGS units (*Smithsonian Physical Tables*, ninth revised edition, 1956, The Smithsonian Institution, pp. 729, 730, 731, and 734) [2]:

$M_T = 5.975 \times 10^{27}$ $\qquad r_T = 6.3712 \times 10^8$

$M_s = 1.987 \times 10^{33}$ $\qquad d_s = 149.5 \times 10^{11}$

$M_l = 7.343 \times 10^{25}$ $\qquad d_l = 384.41 \times 10^8$

$\mu = 6.670 \times 10^{-8}$ $\qquad g = \mu\frac{M_T}{r_T^2} = 981.8$

$$C_s = \frac{M_s}{M_T}\frac{r_T^2}{d_s^3} g = 0.396 \times 10^{-13}$$

$$C_l = \frac{M_l}{M_T}\frac{r_T^2}{d_l^3} g = 0.862 \times 10^{-13}$$

The dimensions of $C_s$ and $C_l$ are $\left[T^{-2}\right]$ (inverse squares of time).

## An unarguable impossibility

7 - Thus, we can answer the second fundamental question posed at the start of this Section in the negative *with absolute certainty. It is totally impossible to explain the observed diurnal lunisolar effects within the framework of currently accepted theories*, and in particular to explain the observed amplitude *2R* of the lunar periodicity of 24h50m, which is of the order of 5 grads.[16]

The theoretical lunisolar influences upon the azimuths of the paraconical pendulum according to the current theory of gravitation are *so small*, and the observed influences relatively *so great*, that, during the visits to my laboratories at Saint-Germain and Bougival, no specialist in the theories of mechanics and astronomy controverted the *total* impossibility of such an explanation.[17]

---

[16] The correction given by the theory of general relativity is *utterly negligible*.

According to Schwarzschild's formulation, the correction to be made to the Newtonian potential in fact corresponds to the coefficient

$$\lambda = 1 - \frac{2\mu M_l}{c^2 d_l}$$

where $\mu$ is the gravitational universal constant, $M_l$ is the mass of the Moon, $c$ is the speed of light, and $d_l$ is the distance from the Moon to the Earth (Darmois, *La théorie einsteinienne de la gravitation. Les vérifications expérimentales* (The Einsteinian theory of gravitation. Experimental verifications), Hermann, 1932, p. 13 [99]). We have (for $c = 3 \times 10^{10}$):

$$\frac{2\mu M_l}{c^2 d_l} = \frac{2 \times 6.67 \times 10^{-8} \times 7.343 \times 10^{25}}{9 \times 10^{20} \times 384.41 \times 10^{8}} = 2.83 \times 10^{-13}$$

[17] The above calculations are founded on Lagrange's method of variation of constants. The movement of the paraconical pendulum can be considered as being effectively an elliptical movement that is disturbed by various perturbing forces. In my general theory of the paraconical pendulum of 1956, the method that seemed to me to be both most convenient and most expeditious was that used by Lagrange, which leads to a system of differential equations that can be integrated by successive approximations.

On this method, see Allais, 1956, *Théorie du Pendule Paraconique* (Theory of the Paraconical Pendulum), Part I, Section D, pp. I 53-I 98 [8]. See also particularly Tisserand (*Traité de Mécanique Céleste* (Treatise on Celestial Mechanics), Vol. I, *Perturbations des planètes d'après la méthode de la variation des constantes arbitraires* (Perturbations of the planets according to the method of variation of arbitrary constants), Gauthier-Villars, 1889, p. 173-188) [265].

*TABLE IV*

## FORCES ACTING UPON THE FOUCAULT PENDULUM RELATIVE TO AXES LINKED TO THE EARTH

### *Classical theory*
### *in the case of a rectilinear [1] or elliptical [2] trajectory*

*Notation*

**$\omega$ = speed of rotation of the Earth**

**S : suspension point**

**S x y z : axes linked to the Earth**

**L : latitude of location of observation**

**G : center of gravity of the pendulum**

**Orientation of the axes: Sx: South; Sy: East; Sz: Zenith**

**l = length of equivalent simple pendulum**

$m = x / l \qquad n = y / l$

$p = 2\pi / T = \sqrt{g/l} \qquad l = SG$

**$U_T$ : gravitational potential of the Earth**

$\vec{g} = \nabla_G U_T$

**$\vec{N}$ : tension in the wire**

**M : mass of the pendulum**

*Vectorial equations [1]*

$$(1) \qquad \vec{F} = M\vec{\gamma}$$

$$(2) \qquad \vec{F} = M \nabla_G U_T + \vec{N}$$

$$(3) \qquad \vec{\gamma} = \frac{d^2 \vec{SG}}{dt^2} + 2\vec{\omega} \wedge \frac{d\vec{SG}}{dt}$$

whence, from (1)

$$(4) \qquad \frac{d^2 \vec{SG}}{dt^2} = -\vec{g} - 2\vec{\omega} \wedge \frac{d\vec{SG}}{dt} + (\vec{N}/M) \qquad \vec{g} = \nabla_G U_T$$

Equation (1) is only supposed to be valid with respect to Copernican axes, up to uniform translation.

*Equations of movement of the horizontal projection of the center of gravity [1]*

$$(5) \qquad \ddot{m} + \frac{g}{l} m = -2\omega \sin L\, \dot{n}$$

$$(6) \qquad \ddot{n} + \frac{g}{l} n = 2\omega \sin L\, \dot{m}$$

*Speed of rotation of the azimuth $\phi$ of the pendulum (derived from (5) and (6))*

$$(7) \qquad \dot{\phi} = -\omega \sin L \qquad \text{(elliptical trajectory)}^1$$

$$(8) \qquad \dot{\phi} = -\omega \sin L + \frac{3}{8} p \alpha \beta \qquad \text{(elliptical trajectory)}^2$$

-----

(1) Paul Appell, 1953, *Traité de Mécanique Rationnelle* (Treatise on Rational Mechanics), Gauthier-Villars, Vol. II, pp 293-296; G. Bruhat and A. Foch, 1967, *Mécanique* (Mechanics) Masson, pp. 153-156.

The two equations (5) and (6) result from the fact that, *to the first approximation*, we can take N ~ g , z ~ - l. (Appell, id., p. 293).

(2) H. Resal, *Traité de Mécanique Générale* (Treatise on General Mechanics), 1895, Gauthier-Villars, Vol. I, p. 130.

See also Jules Haag, *Les mouvements vibratoires* (Vibratory Movements), Volume II, French University Press, 1955, §194, pp. 194-196. The precession (3/8) $p\,\alpha\,\beta$ is due to Airy (1850).

*TABLE V*

## FORCES ACTING UPON THE FOUCAULT PENDULUM RELATIVE TO AXES LINKED TO THE EARTH UNDER THE ACTION OF A CELESTIAL BODY i

*Notation*

$M_i$ : mass of celestial body i

I : center of body i

$U_i$ : potential of attraction of body i

$d_i = SI$ $\quad l = SG$

$\vec{N}$ = force exerted by pendulum support

T = center of the Earth

$\mu$ : coefficient of universal gravitation

$r_T$ = radius of the Earth

*Vectorial equations* [1]

(1) $\vec{F} = M\vec{\gamma}$

(2) $\vec{F} = M\,\mathrm{grad}_G\,U_T + M\,\mathrm{grad}_G\,U_i + \vec{N}$

(3) $\vec{\gamma} = \dfrac{d^2\vec{SG}}{dt^2} + 2\,\vec{\omega} \wedge \dfrac{d\vec{SG}}{dt} + \mathrm{grad}_T\,U_i$

whence

(4) $\dfrac{d^2\vec{SG}}{dt^2} = -\vec{g} - 2\,\vec{\omega} \wedge \dfrac{d\vec{SG}}{dt} + \mathrm{grad}_G\,U_i - \mathrm{grad}_T\,U_i + (\vec{N}/M)$

(5) $\dfrac{d^2\vec{SG}}{dt^2} = -\vec{g} - 2\,\vec{\omega} \wedge \dfrac{d\vec{SG}}{dt} + (\mathrm{grad}_S\,U_i - \mathrm{grad}_T\,U_i)$

$+ (\mathrm{grad}_G\,U_i - \mathrm{grad}_S\,U_i) + (\vec{N}/M)$

*Interpretation*

(6) $\mathrm{grad}_S\,U_i - \mathrm{grad}_T\,U_i$ = deviation of the vertical

*(This term is not concerned in the movement of the pendulum)*

(7) $\mathrm{grad}_G\,U_i - \mathrm{grad}_S\,U_i$ = effective acceleration exerted upon the center of gravity G of the pendulum

*Order of magnitude of the angular acceleration exerted on the pendulum by the celestial body i*

(8) $\vec{\gamma}_i / l = (\mathrm{grad}_G\,U_i - \mathrm{grad}_S\,U_i)/l = \mu\,\dfrac{M_i}{l}\left[\dfrac{\vec{GI}}{GI^3} - \dfrac{\vec{SI}}{SI^3}\right]$

(9) $|\vec{\gamma}_i / l| \sim \mu\,\dfrac{M_i}{d_i^3}$ $\qquad g = \mu\,\dfrac{M_T}{r_T^2}$

(10) $|\vec{\gamma}_i / l| \sim C_i = \dfrac{M_i}{M_T}\,\dfrac{r_T^2}{d_i^3}\,g$

Sun : $C_s = 0.396 \times 10^{-13}$ $\qquad$ Moon : $C_l = 0.862 \times 10^{-13}$

-----

(1) On equations (1), (2), and (3), see the references of note (1) of *Table IV*.

*TABLE VI*

## INFLUENCES OF THE SUN AND THE MOON ON THE MOVEMENT OF THE PARACONICAL PENDULUM

***Angular acceleration due to celestial body i (Table V)***

(1) $$\vec{\gamma}_i/l = \mu \frac{M_i}{l} \left[ \frac{\vec{GI}}{GI^3} - \frac{\vec{SI}}{SI^3} \right] \qquad SI = d_i$$

(2) $$\vec{\gamma}_i/l = \frac{M_i}{M_T} \frac{r_T^2}{d_i^3} \frac{g}{l} \left[ \left(\frac{SI}{GI}\right)^3 \vec{GI} - \vec{SI} \right]$$

***Equations of movement***

**The influence of the celestial body i upon the movement of the pendulum can be calculated by introducing terms corresponding to the horizontal projections of $\vec{\gamma}_i / l$ (equation 2) into the right-hand sides of equations (5) and (6) of *Table IV*, whence to the first approximation**

(3) $$\ddot{m} + \frac{g}{l} m = -2\,\omega \sin L\, \dot{n} + K_i\, (m \cos 2 A_i + n \sin 2 A_i)$$

(4) $$\ddot{n} + \frac{g}{l} n = 2\,\omega \sin L\, \dot{m} + K_i\, (m \sin 2 A_i - n \cos 2 A_i)$$

(5) $$K_i = \frac{3}{2} C_i \sin^2 z_i \qquad C_i = \frac{M_i}{M_T} \frac{r_T^2}{d_i^3}\, g$$

**$z_i$, $A_i$ : zenith and azimuth distance of the body i**

$$C_s = 0.396 \times 10^{-13} \qquad C_l = 0.862 \times 10^{-13}$$

***Variation of the azimuth and the minor axis of the elliptical trajectory of the pendulum***

**Taking account of equation (8) of *Table IV*, we deduce, from (3) and (4)**

(6) $$\dot{\phi} = -\omega \sin L + \frac{3}{8} p\, \alpha\, \beta + \frac{1}{p} \frac{\alpha\beta}{\alpha^2 - \beta^2} K_i \cos 2\,(A_i - \phi)$$

(7) $$\dot{\beta} = \frac{\alpha}{2p} K_i \sin 2\,(A_i - \phi)$$

**In view of the orders of magnitude, and of the fact that in each 14-minute experiment we have $\beta < 1/1000$, equations (6) and (7) give to the first approximation (note 10 of §2.5 above):**

(8) $$\dot{\phi} = -\omega \sin L + \frac{3}{8} p\, \alpha\, \beta$$

(9) $$\dot{\beta} = \frac{\alpha}{2p} K_i \sin 2\,(A_i - \phi)$$

-----

*Sources for Tables V and VI:* Allais, September 1966, *Théorie du pendule paraconique* (Theory of the Paraconical Pendulum); and May 1996, *Sur les périodicités lunisolaires du pendule paraconique* (On the lunisolar periodicities of the paraconical pendulum).

The terms in $K_i$ in equations (6) and (7) are deduced from the terms in $K_i$ of equations (3) and (4), and from *Table II* of *Appendix I* of my *Théorie du pendule paraconique* (Theory of the Paraconical Pendulum) of September 1986 (my note of May 1996, §9, p. 12).

*TABLE VII*

## CALCULATED AND OBSERVED VALUES OF THE INFLUENCE OF THE MOON UPON THE MOVEMENT OF THE PARACONICAL PENDULUM

***for the diurnal lunar period T = 24.84h = 24h50m***

***Observed value***

**From *Table I* (§A.3) for the period T = 25h which is representative of the period T = 24h50m, the average amplitude is**

$$2\,R = 5.3 \text{ grads} = 5.3\,\pi / 200 = 0.0833 \text{ radians}$$

**The corresponding average observed variation during a semi-period 24.84 / 2 = 12.42h is thus**

$$(1) \qquad \dot{\phi}_o = 0.0833 / 12.42 \times 3600 = 0.186 \times 10^{-5} \quad \text{radians per second}$$

***Theoretical value***

**We have (*Table VI*, equations 8 and 9)** [1]

$$(2) \qquad \dot{\phi} = -\,\omega \sin L + \frac{3}{8}\, p\, \alpha\, \beta \qquad\qquad (p = \sqrt{g/l} = 3.44)$$

$$(3) \qquad \dot{\beta} = \frac{\alpha}{2p}\, K_l \sin 2\,(A_l - \phi)$$

**At the end of an experiment of $\Delta t$ = 14 minutes = 840 seconds, we have** [2]

$$(4) \qquad \beta \sim \frac{\alpha}{2p}\, K_l\, \Delta t\, \overline{\sin 2\,(A_l - \phi)}$$

**so that for the average value $\bar{\beta}$ of $\beta$ ( $\bar{\alpha}$ = average value of $\alpha$ ) we have**

$$(5) \qquad |\bar{\beta}| \sim \frac{\bar{\alpha}}{2p}\, K_l\, \frac{\Delta t}{2}\, |\overline{\sin 2\,(A_l - \phi)}| < \frac{\bar{\alpha}}{2p}\, K_l\, \frac{\Delta t}{2}$$

**For the theoretical average value $|\dot{\phi}_t|$ of the influence of the Moon, we thus have**

$$(6) \qquad |\dot{\phi}_t| < \left(\frac{3}{8}\, p\, \bar{\alpha}\, \left(\frac{\bar{\alpha}}{2p}\, K_l\, \frac{\Delta t}{2}\right)\right) = \frac{3}{16}\, \bar{\alpha}^2\, \frac{\Delta t}{2}\, K_l$$

*i.e.* for $\bar{\alpha} \sim 1/10$

$$(7) \qquad \dot{\phi}_t < 0.788\, K_l \qquad K_l = \frac{3}{2} \sin^2 \delta_l\, C_l \qquad C_l = 0.862 \times 10^{-13}$$

*i.e.*

$$(8) \qquad \dot{\phi}_t < 0.788\, \frac{3}{2}\, C_l = 1{,}18 \times 0.862 \times 10^{-13} = 1{,}018 \times 10^{-13}$$

***Observed value / Theoretical value*** [3]

$$(9) \qquad \frac{\dot{\phi}_o}{\dot{\phi}_t} > \frac{0.186 \times 10^{-5}}{1.018 \times 10^{-13}} = 18.3 \times 10^{6}$$

-----

(1) The index l corresponds to the Moon.

(2) The average value of a quantity h over the 14 minutes of pendulum motion is designated by the notation $\bar{h}$. The absolute value of h is designated by the notation |h|.

(3) We may remark that we have $r_T / l = 6.37 \times 10^8 / 83 = 7.67 \times 10^6$ where $r_T$ is the radius of the Earth and l is the length of the equivalent single pendulum (§A.1.1 above).

## *B.3 Periodic components not explicable in terms of known phenomena*

If the series of observations of the paraconical pendulum do indeed contain *statistically significant* diurnal periodic components having periods of 24h and 24h50m, and if the periodic effects thus detected *cannot be considered* as resulting from the current theory of gravitation, *is this because these periodic effects can be attributed to an influence of a known periodic phenomenon, direct or indirect*? This is the third fundamental question.

### Comparative harmonic analysis of different phenomena

1 - In order to answer this question, I proceeded to perform harmonic analysis for the same periods of 24h and 24h50m of the following phenomena: the temperature in the laboratory and at Le Bourget, the atmospheric pressure in the laboratory and at Le Bourget, the magnetic declination, the $K$-index of Earth magnetism by Bartels, and the Wolf numbers (solar activity), and I compared the results of this analysis with those corresponding to the azimuth of the paraconical pendulum, both from the point of view of the amplitude and the phases.

If any one of these phenomena could be considered as being the cause of the observed movements of the paraconical pendulum, one would observe: 1° - *an agreement in phase*, at least approximate, between the cause and the effect; and 2° - *the same periodic structure*, and in particular the same ratio of amplitudes for the 24h period and the 24h50m period (for which it would be legitimately possible to substitute a 25h period, in the first approximation, for convenience of calculation).

However, *this double circumstance did not appear for any of the above phenomena that I analyzed.* Besides, it must be underlined that the Graphs representing these phenomena do not present any visible connection with the Graphs representing the azimuths of the plane of oscillation of the paraconical pendulum.

By way of illustration, *Graphs XII and XIII* show the results of Buys-Ballot analysis for the periodicities of 24 and 25 hours compared together for the period of June-July 1955, on the one hand for the azimuth of the paraconical pendulum and *the temperature measured at Le Bourget* and on the other hand for the paraconical pendulum and the *magnetic declination* registered at Chambon-la-Forêt.[1]

*Graph XII* shows on the one hand that differences in phase are present between the fitting sinusoids for the periods of 24 and 25 hours of the azimuth and of the temperature, and on the other hand, above all, that *the 25 hour wave has a much greater relative amplitude for the azimuth than for the temperature.*

And *Graph XIII* demonstrates *a remarkable coincidence of phase* between the azimuth of the pendulum and the magnetic declination, both for the period of 24 hours and for the period of 25 hours. On the other hand, *the relative amplitudes are very different* for these two periods. *Table VIII* shows that this is even more the case for the period of November-December 1954.[2] Such agreements in phase cannot be attributed to chance.[3]

In fact, *for none of the phenomena examined*, does the lunar component of period 25h (representative of the 24h50m wave) have an amplitude comparable to that of the 24h wave.[4] *This very particular periodic structure of the azimuths of the paraconical pendulum is sufficient to eliminate all known geophysical phenomena as explanations.*[5]

---

[1] These are the observatories closest to the IRSID laboratory in Saint-Germain.

[2] Allais, 1958, *Anomalies du mouvement du pendule paraconique à support anisotrope* (Anomalies of the movement of the paraconical pendulum with anisotropic support), pp. 31-32 [19].

[3] It is very remarkable that this agreement of the phases is not reproduced for the fortnights and the weeks, *as would be expected* according to the hypothesis that this phase agreement corresponds to a real phenomenon, because the different relative amplitudes of the waves corresponding to the azimuths and to the magnetic declination for 24h and 25h ought to yield differences in phase when the Buys-Ballot harmonic analysis is applied to periods that are too short for the waves to be separated.

[4] See particularly *Table I* du §A.3.1. above. For the seven month-long series, the average of the ratios $R_{25}/R_{24}$ is equal to 1.39.

[5] See my *Note* of 16 December 1957 to the Academy of Sciences, *"Théorie du pendule paraconique et influence lunisolaire"* ("Theory of the paraconical pendulum and the lunisolar influence") [17].

See in particular my memoir of 21 April 1958, *Anomalies du mouvement du pendule paraconique à support anisotrope* (Anomalies of the movement of the paraconical pendulum with anisotropic support), pp. 23-33 [19].

*GRAPHS XII*

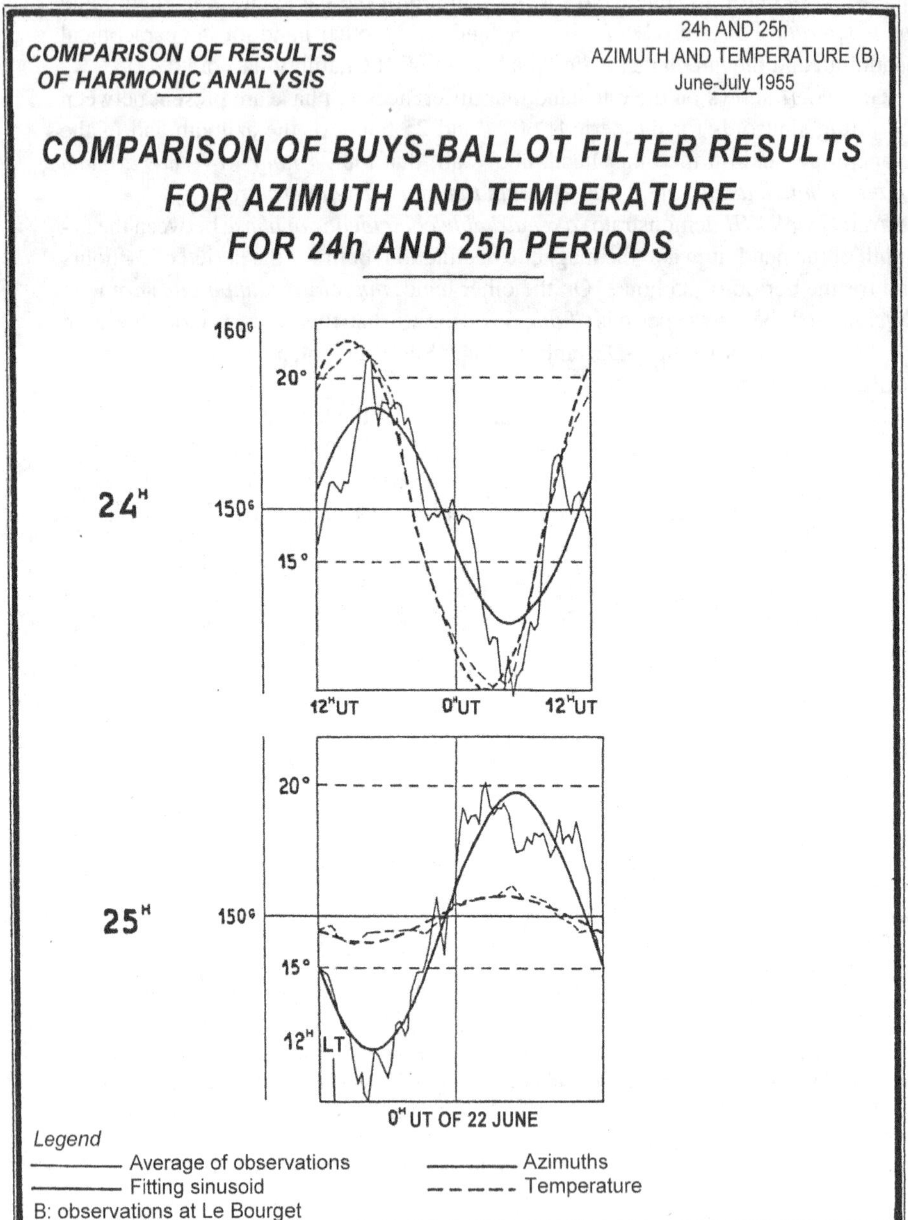


*Source: Graph VA.3.a* of my Conference of 22 February 1958.

*GRAPHS XIII*

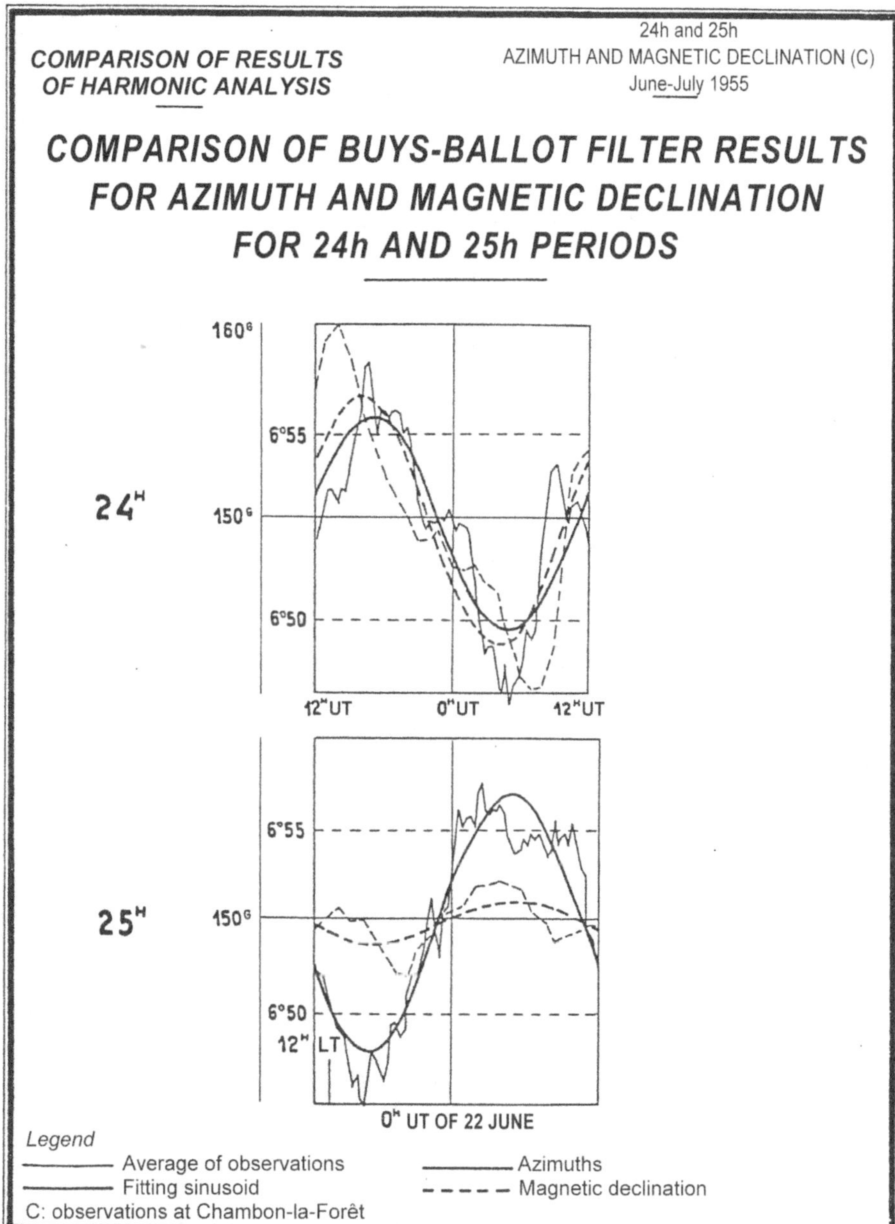


*Source: Graph VA.3.c* of my Conference of 22 February 1958.

*TABLE VIII*

***AZIMUTHS OF PARACONICAL PENDULUM AND MAGNETIC DECLINATION***

***Observations of November-December 1954 and June-July 1955***

***Periodicities of 24 and 25 hours***

| Periods | | Azimuths in grads | | | Declination in sexagesimal minutes | | | $\frac{2R'}{D'} / \frac{2R}{D}$ | H - H' |
|---|---|---|---|---|---|---|---|---|---|
| | | 2 R | 2 R/D | H | 2 R' | 2 R'/D' | H' | | |
| 25 h. | **Nov-Dec 1954** | 12.87 | 0.080 | 5h 52 m | 0.578 | 0.037 | 4h 33 m | 0.461 | 1h 19 m |
| | **June-July 1955** | 14.01 | 0.155 | 15h 58 m | 1.205 | 0.056 | 16h 02 m | 0.362 | - 0h 04 m |
| 24 h. | **Nov-Dec 1954** | 10.34 | 0.064 | 13h 17 m | 2.67 | 0.17 | 11h 36 m | 2.65 | 1h 41 m |
| | **June-July 1955** | 11.66 | 0.129 | 16h 54 m | 6.96 | 0.32 | 15h 56 m | 2.51 | 0h 58 m |

*Legend:* 2R = amplitudes of periodic components in grads; 2R' = amplitudes of periodic components in sexagesimal minutes.
D and D'=differences between greatest and least values of series in question.
H and H' represent the times of the wave peaks in lunar and solar time. See *Graphs XIII*.

*Source:* My memoir of 21 April 1958 (revised on 28 January 1960), *Anomalies du mouvement du pendule paraconique à support anisotrope* (Anomalies of the movement of the paraconical pendulum with anisotropic support), p. 32.

## Lunisolar variations of the vertical and the movement of the paraconical pendulum

2 - Lunisolar variations of the vertical have often been advanced as a possible explanation of the detected effects.[6] In fact, they provide no such explanation. *By themselves*, as I have detailed, *variations of the vertical cannot have any influence on the variations of the azimuth of oscillation of the paraconical pendulum.*[7]

---

[6] At the suspension point $S$, the component of the direction of the vertical corresponding to a celestial body $i$ is represented by the difference

(1) $\quad \vec{f_i} = \nabla_S U_i - \nabla_T U_i$

of the gradients of the potential $U_i$ at the point $S$ and at the center of the Earth (equation 6 of *Table V* above). If it is supposed that the Earth cannot be deformed, then we can show that the horizontal components $f_{ih}$ and vertical components $f_{iv}$ of $f_i$ may be written as follows:

(2) $\quad f_{ih} = \frac{3}{2}\frac{M_i}{M_T}\frac{r_T^3}{d_i^3} g \sin 2z_i$

(3) $\quad f_{iv} = 3\frac{M_i}{M_T}\frac{r_T^3}{d_i^3} g \left(\cos^2 z_i - \frac{1}{3}\right)$

$M_i$ and $M_T$ are the masses of the celestial body $i$ and of the Earth. $r_T$ is the radius of the Earth and $d_i$ is the distance from the Earth to the body $i$. $g$ is the acceleration of gravity, and $z_i$ is the zenith distance of the celestial body $i$.

But if account is taken of the deformation of the Earth, then the deviation of the vertical is equal to the ratio

(4) $\quad \delta_i = \frac{f_{ih}}{g} = \frac{3}{2}\left(1+k-h\right)\frac{M_i}{M_T}\frac{r_T^3}{d_i^3} \sin 2z_i \qquad 1+k-h \sim \frac{2}{3}$

The coefficients $k$ and $h$ are the *Love* numbers corresponding to the deformation of the Earth by the action of the celestial body $i$. We therefore have:

(5) $\quad \frac{\delta_i}{\sin 2z_i} = \frac{M_i}{M_T}\frac{r_T^3}{d_i^3} = C_i\frac{r_T}{g} \qquad C_i = \frac{M_i}{M_T}\frac{r_T^2}{d_i^3} g$

The values of $\delta_i / \sin 2z_i$ for the Sun and the Moon are, in radians and in sexagesimal seconds (§2.5 and note 14 of §2.6 above)

(6) $\quad \frac{\delta_s}{\sin 2z_s} = C_s\frac{r_T}{g} = \frac{0.396\times10^{-13}\times6.371\times10^8}{981.8} = 2.570\times10^{-8}\text{rad} = 5.30\times10^{-3}\text{sec}$

(7) $\quad \frac{\delta_l}{\sin 2z_l} = C_l\frac{r_T}{g} = \frac{0.862\times10^{-13}\times6.371\times10^8}{981.8} = 5.590\times10^{-8}\text{rad} = 11.54\times10^{-3}\text{sec}$

See particularly Schureman, 1941, *Manual of Harmonic Analysis and Prediction of Tides*, p. 14 [244], and Stoyko, 1949, *L'attraction lunisolaire et les pendules*, Bulletin astronomique (The lunisolar attraction and pendulums, Astronomical Bulletin), Vol. XIII, (pp. 1-36), p. 3, equation 2, p. 6, and p. 30 [254].

It should be underlined that, for the Sun and the Moon, the agreement between the observed and calculated values of the deviations of the vertical is *very remarkable* (see particularly Stoyko, 1947, i.d., p. 31).

[7] See above §B.2.4, pp. 120-121.

## Variations of the horizontality of the support and the movement of the paraconical pendulum

3 - At the beginning of May 1957, it was maintained that the horizontality of the surface supporting the pendulum could change during the experiments, particularly due to thermal deformation of the building following the modification of insolation during the course of the day,[8] and that, if the surface of the support assumed a slight inclination, the plane of oscillation of the pendulum would tend towards the position of the vertical plane containing the line of greatest slope of the surface supporting the pendulum.

One of my correspondents, in fact very competent in the field, wrote to me on 2 May 1957: "*How can it be supposed that the variation of the position of your support is not as large as one minute, considering the installation you have built?*"

In fact eighteen months earlier, i.e. on 15 and 16 December 1955, I had already made experiments on the effects of inclination of the support upon the movement of the paraconical pendulum, which showed that these effects were completely negligible. In view of the high quality of my correspondent, I immediately made new experiments for checking upon the effects of variations of horizontality of the support in May and June 1957, which confirmed the results of my experiments in 1955.

• In order to determine the influence of inclination of the support, and in order to eliminate any systematic periodic influence of the balls, I performed successive experiments with the surface of the support being alternatively horizontal and inclined.[9]

---

[8] It should be remarked that the only dependence of the paraconical pendulum upon the exterior is via its suspension upon the bearing surface. No influence from the exterior can therefore act, except via inclination of that surface.

[9] See my very detailed *Note* of 19 January 1959 to the Academy of Sciences, "*Détermination expérimentale de l'influence de l'inclinaison de la surface portante sur le mouvement du pendule paraconique à support anisotrope*" ("Experimental Determination of the Influence of Inclination of the Bearing Surface upon the Movement of the Paraconical Pendulum with Anisotropic Support") [31].

Let $i$ be the inclination of the support in sexagesimal seconds and $J$ be the angle that the projection of the line of greatest slope in the downward direction makes with the North-South vector. The azimuth of the plane of oscillation is designated by $\phi$ (*Graph XIV, 2*)

*Graphs XIV, I; XV, III and XV, IV* represent the observations performed on *15 and 16 December 1955* with the paraconical pendulum[10] for 48 14-minute experiments, the odd-numbered experiments corresponding to a horizontal surface and the even-numbered experiments to an inclined surface and the ball being changed after each experiment, for an inclination $i = 2064$" and an angular spacing $\phi - J = -50$ grads.

Although the various curves are perceptibly different, in particular due to the perturbing influence of the balls (see *Graph XIV, I* which represents the results corresponding to the experiments 5 to 12), the average curves of the 24 odd-numbered experiments and of the 24 even-numbered experiments are not much different (*Graph XV, III*). The same is true for the averages calculated for 15 and 16 December (*Graphs XV, IVA and IVB*).

However, if we consider the average curves for all the experiments in the morning and in the evening, without considering the inclination, we find two average curves that are very different (*Graph XV, IVC*).

Analogous results were obtained *on 22 and 24 May 1957* for $i = 1032$" and $\phi - J = 25$ grads.

These results show that the systematic influence of the epoch upon the movement, at a few hours spacing, is much greater than the influence due to an inclination of the order of 2000 or 1000 sexagesimal seconds. It is very remarkable that in all cases the tangents at the starts of the average curves correspond exactly to the Foucault effect.

• Furthermore, *from 19 to 23 June 1957*, day and night, I performed a continuous series of *doubly enchained* experiments with an elementary duration of 14 minutes. For the odd-numbered experiments the surface was horizontal and for the even-numbered experiments we set $i = 1032$" and $J = 396$ grads. The ball was changed for each experiment. Each odd-numbered experiment of 14 minutes started in the azimuth at which the previous odd-numbered experiment ended, and the same for the even-numbered experiments. There were thus two independent series of enchained experiments (*Graph XV, V*). The differences of the azimuths that they present are due partly to a slight systematic influence of the inclination of the surface (1.72 grads for 1032" and for $\phi - J = -29.7$ grads) and also partly to a preponderant accidental effect of the balls.

[10] The pendulum that was used during the continuous experiments of November-December 1954 and June-July 1955.

For these two series, the Buys-Ballot filter (*Graph XV, VI*) for 24h gives two curves of the same general appearance, the effect of the balls being virtually eliminated, and the amplitude of the periodic effect increasing with the inclination.

On the other hand, eight doubly enchained experiments performed in *May-June 1957*, each of ten hours duration, demonstrated a systematic influence of 2.77 grads for a double inclination of 2064" and $\phi - J = -33.77$ grads.

From all these experiments, and taking accidental errors into account, we may conclude that the influence of the inclination is approximately proportional to $i$, and is about 1.5 grads for 1000" with $\phi - J$ little different from -30 grads.

• These numbers can be usefully compared with the results obtained at Bougival and Saint-Germain in July 1958. The amplitudes $2R$ of the effets corresponding to the filter of 24h50m were of the order of 2.15 grads, while the daily variations of the inclinations were never greater than 8" (sexagesimal seconds).[11, 12]

• From all these indications, we see that comparison of the experimental results obtained by varying the inclination of the ball support surface of the paraconical pendulum, both during releases at a given azimuth and during doubly enchained experiments, shows that *the periodic variations of azimuth over time cannot be considered as resulting from variations of the inclination of the support with respect to the vertical.*

[11] During the experiments in July 1958 at Bougival and Saint-Germain (see *Section C* below, pp. 142-161), the inclinations of the bearing surfaces were checked every twenty minutes with precision levels of the *Institut Géographique National.*

[12] I remind the reader that the theoretical deviations of the vertical due to the lunisolar action are of the order of a hundredth of a sexagesimal second (§3.2 above).

## GRAPHS XIV

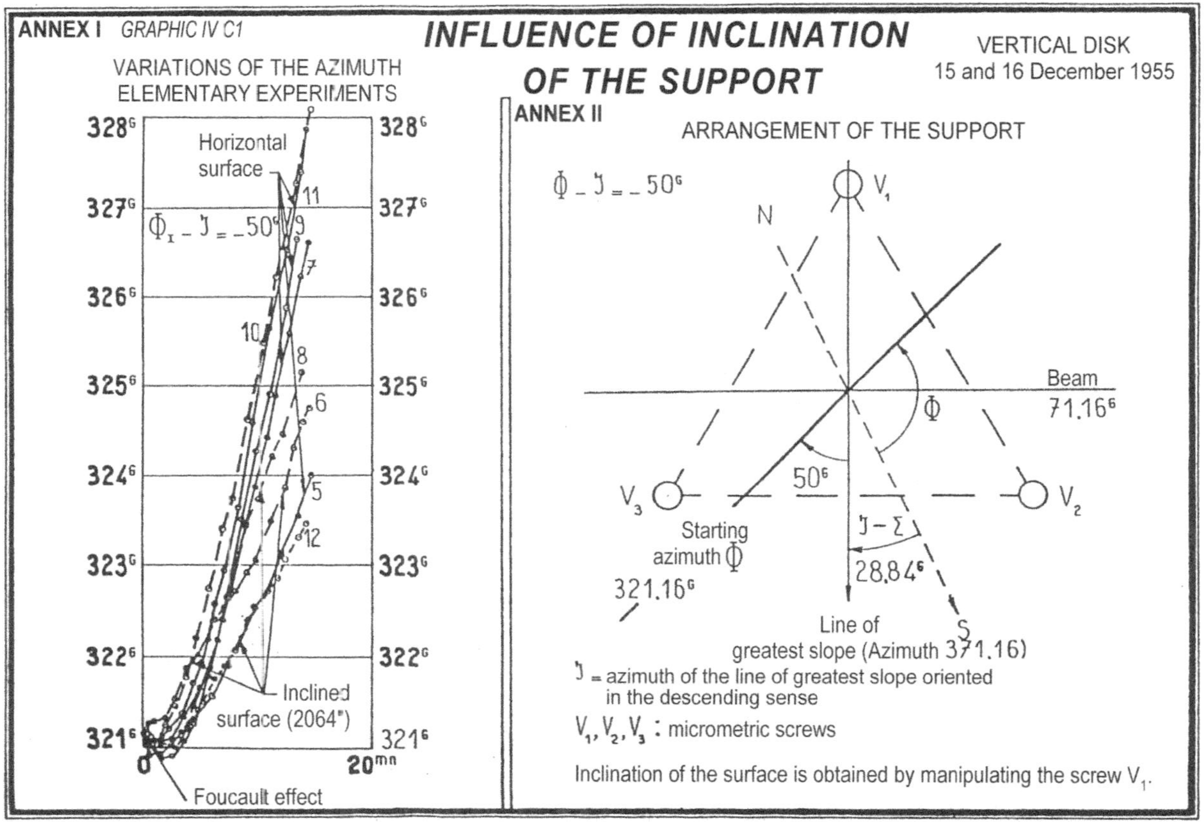


*Source:* My Note of 19 January 1959, Comptes Rendus de l'Académie des Sciences, Vol. 248, pp. 359-362 (photographic reproduction).

## GRAPHS XV

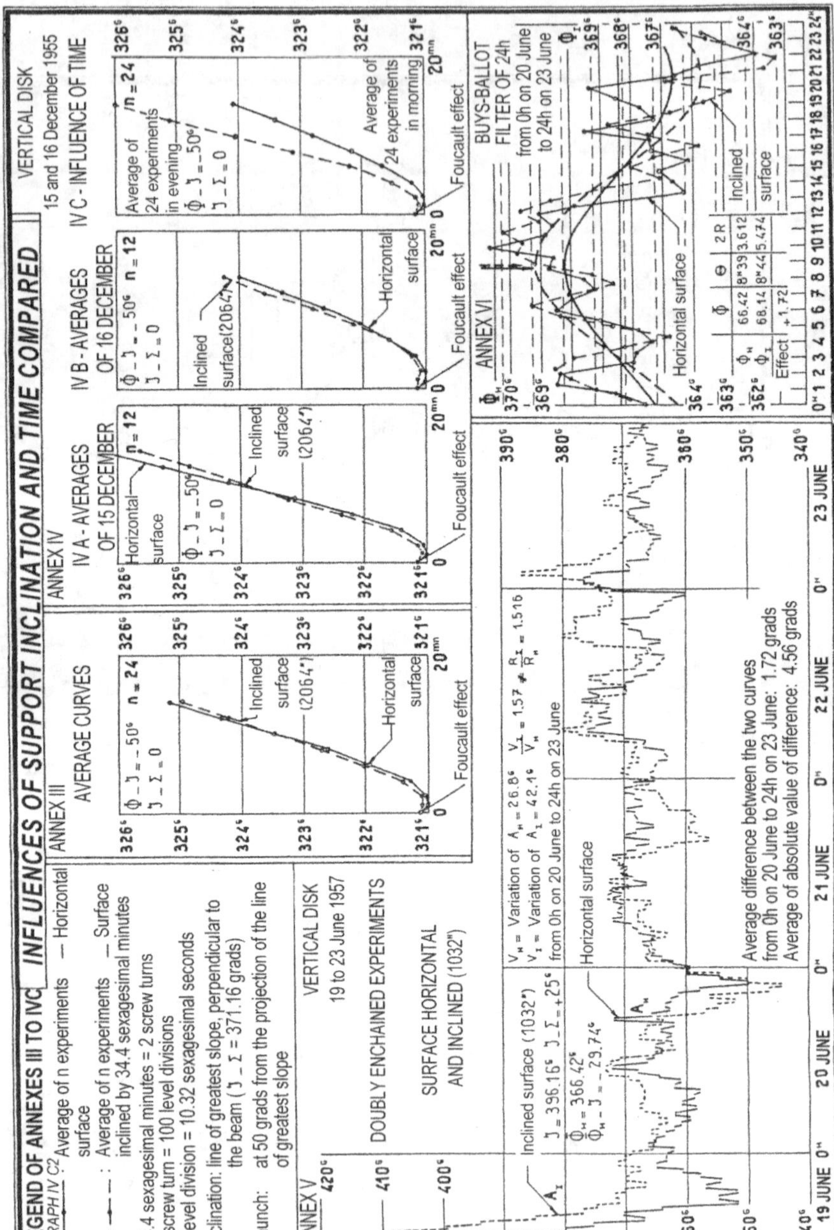


Source: My Note of 19 January 1959, Comptes Rendus de l'Académie des Sciences, Vol. 248, pp. 359-362 (photographic reproduction).

## No direct relation between the azimuths of the paraconical pendulum and known phenomena

4 - Overall, we must conclude that the lunisolar components of 24h and 24h50m, and particularly the latter, whose existence must be considered *as completely real*, and that are *totally inexplicable* by the current theory of gravitation, *cannot be viewed as resulting from the direct or indirect action of any known phenomenon; and this conclusion is utterly certain.*

Thus, the third fundamental question posed at the start of this Section B[13] must be answered *with complete certainty* in the negative. *We are concerned with a new phenomenon that is completely inexplicable in the framework of the currently accepted theory of gravitation.*

[13] p. 102 above.

# C THE CRUCIAL EXPERIMENTS OF JUNE-JULY 1958 IN SAINT-GERMAIN AND BOUGIVAL

## *C.1 The two laboratories at Saint-Germain and Bougival*

At the end of 1957, with the agreement of Albert Caquot and Pierre Tardi, I established a program of research for the first semester of 1958, with the object of obtaining funding from the *Centre National de la Recherche Scientifique* (National Center for Scientific Research).

This program aimed particularly at repeating the experiments on the paraconical pendulum in a laboratory quite deep below the Earth's surface, either in an underground quarry or in a mine, so as to eliminate any possible superficial perturbation due to wind or temperature.

Finally a site was found at the beginning of 1958 in a disused subterranean quarry at Bougival, the "*Blanc Minéral*" quarry.

*The proposed combination of experiments was to be crucial for confirming or disproving the existence of a lunar component of period 24h50m having amplitude inexplicable in the framework of the theory of gravitation, in the azimuths of the paraconical pendulum with anisotropic support.*

Pursuant to this project, and with the help of the *Centre National de la Recherche Scientifique* (National Center for Scientific Research - CNRS) and the *Comité d'Action Scientifique de la Défense Nationale* (National Defense Committee for Scientific Action - CASDN), in June-July 1958, I proceeded to perform two simultaneous series of continuous experiments with two *identical* sets of apparatus in my laboratory at Saint-Germain and in a new laboratory at Bougival, installed in a gallery in a currently abandoned part of the "*Blanc Minéral*" (White Mineral) underground quarry having no previous facilities, overlaid with 57 meters of clay and chalk, about 6.5 km distant from Saint-Germain. The horizontal distance from the open surface was of the order of 800 m.[1]

---

[1] The installation in the Bougival laboratory was performed very efficiently by the IRSID workshop by using prefabricated elements and installing a rigid metallic pendulum support identical to that at IRSID (see §E.3 below). The two pendulums were both *identical* to that described in §A.1 above.

The support was installed upon massive reinforced concrete pedestals that were criss-crossed by a system of metallic beams.

The anisotropy of the support proved to be *entirely comparable* to that at IRSID (see §E.3.3 and *Table X* below, p. 180).

After a trial period of a month in June,[2] two simultaneous continuous experiments were performed in the two laboratories of Bougival and Saint-Germain over a duration of 30 days from 0h UT on 2 July 1958 to 23h40m UT on 31 July.[3]

These two *absolutely crucial* experiments demonstrated that the previously demonstrated anomalies *were again observed*, and that, in the two laboratories, they presented periodic structures that corresponded in a remarkable manner.[4]

---

[2] In particular, it was necessary to organize two teams of seven observers in the two laboratories, operating in relays day and night.

[3] I experienced a certain apprehension as I proceeded to set up these two crucial experiments, because, although I was absolutely certain of the existence at IRSID, two meters below ground level, of a diurnal lunar periodicity of 24h50m having an amplitude incompatible with the accepted theory of gravitation, I was not able to anticipate what effects might be found at a depth of 57 meters.

[4] The reader may usefully refer to the two following *Notes* that I presented to the Academy of Sciences on 3 November and 22 December 1958.

- *Nouvelles expériences sur le Pendule Paraconique à support anisotrope* (New experiments on the Paraconical Pendulum with anisotropic support). CRAS, Vol. 247, 1958, pp. 1428-1431 [25].

- *Structure périodique des mouvements du Pendule Paraconique à Bougival et Saint-Germain en juillet 1958* (Periodic structure of the movements of the Paraconical Pendulum at Bougival and Saint-Germain in July 1958). CRAS, Vol. 247, 1958, pp. 2284-2287 [27].

## C.2 *The periodic diurnal structure of the two series of enchained experiments of Bougival and Saint-Germain*

*Graphs XVI to XXIII* and *Table IX* below represent the essential results of harmonic analysis of the two series of enchained experiments at Bougival and Saint-Germain, as they were published in my two *Notes* of 3 November and 22 December 1958 to the Academy of Sciences.

### The periodic component of 24h50m

1 - *Graphs XVI and XVII* represent the results of harmonic analysis obtained by the Buys-Ballot filter method for periods of 25h and 24h50m of the continuous enchained experiments performed *simultaneously* in the two laboratories day and night over a month, from 0h UT on 2 July to 23h40m UT on 31 July, in conditions *identical* to those of my previous experiments of June-July 1955.[1], [2]

*Graphs XVI and XVII* make it possible to compare the results obtained in the two laboratories of Saint-Germain and Bougival. In the case both of the 25h filter and the 24h50m filter, the two fittings have amplitudes that differ very little, and exhibit *very remarkable* agreements of phase.

In fact, these agreements are generally *more marked* for *Graph XVII* which corresponds to the period of 24h50m than for *Graph XVI* which corresponds to the period of 25h.

---

[1] §A.1 above.

[2] The symbol UT corresponds to universal time. The symbol LT corresponds to lunar time. Thus the Moon passes the meridian at 12 hours LT.

*Graph XVI* was published in my *Note of 3 November 1958* to the Academy of Sciences. The results corresponding to *Graph XVII* were published in my *Note of 22 December 1958*, and also in the *Complementary Note* to my memoir of 1958, *Doit-on Reconsidérer les Lois de la Gravitation ?* (Should the Laws of Gravitation be Reconsidered?) [23].

Here I should remind the reader that at the time we had none of the means for calculation that are available today. Calculations were made by hand with an electrically powered machine.

With data available every twenty minutes, it was naturally much faster to apply the Buys-Ballot filter for a period of 25h. Accordingly in my *Note of 3 November 1958* I presented the results immediately by considering a period of 25h.

In the following weeks I applied the Buys-Ballot filter for a period of 24h50m by interpolating the data every ten minutes.

*The comparison of the results obtained by the two calculations only became more significant.* The agreement of the results is indeed better when the period of 24h50m is considered instead of the approximate period of 25h.

GRAPH XVI

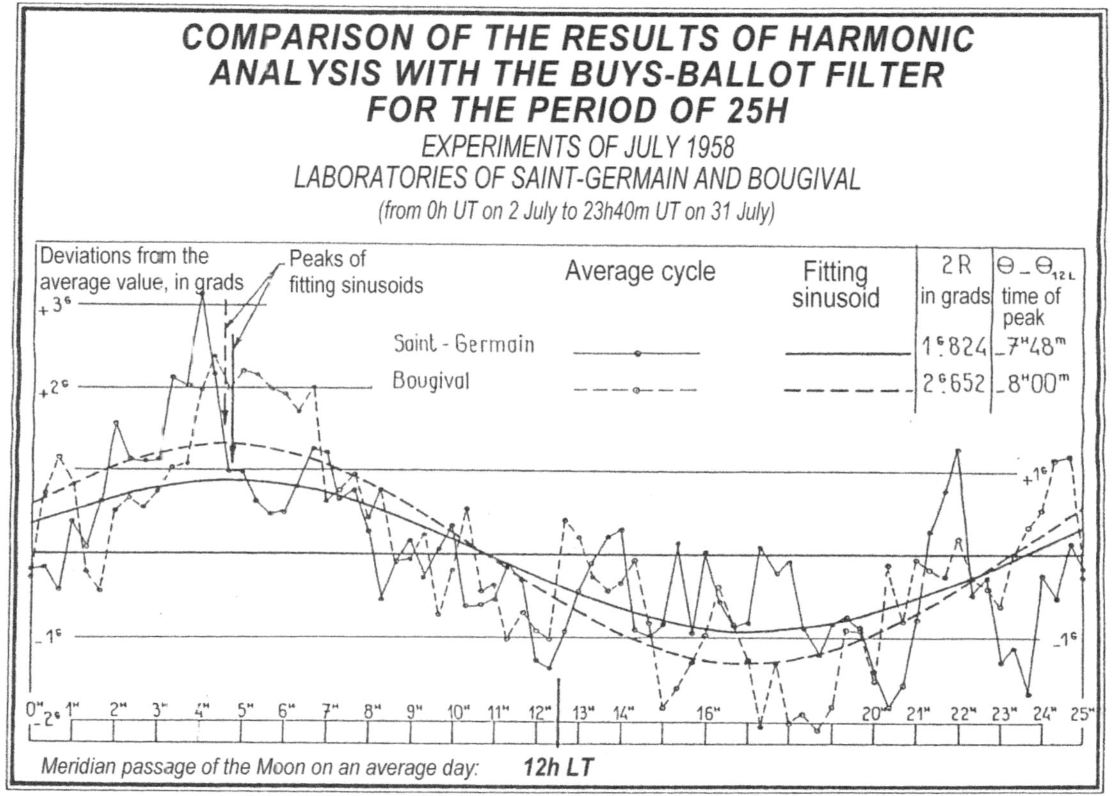


*Source:* Note of 3 November 1958 to the Academy of Sciences, *Nouvelles experiences sur le pendule paraconique à support anisotrope* (New Experiments on the Paraconical Pendulum with Anisotropic Support), CRAS, Vol. 247, p. 1429.

*GRAPH XVII*

EXPERIMENTS OF JULY 1958 AT BOUGIVAL AND SAINT-GERMAIN

***Results of the Buys-Ballot filter for a period of 24h50m***

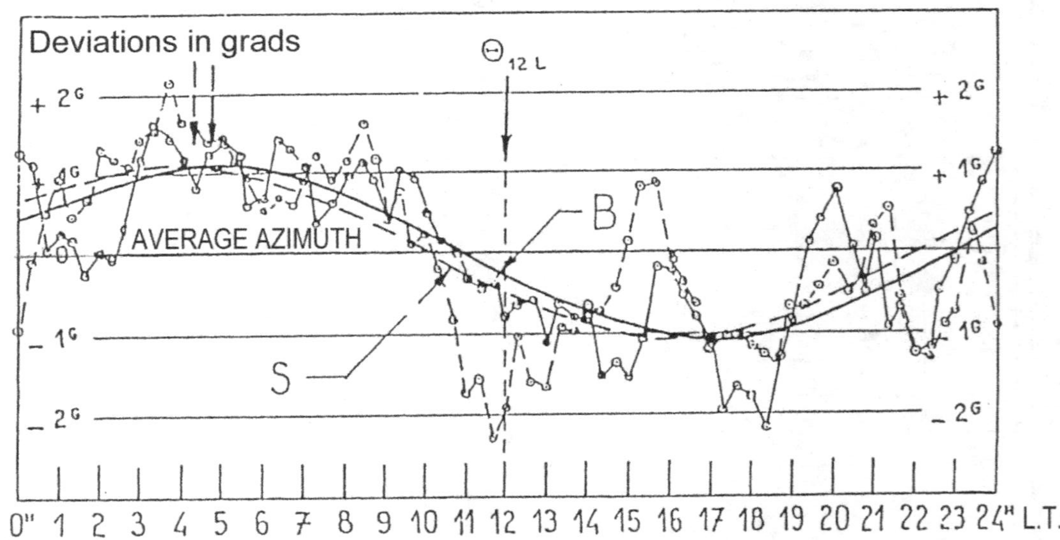


*Legend*

| | 2 R in grads | $\Theta - \Theta_{12\,L}$ |
|---|---|---|
| BOUGIVAL | 2.17 | – 7 h. 23 m. |
| SAINT-GERMAIN | 2.10 | – 7 h. 55 m. |

———— : BOUGIVAL — — — — — : SAINT-GERMAIN

Fine lines: average cycles Coarse lines: fitting sinusoids

## The periodic structure of the two series of enchained experiments of Bougival and Saint-Germain

2 - *Table IX* gives the overall numerical results for the analyses performed. These are displayed in *Graphs XVI to XXIII.*

The symbols B and S respectively correspond to the two pendulums at Bougival and Saint-Germain. The symbols B+S and B-S correspond to the half-sum and to the half-difference of their observed azimuths.

The 30 days of observation were divided into four elementary periods of 7, 8, 8, and 7 days, designated by the numbers 1, 2, 3, and 4. The symbols 1+2 and 3+4 denote the first and the second fortnight of the month, and the symbol 1+2+3+4 denotes the entire month.

$\Theta$ represents the time of the peak of the fitting sinusoid in UT, and $\Theta - \Theta_{12L}$ represents the time of the peak of the fitting sinusoid in lunar time, taking the hour of meridian passage of the Moon as origin $\Theta_{12L}$.

It is very remarkable that on average the amplitude of the wave of 24h50m was about double that of the wave of 24h at Bougival and at Saint-Germain (*Table IX*).[3]

We see that, *during the month of July 1958, the effects having period 24h50m were substantially the same in the two laboratories. The effects having periods 24h and 12h were, by contrast, opposite in sign.*

*Graphs XIX to XXIII* show that this structure is also present in each fortnight considered in isolation.[4, 5, 6]

---

[3] The average of the two ratios $2.174/1.394 = 1.559$ and $2.106/0.776 = 2.714$, *i.e.* 2.136, is substantially equal to the ratio $C_l/C_s = 2.177$ (§B.2.5 above, p. 123).

[4] From *Table IX*, if we consider the wave of 24h50m, the half-sum $(B+S)/2$ has amplitude 2.140 grads, while the half-difference $(B-S)/2$ has amplitude 0.034 grads.

For the wave of 24h, the half-difference $(B-S)/2$ and the half-sum $(B+S)/2$ have respective amplitudes 1.086 grads and 0.310 grads.

We see that *the half-sum $(B+S)/2$ has no substantial component of period 24h, while the half-difference $(B-S)/2$ has no substantial component of period close to 24h50m.*

Application of a Buys-Ballot filter of 25h to the half-difference $(B-S)/2$ is accordingly equivalent to application of a filter of 25h to a wave of 24h.

[5] One circumstance is *totally unexplained* even today: *the oppositions in phase for the waves of 24h and 12h at Saint-Germain and at Bougival.*

[6] These *average* agreements are all the more remarkable because the *daily* azimuth curves at Bougival and Saint-German *are quite substantially different.*

**Overall view**

3 - These *completely fundamental* results thus demonstrate:

a. that the existence of a periodicity of period 24h50m *is not due to any accidental cause*;

b. that this periodicity still appears when there is *no substantial thermal variation*, as was the case at Bougival;

c. that this periodicity still appears when the external structure of the laboratory is *sheltered from all external perturbations*, as was the case at Bougival.

*These two crucial and absolutely fundamental series of experiments totally swept aside all previously presented objections to my experiments on the paraconical pendulum, as far as concerned the existence of a diurnal lunar periodicity of 24h50m having an amplitude totally inexplicable in the framework of the current theory of gravitation.*[7]

*They totally confirmed that anomalies of periodic character, inexplicable in the framework of the current theory of gravitation, are present in the movement of the asymmetric paraconical pendulum with anisotropic support, and that, in the present state of the discussion, these anomalies cannot be linked to any known phenomenon.*

---

[7] By way of illustration, see the *Note* of Jean Goguel to the Academy of Sciences, "*Observations à propos du pendule dit paraconique*" ("Observations in connection with the so-called paraconical pendulum"), (CRAS, Vol. 246, no. 16, 21 April 1958, pp. 2340-2342) [152].

This *Note*, which essentially attributed the observed effects to oscillations of the building, is typical of the *completely unfounded* objections that were presented against my work.

It is based upon hypotheses formulated in a very vague form, whose precise meaning is difficult to grasp. Moreover no theoretical calculation or numerical application was presented to justify them.

But this is only one example among many others. For example one of my most eminent interlocutors, Henri Villat (1879-1972), professor of fluid mechanics at the Faculty of Sciences and president of the Mechanics Section of the Academy of Sciences, declared to me on 3 December 1956: "*Replace your steel balls with balls of agate, and you will see: everything will be different.*"

Nevertheless, in answer to my invitation to visit my laboratory at IRSID, he answered: "*Well, you see... It would be of little use for me to trouble to travel to Saint-Germain, because I am not an experimenter.*"

*TABLE IX*

EXPERIMENTS OF JULY 1958 AT BOUGIVAL AND SAINT-GERMAIN

***Overall results***

| PERIOD OF FILTER | PENDULUM | TIME PERIOD | 2R | $\Theta$ | $\Theta - \Theta_{12L}$ | PERIOD OF FILTER | PENDULUM | TIME PERIOD | 2R | $\Theta$ | $\Theta - \Theta_{12L}$ |
|---|---|---|---|---|---|---|---|---|---|---|---|
| $24^h50^m$ | B | 1 + 2 + 3 + 4 | 2.174 | | $-7^h23$ | $25^h$ | B + S | 1 + 2 | 1.804 | | $-7^h45$ |
| | S | " | 2.106 | | $-7^h55$ | | " | 3 + 4 | 2.880 | | $-8^h01$ |
| $25^h$ | B | 1 + 2 + 3 + 4 | 2.562 | | $-8^h01$ | $24^h$ | B − S | 1 + 2 | 0.502 | $13^h17$ | |
| | S | " | 1.840 | | $-7^h47$ | | " | 3 + 4 | 1.708 | $13^h49$ | |
| $24^h$ | B | 1 + 2 + 3 + 4 | 1.394 | $13^h56$ | | $12^h$ | B − S | 1 + 2 | 1.502 | $5^h26$ | |
| | − S | " | 0.776 | $11^h40$ | | | " | 3 + 4 | 1.986 | $6^h31$ | |
| $12^h$ | B | 1 + 2 + 3 + 4 | 1.890 | $5^h41$ | | $25^h$ | B − S | 1 + 2 | 1.388 | | $7^h23$ |
| | − S | " | 1.540 | $6^h30$ | | | −(B − S) | 3 + 4 | 1.880 | | $7^h54$ |

*Legend:* $\Theta$ = time of peak of the fitting sinusoid at the middle of the period.
$\Theta_{12L}$ = time of meridian passage of the Moon at the middle of the period.
At Bougival we have: 2R (24h50m) / 2R (24h) = 2.174 / 1.394 = 1.563
At Saint-Germain we have: 2R (24h50m) / 2R (24h) = 2.106 / 0.776 = 2.714
The average ratio of the amplitudes is accordingly: (1.563 + 2.714) / 2 = 2.138

*Source: Note* to the Academy of Sciences of 22 December 1958, *Structure périodique des mouvements du pendule paraconique à Bougival et à Saint-Germain en juillet 1958* (Periodic structure of the movements of the paraconical pendulum at Bougival and at Saint-Germain in July 1958), CRAS, Vol. 247, 1958, p. 2285.

*GRAPH XVIII*

EXPERIMENTS OF JULY 1958 AT BOUGIVAL AND SAINT-GERMAIN

***Results of the Buys-Ballot filter for a filter of 24h***

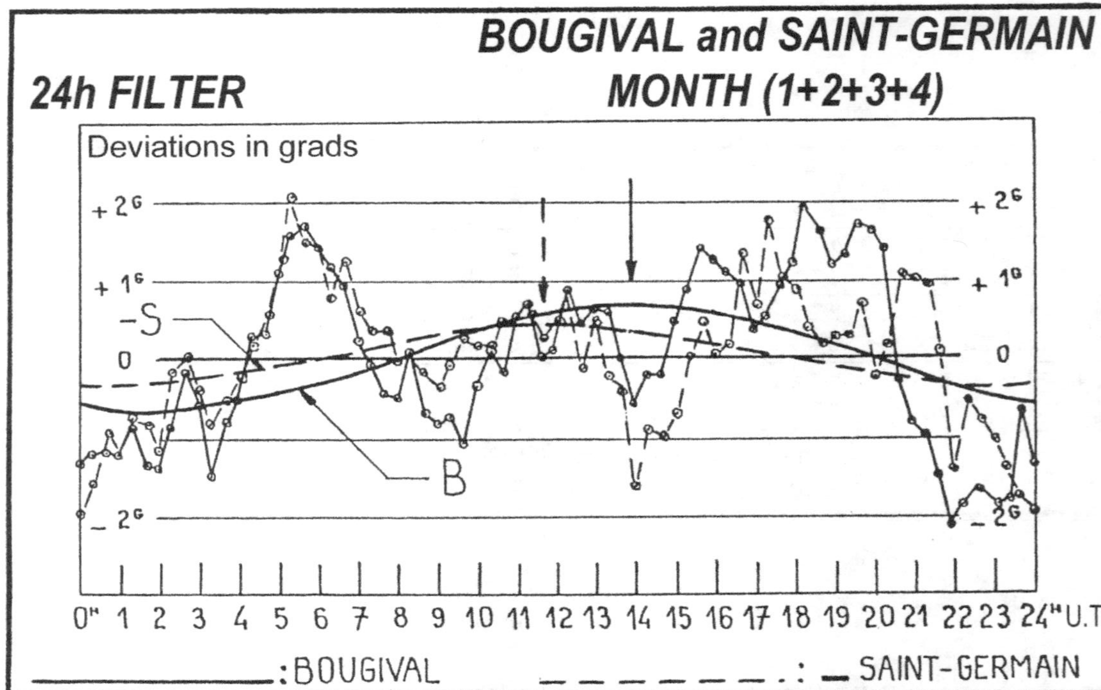


*Source: Note* to the Academy of Sciences of 22 December 1958, *Structure périodique des mouvements du pendule paraconique à Bougival et à Saint-Germain en juillet 1958* (Periodic structure of the movements of the paraconical pendulum at Bougival and at Saint-Germain in July 1958), CRAS, Vol. 247, 1958, p. 2285.

***GRAPHS XIX, XX, and XXI***

EXPERIMENTS OF JULY 1958 AT BOUGIVAL AND SAINT-GERMAIN

***Results of Buys-Ballot filter with filters of 12h and 24h***

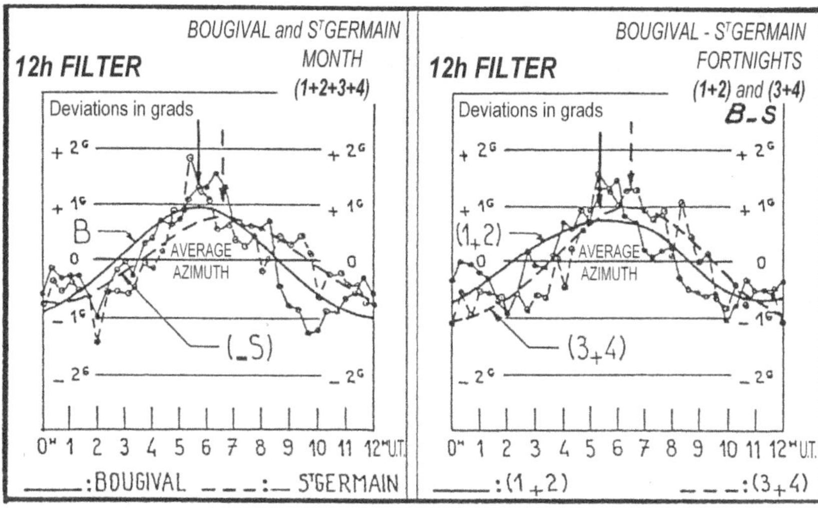


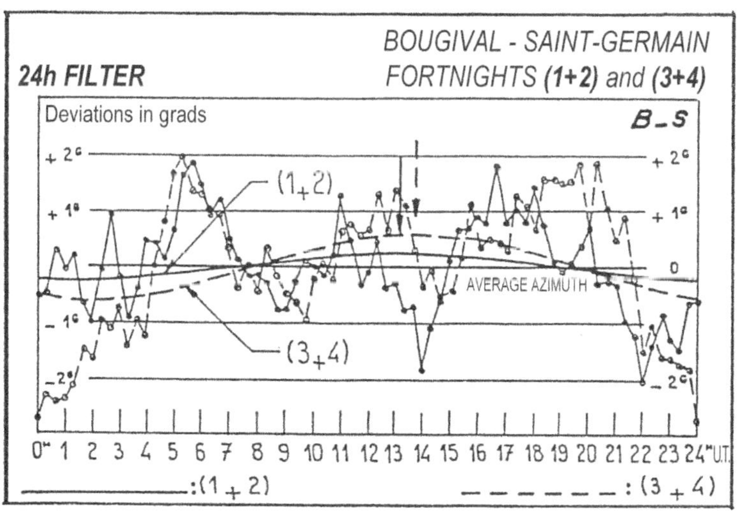


*Source: Note* to the Academy of Sciences of 22 December 1958, *Structure périodique des mouvements du pendule paraconique à Bougival et à Saint-Germain en juillet 1958* (Periodic structure of the movements of the paraconical pendulum at Bougival and at Saint-Germain in July 1958), CRAS, Vol. 247, 1958, p. 2286.

***GRAPHS XXII and XXIII***

EXPERIMENTS OF JULY 1958 AT BOUGIVAL AND SAINT-GERMAIN

***Results of Buys-Ballot filter for a filter of 25h***

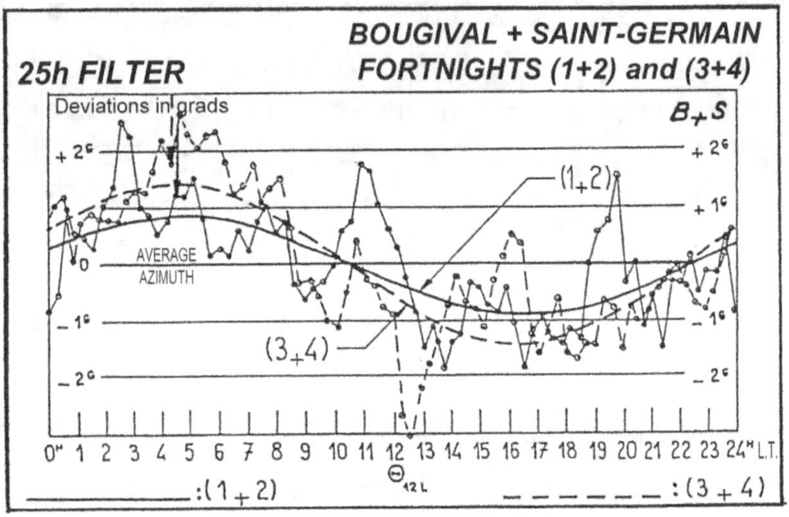


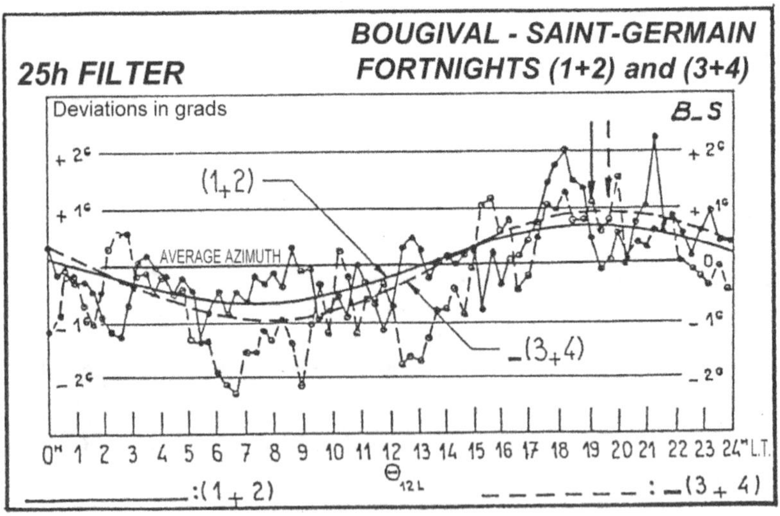


*Source: Note* to the Academy of Sciences of 22 December 1958, *Structure périodique des mouvements du pendule paraconique à Bougival et à Saint-Germain en juillet 1958* (Periodic structure of the movements of the paraconical pendulum at Bougival and at Saint-Germain in July 1958), CRAS, Vol. 247, 1958, p. 2286.

## Frequency graph of the Bougival series of July 1958

4 - Consideration of the frequency graph of the month-long Bougival series is particularly significant, and it *totally* confirms the above conclusions.[8]

*Graph XXIV* shows the series of 721 hourly values of the paraconical pendulum observed at Bougival from 0h UT on 2 July to 23h40m UT on 31 July.[9] This graph has an appearance very comparable to that of *Graph II* which represents the azimuths of June-July 1955.[10]

*Graph XXV* shows the cycle of 24h50m obtained by Buys-Ballot analysis from the hourly values of *Graph XXIV*,[11] having an amplitude $2R = 2.13$ grads.

*Graph XXVI* is a frequency graph of the series of 721 hourly values of *Graph XXIV*, with lines of equal significance being shown. The level of significance of the period of 24h50m is 0.07%, corresponding to a probability of less than one in a thousand.[12]

The period of 24h50m has an amplitude about 1.68 times greater than that of the period of 24h, whose level of significance is only about 14%.[13]

---

[8] This analysis was presented as an illustration in my Communication of 1961 to the *International Statistical Institute* (note 10 of §B.1.3 above, p. 113) [40].

[9] *Annex II B* of my Communication of 1961 to the *International Statistical Institute*.

[10] *Graph II* of §A.2 above, p. 89.

[11] *Annex III C* of my Communication of 1961 to the *International Statistical Institute*.

[12] My Communication of 1961, §18. The value of $\mu$ corresponding to the period of 24h50m is 7.28, with $P = e^{-\mu} = 0.0007$.

See §B.1.3 above.

[13] By way of comparison, see the frequency graph for the series of 721 hourly values of November-December 1954 (*Graph XI* of §B.1.3 above).

*GRAPH XXIV*

OBSERVATIONS OF JULY 1958 AT BOUGIVAL

***Azimuths of plane of oscillation***

*N = 721 hourly values*

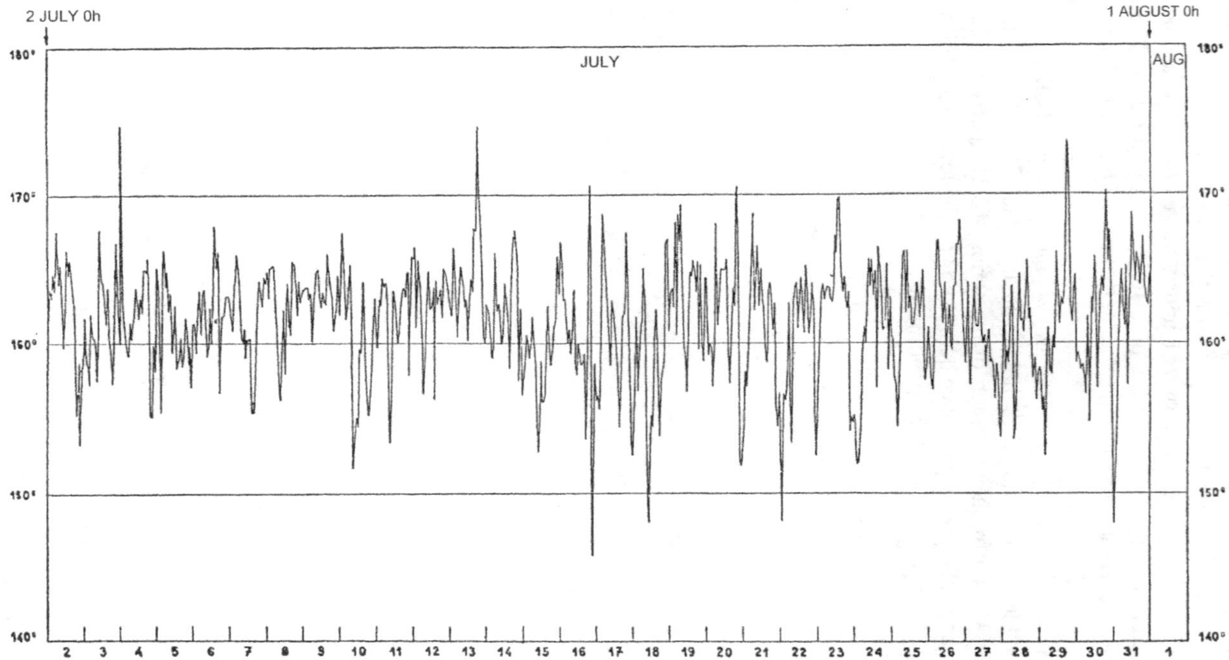


*Source: Annex II B* of my Communication of 1961 to the *International Statistical Institute* "*Test de périodicité. Généralisation du test de Schuster au cas de séries temporelles autocorrélées dans l'hypothèse d'un processus de perturbations aléatoires d'un système stable*" (Test of periodicity. Generalization of the Schuster test to the case of autocorrelated time series, under the hypothesis of a process of random perturbations of a stable system). The 721 hourly values of the azimuths of the paraconical pendulum at Bougival are given in *Annex II A* of my Communication of 1961 to the *International Statistical Institute*.

*GRAPH XXV*

OBSERVATIONS OF JULY 1958 AT BOUGIVAL

***Azimuths of plane of oscillation***

***Application of the Buys-Ballot method to the period of 24h50m***

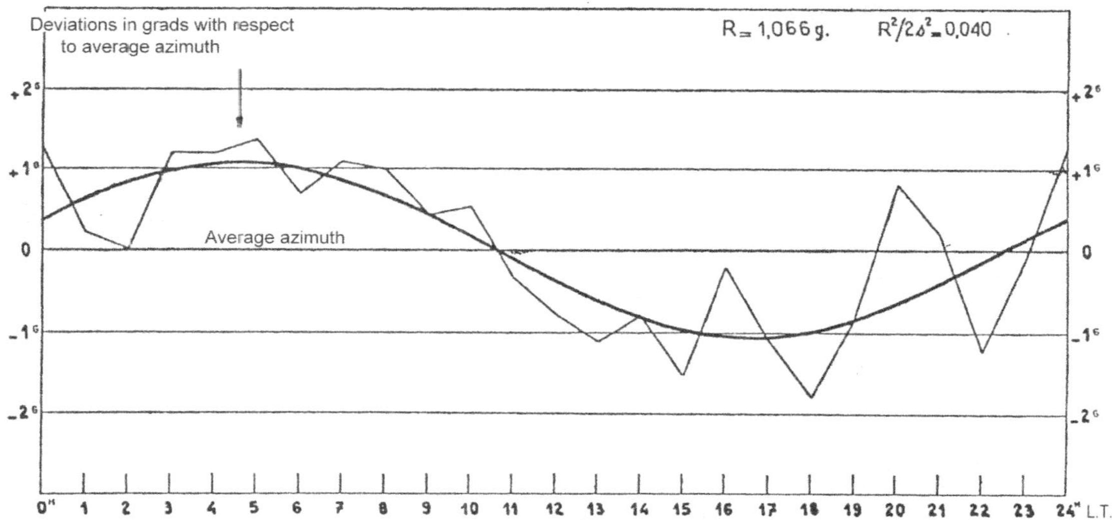


*Legend:* —— observed values ▬▬ fitting sinusoid

R = 1.066 g. $R^2/2\sigma^2 = 0.040$

*Source: Annex III C* of my Communication of 1961 to the *International Statistical Institute* "*Test de périodicité. Généralisation du test de Schuster au cas de séries temporelles autocorrélées dans l'hypothèse d'un processus de perturbations aléatoires d'un système stable*" (Test of periodicity. Generalization of the Schuster test to the case of autocorrelated time series, under the hypothesis of a process of random perturbations of a stable system).

***GRAPH XXVI***

## OBSERVATIONS OF JULY 1958 AT BOUGIVAL

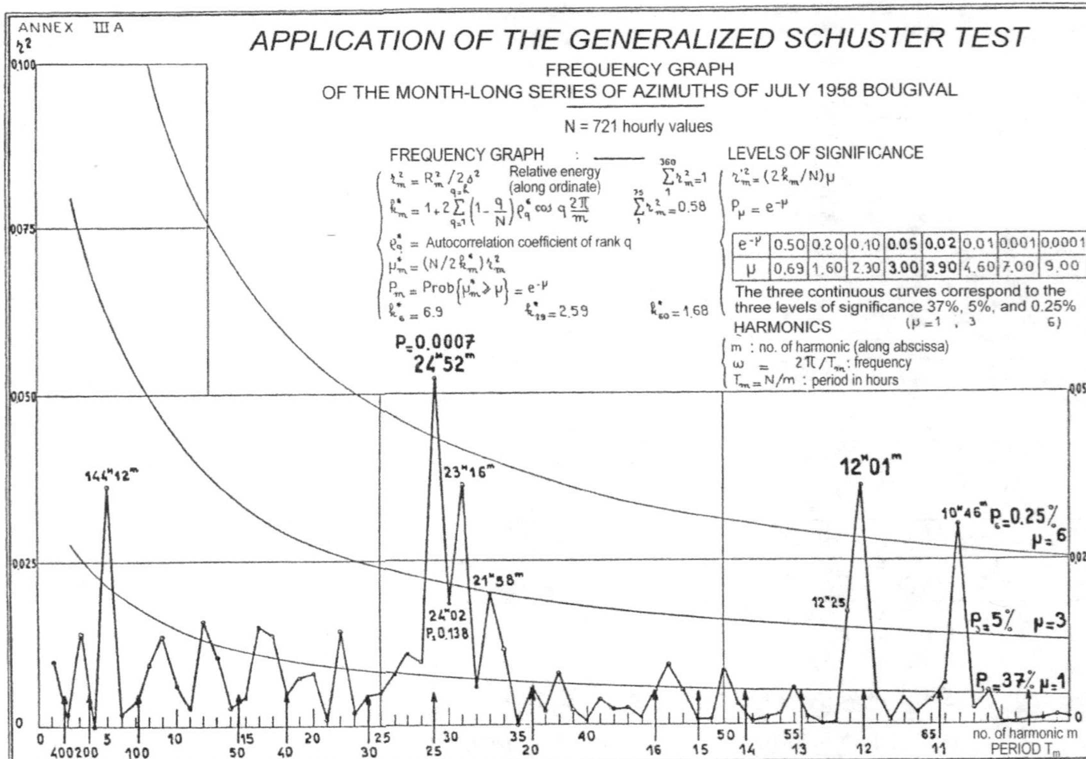


*Legend:* On the formulation of the test, see §B.1.3 above and the *Legend* of *Graph XI*.

*Source: Annex III A* of my Communication of 1961 to the *International Statistical Institute* (see *Source* of *Graph XXII*).

## *C.3 Monthly sidereal lunar periodicity*

*Graphs XXVII and XXVIII* show the fittings of the half-sums $(B+S)/2$ and of the half-differences $(B-S)/2$ of the *daily averages* of the azimuths $B$ and $S$ of the paraconical pendulums at Bougival and at Saint-Germain from 1 July to 31 July 1958,[1] taking account both of linear trends[2] and of sinusoids of period 27.322 days, which is *equal to the sidereal period of the Moon*.[3, 4]

We see that the two fittings of *Graphs XXVII and XXVIII* are virtually in phase. The peak of the fitting to $(S+B)/2$ is at 20h on 24 July, and the peak of the fitting to $(S-B)/2$ is at 0h on 24 July. The two trends of -0.149 and -0.147 grads per day of *Graphs XXVII and XXVIII* are virtually identical.

---

[1] *In contrast to the above calculations, the following fittings were performed recently, in January 1996.*

On *Graph XXVII*, the azimuths are reckoned *in grads from the North* in the anticlockwise sense. I remind the reader that we have

(1) $\sigma^2/\Sigma^2 = 1 - R^2$

The difference $1-R^2$ thus represents the fraction of the variance that is not explained by the multiple correlation (see above, p. 101, note 12).

[2] These linear trends correspond to fluctuations of longer period, and particularly to fluctuations having period of six months (see *Chapter V*, Section B, below).

[3] In fact the two periods of the Moon, *i.e.* its sidereal period and its synodic period, give almost exactly the same coefficients of correlation. The reason is that, for a total duration of 30 days (*i.e.* 720 hourly values), these two sidereal and synodic periods are indistinguishable.

I considered the *sidereal period* in order to facilitate eventual comparison with the results of Esclangon and Miller, which correspond to sidereal time.

In fact, this choice was entirely justified by the analyses of the series of observations of the paraconical pendulum with isotropic support (see below *Chapter II*, §E.2 and F.3).

[4] We have the relation

(1) $\frac{1}{T_1} - \frac{1}{T_2} = \frac{1}{T}$

between the sidereal period $T_1$ and the synodic period $T_2$ of the Moon and the annual period $T$ of the Earth, expressed in average days, with:

$T_1 = 27.32166 \qquad T_2 = 29.53059 \qquad T = 365.25636$

*GRAPH XXVII*

HALF-SUM (S+B)/2 OF THE AVERAGE DAILY VALUES
OF THE AZIMUTHS AT BOUGIVAL AND SAINT-GERMAIN
Fitting to the sidereal period of the Moon of 27.322 days
with linear trend accounted for
12h on 2 July - 12h on 31 July 1958

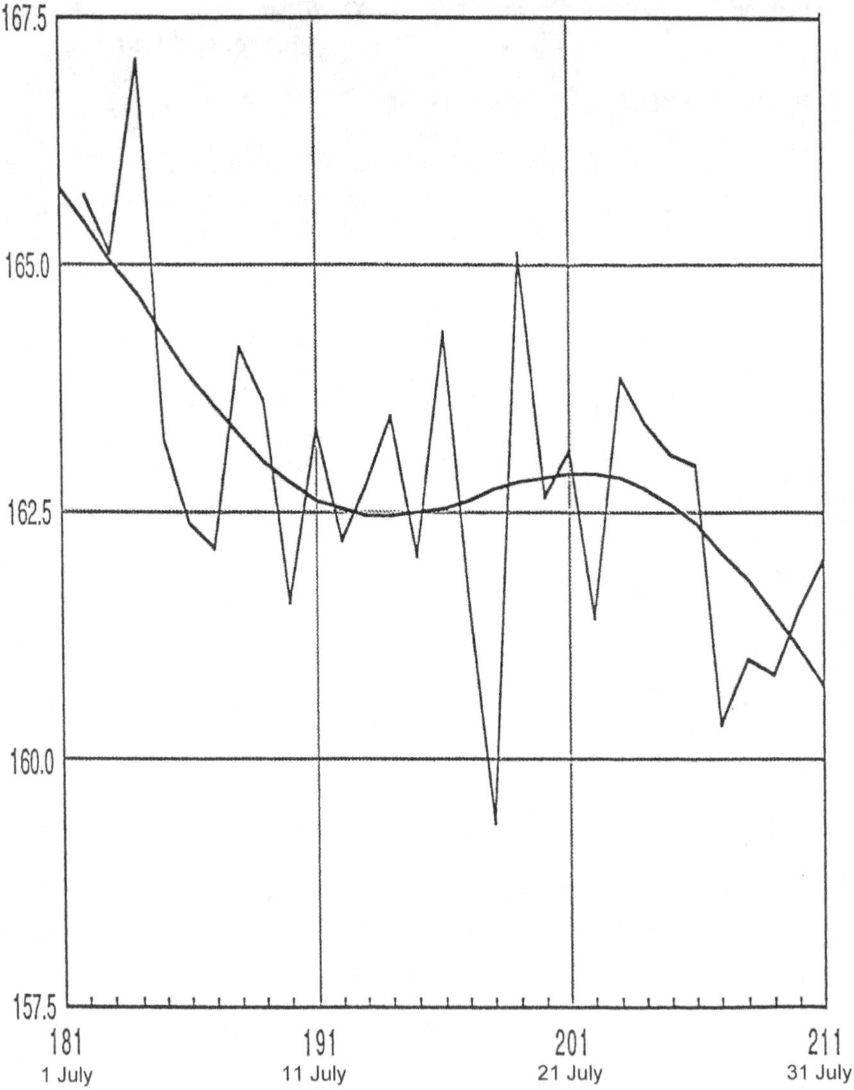


*Legend:* $\Sigma = 1.61$; $R = 0.629$, $1\text{-}R^2 = 0.604$; m = 162.9 grads, trend = -0.149 grads per day, r = 1.02 grad, $\sigma = 1.25$ grad; N = 30 days.

$\Sigma$ = standard deviation of the data; R = coefficient of multiple correlation; r = semi-amplitude of fitting sinusoid; $\sigma$ = standard deviation of residues.

Dates of maximum and minimum of the sinusoid: 20h on 24 July ($s_0 = 204.85$) and 4h on 11 July ($s'_0 = 191.19$). The dates s are reckoned in days from 0h on 1 January 1958.

***GRAPH XXVIII***

HALF-DIFFERENCE (S - B)/2 OF THE AVERAGE DAILY VALUES OF THE AZIMUTHS AT BOUGIVAL AND SAINT-GERMAIN
Fitting to the sidereal period of the Moon of 27.322 days
with linear trend accounted for
2 juillet 12 h - 31 juillet 12 h 1958

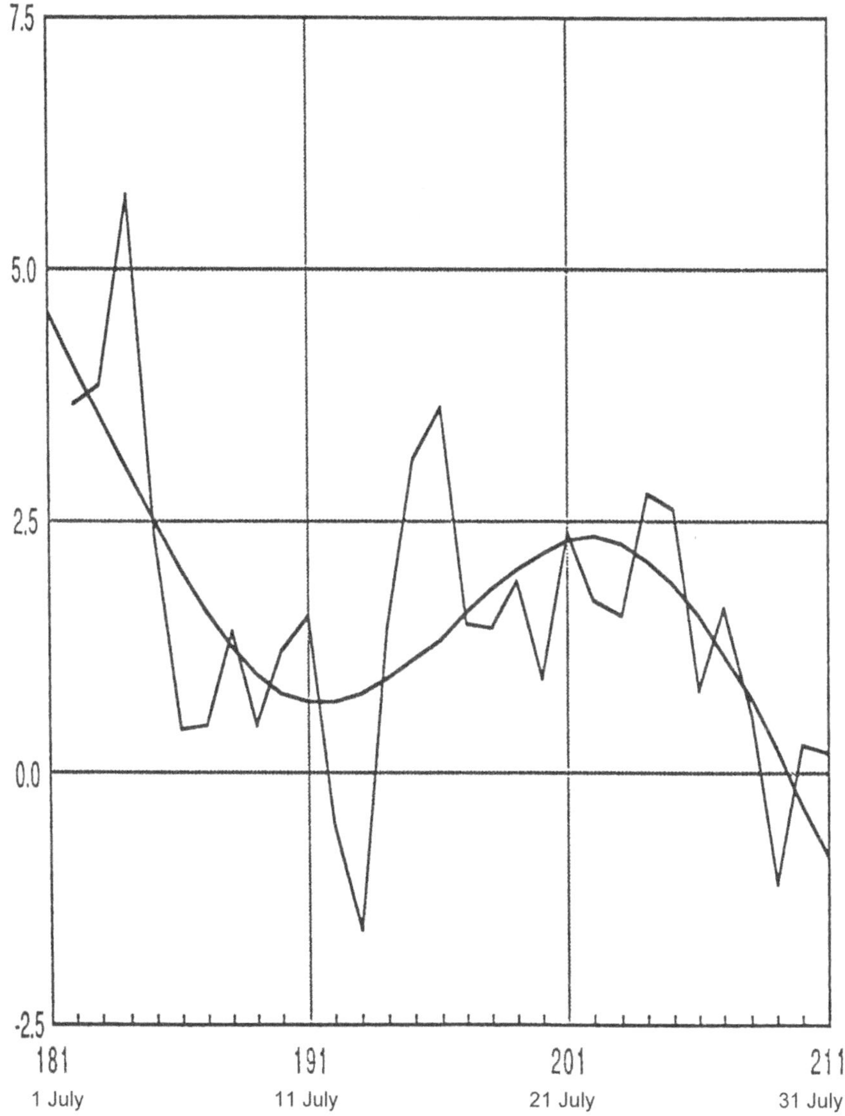


*Legend:* $\Sigma = 1.51$; $R = 0.680$, $1-R^2 = 0.537$; $m = 1.55$ grads, trend = -0.147 grads per day, $r = 1.70$ grad, $\sigma = 1.11$ grad; $N = 30$ days (See *Legend* of Graph XXVII).

Dates of maximum and minimum of the sinusoid: 0h on 24 July ($s_0 = 204.03$) and 8h on 10 July ($s'_0 = 190.37$). The dates s are reckoned in days from 0h on 1 January 1958.

*Source:* Calculation 948*, Graph 13665 (26 June 1996).

## *C.4 Overall view*

The *crucial* experiments made *at the same time and in the same conditions* at Saint-Germain and at Bougival made it possible to apply *decisive* factors to my previous analyses of 1954 and 1957, in order to eliminate with *complete certainty* the totality of the pseudo-explanations of the observed periodic effects put forward by those who contradicted me. Thus, for example:

- The *virtual identity* of the periodic components of 24h50m observed at Saint-Germain and at Bougival, *itself alone, makes it possible to eliminate any explanation by a fortuitous cause.*
- Similarly, the *virtually invariable* temperature conditions obtained in the Bougival laboratory *make it possible to eliminate any thermal effect.*
- The *parallelism* of the periodic effects detected at Bougival and at Saint-Germain *makes it possible to eliminate* any influence based upon the influence of the building at IRSID, or based upon any superficial cause.

In fact, these two crucial experiments *swept away all the previously presented objections* to my experiments with the paraconical pendulum, in particular as to the existence of a diurnal lunar periodicity of 24h50m[1] having an amplitude *totally inexplicable* in the framework of the theory of gravitation.

---

[1] In my Conference of 7 November 1959 under the aegis of the Alexandre Dufour Circle, I recalled what had been previously said to me:

> *"For months I have been told: "So, then, perform experiments at the same time in two different places, and if you obtain analogous results, this will be decisive. The question will be heard, it will be judged, and it will be proved that you are right." A very competent personality even said to me: "It is not even necessary that the results should be the same. If your pendulums exhibit analogous movements, that will be an important result."*

*In fact the identity of the periodic components of 24h50m was much higher than that which, before the experiments of July 1958 at Bougival and at Saint-Germain, was considered would be the decisive level.*

The fact that, after *such a decisive and shattering success*, the responsible scientific authorities did not furnish me with the financial means needed for pursuit of my experiments, remains for me today *totally incomprehensible, so scientifically aberrant was their attitude*.[2]

[2] See below, *Overall View* (*Section G* of this Chapter).

In fact opinions were very divided. *At least nine members of the Academy of Sciences thought that my experiments should be pursued, but apparently they were in the minority* (§G.6 below, note 3, p. 233).

# D DEVIATIONS OBSERVED DURING TWO TOTAL SOLAR ECLIPSES

During the two total solar eclipses of 30 June 1954 and 2 October 1959, an abnormal lunisolar influence manifested itself in the form of remarkable perturbations of the movement of the azimuth of the paraconical pendulum.

## *D.1 The total solar eclipse of 30 June 1954*

During the 32-day series of enchained experiments of June-July 1954,[1] and at the moment of the total solar eclipse of 30 June 1954, the plane of oscillation of the paraconical pendulum was abruptly displaced through about 15 grads.

The two *Graphs XXIX and XXX*[2] give the angular displacement of the plane of oscillation (shown along the ordinate) as a function of time (shown along the abscissa). Each point represents the starting azimuth corresponding to each 14-minute long series of observations, which was equal to the azimuth of the plane of oscillation at the end of the previous 14-minute long experiment.

[1] The pendulum consisted of a vertical bronze disk and two horizontal disks (see §A.3.1 above).

[2] See my Note of 18 November 1957: "*Mouvement du pendule paraconique et éclipse totale de Soleil du 30 juin 1954*" ("Movements of the paraconical pendulum and the total solar eclipse of 30 June 1954"), CRAS, Vol. 245, 4 December 1957, pp. 2001-2003 [11].

The displacement of the azimuth of the plane of oscillation of the pendulum, *which was located in the basement* (§A. 1.2 above), was so abrupt that it took the observer Jacques Bourgeot (my laboratory chief) totally by surprise, and he immediately called me on the telephone. He had never seen such a displacement before.

*Graph XXIX* represents the curve of the observed azimuths from 20h on 28 June 1954 to 4h on 1 July 1954, together with, on the right side of the vertical line corresponding to 0h on 30 June, the mirror image (dotted line) with respect to that vertical line of the left part of the curve representing the azimuths. *Graph XXX* shows the detailed curve of the azimuths observed from 9h on 30 June to 15h on 30 June (both UT).

The eclipse started at 11h21m and terminated at 13h55m. At the *exact* moment of the start of the eclipse, the azimuth of the plane of oscillation brutally increased by 5 grads away from the trend that had previously characterized its movement. This deviation reached a maximum of 15 grads twenty minutes before the maximum of the eclipse which took place at 12h40m, and then it diminished progressively but more abruptly than it had risen, so that the deviation was only 1 grad and 20 centesimal minutes before the end of the eclipse.

Remarkably, as far as one can judge, after the eclipse, the displacement of the plane of oscillation recommenced a movement CD that was analogous to its movement AB observed before the start of the eclipse (*Graph XXIX*).

*Graph XXIX* shows a rough symmetry of the azimuth curve with respect to the vertical line corresponding to 0h on 30 June. This symmetry, which can be attributed to the almost periodic structure of the azimuths,[3] appears during around 28 hours on both sides of this axis of symmetry. Supposing - which is plausible - that this symmetry corresponds to some physical reality which is independent of the perturbations caused by the contact between the ball and the bearing surface, it is conspicuous that nothing in the part of the azimuth curve before the time of the center of symmetry corresponds to the very pronounced deviation observed during the eclipse.

[3] On the almost periodic structure of a series, see §A.5.4 above, p. 101.

It should also be underlined that, during all the periods of continuous observation our team performed previously, I had never observed any variation of the azimuth curve analogous to this portion BC corresponding to the solar eclipse of 30 June 1954.[4]

Finally, one can remark that the rapid variations of azimuth observed from 11h20m to 12h and from 12h20m to 13h correspond to angular speeds of the order of $0.62 \times 10^{-4}$ and $0.79 \times 10^{-4}$ radians per second, *i.e.* respectively 1.13 and 1.43 times the Foucault effect (which is $0.55 \times 10^{-4}$ radians per second at the latitude of the Saint-Germain laboratory). The forces in play during the observed perturbations accordingly are of the same order of magnitude as those that act during the Foucault effect.

[4] It is remarkable that the maximum of the apparent deviation due to the eclipse took place twenty minutes before the maximum of the eclipse. Thus there is a certain dissymmetry in the observed effect. An analogous dissymmetry, but in the reverse sense with the maximum of the effect being later than the eclipse maximum, has been observed for terrestrial magnetism (Lion, *CRAS*, 33, 1851, p. 202 ; 34, 1852, p. 207 [179, 180]; Lion and Diamilla-Muller, *CRAS*, 74, 1872, p. 199 [181]) and for the terrestrial electrical field (Nordmann, *CRAS*, 142, 1906, p. 40 [205]; Chevrier, *CRAS*, 197, 1933, p. 1143 [93]; Rouch, *CRAS*, 239,1954, p. 465 [234]).

## GRAPH XXIX

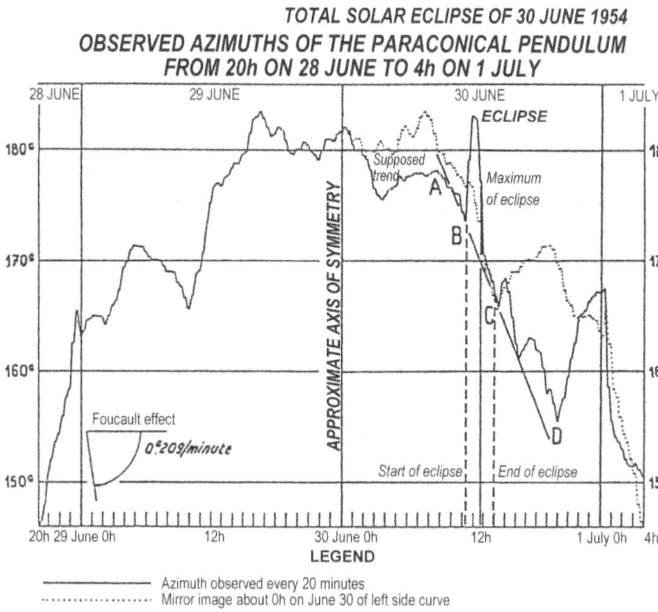


## GRAPH XXX

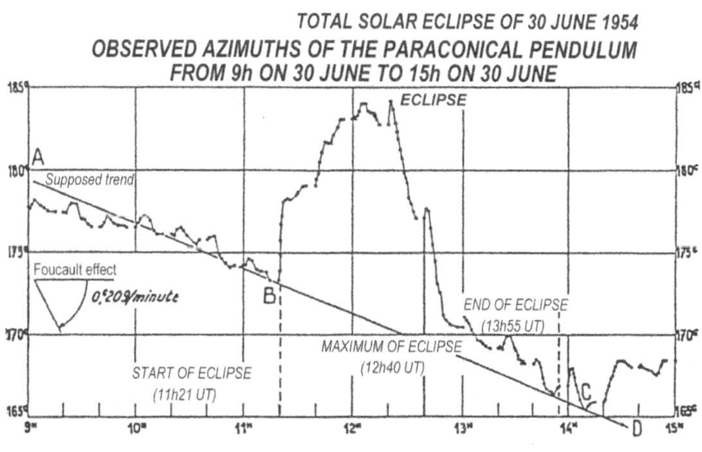


*Source: Note of 4 December 1957* to the Academy of Sciences, *Mouvements du pendule paraconique et éclipse totale de Soleil du 30 juin 1954* (Movements of the paraconical pendulum and the total eclipse of the Sun on 30 June 1954), CRAS, vol. 245, pp. 2001-2003 (photographic reproduction).

## *D.2 The total solar eclipse of 2 October 1959*

A similar perturbation of the order of 10 grads was observed during the total solar eclipse of 2 October 1959, *which was only partial at Paris.*[1]

The movement of the paraconical pendulum was observed during a period of three days, from 20h UT on 30 September to 4h UT on 4 October. The circumstances of the movement are represented in *Graphs XXXI and XXXII.*

The experimental conditions (laboratory, support, pendulum, use of the same ball during the series of enchained experiments) were *exactly the same* as during the eclipse of 30 June 1954.

While on 30 June 1954 the eclipse took place upon a regularly descending portion of the azimuth curve and the observed deviation exhibited a form never previously seen, by contrast the deviation observed on 2 October 1959 - if deviation there was - took place near a peak and its form had nothing exceptional, so that absolutely certain interpretation became rather difficult.

However if we admit that, in the absence of the eclipse, the general observed movement would have been represented by the line ABCD, then the deviation corresponding to the eclipse can be determined as indicated on *Graph XXXII.*[2]

---

[1] See my unpublished *Note* of 10 November 1959: "*Mouvements du pendule paraconique et Eclipse totale de soleil du 2 octobre 1959*" ("Movements of the paraconical pendulum and the total solar eclipse of 2 October 1959").

[2] For the influence of the total solar eclipse of 2 October 1959 on the paraconical pendulum with isotropic support, see *Chapter II* below, Section H.

## *GRAPH XXXI*

**TOTAL SOLAR ECLIPSE OF 2 OCTOBER 1959**
**OBSERVED AZIMUTHS OF THE PARACONICAL PENDULUM**
**FROM 15h ON 1 OCTOBER TO 23h ON 3 OCTOBER**

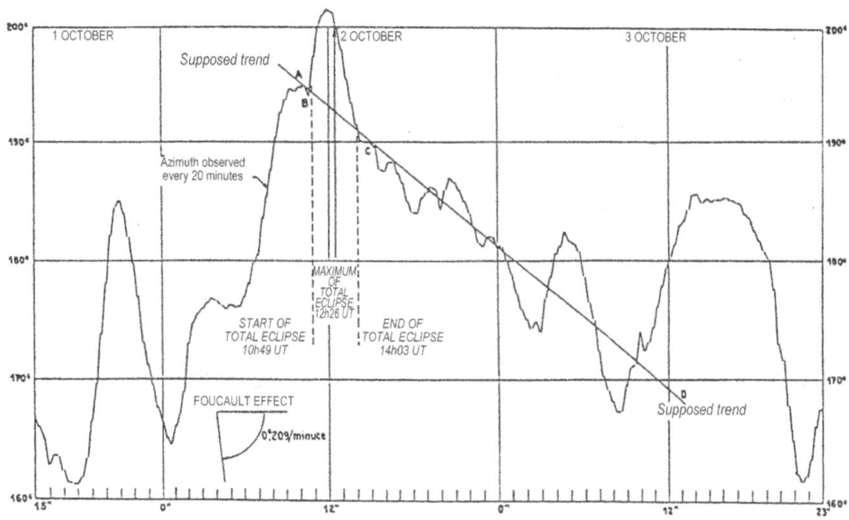


## *GRAPH XXXII*

**TOTAL SOLAR ECLIPSE OF 2 OCTOBER 1959**
**OBSERVED AZIMUTHS OF THE PARACONICAL PENDULUM**
**FROM 9h ON 2 OCTOBER TO 15h40m ON 2 OCTOBER**

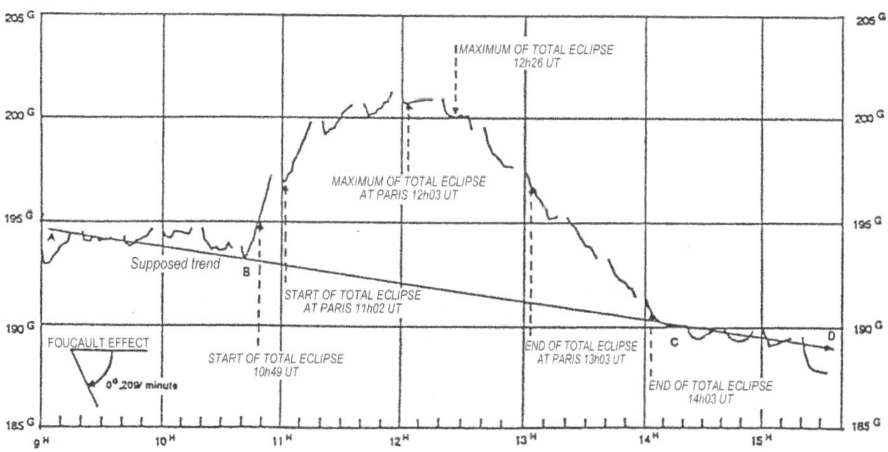


*Source:* Allais, unpublished *Note* of 10 November 1959, *Mouvement du Pendule paraconique et Eclipse totale de soleil du 2 octobre 1959* (Movement of the paraconical pendulum and the total solar eclipse of 2 October 1959) (photographic reproduction).

### *D.3 Comparison of the observed perturbations during the two solar eclipses of 30 June 1954 and 2 October 1959*

If the observed deviations from the supposed trends for the two eclipses during the period of time spanning the eclipses are displayed upon the same Graph, if the maxima of the total eclipses are taken as the common origin, and if the scales of the abscissas are taken so that the two durations of 3h14m for 1959 and 2h51m for 1954 are represented by the same length, then the supposed perturbations appear very similar in the two cases (*Graph XXXIII*).

Although only two experiments are involved, and although the supposed tendencies of the movement of the plane of oscillation of the pendulum before and after the eclipses are not absolutely certain, the perturbations considered in the two cases present *quite remarkable similarity.*[1]

In both cases, the observed angular deviations per unit time are of the same order of magnitude as those corresponding to the Foucault effect. We may deduce from this that *the forces in play during the observed perturbations are of the same order of magnitude as those that act in the Foucault effect.*

*In both cases, the deviations had the effect of bringing the plane of oscillation of the pendulum towards the meridian.* These deviations are totally inexplicable in the framework of the currently accepted theory of gravitation.

---

[1] It must be observed that this similarity only appears if we consider the starts and ends of the total eclipses *for the entire Earth* as given by *Connaissance des Temps* (French astronomical ephemerides, published annually since 1679), and not the starts and ends observed *locally* at Paris (on the Graphs of my *Note* to the Academy of Sciences of 1957 relating to the eclipse of 1954, the various times indicated are relative to the eclipse observed at Paris).

This interpretation also appears to be justified by the fact that the amplitudes of the supposed deviations are of the same order of magnitude in both cases, whereas the portion of the solar surface eclipsed *at Paris* in 1959 only represented 36.8% of the surface eclipsed in 1954. The somewhat greater deviation observed in 1954 can be explained by the fact that, at the start of the eclipse, the pendulum was about 27 grads away from the North-South azimuth (towards which it appears that the plane of oscillation is pulled at the moment of the eclipse), while it was only 7 grads away at the corresponding instant in 1959.

For lack of sufficient finance, similar observations could not be performed in my laboratory at Bougival, which I had to dismantle in 1960.

*Whatever may be the intrinsic scientific importance of the anomalies of the para-conical pendulum corresponding to eclipses - certainly major*[2] - their *relative* scientific significance is in fact much less than that of the observed periodic lunisolar anomalies, whose existence is *completely established* by the hundreds of thousands of observations that we performed from which they have been deduced.[3]

[2] Because these anomalies are *totally inexplicable in the framework of the currently accepted theory of gravitation.*

[3] In fact, the observed anomalies were confirmed by the experiments of Saxl and Allen during the solar eclipse of 7 March 1970 (*1970, Solar Eclipse as seen by a Torsion Pendulum*, Erwin J. Saxl and Mildred Allen, Physical Review, D, Vol. 3, Number 4, 15 February 1971 [242]). The article of Saxl and Allen explicitly refers to my own experiments.

## GRAPH XXXIII

### COMPARISON OF OBSERVED AZIMUTHS DURING THE TWO ECLIPSES OF 30 JUNE 1954 AND 2 OCTOBER 1959

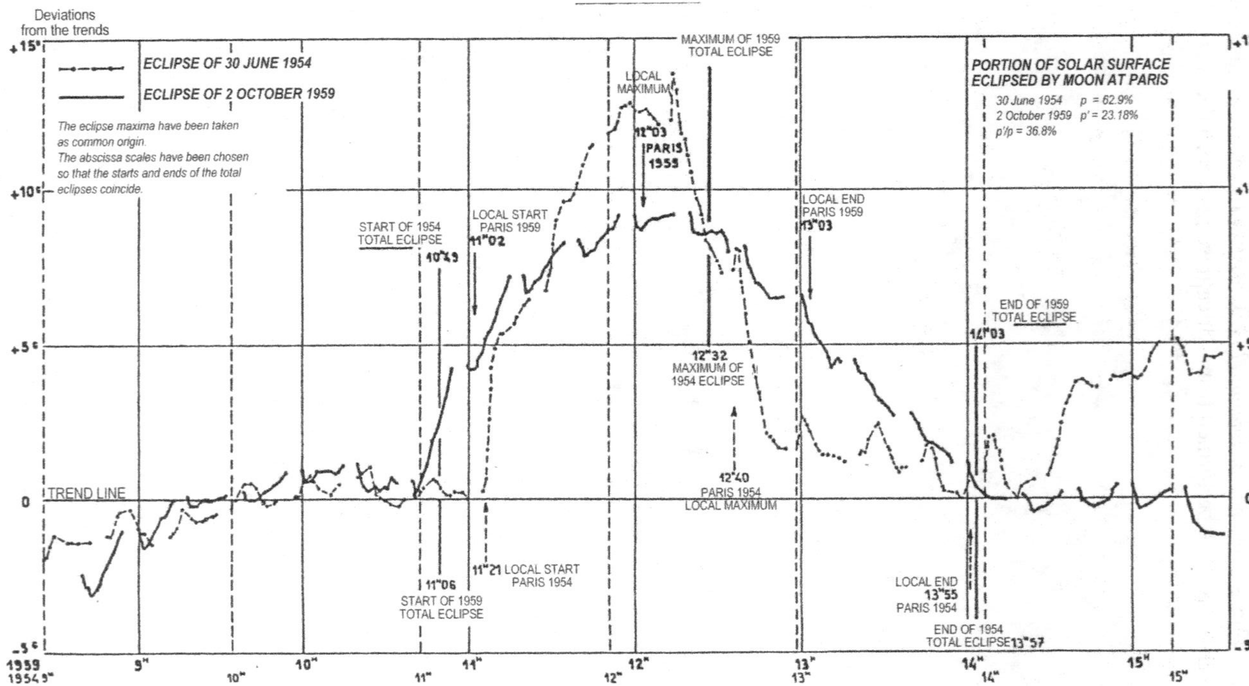


*Source:* Allais, unpublished *Note* of 10 November 1959, *Mouvement du Pendule paraconique et Eclipse totale de soleil du 2 octobre 1959* (Movement of the paraconical pendulum and the total solar eclipse of 2 October 1959) (photographic reproduction).

# E ESSENTIAL FACTORS IN THE MOVEMENT OF THE PARACONICAL PENDULUM WITH ANISOTROPIC SUPPORT

## *E.1 An extremely complex movement*

The totality of the experiments I conducted from 1954 to 1960 demonstrated that the movement of the paraconical pendulum with anisotropic support is an *extremely complex phenomenon that is very difficult to analyze*, and over many months this subject continued to pose many riddles.

*In fact it was only in September 1955, after the third month-long series of observations of June-July 1955, that I became certain of the real existence of a diurnal lunar component of 24h50m in the various month-long series of observations of the paraconical pendulum, the amplitude of this component being completely inexplicable in the framework of current theories.*[1]

It was from that date that I took the decision to submit my results to the examination of outside persons. In the three years following, 127 persons of importance - *among whom more than fifty were specialists in mechanics and geophysics* - came to visit my laboratory at IRSID, and subsequently that at Bougival.

[1] See my "*Note sommaire sur les mouvement du pendule paraconique*" ("Short note on the movements of the paraconical pendulum"), September 1955, 18 p.

• In December 1955 I had formulated a first overall version of the theory of the conical pendulum,[2] and in September 1956 I was able to evolve an overall version of the theory of the paraconical pendulum.[3, 4]

*Since there was no test in the literature for autocorrelated time series*, I had to create one. In February 1957,[5] after months of work, I arrived at my *Generalization of the Schuster Test*, which was the object of my *Note* of 13 May 1957 to the Academy of Sciences [16].[6] This test enabled me totally to confirm my previous conclusions as to the reality of the existence of the lunar periodicity of 24h50m *having an amplitude totally inexplicable in the framework of the current theory of gravitation.*

• It is evidently impossible to analyze in an exhaustive fashion all the factors that determine the movement of the paraconical pendulum with anisotropic support. Therefore I shall limit myself in this section to examining *the four factors that are really absolutely essential*: the Foucault effect, the anisotropy of the support, the random influence of the balls, and the periodic influences exerted upon the pendulum.

By way of illustration, I shall complete this analysis with examination of several claimed explanations of the observed phenomena.

Finally I shall briefly present a few observations in connection with the existence of a limit plane, variable with time, towards which the plane of oscillation of the pendulum tends to converge during each 14-minute experiment.

---

[2] *Théorie du pendule conique* (Theory of the conical pendulum) (provisional edition), December 1955, 50 p [5].

[3] *Théorie du pendule paraconique* (Theory of the paraconical pendulum), September 1956, 441 p [8].

[4] In particular, that theory enabled me to calculate exactly the lunisolar influence upon the movement of the paraconical pendulum (see §B.2 above).

[5] "*Note sur l'interprétation à donner aux expériences sur le pendule paraconique*" ("Note on the interpretation to be given to the experiments with the paraconical pendulum"), 25 April 1957, 18 p.

[6] See §B.1.3 above.

## *E.2 Foucault Effect*

### The paraconical pendulum and the Foucault effect

1 - It is very significant that *during all my experiments, as long as the oscillation of the pendulum remained planar, the displacement in azimuth of the plane of oscillation of the paraconical pendulum reduced to the Foucault effect.*[1]

From this it results that my experiments are in no way in contradiction with the general result of Foucault's experiment, as has only too often been suggested. *They absolutely do include the Foucault effect.*

### The Foucault effect and generation of ellipses

2 - Any cause acting upon the pendulum, other than the Foucault effect, can either act *directly* by modifying the speed of rotation $\dot{\phi}$ of the azimuth $\phi$ of the pendulum, or *indirectly* by creating an ellipse, and thus, due to the Airy effect,[2] causing a precession at the rate

$$\dot{\phi} = (3/8)\, p\alpha\beta \qquad p = 2\pi/T$$

of the plane of oscillation of the pendulum, where $\alpha$ and $\beta$ represent the major axis and the minor axis of the elliptical trajectory of the pendulum in radians, and $T$ is its period of oscillation.

Since the Foucault effect always appears when the oscillation is planar, we can conclude that, *at least in the first approximation, any cause other than the Foucault effect acting upon the pendulum works indirectly by the creation of ellipses.*

---

[1] §A. 4 above, pp. 93-95.

At the latitude of IRSID ($L = 48.90°$) the speed of rotation of the plane of oscillation corresponding to the Foucault effect is $-\omega \sin L = -0.550 \times 10^{-4}$ rad/sec, where $\omega$ represents the speed of rotation of the Earth ($\omega = 0.729 \times 10^{-4}$ rad/sec).

If no other effect were to intervene, the plane of oscillation of the pendulum would make a complete turn in $2\pi/\omega \sin L = 11.42 \times 10^4$ seconds, *i.e.* in 31.76 hours.

[2] §B.2.3 above, p. 120, and §B.2.5, note 11, p. 122.

## The classical experiments on the Foucault pendulum, and the paraconical pendulum

3 - It is a fact that the Foucault experiment gave rise to spectacular demonstrations, such as that performed in 1852 in the Pantheon at Paris with a pendulum 67 meters long and weighing 28 kgs. The oscillation remained almost rectilinear, with an amplitude of 0.06 radians.[3]

Here it is sufficient to underline *the essential differences* between the experimental conditions of the paraconical pendulum and those during the classical experiments of Foucault.

1 - The paraconical pendulum I utilized was *a short pendulum* whose length was of the order of a meter, as contrasted with several meters or more, and usually several tens of meters, in the experiments of Foucault and his successors.

It is a well-known fact that it is very difficult to obtain the Foucault effect with short pendulums. Anomalies almost always appear.[4]

---

[3] See in the *Second Part* of this work (*Chapter II*, Section C) an overall analysis of the *very numerous* experiments on the Foucault pendulum.

The bibliography of experimental researches on the Foucault pendulum is *very considerable*, but, upon investigation, one cannot fail to be struck by two facts: the rarity of serious work, and the paucity of numerical data upon the results obtained.

For the experiments on the Foucault pendulum, see in particular the *Bibliographie du Pendule* (Bibliography of the Pendulum), which is very extensive, published in 1889 by the *French Society of Physics* (*Collection de mémoires relatifs à la Physique* (Collection of Memoirs Relating to Physics), Vol. IV, Gauthier-Villars [1]), and my memoir of 1958, "*Doit-on reconsidérer les lois de la gravitation ?*" ("Should the Laws of Gravitation be Reconsidered?"), pp. 99-100 [23].

[4] In a very interesting article (*On the Irregularities of Motion of the Foucault Pendulum*, The Physical Review, April 1919, Vol. XIII, no. 4, pp. 241-258 [183]), A.C. Longden wrote:

> *"More than a score of well-known physicists and astronomers are on record as affirming that the Foucault Pendulum must be very long and very heavy in order to give satisfactory results."*

2 - The paraconical pendulum employed *can rotate around itself*, while the Foucault pendulum is attached to the wire that suspends it.[5, 6]

3 - The movement of the paraconical pendulum utilized was observed *without any discontinuity*, day and night, during periods of the order of a month. *This was never done before in any previous experiments upon the Foucault pendulum.*

From all these indications, it clearly follows that *nothing in the results of my experiments contradicts previously acquired results in any way.* On the contrary, anomalies have constantly appeared in all previous results, and it has certainly been a very great mistake to neglect the study of these anomalies.[7]

---

[5] As far as I know, the only experimenter who ever tried a suspension resting upon a ball was Longden (see note 4 above). He described his abandonment of the use of this suspension in the following forthright terms (p. 249):

> *"I decided not to use the ball and plane support on account of its tendency to rotate at the upper surface of the ball".*

The sphere was fixed, and it was the surface that rolled on the sphere.

[6] It is appropriate to underline here that no memoir on the theory of the pendulum has ever studied the movement of the central trihedral of inertia of the pendulum when the pendulum is suspended upon a point or upon a ball.

I have filled this gap in my memoir of 1956, *Théorie du pendule paraconique* (Theory of the Paraconical Pendulum, note 3 of §E.1 above) [8].

In the *Second Part* of this work (*Chapter I*, section B) I shall describe the essential elements of this movement.

[7] Among the anomalies detected, *undoubtedly the most curious* is that described by Abbot Panisetti: *a pendulum starts moving by itself in the East-West direction.* His experiments were performed with pendulums from 1 to 16 meters in length. (Revue Cosmos, 1856, Vol. VIII, pp. 503-504 [211], and 1857, Vol. IX, pp. 638-639).

These experiments may be compared with the similar experiments of Surgeon General Félix Pasteur (see my memoir of 1958, *Doit-on reconsidérer les lois de la gravitation ?* (Should the Laws of Gravitation be Reconsidered?), p. 101 [23]).

## *E.3 Anisotropy of the support*

### Experimental procedure

1 - In order to bring out the influence of the anisotropy of the support upon the movement of the paraconical pendulum during an experiment of duration $\theta = 14$ minutes, I performed successive releases at azimuths spaced apart by $q$ grads, as follows.[1]

$p$ releases were made in each azimuth, so that $N = (200/q) \times p$ experiments were performed. In order to eliminate any systematic influence of the time, the order of succession of starting azimuths was determined with the help of a table of random numbers.

During any given experiment of duration $\theta = 14$ minutes, the plane of oscillation of the paraconical pendulum shifted through an angle $\Delta\phi$ from the initial azimuth in question $\phi_0$. Graphing the $\phi_0$ along the abscissa and the average variations *in grads per minute* along the ordinate,

(1) $$\overline{\dot{\phi}} = \Delta\phi/\theta$$

we obtain a correlation graph between the average rate of angular shifting per minute and the azimuth of release.

### Empirical representation

2 - If we average the $\overline{\dot{\phi}}$ for each azimuth, we obtain a curve that looks like a sinusoid, whose period is equal to 200 grads. By the method of least squares, we can therefore calculate the fitting sinusoid

(2) $$\overline{\dot{\phi}} = a_0 + a_1 \sin 2(\phi - \Sigma_1)$$

that best represents the observations as a whole.

[1] See my *Note* of 9 February 1959 to the Academy of Sciences, *Détermination expérimentale de l'influence de l'anisotropie du support sur le mouvement du pendule paraconique* (Experimental determination of the influence of anisotropy of the support on the movement of the paraconical pendulum) [30].

We then observe that the residues of the correlation also have a sinusoidal appearance whose period is a half of the previous one, so that finally we are led to the *empirical representation*

(3) $$\overline{\phi} = a_0 + a_1 \sin 2(\phi - \Sigma_1) + a_2 \sin 4(\phi - \Sigma_2)$$

We can similarly observe the value in centimeters at the end of 14 minutes of the minor axis $2b$ of the ellipse described by the point of the needle that extends from the bottom of the pendulum, and the correlation obtained can likewise be represented by an expression of the type[2]

(4) $$2b = 2b_0 + 2b_1 \sin 2(\phi - \Sigma_1') + 2b_2 \sin 2(\phi - \Sigma_2')$$

*The expressions (3) and (4) represent the combined action of the support and the Foucault effect.*

## Estimation of the effects of the anisotropy of the support

3 - *Table X* shows the results obtained for different values of $p$ and $q$ with the two identical paraconical pendulums $P_1$ and $P_2$ and the two virtually identical suspensions $S_S$ and $S_B$ that I used in my two laboratories at Saint-Germain and at Bougival during the experiments of 1955, 1956, and 1958.[3]

The two Graphs *XXXIV* represent, for $q = 10$ grads and $p=5$, the results obtained for the angular variation $\overline{\phi}$ and for the minor axis $2b$ in the case of the pendulum and the suspension in the laboratory of Saint-Germain, from 4 to 10 March 1955.

The angles $\Sigma_1$ and $\Sigma_1'$ on the one hand, and $\Sigma_2$ and $\Sigma_2'$ on the other hand, correspond in a remarkable manner. The azimuths $\Sigma_1$ are very close to the azimuth orthogonal to the support, which was equal to 371.16 grads, so that this azimuth may be considered as representing the azimuth of anisotropy of the support at IRSID.

In *Graph XXXIV*, the theoretical displacement corresponding to the Foucault effect is represented relative to the displacement of the plane of oscillation.

[2] In order to obtain the minor axis in radians, its measure in centimeters must be divided by the distance $l' = 105$ cm from the point of the needle of the pendulum to the center of the ball (see note 6 of §A.1.2 above, p. 84).

[3] Some mistakes in the *Table* in my *Note* of 9 February 1959 have been corrected in *Table X* below.

The comparison of the results of the correlations of 13 August 1958 for Bougival and Saint-Germain (*Table X*) makes it possible to verify that the two supports in the two laboratories indeed did exert practically *the same influence* upon the movements of the two pendulums.

Considering only the principal effect on the azimuth as represented by equation (2), the average *in grads per minute* is

(5) $\dot{\phi}_1 = -0.127 - 0.677 \sin 2(\phi - 372.11)$

and *in radians per second*[4]

(6) $\dot{\phi}_2 = 0.262 \times 10^{-3} \dot{\phi}_1 = -0.333 \times 10^{-4} - 1.772 \times 10^{-4} \sin 2(\phi - 372.11)$

*The effect of the anisotropy of the support is thus of the same order of magnitude as the Foucault effect which is equal to* $-0.550 \times 10^{-4}$, *i.e.* to $-0.21$ grad/minute.[5, 6]

## Enchained experiments

4 - We thus see that when, during a series of *enchained* experiments, the azimuth of the pendulum establishes itself stably in an azimuth separated from the azimuth $\Sigma = 371$ grads, *this is because there is a cause C* acting that counterbalances the restoring torque effect of the suspension. As far as appears from the experiments analyzed above, this restoring torque effect would rapidly bring the plane of oscillation towards the direction of the plane $\Sigma$, which constitutes a direction of stable equilibrium, taking into account the combined influence of the support and the balls. For a deviation of 50 grads, this cause $C$ is equivalent to approximately three times the Foucault force.[7]

*The fluctuations of the position of equilibrium of the plane of oscillation of the pendulum due to cause C correspond to the anomalies of the paraconical pendulum with anisotropic support.*

---

[4] $(\pi/200)/60 = 2.618 \times 10^{-4}$. The constant term $-0.127$ is of the order of half the Foucault effect which is equal to $-0.21$ grad/minute.

[5] Note (1) of §2.1 above, p. 173.

[6] By putting, with the notation of *Table VI* (p. 128)

$$\dot{\beta} = \mu_S \sin 2(\Sigma - \phi)$$

we have (Table X, p. 180)

$$\mu_S = (0.174/2)/(103 \times 840) = 1.01 \times 10^{-6} \text{ radians/second.}$$

[7] $1.772 \times 10^{-4}/0{,}550 \times 10^{-4} = 3.22$.

## Other observations

5 - *Graphs XXXV* make it possible to compare the results obtained by the same analysis *at different epochs and for pendulums of different types*. The second *Graph XXXV* corresponds to the observations of *Table X* from 4 March to 10 March 1955.

The *algebraic* mean of the four series of observations gives, *in grads per minute*

(7) $\overline{\dot{\phi}} = -0.047 - 0.897 \sin 2(\phi - 374.95)$

and *in radians per second*

(8) $\overline{\dot{\phi}} = -0.123 \times 10^{-4} - 2.348 \times 10^{-4} \sin 2(\phi - 374.95)$

As for the amplitude and the phase of the fitting sinusoid, these results are completely analogous to the results above.[8, 9]

## Effect of the anisotropy of the support

6 - In total, the anisotropy of the support exerted a restoring torque effect towards the azimuth of 371 grads (counted from the South in the anticlockwise sense, *i.e.* 171 grads counting the azimuths in the anticlockwise sense from the North) corresponding to the perpendicular to the beam supporting the pendulum.

From this it results that, for *the seven month-long series* performed from 1954 to 1960, the azimuths of the plane of oscillation (*measured from the North in the anticlockwise sense*) were always in the interval between 93 and 268 grads. The average azimuths $\overline{\phi}$ of the seven month-long series were always in the interval between 150 and 174 grads. The average value of the average azimuths $\overline{\phi}$ was around 164 grads.[10]

---

[8] The quantity represented in *Graph XXXV* is $y = y_0 + a \sin 2(\phi_0 - \phi)$, while in *Graph XXXIV* the quantity represented is $\dot{\phi} = a_0 + a_1 \sin 2(\phi - \Sigma_1)$.

The correspondence of notation is thus $y_0 = a_0$, $a = -a_1$.

[9] For the experiments of August 1954 we had $y_0 = 0.18$ grads/minute, a positive value, while all the other values are negative. In this series of experiments the weight of the pendulum was much greater (see §A.3.1 above).

[10] *Table I* of §A.3.1 above.

In the *Théorie Générale du Pendule* (General Theory of the Pendulum) [8] which I elaborated in 1956, I reckoned the angles *positively in the anticlockwise sense from the South*, but for calculations to be applied elsewhere, it seems more simple to calculate them *in the anticlockwise sense from the North*.

*TABLE X*

## ANISOTROPY OF THE SUPPORT

### CHARACTERISTICS OF THE CORRELATION OF THE MOVEMENT WITH THE STARTING AZIMUTH

| SUSPENSION | PENDULUM USED | q IN GRADS | n | TOTAL NUMBER OF EXPERIMENTS | PERIODS OF OBSERVATION | DISPLACEMENT OF AZIMUTH IN GRADS/MINUTE $\Phi' = a_0 + a_1 \sin 2(\Phi - \Sigma_1) + a_2 \sin 4(\Phi - \Sigma_2)$ | | | | | MINOR AXIS OF ELLIPSE AT END OF EXPERIMENT IN CM $2b = 2b_0 + 2b_1 \sin 2(\Phi - \Sigma'_1) + 2b_2 \sin 4(\Phi - \Sigma'_2)$ | | | | |
|---|---|---|---|---|---|---|---|---|---|---|---|---|---|---|---|
| | | | | | | $a_0$ | $a_1$ | $\Sigma_1$ | $a_2$ | $\Sigma_2$ | $2b_0$ | $2b_1$ | $\Sigma'_1$ | $2b_2$ | $\Sigma'_2$ |
| Saint-Germain | $P_1$ | 10 | 5 | 100 | 4 to 10 March 1955 | -0.114 | -0.838 | 370.67 | +0.089 | 384.00 | +0.003 | -0.166 | 370.90 | +0.015 | 375.78 |
| " | $P_1$ | 20 | 2 | 20 | 4 January 1956 | -0.131 | -0.607 | 383.58 | -0.089 | 397.71 | +0.030 | -0.141 | 388.95 | -0.047 | 391.24 |
| " | $P_1$ | 20 | 2 | 20 | 21 May 1958 | -0.157 | -0.781 | 371.54 | +0.152 | 373.08 | +0.016 | -0.200 | 372.74 | +0.065 | 373.05 |
| " | $P_2$ | 20 | 2 | 20 | 13 August 1958 | -0.153 | -0.541 | 365.51 | -0.166 | 390.02 | -0.010 | -0.174 | 371.69 | -0.065 | 402.63 |
| Bougival | $P_1$ | 20 | 2 | 20 | 13 August 1958 | -0.078 | -0.616 | 369.24 | +0.087 | 382.09 | +0.044 | -0.188 | 369.19 | +0.047 | 380.61 |
| | | | | | AVERAGES | -0,127 | -0,677 | 372,11 | +0,015 | 385.38 | +0,017 | -0.174 | 374.70 | +0.003 | 384.66 |

PENDULUMS

$P_1$ : Pendulum used in the continuous month-long experiments in Nov-Dec 1954 and June-July 1955 at St.-Germain and June-July 1958 at Bougival

$P_2$ : Pendulum used in June-July 1958 at St.-Germain

FOUCAULT EFFECT

$\Phi' = -0.21$ grad /minute

CHARACTERISTICS OF SUSPENSIONS USED

Azimuth of perpendicular to support: $\Sigma_0 = 371.16$ grads

CHARACTERISTICS OF PENDULUMS $P_1$ AND $P_2$ USED (vertical disks)

Mass $M = 12$ kg

Principal moments of inertia around center of suspension ball: $A = 82.89\ 10^6$, $B = 83.11\ 10^6$, $C = 0.325\ 10^6$

Period $T = 1.85$ second

Distance from center of gravity to center of ball $OG = \ell = 83$ cm

Radius of ball $\rho = 0.325$ cm

CORRESPONDENCE BETWEEN AZIMUTHS $\Sigma$ AND COEFFICIENTS a AND b

| PERIODS OF OBSERVATION | $\Sigma_1 - \Sigma'_1$ | $\Sigma_2 - \Sigma'_2$ | $\frac{2b_1}{a_1}$ | $\frac{2b_2}{a_2}$ |
|---|---|---|---|---|
| 1 | -0.23 | +8.22 | +0.198 | +0.169 |
| 2 | -5.37 | +6.47 | +0.232 | +0.528 |
| 3 | -1.20 | +0.03 | +0.256 | +0.428 |
| 4 | -6.18 | -12.61 | +0.322 | +0.392 |
| 5 | +0.05 | +1.48 | +0.305 | +0.540 |
| AVERAGES | -2.59 | +0.72 | +0.263 | +0.411 |

ARRANGEMENT OF SUPPORT

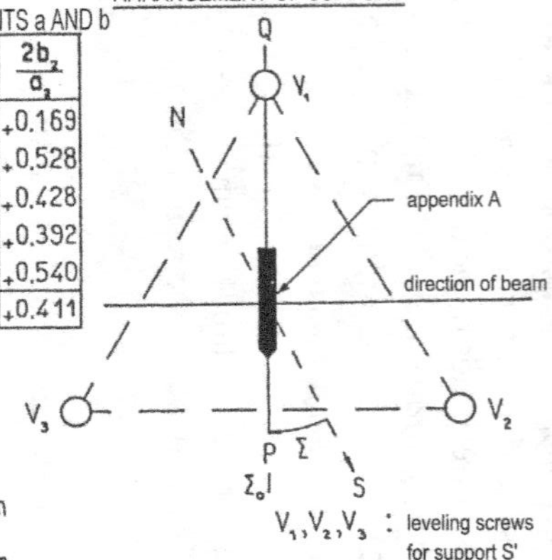


*Source: Graph IV.B.2* of my Conference of 7 November 1959, and my *Note* of 9 February 1957 to the Academy of Sciences, *Détermination expérimentale de l'influence de l'anisotropie du support sur le mouvement du pendule paraconique* (Experimental Determination of the Influence of Anisotropy of the Support on the Movement of the Paraconical Pendulum).

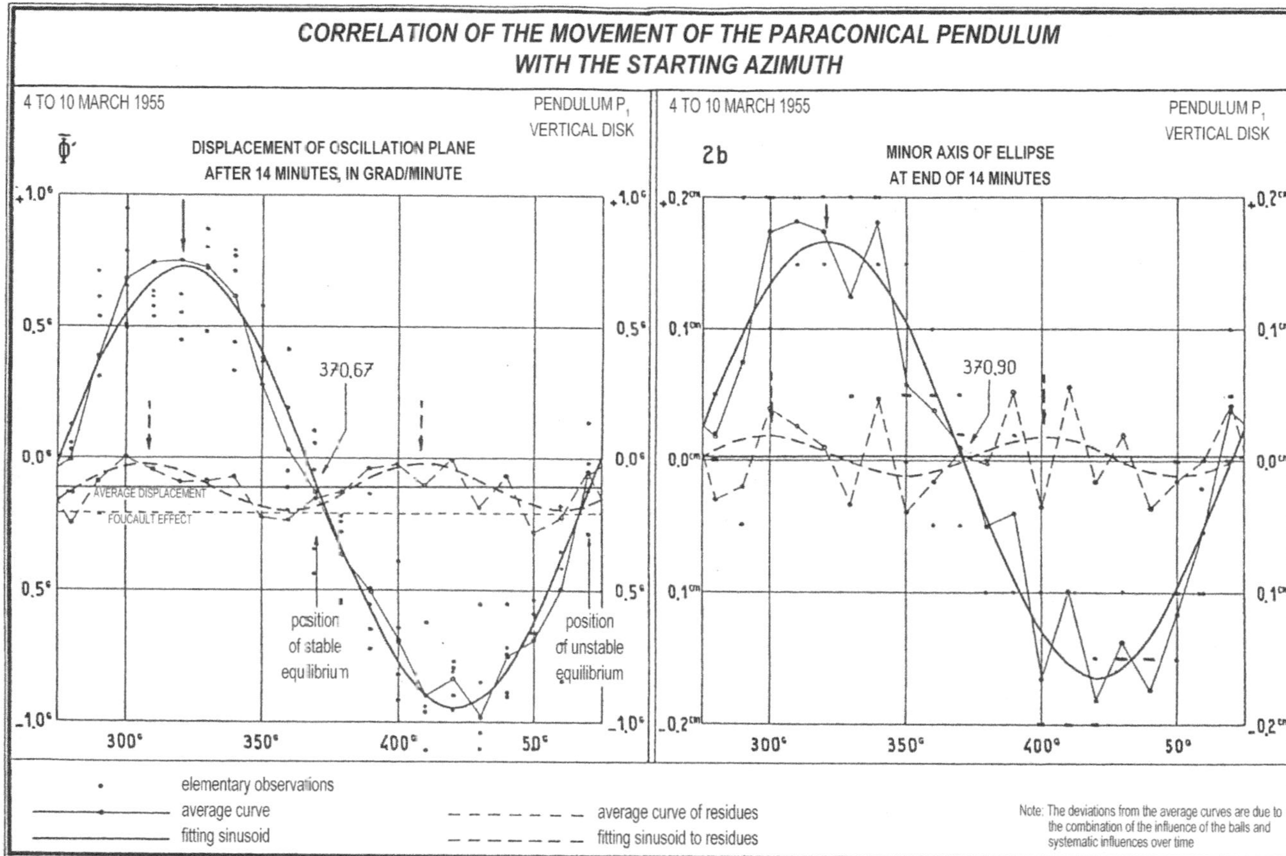


*Source:* *Graph IV.B.1* of my Conference of 7 November 1959, and my *Note* of 9 February 1957 to the Academy of Sciences, *Détermination expérimentale de l'influence de l'anisotropie du support sur le mouvement du pendule paraconique* (Experimental Determination of the Influence of Anisotropy of the Support on the Movement of the Paraconical Pendulum).

*GRAPH XXXV*

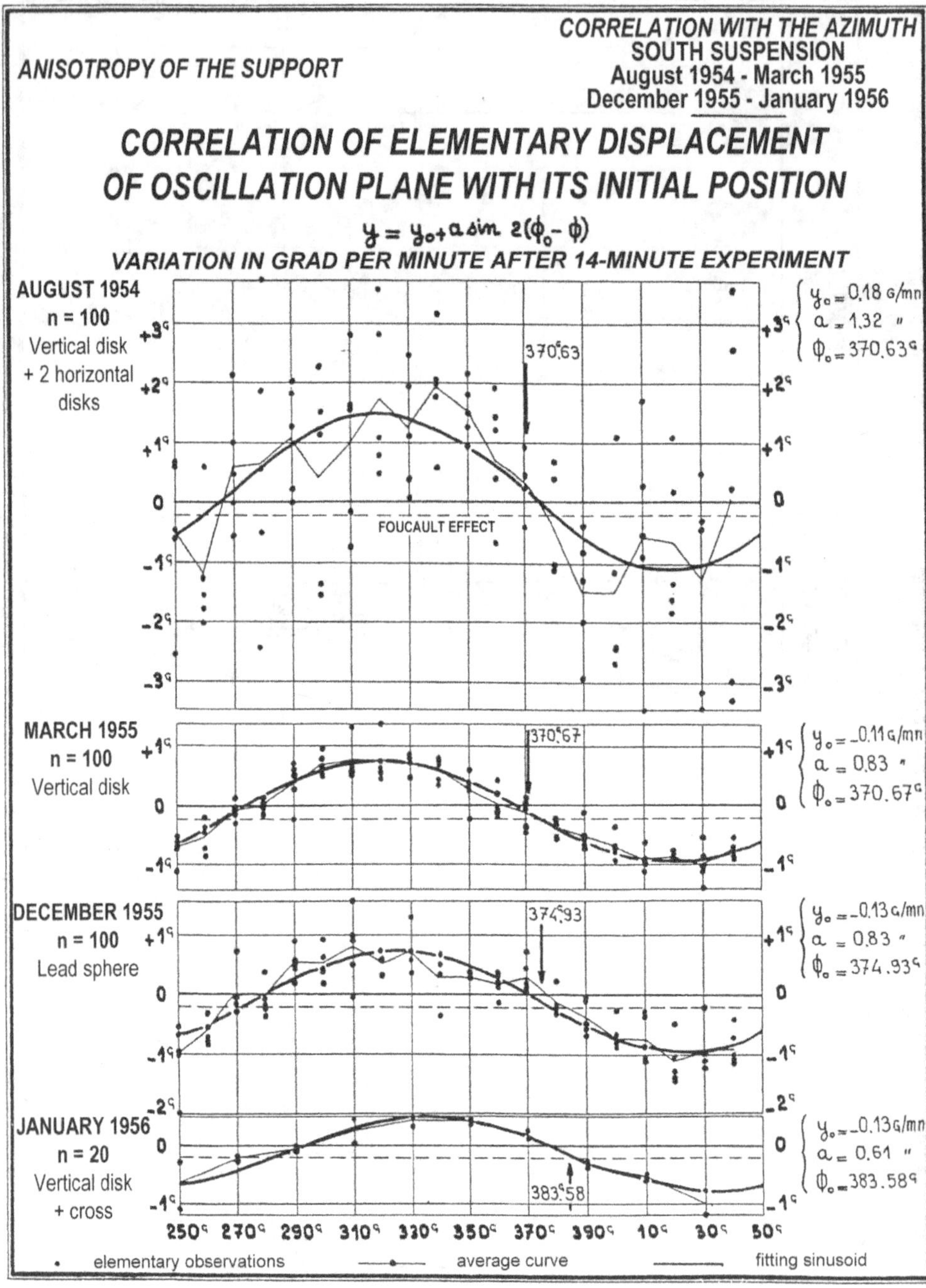


*Legend:* The Foucault effect corresponds to -0.21 grad/minute.
*Source: Graph IV.B.2* of my Conference of 22 February1958

## *E.4 Influence of the balls*

*The third important influence* acting upon the movement of the paraconical pendulum is that of the balls.

Neither the balls nor the bearing surfaces could be considered as being perfect.[1] In fact, the experiments performed showed that, for each elementary experiment of 14 minutes, the balls had an average random influence of 2.5 grads, of the same order of magnitude as the Foucault effect,[2] but that this could not explain the considerable variations of the observed azimuth, as shown by the triply enchained experiments performed in May 1957.[3]

In any case, *the random effect of the balls could not explain the very significant periodicities that were observed, in particular the diurnal lunar periodicity of 24h50m.*

[1] There was thus a merit in changing the balls as often as possible in order to eliminate their effects by the action of averages, a merit that did not appear to me during my first experiments in June-July 1954.

In fact, from November 1954, the ball was changed at the start of each experiment, and on average the bearing surface was changed each week.

[2] See §B.1.1 above.

An influence of 2.5 grads over 14 minutes = 840 seconds is equivalent to an average influence of $2.5\pi/(200 \times 840) \sim 0.47 \times 10^{-4}$ rad/sec, while the Foucault effect is $0.55 \times 10^{-4}$ rad/sec. The two effects are accordingly of the same order of magnitude.

[3] §B.1.1 above.

## *E.5 Periodic astronomical influences*

In *all* the time series formed by the observed azimuths during the month-long periods of observation, *periodic diurnal and semi-diurnal lunisolar influences that represent one of the most remarkable aspects of the movement of the paraconical pendulum with regard to their amplitudes* have been brought to light by different techniques of harmonic analysis, *whose results have been in remarkable mutual agreement*: the Buys-Ballot filter, fitting to a given group of waves by Darwin's method or by the method of least squares, periodogram, and correlogram.

*The existence of these periodic influences, and in particular of the periodic diurnal lunar component of 24h50m, is a certainty.*[1]

### The structure of almost periodic functions

1 - In fact, all the Graphs that can be deduced from the various observational series present *very numerous* morphological similarities, symmetries and double symmetries, and local periodicities, and these series present *all the characteristics of almost periodic functions.*[2]

With regard to their periodic components which, as far as they can be identified, are relatively numerous and of incommensurable periods, it is certain that the series of azimuths of the paraconical pendulum are *effectively* almost periodic functions to which have been added random components, due essentially to the influence of the balls.

---

[1] Naturally here we are not concerned with metaphysical certainty which would insist upon the availability of an infinite set of observations, but with a practical and human certainty *in the sense of the calculation of probabilities.*

[2] See §A.5.4 above, p. 101.

## Orders of magnitude

2 - *The angular speeds of variation of azimuth* corresponding to the amplitudes of the two most important periodic components detected in the seven month-long series of observations of the paraconical pendulum with anisotropic support, whose periods are respectively 24h and 24h50m, are both of an order of magnitude equal to a *thirtieth* of the Foucault effect.[3] The sum of the amplitudes of all the periodic components seems to be of the order of the Foucault effect.

*In fact, no experimenter had ever detected lunisolar effects* in the movement of the Foucault pendulum,[4] no doubt because of the small scale of the angular amplitudes of the oscillations and because of the great length of the pendulums generally employed, but also, and above all, because of the timespans covered by their experiments with the pendulum, which were always very limited.

*On the theoretical plane*, no author ever has really studied the lunisolar action upon the movement of the pendulum. All previous authors had in fact considered such an action to be so small as to be experimentally indiscernible, and thus useless to calculate.

## Lunisolar effects and coefficients of the general theory of Newtonian potentials

3 - With regard to the general theory of Newtonian potential in the literature,[5] it is essential to compare the *relative* orders of magnitude of the different periodic lunisolar components corresponding to the paraconical pendulum with the gravitational forces deduced from the lunisolar potentials.

---

[3] In fact, a variation of the order of $0.2 \times 10^{-5}$ radians per second corresponds to the diurnal lunar period of 24h50m, while the Foucault effect is $0.550 \times 10^{-4}$ radians per second. We therefore have $2 \times 10^{-6}/0.550 \times 10^{-4} \sim 0.36 \times 10^{-1} = 1/27.8$ (§A5.3. above).

[4] As far as I know, Dejean de Fonroque was the only experimenter to consider lunisolar effects. Even in 1879, long before Miller (*Chapter IV* below, Sections C and F), he underlined the existence of an effect corresponding to the movement of the Earth along its orbit, and the existence of an effect corresponding to the movement of the solar system towards the constellation of Hercules. However *his analyses are only qualitative, and really lead to nothing*. They are virtually unusable (see note 8 of §B.2.6, p. 50 of the *Introduction* above).

A detailed analysis of the memoirs of Dejean de Fonroque is presented in the *Second Volume* of this work (*Chapter II*, Section C).

[5] See particularly: Paul Schureman, *Manual of Harmonic Analysis and Prediction of Tides*, U.S. Department of Commerce, Washington, 1941 [244].

For simplification, I shall limit myself here to the diurnal lunisolar effects. *Table XI* presents an analysis of these effects for the four most important diurnal waves $K_1$ (23.93h), $M_1$ (24.84h), $O_1$ (25.82h), and $Q_1$ (26.87h).[6]

We see that the *relative* structure of the amplitudes is *totally different* for the paraconical pendulum and for the gravitational theory of the lunisolar forces. Thus, for example, the lunar component $M_1$ of 24h50m (24.84h) is *relatively four times greater* for the paraconical pendulum than for the theory of tides, while the solar component of 24 hours is *relatively half as strong as* for the gravitational force, *so that there is a relative difference of eight to one.*

It is this *absolutely fundamental* structural difference that particularly allows us to assert that the anomalies of the paraconical pendulum constitute *an entirely new phenomenon*, completely distinct from the phenomena deduced from the Newtonian theory of gravitation. *In fact, the observed periodic structure cannot be considered as being derived from any of the phenomena resulting from the gravitational potential of the Moon and the Sun.*

---

[6] A detailed analysis is presented in the *Second Volume* of this work (*Chapter VI*, Section A, see above, p. 30).

The coefficients deduced from the theory of Newtonian potentials are valid both for deviations of the vertical and for the theory of tides (see in particular note 5 above).

It is at least curious to realize the extent to which the orders of magnitude of the power series coefficients of the lunisolar potentials (*Table XI*) are *very widely* ignored in scientific circles, even among geophysical specialists.

*TABLE XI*

DIURNAL LUNISOLAR EFFECTS
PARACONICAL PENDULUM
AND COEFFICIENTS OF LUNISOLAR FORCES

| Periodicities | $K_1$ et $P_1$ 24 h. | $M_1$ 24.84 h. | $O_1$ 25.82 h. | $Q_1$ 26.87 h. | Total of amplitudes |
|---|---|---|---|---|---|
| Paraconical Pendulum Amplitude of wave 2R in grads | | | | | |
| June-July 1954 [1] | 2.34 | 4.12 | 2.56 | 6.76 | 15.98 |
| Nov-Dec 1954 [2] | 11.26 | 11.54 | 4.64 | 6.18 | 33.62 |
| June-July 1955 [3] | 13.00 | 10.46 | 4.78 | 7.78 | 36.02 |
| Averages | 8.87 | 8.71 | 3.99 | 6.91 | 28.54 |
| Relative values: a | 31.08 | 30.52 | 13.98 | 24.21 | 100 |
| Theory of lunisolar forces | | | | | |
| Coefficients | 0.7060 | 0.0977 | 0.3771 | 0.0730 | 1.2538 |
| Relative values: b | 56.31 | 7.79 | 30.08 | 5.82 | 100 |
| Ratios of relative values | | | | | |
| Ratios a/b | 0.552 | 3.92 | 0.465 | 4.16 | 1 |

*Legend:* 1) Series of 721 hourly values centered upon 12h20m on 23 June 1954.
2) Series of 721 hourly values centered upon 12h20m on 3 December 1954.
3) Series of 721 hourly values centered upon 12h20m on 22 June 1955.

*Source:* 1) *Paraconical Pendulum*: Calculations of the Hydrographic Institute of Hamburg; my *Note* of 4 December 1956, *Analyse harmonique des mouvements du pendule paraconique. Compléments* (Harmonic analysis of the movements of the paraconical pendulum. Additions) (3 p.).
2) *Coefficients of the theory of lunisolar forces*: Schureman, *Manual of Harmonic Analysis and Prediction of Tides*, 1941, pp. 164-165.
The amplitudes of closely neighboring periods are grouped together.

## *E.6 Other factors*

In view of the detected periodicities, and particularly in view of their amplitudes, supposed explanations and objections were repeatedly advanced. I think it is necessary to examine some of them, *by way of illustration.*[1]

**The apparatus was imperfect, and therefore no valid conclusion could be drawn from my experiments**

1 - It is certain that, whatever was the precision with which the various apparatuses were constructed,[2] they still suffered from constructional defects.

For example, the center of gravity of the pendulum could have been slightly off-center with respect to the principal axis of inertia passing through the center of the ball,[3] or the pieces of metal used might not have been perfectly homogenous. And, as far as the balls and the bearing surfaces were concerned, they were not perfect and they were deformable to a greater or lesser extent.

However, *whatever the defects of construction might have been, they could never have caused the appearance of real periodic effects in any way.*

**The observed effects were due to magnetism of the small conical steel turret (about two hundred grams) connected to the pendulum, in which the ball was seated**

2 - This magnetic action may probably have existed, but it was *very weak*, and it is possible to assert without any calculation that such magnetism would have had the effect of orienting the pendulum in an azimuth that only varied by a few centesimal minutes per day, *which was not the case.*

---

[1] See my *Note* of 25 April 1957, *Interprétation à donner aux expériences sur le pendule paraconique* (Interpretation to be given to the experiments with the paraconical pendulum), 18 p.

[2] All the pendulums employed were constructed with a precision of one hundredth of a millimeter (see my *Note* of 15 January 1957, *Note sommaire sur les recherches sur le pendule paraconique* (Short note on researches with the paraconical pendulum), 13 p., p. 5).

[3] It is in fact extremely difficult to produce a perfectly straight rod made from a material such as brass, which however was chosen for its non-magnetic qualities.

**The observed effects were due to Foucault currents induced in the pendulum by its motion in the Earth's magnetic field**

3 - If such an effect existed, its effect would have been to orient the pendulum in a privileged direction which would not vary more than a few centesimal minutes a day, *which was not the case.*

**The displacements of the plane of oscillation were due to air movements in the laboratory**

4 - If such an action was the explanation looked for, the movement of the pendulum would be a result of a set of random actions. In that case, *statistically significant periodicities* of the azimuths would not appear in the harmonic analysis.

**The detected periodicities were real, but they were due to elastic reactions of the building**

5 - However, the building by itself could not exert any periodic action. If it exerted an action, that would actually be a transmitted action, and, in that case, the observed phenomena would be due to the amplification of an already known geophysical phenomenon. But in fact all known geophysical phenomena have periodic structures *very different* from those of the movement of the paraconical pendulum.[4]

**The equations of movement of the paraconical pendulum included periodic solutions whose periods were very close to 24h and 24h50m**

6 - First of all, it should be underlined that such an argument completely neglects the other periodicities observed.[5]

In any case, the argument would only be valid if the paraconical pendulum had oscillated in a continuous manner.

[4] See §B.3.1 and §B.3.3 above.

[5] See in particular *Table II* of §A.5.2 above and *Table XI* of §5 above.

In fact, the pendulum was re-launched every 20 minutes from a stationary starting position, so that its movement could only depend upon its starting azimuth, and ought always to have been the same. However, for all the azimuths in the interval ($370 \pm 50$ grads), it was observed that the variation of azimuth during any given experiment of 14 minutes was sometimes positive and sometimes negative, according to the epoch.

In any case, such periodicities of 24h and 24h50m could not appear during the course of experiments of 14 minutes duration.

### The distribution of the observed anomalies according to the normal law shows that they were of a random nature

7 - It is indeed certain that chance can imitate periodicity, but periodicity can equally imitate chance.

In fact, starting from 1954, I was able to verify that the sum of the ordinates of 13 sinusoids obtained in the analysis of my numerical series by the method of least squares *was distributed according to the normal law*. However this value was really the sum of 13 sinusoids, and thus its magnitude was completely determinate and *non-random*. This implies that the fact that a numerical series has a *random* appearance does not in any way exclude the possibility that it may represent an almost periodic phenomenon that is in no way random.

This understanding led me in 1981 to prove a theorem that I called *Theorem T*.[6]

### No analogous phenomenon has ever been observed in experiments upon gravitation

8 - In fact, such an argument has no value by itself. *The very meaning of the discovery of any new phenomenon is that it has not been demonstrated before.*[7]

---

[6] See my memoir of 1982, *Fréquence, Probabilité et Hasard* (Frequency, Probability, and Chance), with two Appendices : 1.-*Fréquences empiriques et fréquences mathématiques - Illustration* (Empirical frequencies and mathematical frequencies - Illustration); 2. *Le Théorème T - La simulation du hasard par des fonctions presque périodiques* (Theorem T - The simulation of chance by almost periodic functions) [43, 44]. This memoir is attached to the *Second Volume* of this work as *Appendix E*.

In fact, if a phenomenon results from sufficiently numerous incommensurable periodic influences, its values over time are distributed according to the normal law.

[7] In reality the crowd that contradicted me were animated by *the same single conviction*. Accepted theories, perfectly verified, could not be brought into question. It was totally impossible that a non-professional could have done that, and accordingly his experiments must be worthless.

### *A priori* assertions

9 - Starting in December 1956, Henri Villat, President of the Mechanical Section of the Academy of Sciences, repeatedly asserted over and over that I had not formulated a theory of the apparatus, and that if the theory were correctly elaborated everything would be explained.

In his letter of 26 March 1956 he wrote to me:

> *"But the anomalies, or rather what you consider to be anomalies in your experiments, will be explained in the simplest way in the world when you have made the necessary calculations."*

And in his letter of 4 June 1958 to our common friend G. Varlan he also wrote:

> *"Once the equations (of the apparatus) have been integrated or at least interpreted properly, there will be no further mystery."*

In fact, on 6 December 1956, I had sent to him my general memoir of September 1956, "*Theory of the Paraconical Pendulum*" [8].[8]

But this is not the real question. The problem is in the orders of magnitude. As I indicated to Henri Villat in my letter of 24 July 1958:

> *"I think I should attract your attention to the fact that the periodic lunisolar effects upon the movement of the paraconical pendulum that can be calculated from the classical theory of gravitation are of the order of* $10^{-13}$*, and thus are not appreciable."*

[8] This memoir of 441 pages explained the general theory of the paraconical pendulum. In order to determine the elliptical trajectory of the paraconical pendulum, this memoir used Lagrange's method of variation of constants, and in its synthesizing Seventh Part (pp. VII.1 to VII.29) it specified all the formulas that, to the first approximation, correspond to the movement of the paraconical pendulum, and to all the factors affecting this movement.

On the influence of the experimental arrangements, see §6.1 to §6.6 above.

*"After that realization, the rigorous theoretical calculation of the movement seemed to me to have only a purely academic interest, because the order of magnitude of the effects thus calculated is about a hundred million times smaller than the order of magnitude of the observed effects.*

*"That, and only that, seems to me to be the important question. I have observed effects of rotation of the plane of oscillation that have a periodic character and whose order of magnitude is around* $10^{-5}$ *radians per second, while, under the hypothesis that the observed movements are due to lunisolar action, theory can only explain effects of the order of* $10^{-13}$ *radians per second. It is only when viewed in this way that the theory of the apparatus seems to me to be interesting and indispensable...*

*"You say, at the end of your letter of 4 June 1958, that after the calculations will have been made and correctly interpreted, there will be no more mystery.*

*"I admit that I do not understand your point of view very clearly. It is easy to see that the effects on the rotation due to the lunisolar influence are of the order of* $10^{-13}$ *radians per second,* while the observed effects are of the order of* $10^{-5}$ *radians per second. The difference of orders of magnitude is such that an explanation of the observed effects by the classical lunisolar influence appears to me to be impossible."*

* As far as I know, this order of magnitude has never been contested by anyone who takes the currently accepted theory of relative movement as his starting point.

In fact, Henri Villat never made any answer to my letter of 24 July 1958.[9, 10]

[9] In view of the impact of the *a priori* assertions of Henri Villat who was at the very heart of the Academy of Sciences, I reproduce the whole of the correspondence relative to Henri Villat's assertions, which he incessantly repeated, in *Annex I.A* in the *Second Volume* of this work

In fact, Henry Villat always refused to come to visit my laboratory at IRSID, and I was only able to meet him once, on 3 December 1956 (see above §C.2.3, note 7, p. 148).

[10] In his letter of 1 June 1960, Jean Leray, member of the Mechanics Section of the Academy of Sciences, wrote to me:

*"Your theoretical considerations... do not lead to numerical conclusions; you do not compare the magnitudes that you have measured with their theoretically calculated values. Therefore you have not formulated the theory of your pendulum."*

Thus, Jean Leray reproduced *almost word for word* the arguments presented two years earlier by Henri Villat, and, like him, took no account of my memoir of September 1956, *Théorie du pendule paraconique* (Theory of the Paraconical Pendulum) [8] which I had transmitted to him, nor of the orders of magnitude *which are completely essential in this matter.*

In fact Jean Leray took no account of the crucial experiments of July 1958 (Section C above).

## *E.7 Existence of a limit plane, variable with time, in the enchained series of the paraconical pendulum with anisotropic support*

In fact what observation shows is that *everything happens as though*, during each 14-minute experiment, there exists a limit plane, resulting in particular from the combination of the action of the support and astronomical influences such as the lunisolar action, towards which the plane of oscillation of the pendulum tends to converge. *This limit plane varies continually with the passage of time.*

### Empirical representation

1 - With regard to the results obtained during analysis of the influence of the anisotropy of the support,[1] it is helpful to suppose that *during each 14-minute experiment*, for the average value $\overline{\dot{\phi}}$ of the variation of the azimuth per unit time, we have

(1) $$\overline{\dot{\phi}} = -\omega \sin L + k \sin 2\left(\overline{X} - \overline{\phi}\right) + K \sin 2\left(\Sigma - \overline{\phi}\right) + \varepsilon$$

with[2]

(2) $$k \sin 2\left(\overline{X} - \overline{\phi}\right) = \sum_i k_i \sin 2\left(\overline{X_i} - \phi\right)$$

In these equations, $-\omega \sin L$ represents the Foucault effect, $\Sigma$ the direction of anisotropy of the support, and $\overline{X_i}$ the average azimuth corresponding to the astronomical influence *i during the 14-minute period under consideration*, and the coefficients $k_i$ and $k$ are coefficients that are variable with time. The term $\varepsilon$ represents the random influence of the balls.

$\overline{\phi}$ represents the average azimuth of the plane of oscillation of the pendulum during the 14-minute experiment under consideration, and $\overline{\dot{\phi}}$ represents its average variation per unit time during that period, with $\dot{\phi} = \mathrm{d}\phi/\mathrm{dt}$.

---

[1] §E.3 above, p. 176.

[2] Naturally we have

(1) $$k \sin 2\overline{X} = \sum_i k_i \sin 2\overline{X_i} \qquad k \cos 2\overline{X} = \sum_i k_i \cos 2\overline{X_i}$$

(2) $$k = \sqrt{\sum k_i^2 + 2\sum k_i k_j \cos 2\left(\overline{X_i} - \overline{X_j}\right)}$$

(3) $$\tan 2\overline{X} = \left(\sum_i k_i \sin 2\overline{X_i}\right) / \left(\sum_i k_i \cos 2\overline{X_i}\right)$$

**Expression of the azimuth of the limit plane**

2 - Equation (1) can also be written

(3) $\dot{\overline{\phi}} = -\omega \sin L + f \sin 2\left(\overline{Y} - \overline{\phi}\right) + \varepsilon$

by putting[3]

(4) $f \sin 2\left(\overline{Y} - \overline{\phi}\right) = k \sin 2\left(\overline{X} - \overline{\phi}\right) + K \sin 2\left(\Sigma - \overline{\phi}\right)$

$\overline{Y}$ represents the average azimuth of the limit plane towards which, if *f has a sufficiently high value*, the plane of oscillation of the pendulum tends to converge during the 14-minute experiment in question.

We naturally have

(5) $\overline{Y} = \overline{Y}(t) \quad \overline{X} = \overline{X}(t) \quad k = k(t) \quad f = f(t) \quad \varepsilon = \varepsilon(t)$

The direction $\overline{Y}$ corresponds to the combined effect of astronomical influences, which have average direction $\overline{X}$, and of the influence of the support anisotropy, which has direction $\Sigma$.

---

[3] Naturally we have

(1) $f \sin 2\overline{Y} = k \sin 2\overline{X} + K \sin 2\Sigma \qquad f \cos 2\overline{Y} = k \cos 2\overline{X} + K \cos 2\Sigma$

(2) $f = \sqrt{k^2 + K^2 + 2kK \cos 2\left(\overline{X} - \Sigma\right)}$

(3) $\tan 2\overline{Y} = \dfrac{k \sin 2\overline{X} + K \sin 2\Sigma}{k \cos 2\overline{X} + K \cos 2\Sigma}$

We can also write

(4) $\tan 2\left(\overline{Y} - \Sigma\right) = \dfrac{k \sin 2\left(\overline{X} - \Sigma\right)}{K + k \cos 2\left(\overline{X} - \Sigma\right)}$

## The limit plane and experiment

3 - Equation (3) exactly corresponds to what is suggested by the experiment on three enchained series that was performed from 6 to 10 May 1957.[4]

Naturally what is observed during an enchained experiment is the azimuth $\phi$ of the paraconical pendulum, and this azimuth is related simultaneously, both to the direction $\Sigma$ of anisotropy of the support, and also to the direction $\overline{X}$ which is representative of the totality of astronomical influences.

In representation (1), the attraction of the plane of oscillation $\overline{\phi}$ towards the direction $\overline{X}$ is stronger the greater the coefficient $k$ is, and the attraction of the plane of oscillation towards the direction of anisotropy $\Sigma$ of the support is stronger the greater the coefficient $K$ is.

As far as can be judged, the fact that the plane of oscillation is constantly veering away from the direction $\Sigma$ of anisotropy of the support shows that the order of magnitude of the coefficient $k$ is *comparable* to that of the coefficient $K$.[5]

## Determination of the direction X of anisotropy of space

4 - The determination of the direction $X$ of anisotropy of space due to astronomical influences naturally presents *considerable* theoretical and practical interest.

*This determination implies the implementation of an isotropic support* for which the coefficient K of equation (1) should be null, and this is the object of *Chapter II* below.

---

[4] §B.1.1. above, p. 103.

[5] This deduction appears to be confirmed by analysis of the enchained series of 28 September to 4 October 1959 with the paraconical pendulum and the isotropic support (see below, *Section H* of *Chapter II*).

## *E.8 Overall view*

From the empirical and theoretical analysis of the observations performed on the movement of the paraconical pendulum, the result is that we may consider that the movement of the azimuth of the paraconical pendulum with anisotropic support essentially results from four cooperating effects: the Foucault effect, a restoring torque effect of the anisotropic suspension, the random influence of the balls, and periodic influences of astronomical origin.

Both in the case of lunisolar effects and in the case of effects of the anisotropy of the support, *to the first approximation*, the observed effects *result from ellipticity of the trajectories*, engendered on the one hand by the anisotropy of the support and on the other hand by astronomical influences.

In fact, *as long as the oscillation of the pendulum remains planar, there is no detectable effect due to the anisotropy of the support or to the periodic lunisolar influence, and the movement of the paraconical pendulum reduces to the Foucault effect.*[1]

At each instant there exists *a limit plane* towards which the plane of oscillation of the pendulum tends to converge, resulting from the Foucault effect and the combined action of the support and periodic influences of astronomical origin. This limit plane varies over time due to the periodic influences of astronomical origin.

---

[1] *Graphs III and IV* of §A.4 above, pp. 94-95.

# F THE HYPOTHESIS OF ANISOTROPY OF INERTIAL SPACE

## *F.1 Theory and experiment*

### Incompatibility of the results of the observations of the paraconical pendulum with anisotropic support with the classical theory of mechanics

1 - From all the analysis above, one *absolutely unarguable* conclusion follows. The amplitudes of the periodic lunisolar components of the movement of the paraconical pendulum with anisotropic support are *totally inexplicable* in the framework of the current theory of gravitation.

For the periodicity of 24h50m in particular, that theory basically leads to diurnal variations of the azimuth of the plane of oscillation of the paraconical pendulum of the order of $10^{-13}$ radians per second, while the observed effects are of the order of at least $10^{-5}$ radians per second in the case of the anisotropic suspension.[1]

The difference of orders of magnitude between the observed effects and the calculated effects is so great than *not one of the persons of importance* who came to visit my laboratory at IRSID ever controverted the *total impossibility* of explaining the observed movements of the paraconical pendulum in the framework of the current theory of gravitation and of relative movements.

[1] See §I.B.2.5 above, pp. 121-123.

The observed effects are at least of the order of $10^{-4}$ radians per second in the case of the isotropic suspension (*Chapter II*, §F.2).

**The postulates of the current theory of gravitation**

2 - Since, in the case of the periodic lunisolar components of the movement of the paraconical pendulum with anisotropic support, the current theory of mechanics leads to orders of magnitude that are *completely incompatible with the observational data*, it seems proper to re-examine the hypotheses from which these theoretical orders of magnitude have been obtained. These hypotheses are essentially the following:

1. The forces of gravitation are *supposed to be inversely proportional to the square of the distance* and proportional to the *gravitational masses*. The coefficient of proportionality is supposed to be *the same* for all masses and for all distances.
2. Transmission of gravitational forces is supposed to be *instantaneous* and to take place in a *straight line* in a space that is supposed to be *Euclidean and isotropic*.
3. The force exerted upon a point $M$ whose inertial mass is $m$ is supposed to follow the law
   (1) $\quad \vec{F} = m\vec{\gamma}$
   It is thus supposed to be proportional to the inertial mass.
4. The *gravitational mass* corresponding to the forces of gravitation is supposed to be equal to the *inertial mass*.
5. The law $\vec{F} = m\vec{\gamma}$ is supposed to be only applicable with respect to a system of axes $O'x'y'z'$, so-called *Galilean axes*, which are moving with a uniform movement of translation with respect to a system of axes, so-called Copernican axes, whose origin is the center of gravity of the solar system and whose directions connect this center of gravity to three fixed stars.
6. Let $Oxyz$ be a system of axes linked to the Earth. We have
   (2) $\quad \vec{\gamma} = \vec{\gamma_r} + \vec{\gamma_e} + \vec{\gamma_c}$
   (3) $\quad \vec{F} = m\vec{\gamma_r} + m\vec{\gamma_e} + m\vec{\gamma_c}$

$\overrightarrow{\gamma_r}$ designates the relative acceleration of the point $M$ with respect to the reference frame $Oxyz$ linked to the Earth; $\overrightarrow{\gamma_e}$ designates the induced acceleration due to the movement of the reference frame $Oxyz$ with respect to the reference frame $O'x'y'z'$; and $\overrightarrow{\gamma_c}$ designates the complementary acceleration, the so-called Coriolis acceleration.
We have
(4) $\overrightarrow{\gamma_c} = 2\overrightarrow{\omega} \wedge \overrightarrow{v_r}$
where $\overrightarrow{v_r}$ denotes the velocity of the point $M$ with respect to the axes $Oxyz$, and $\overrightarrow{\omega}$ denotes the instantaneous rotation of the reference frame $Oxyz$ with respect to the Galilean reference frame $O'x'y'z'$.
Equation (2) is a pure mathematical identity.

7. We have
(5) $\overrightarrow{F} = \overrightarrow{F_1} + \overrightarrow{F_2}$
$\overrightarrow{F_1}$ represents the force of attraction of the Earth and the other celestial bodies, and $\overrightarrow{F_2}$ represents the resultant of the other applied forces. The force $\overrightarrow{F_1}$ is supposed to be proportional to the gravitational mass of the point $M$.
In the case of the Foucault pendulum, $\overrightarrow{F_2}$ represents the tension $\overrightarrow{N}$ of the suspension wire. In the case of the paraconical pendulum, $\overrightarrow{F_2}$ represents the force $\overrightarrow{N}$ exerted by the support upon the ball of the pendulum.
8. It particularly results from these hypotheses that the space corresponding to a Copernican reference system is considered *as being perfectly Euclidean and isotropic throughout.*

All these hypotheses are classical,[2] but it is proper to remind the reader of them here. With regard to the observations of the movement of the paraconical pendulum, *some of these hypotheses are certainly effectively invalidated by experiment.*

### Verifications of the current theory of gravitation

3 - In fact, the current theory of gravitation appears to be well verified:

- in the case of the movements of the celestial bodies (celestial mechanics);[3]
- in the case of falling bodies at the surface of the Earth;[4]
- in the case of the Foucault pendulum, for rectilinear oscillations;[5]
- in the case of the lunisolar deviations of the vertical, if account is taken of the deformation of the terrestrial spheroid by the lunisolar action.[6]

On the other hand, the theory of gravitation *receives a setback* when it is applied to the case of the influence of the attraction of the Sun and the Moon upon the movement on the paraconical pendulum, both in the case of the amplitudes of the periodic lunisolar components and in that of the anomalies observed during total solar eclipses.

---

[2] See for example Paul Appell, 1953, *Traité de Mécanique Rationnelle* (Treatise on Rational Mechanics), Vol. II, Chapter XXII, pp. 267-302 ; and A. Foch, 1967, *Mécanique* (Mechanics), Masson, pp. 149-156 [88].

[3] In his *Cours de Mécanique* (Lecture on Mechanics) (Gauthier-Villars, 1930, Vol. I, p. 387) Paul Painlevé wrote [209]:

> *"By successive approximations, the equations which give the solutions of the problem of three bodies in the case in which only three bodies attract one another allow us to calculate ephemerides over a very long interval of time; the tables of Le Verrier give them for three centuries (and also for three centuries in the past). If we compare the calculated positions with the observed positions, the agreement is an astounding confirmation of Newton's Laws: the celestial body that strays furthest from its anticipated position is the Moon, which sometimes is a little ahead of its calculated position and sometimes a little behind it, without this advance or delay becoming greater than one second of time in a century."*

[4] See for example Foch, 1967, id., pp. 151-152 [88].

[5] See for example Foch, 1967, id., pp. 155-156 [88], and §A.4 above.

[6] See notably M. N. Stoyko, *L'attraction lunisolaire et les pendules* (The lunisolar attraction and pendulums), Bulletin astronomique (Astronomical Bulletin), Vol. XIII, 1949, pp. 6, 29-31, and 46 [254]. The *observed* deviation of the vertical is equal to its calculated value, considering the terrestrial spheroid as being non-deformable, multiplied by $1+k-h=0.667$, where $k$ and $h$ are the Love numbers corresponding to the deformation of the terrestrial spheroid by the lunisolar action.

For the lunisolar deviation of the vertical, see §B.2.4 above and *Table V* of §B.2.4, p. 127. See also note 6, p. 135.

## *F.2 Accuracy of the verification of the laws of gravitation*

2. - In view of the *major* anomalies of the paraconical pendulum, it seems proper to examine to what accuracy the laws of gravitation have been verified, both in the astronomical domain and at the surface of the Earth.[1]

Astonishing as it may seem, all the treatises on Mechanics and Astronomy are remarkably silent on this fundamental question. This is a serious gap and an evident deficiency from the point of view of scientific discipline. In fact, no law has any meaning of any kind if one does not know the accuracy to which it has been verified.

### Accuracy of the astronomical verifications of the postulates of mechanics

1 - Since the fundamental laws of mechanics at the surface of the Earth result from extrapolation of the results obtained in astronomy, it is not without interest to examine with what accuracy these laws have been effectively verified in this domain.

Unhappily this discussion is not available anywhere, because it is postulated as a principle that Newton's laws have been rigorously verified. However, without entering into a detailed discussion which would be outside the scope of this work, it is relatively easy to grasp the order of magnitude of this accuracy.

In fact, consideration of the residues of the fittings by the method of least squares which have been used for construction of the tables currently used in astronomy shows that, for angular displacements, the order of magnitude of the detected deviations between observation and theory is at least of the order of one sexagesimal second of arc,[2] which gives a relative error of the order of

(1) $1/(90 \times 60 \times 60) \approx 3 \times 10^{-6}$

[1] Allais, 1958, *Doit-on reconsidérer les lois de la gravitation ?* (Should the Laws of Gravitation be Reconsidered?), pp. 101-102 [23]. See also Allais, 21 April 1957, *Anomalies du mouvement du pendule paraconique à support anisotrope* (Anomalies of the movement of the paraconical pendulum with anisotropic support) (71 p.), pp. 51-56.

See also my *Third Conference* of 18 March 1967, Bulletin no. 121 of the Alexandre Dufour Circle, May 1967, pp. 114-118 (§B.2.9 of the *Introduction*, above).

[2] Ann. de l'Obs. de Paris (Annals of Observatory of Paris): Memoirs, Vol. IV (the Earth); V (Mercury); VI (Mars and Venus); XXIV (Saturn); XXVIII (Uranus and Neptune); XXXI (Jupiter).

The reader can usefully consult the fundamental article of H. Spencer Jones, *The rotation of the earth and the secular accelerations of the Sun, Moon and Planets.*[3]

[3] Monthly Notices of the Roy. Astr. Soc., vol. 99, 1939, 541-558 [252]. A good summary of this study can be found in Danjon, Astronomie Générale (General Astronomy), Sennac, Paris, 1952-1953, pp. 120 to 126 [98].

The Graphs given on pp. 36 and 37 of this study in fact show that, taking as unit the second of arc corresponding to the average movement of the Moon, the Newtonian times determined with the aid of the orbital movements of the Moon, the Sun, Mercury, and Venus are only in agreement with one another to about two seconds, due to irregular fluctuations. These accidental deviations can even reach 75 seconds for Venus. If we consider that the Sun, Mercury, and Venus traverse respectively 1/13.37, 1/3.22, and 1/8.20 seconds of arc along their trajectories while the Moon traverses 1 second of arc along its trajectory, we arrive at the conclusion that the positions along the orbits are certainly not determined to a better precision than one second of arc.

This is also the conclusion at which we arrive if we compare the predictions made by different tables. For example, consider the "*Connaissance des temps 1957*" and the "*American Ephemeris 1957*" for 1 January 1957 at 0h UT:

| | RIGHT ASCENSION | | | |
|---|---|---|---|---|
| | C.D.T. | A.E. | deviation in seconds of time | deviation in arc-seconds |
| *Sun* | 18h 44m 53.89s | 18h 44m 54.02s | −0.13 | −1.95 |
| *Mercury* | 19h 57m 46.40s | 19h 57m 46.41s | −0.01 | −0.15 |
| *Venus* | 16h 56m 13.30s | 16h 56m 13.22s | +0.08 | +1.20 |
| *Mars* | 00h 49m 57.52s | 00h 49m 57.35s | +0.17 | +2.55 |
| *Jupiter* | 12h 07m 17.48s | 12h 07m 17.47s | +0.01 | +0.15 |
| *Saturn* | 16h 31m 32.96s | 16h 31m 32.01s | −0.05 | −0.75 |
| *Uranus* | 08h 33m 58.63s | 08h 33m 58.63s | 0.00 | 0.00 |
| *Neptune* | 14h 02m 51.75s | 14h 02m 51.71s | +0.04 | +0.60 |

*The average of the absolute values of the deviations is around 0.92 arc-seconds.*

It should be noted that it would be an error to take the angle traversed during the total duration of the observations, which is of the order of a century, as the denominator of equation (1), because the unknown constants of the movement are determined precisely in such a way that, on average, there should be no secular deviation between the observed and the calculated azimuths.

We are thus led to the conclusion that Newton's laws have only been verified in astronomy to a relative precision of the order of $3 \times 10^6$, which is after all very remarkable.[4]

This conclusion goes against ideas that are generally admitted, albeit without real discussion; but it must be accepted.

From the above, we must consider that the current laws of gravitation are not in any way perfectly verified laws that are definitive and unchangeable,[5] and upon which we may base assertions that my experimental results are contrary to all experience acquired in the domain of astronomy. *These laws, like all experimental laws, are only verified to a certain approximation.*

### The accuracy to which the postulates of mechanics have been verified at the surface of the Earth

2 - The mechanical experiments at the surface of the Earth that have been performed with the greatest accuracy are those relating to the seconds pendulum. These experiments accept as a fact the well-known equation

$$T = 2\pi\sqrt{I/Mgl}$$

deduced from the postulates of mechanics. $I$ denotes the moment of inertia of the pendulum. We calculate the quotient $I/Ml$ from measures of length; we measure $T$; and then we deduce $g$. The experiments performed by Volet at the Breteuil Pavilion at Sèvres made it possible to measure $g$ directly by photographing falling invar rulers, and provided confirmation of the values deduced from the pendulum observations to approximately $10^{-5}$. This is the order of magnitude of the accuracy to which the principles of mechanics appear to have been verified at the surface of the Earth.

[4] For the secular acceleration of the movement of the Moon, see particularly F. Tisserand, *Traité de Mécanique Céleste* (Treatise on Celestial Mechanics), Vol. III, *Exposé de l'ensemble des théories relatives au mouvement de la Lune* (Explanation of the various theories relating to the movement of the Moon), Gauthiers-Villars, 1894, Chapters XIII, XVIII and XIX [266]. See also W.M. Smart, *Celestial Mechanics*, Longmans, 1953, Chapters 17, 18 and 19 [250].

[5] These laws which were meant to be so perfect that Hegel thought he could demonstrate them by metaphysics (*Philosophy of Nature*, Franch transl. Vera, I, p. 293, Parag. 270 [156]. See commentary by Pareto, *Treatise on General Sociology*, Vol. I, p. 269 [212]).

## Order of magnitude of the observed anomalies in the movement of the paraconical pendulum with anisotropic support

3 - It is interesting to compare these figures with the order of magnitude of the anomalies I have observed. This order of magnitude is around a thirtieth that of the Foucault effect,[6] which itself is of the order of three millionths that of gravity.[7]

*The observed effects are thus of order of magnitude less than or equal to the order of magnitude to which one may consider that the principles of mechanics have been verified, either at the surface of the Earth or in the domain of astronomy.*

---

[6] §A.5.3 above, p. 98.

[7] The two equations that determine the Foucault precession are (§B.2.3, *Table IV*, p. 126)

$$\ddot{m} + p^2 m = -2\omega \sin L\, n' \qquad\qquad p^2 = g/l = 981/83 = 11.82$$
$$\ddot{n} + p^2 n = 2\omega \sin L\, m'$$

Since we have

$$\vec{F} = M\vec{\gamma}$$

and since $\vec{\gamma}$ has $l\ddot{m}$ and $l\ddot{n}$ as components, we see that the Foucault perturbing force has components

$$-2\omega \sin L M l \dot{n}$$
$$2\omega \sin L M l \dot{m}$$

whose order of magnitude is

$$2\omega \sin L\, M l p \alpha \qquad\qquad p = 2\pi/T = \sqrt{g/l}$$

where $\alpha$ is the amplitude. Thus we have the Foucault force

$$\frac{\text{Foucaultforce}}{\text{Gravity}} \sim \frac{2\omega \sin L\, M p l \alpha}{Mg} = \frac{2\omega \sin L\, l p \alpha}{g}$$

In the case of my experiments of June-July 1955 (§A.1 and §A. 4 above)

$$\omega \sin L = 0.55 \times 10^{-4} \qquad l = 83\text{cm} \qquad p = 3.44 \qquad \alpha = 0.1$$

whence

$$\frac{\text{Foucaultforce}}{\text{Gravity}} = \frac{2 \times 0.55 \times 10^{-4} \times 3.44 \times 83 \times 0.1}{981} = 3.20 \times 10^{-6}$$

Therefore the order of magnitude of the anomalies of the paraconical pendulum with anisotropic support, equal to a thirtieth of the Foucault effect, *is about one ten-millionth of gravity.*

It should also be underlined that, due to their periodic structure, the detected anomalies *are zero on average*. Thus, if new forces are to be considered, the corresponding diurnal lunisolar anomalies only manifest on the scale of the solar day, of the sidereal day, or of the lunar day. On the astronomical scale of the movement of the planets, we must therefore provide corresponding forces whose integrals are null over the entire trajectory of a planet. Their order of magnitude is thus quite comparable to the order of magnitude with which we may consider that Kepler's laws are verified over one planetary revolution.[8]

*From the above, it follows that the anomalies I have detected are not in any way in contradiction with the previous experimental data, neither at the surface of the Earth nor in the domain of astronomy.*

[8] Otherwise put, if, to the Newtonian actions, we were to add actions $10^{-6}$ smaller and on average null during the revolution of a planet, these actions would probably remain undiscovered.

## F.3 The hypothesis of anisotropy of inertial space, and its implications

### Explanation of the anomalies of the paraconical pendulum by an anisotropy of inertial space

1 - *As soon as 1955, I have been able to show that a difference of inertial mass in two orthogonal directions of the order of* $10^{-6}$ could explain the effects observed.[1] In fact, due to their minuscule nature, the effects of such an anisotropy of inertial space could easily have escaped experimental observation until now.

The reason that I did not pose this hypothesis in my *Notes* to the Academy of Sciences from 1957 to 1959 was in order to avoid exciting the general opposition of all the partisans of the theory of relativity.[2] I thought - wrongly - that the anomalies of the paraconical pendulum *demonstrated* by my experiments would be sufficient by themselves, following the principle of Planck:[3]

> *"The scientific value of precise experiments is independent of their theoretical interpretation."*

Truthfully speaking, if we consider that the precision with which the present laws of gravitation have been verified is with a relative error of the order of $10^{-6}$, then *the hypothesis of the anisotropy of inertial space seems quite compatible with all the observational data.*[4]

### Equations of movement of the paraconical pendulum, under the hypothesis of anisotropy of inertial space

2 - We can effectively explain the orders of magnitude of the periodic lunisolar components detected in the movement of the paraconical pendulum with anisotropic support by an anisotropy of inertial space of the order of $10^{-6}$, caused by the influence of the celestial bodies.[5]

---

[1] Allais, 12 August 1955, *Mouvements Périodiques du Pendule Conique* (Periodic Movements of the Conical Pendulum) (12 p. with four appendices), p. 7.

[2] See above, note (3) of §C.1.3 of the *Introduction*, p. 57.

[3] Planck, *Initiations à la Physique* (Initiations to Physics), 1941, Flammarion, p. 256 [218].

[4] See §F.1.3 above.

[5] The same explanation holds for the paraconical pendulum with isotropic support (see *Chapter II* below, Section I).

• *Table XII* below gives the essential equations relating to the influence of an anisotropy of inertial space caused by the influence of the celestial bodies.

It is supposed that the inertial mass $M_i$ of the pendulum is relatively increased with respect to its gravitational mass $M_g$ by a factor $1+\varepsilon_i$ in the direction of the celestial body $i$. Thus, the coefficient $\varepsilon_i$ represents the influence of the body $i$.

Equations (1) and (2) of *Table XII* show what happens to *the first terms* of the equations (5) and (6) of *Table IV* of Section B.2 above, in the case that the axis $Om$ is oriented towards the celestial body $i$. These equations are equivalent to equations (3) and (4) of *Table XII*, where the second term of equation (3) can be considered *as being a perturbation.*

An elliptical trajectory of the paraconical pendulum corresponds to these equations (3) and (4), for which equations (5) and (6) define the variations $\dot{\phi}$ of the azimuth of the ellipse and $\dot{\beta}$ of its minor axis.[6]

For an arbitrary orientation of the azimuth $X_i$ of the celestial body $i$, the equations (5) and (6) should be replaced by equations (7) and (8).

The body $i$ exerts a *direct* influence upon $\dot{\phi}$ represented by equation (7), and an *indirect* influence upon $\dot{\phi}$ via $\dot{\beta}$ and the Airy precession as intermediaries. In fact, the direct influence is *relatively negligible* by comparison to the indirect influence.[7]

*Formally*, equations (7) and (8) are completely analogous to the expressions for $\dot{\phi}$ and $\dot{\beta}$ corresponding to the current theory of gravitation.[8] *Only the coefficients are different.*

Naturally, we can define an average direction of the anisotropy of inertial space by putting[9]

(1) $\varepsilon \sin 2(X-\phi) = \sum \varepsilon_i \sin 2(X_i - \phi)$

• *Table XIII* gives, *to the first approximation*, expressions for $\dot{\phi}$ and $\dot{\beta}$ in which account is taken, all together, of the Foucault effect, the Airy precession,[10] the anisotropy of the support,[11] and the anisotropy of inertial space caused by the various celestial bodies.

The expressions (1) and (2) for $\dot{\phi}$ and $\dot{\beta}$ in *Table XIII* enable us to estimate the order of magnitude of the variation of $\dot{\phi}$ corresponding to the influence of the Moon (equations 3 and 4).

The order of magnitude of the observed value $\dot{\phi}_0$ of $\dot{\phi}$ has already been calculated for the lunar periodicity of 24h50m in the case of the paraconical pendulum with anisotropic support.[12]

---

[6] Allais, 1956, *Théorie du Pendule Paraconique* (Theory of the Paraconical Pendulum), Part III, no. 1103, *Table II*, p. AI,6 [8].

[7] See note (11) of §B.2.5 above, p. 122.

[8] Note (3) of *Table XII*.

[9] For the implications of this relationship, see the case of equation (2) of §E.7.1 analyzed above, p. 193.

[10] §B.2.3 above, p. 120.

[11] Allais, 1956, *Théorie du Pendule Paraconique* (Theory of the Paraconical Pendulum), Table T1 of *Attached Note* 28 [8].

[12] §B.2.6, p. 123.

From the equality $\overline{\dot{\phi}_t} = \overline{\dot{\phi}_0}$, we deduce the estimate $\varepsilon_l = 0.20 \times 10^{-6}$ of $\varepsilon_l$ (*Table XIII*, equation 6).

We thus see that *it is possible to explain the order of magnitude of the lunisolar components in the movement of the paraconical pendulum effectively by an anisotropy of inertial space of the order of* $10^{-6}$ *for the influence of each celestial body.*

**Effects of anisotropy of inertial space**

3 - The more marked the anisotropy is, the more the plane of oscillation of the pendulum tends to align itself towards the direction of the celestial body in question, the effect of the body being to increase inertia in the direction of that body.

*An action at a distance is thus replaced by a local anisotropy of inertial space.*[13]

• If we compare the equations of movement of the pendulum corresponding to an anisotropy of inertial space with the equations corresponding to the classical theory, we see that the effect corresponding to the anisotropy of inertial space is proportional to the square of the amplitude of the pendulum and inversely proportional to its length, whereas the effect corresponding to the current theory of gravitation is also proportional to the square of the amplitude of the pendulum, but is independent of its length.[14]

*Thus, according to the current theory of gravitation, the influence of the celestial bodies is independent of the length of the pendulum, whereas, under the hypoth-*

---

[13] In reality, the postulate of instantaneous propagation to a distance of the forces of gravitation *cannot be retained.*

According to Maxwell's representation of local actions by tensors, the opinion is generally held that the actions of gravitation are propagated, and that they can be represented by local properties of space (see *Chapter VI* below).

[14] From *Table XIII* (equation 7), the effect corresponding to the anisotropy of inertial space is proportional to $\alpha^2/l$, while the effect corresponding to the current theory of gravitation is proportional to $\alpha^2$ (*Table XIII*, note 2).

This explains why the Foucault effect is the less perturbed, the longer is the pendulum.

In the case of the Foucault pendulum in the Pantheon in 1852, the length of the pendulum was 67 meters, and the amplitude of the oscillations was 0.06 radians (§E.2.3 above). *Under the hypothesis of anisotropy of inertial space*, the theoretical effect was around 220 times smaller than in the case of the paraconical pendulum. In fact, we have

$$(0.06^2/67)/(0.1^2/0.83) = 1/224$$

In the case considered by Dejean de Fonroque (*Introduction*, §B.2.6, note 7 above), the pendulum was suspended with a wire around one meter long, and the initial amplitude $\alpha$ of the oscillations was of the order of 45°, *i.e.* 0.785 radians, whereas in the case of the paraconical pendulum these figures were $l = 83$ cm and $\alpha = 0.1$ radian (around 6°). The ratio $\alpha^2/l$ was accordingly about fifty times greater than in the case of my own experiments $(0.78^2/100)/(0.1^2/83) = 51.2$.

*esis of anisotropy of inertial space, the corresponding effect is considerably more marked when the pendulum is short.*[15, 16]

## Determination of the coefficients of anisotropy $\varepsilon_i$

4. - With regard to the *very limited* data that we possess at the moment, precise determination of the coefficients of anisotropy $\varepsilon_i$ as functions of the characteristic parameters of the celestial bodies and their variations over time is completely impossible.[17]

---

[15] From the equations (1) and (2) of *Table XIII*, and putting $\lambda = 3p\alpha/8$, $\mu_S = p^3\alpha\sigma/4$, $\mu_i = p\alpha\varepsilon_i/2$, we can deduce as a first approximation the equation

(1) $\ddot{\phi} = \lambda\,[\mu_S \sin 2(\Sigma - \phi) + \sum \mu_i \sin 2(X_i - \phi)]$

The very interesting implications of this second order differential equation will be examined in *Chapter I* of the *Second Volume* of this work. Equation (1) can be written

(2) $\ddot{\phi} = \lambda f \sin 2(Y - \phi) \qquad \lambda = (3/8)\,p\alpha \sim (3/8) \times 3.44 \times 0.1 = 0.129$

if the azimuth of the limit plane is denoted by $Y$. In a case in which the azimuth $\phi$ of the plane of oscillation is close to the limit plane $Y$ that results from the combined action of the support and the celestial bodies, equation (1) reduces to the linear differential equation

(3) $\ddot{\phi} + 2\lambda f \phi = 2\lambda f Y$

If $Y(t)$ varies relatively little during a 14-minute experiment (= 840 seconds), the general integral of (3) is

(4) $\phi = Y + A\cos\Omega t + B\sin\Omega t \qquad \Omega = \sqrt{2\lambda f}$

For the initial conditions

(5) $\phi = \phi_0 \qquad \dot{\phi} = -\omega\sin L \qquad \text{for } t = 0$

equation (4) becomes

(6) $\phi = Y + (\phi_0 - Y_0)\cos\Omega t - \frac{\omega\sin L}{\Omega}\sin\Omega t$

Taking into account the restoring torque effect of the anisotropic suspension, for which we have the order of magnitude $\mu_S = 10^{-6}$ (Note 6, p. 178, of §E.3.3 above), we can estimate (*Second Volume* of this work, *Chapter I*, Section B) that we have approximately: $\mu = \sqrt{2}\mu_S = 1.41 \times 10^{-6}$, and thus

(7) $\Omega = 0.611 \times 10^{-3} \qquad \Theta = 2\pi/\Omega = 10284" \sim 2.86$ hours

From this, it follows that the azimuth $\phi$ oscillates about the limit plane with a period of the order of three hours. For experiments having a duration of 14 minutes, such an oscillation does not take place, and during each 14-minute experiment the azimuth of the plane of oscillation approaches towards the azimuth of the limit plane.

(See also the calculation of note 2 of §II.I.2, and *Table X* of *Chapter II*, pp. 324-325).

During a 14-minute experiment, the third term of equation (6) corresponding to the Foucault effect gives an average effect

(8) $\overline{\dot{\phi}_f} = -\omega\sin L\,\overline{\cos\Omega t}$

Since the average of $\cos\Omega t$ during a 14-minute experiment is smaller than unity, we have $\left|\overline{\dot{\phi}_f}\right| < \omega\sin L$.

[16] For the effects of the anisotropy of inertial space in the case of the paraconical pendulum with isotropic support, see *Chapter II* below, Section I, pp. 320-325.

[17] If we consider the respective actions of the Sun and the Moon upon the diurnal components of the azimuths of the plane of oscillation of the paraconical pendulum with anisotropic support, and the available empirical data (*Table II* of §A.5 above), we may think that the coefficients of anisotropy for the Sun and the Moon are of the same order of magnitude, as are the coefficients $C_s$ and $C_l$ corresponding to the current theory of gravitation (§B.2.5 above, p. 123).

## *TABLE XII*

### INFLUENCE OF THE CELESTIAL BODY i ON THE AZIMUTH AND THE MINOR AXIS OF THE ELLIPTICAL TRAJECTORY OF THE PARACONICAL PENDULUM UNDER THE HYPOTHESIS OF ANISOTROPY OF INERTIAL SPACE

*Notation:*

Axis Om = direction of greatest inertia [1]

$\varepsilon_i$ = coefficient of anisotropy induced by celestial body i [2]

*Differential equations in m and n:*

(1) $(1 + \varepsilon_i)\ddot{m} + p^2 m = 0$ $\qquad p^2 = g/l$

(2) $\ddot{n} + p^2 n = 0$ $\qquad \varepsilon_i << 1$

*To the first approximation:*

(3) $\ddot{m} + p^2 m = \varepsilon_i p^2 m$

(4) $\ddot{n} + p^2 n = 0$

*Solution:*

(5) $\dot{\phi}_i = \varepsilon_i p \dfrac{\alpha\beta}{\alpha^2 - \beta^2} \cos 2\phi$

(6) $\dot{\beta}_i = -\varepsilon_i \dfrac{p}{2} \alpha \sin 2\phi$

*Solution with arbitrary axes Om, On:*[3]

(7) $\dot{\phi}_i = \varepsilon_i \dfrac{p\alpha\beta}{\alpha^2 - \beta^2} \cos 2(X_i - \phi)$ $\qquad$ *$OX_i$ = azimuth of celestial body i*

(8) $\dot{\beta}_i = +\varepsilon_i \dfrac{p}{2} \alpha \sin 2(X_i - \phi)$

-----

(1) Notation of §B.2.3 above and of the corresponding *Table IV*, p. 126.

(2) The coefficient of anisotropy $\varepsilon_i$ *is a dimensionless coefficient.*

(3) These equations should be compared with the equations deduced from the current theory of universal gravitation and corresponding to the attraction of a celestial body i (§B.2.5, *Table VI* above, p. 128).

(1) $\dot{\phi} = \dfrac{1}{p} K_i \dfrac{\alpha\beta}{\alpha^2 - \beta^2} \cos 2(X_i - \phi)$

(2) $\dot{\beta} = \dfrac{1}{2p} \alpha K_i \sin 2(X_i - \phi)$

*TABLE XIII*

# EQUATIONS OF MOVEMENT OF THE PARACONICAL PENDULUM UNDER THE HYPOTHESIS OF ANISOTROPY OF INERTIAL SPACE AND IN THE CASE OF AN ANISOTROPIC SUPPORT

*Equations of the movement (to the first approximation):*

(1) $\dot{\phi} = -\omega \sin L + \frac{3}{8} p \alpha \beta \qquad p = \sqrt{g/l} = 3.44$

(2) $\dot{\beta} = \frac{1}{4} p^3 \alpha \sigma \sin 2(\Sigma - \phi) + \sum_i \frac{p}{2} \alpha \varepsilon_i \sin 2(X_i - \phi)$

*Orders of magnitude of the ε - Illustration in the context of the Moon - Periodicity of 24h50m:*

*Theoretical value of the average component* $\bar{\dot{\phi}}_{tl}$ *of* $\bar{\dot{\phi}}$ during a single experiment of duration Δt =14 min = 840 sec.[1]

(3) $\bar{\dot{\phi}}_{tl} = \frac{3}{8} p \bar{\alpha} \bar{\beta} \qquad \bar{\alpha} = 0.1$

$= \frac{3}{8} p \bar{\alpha} [\frac{1}{2} \Delta t \frac{p}{2} \bar{\alpha} \varepsilon_l \overline{\sin 2(X_l - \phi)}]$

$= 9.32\, \varepsilon_l \overline{\sin 2(X_l - \phi)}$

(4) Order of magnitude of $\bar{\dot{\phi}}_{tl} = 9.32\, \varepsilon_l$ rad/sec.

*Order of magnitude of the average observed value* $\dot{\phi}_{ol}$ (§B.2.6, p. 123):

(5) $\bar{\dot{\phi}}_{ol} = 0.186 \times 10^{-5}$ rad./sec.

*Order of magnitude corresponding to* $\varepsilon_l$ *according to (4) and (5):*

(6) $\varepsilon_l = 0.186 \times 10^{-5} / 9.32 = 0.20 \times 10^{-6}$

*From (1) and (2), the effect of the anisotropy of inertial space upon* $\dot{\phi}_l$ is equal to:[2]

(7) $(\frac{3}{8} p \bar{\alpha})(\frac{p}{2} \bar{\alpha} \varepsilon_i) \frac{\Delta t}{2} \overline{\sin 2(X_i - \phi)} = \frac{3}{16} \frac{g}{l} \bar{\alpha}^2 \varepsilon_i \frac{\Delta t}{2} \overline{\sin 2(X_i - \phi)}$

-----

(1) The ——— mean that these are average values over the duration Δt of the experiment, =14 minutes=840 seconds.

(2) From the current theory of gravitation, the effect of universal attraction upon $\dot{\phi}_i$ is proportional to:

$$(\frac{3}{8} p \bar{\alpha})(\frac{1}{2p} \bar{\alpha} K_i) = \frac{3}{16} \bar{\alpha}^2 K_i$$

(*Table XII*, note 3), and we have, from relation (7):

$$(\frac{3}{16} p^2 \bar{\alpha}^2 \varepsilon_i)/(\frac{3}{16} \bar{\alpha}^2 K_i) = \frac{g}{l} \frac{\varepsilon_i}{K_i}$$

# G OVERALL VIEW OF MY EXPERIMENTS ON THE PARACONICAL PENDULUM WITH ANISOTROPIC SUPPORT 1954-1960

## *G.1 Significance and implications of the experiments on the paraconical pendulum with anisotropic support*

To me, it seems that the essential features of the *necessarily very brief* analysis I have presented above of the anomalies of the paraconical pendulum with anisotropic support can be summarized as follows:

1. The movement of the asymmetrical paraconical pendulum with anisotropic support includes periodic components of *statistically significant* amplitude, in particularly periodic components having periods near 24h and 24h50m, of the order of a thirtieth of the Foucault effect.
2. In particular, the diurnal lunar component of 24h50m cannot be identified with the periodic diurnal lunar component resulting from the theory of gravitation *the way it can be calculated* from the principle of inertia, the principle of universal gravitation, and the theory of relative movements, *whose amplitude is around twenty million times smaller* than in the case of the paraconical pendulum with anisotropic support.
3. The *very particular* periodic structure of the observed phenomena, notably resulting from the relative magnitude of the amplitude of the lunar periodicity of 24h50m, *totally excludes* any explanation based upon any of the already known periodic phenomena that have been suggested as capable of explaining the amplitude of the observed periodicity.
4. My analyses of 1954 to 1957 were confirmed in an *astounding and spectacular* manner by the *similar* results obtained during the two *crucial* experiments performed in July 1958 at IRSID and at Bougival in a gallery 57 meters underground and 6.5 km from Saint-Germain.

5. In fact the observed periodicities, and in particular the lunisolar component of 24h50m, *really exist*. They are *totally inexplicable* in the framework of the current theory of gravitation. *And they cannot be considered as being the results of the direct or indirect action of any known phenomenon.*
6. Anomalies were observed during *the two total solar eclipses* of 30 June 1954 and 2 October 1959. They are *totally inexplicable* in the framework of the current theory of gravitation.
7. *The observed effects, whose order of magnitude is of the order of a millionth of the gravitational force,*[1] *are not actually incompatible with any experimental results deriving from previous experiments*, because the precision with which those results have been obtained was not actually better than a few millionths.
8. *In the current state of the discussion, the observed anomalies can only be explained by supposing the existence of additional terms* in the actions of gravitation. *The simplest hypothesis that can be retained is that of an anisotropy of inertial space.*
9. Naturally, the fact that it is possible to explain the anomalies of the paraconical pendulum by an anisotropy of inertial space *does not prove* the actual existence of such an anisotropy, but it does prove that *everything happens as though inertial space were anisotropic*.
10. The observed anomalies are not isolated anomalies. Many anomalies have also been detected in numerous geophysical phenomena, and it seems probable, even certain, that they can all be ascribed *to one and the same cause*.[2]

---

[1] As follows from the order of magnitude calculation for the $\varepsilon_i$ (§F.2.2 above), and from the calculation of note (7) of §F.2.3 above, p. 204.

[2] See particularly the optical deviations of sightings at marks and at collimators, the optical observations of Esclangon, and the interferometric observations of Miller (*Chapters III and IV* below).

See also the agreements in phase brought to light in *Chapter V*, particularly in §E.1 of that Chapter.

## G.2 *The publication of my work*

### The publication of my work upon the paraconical pendulum with anisotropic support

1 - By triggering a very great debate, the diffusion of my experimental results that started in 1956 and their publication from 1957 onward helped me to obtain the necessary finance to continue my experiments.

In particular, the impact of my Conference of 22 February 1958 at the Ecole Polytechnique under the presidency of Albert Caquot made it possible for me to get financial support for the crucial experiments of July 1958 at Bougival and Saint-Germain.[1]

However, after my Note of 9 February 1959 to the Academy of Sciences relating to the influence of anisotropy of the support,[2] it became impossible for me to publish any further Note, in particular upon the theory of the paraconical pendulum. In particular it was impossible for me to publish five Notes that I had prepared, the first upon the application of Bour's theorem to the case of the Earth's movements, and the other four upon application of Lagrange's method of variation of constants to calculating the influence of all the factors affecting the movement of the paraconical pendulum.[3]

---

[1] See notably §B.2 and §D.2 of the *Introduction*, above.

I was much reproached (by Henri Villat in particular) for having set up my Conference of 22 February 1958 at the Ecole Polytechnique, and in his previous letter of 6 February 1958 Henri Villat did not hesitate to assert in advance that the discussion that was anticipated to follow would be "*absurd*".

It even seemed intolerable that my Conference was organized by the *Alexandre Dufour Circle*, and in my letter to Albert Caquot of 18 February 1958 I was impelled to write:

> *"During recent private conversations, the lack of objectivity of my adversaries became incredible. Thus it seems that the fact that I have accepted the invitation to speak under the aegis of the Alexandre Dufour Circle tends to compromise my own contentions, because, among the members of the Alexandre Dufour Circle, there are certain persons who are anti-relativists and thus judged to be anti-scientific.*
>
> *"One would think that one had returned to the age of Galileo."*

[2] See the *Introduction* above, §B.2.7, note 8, p. 51.

[3] See the *Introduction* above, §B.3.2, note 2, p. 54.

The opposition raised to the publication of my work can be illustrated by the comments of Jean Leray, a member of the Commission on Mechanics of the Academy of Sciences, in his letter of 18 December 1960.[4]

### The non-publication of my results on the movement of the paraconical pendulum with isotropic support

2 - Due to the shutdown of my laboratory at IRSID in June 1960 and the delays involved in exploiting the observations of November-December 1959 and March-April 1960, and due to the definitive refusal of the Academy to publish any note from me after 23 February 1960,[5] *I unfortunately was not able to publish any result corresponding to the paraconical pendulum with isotropic support.*[6, 7]

---

[4] Letter from Jean Leray on 18 December 1960:

> *"I continue to regret that the Academy has published any of your Notes; it cannot, without discrediting itself, continue to publish them: from experimental results which are random because friction plays a leading role in them, and which are without interest because their theory is not developed as far as numerical conclusions, you deduce, using statistical tricks, conclusions which seem unjustified both to me and to more than one of my colleagues, although to you they appear sensational."*

See §E.6 above, note 10, p. 192.

The reader will find all my (very edifying!) correspondence with Jean Leray in Annex 1A of the *Second Volume* of this work.

[5] See *Introduction* above, §D.3.2, note 4, p. 71, and *Chapter III* below, §B.4, pp. 331-340.

[6] See *Introduction* above, §B.2.5, p. 49, and *Chapter II* below.

[7] However, on 5 and 10 November 1959 I prepared two *Notes* presented by René Thiry and André Léauté which unfortunately could not be published by the Academy of Sciences: the first being *Pendule paraconique, Réalisation d'un support aussi isotrope que possible* (The Paraconical Pendulum, Implementation of a Support as Isotropic as Possible) (10 p.), and the second being *Pendule paraconique à suspension isotrope. Détermination des variations au cours du temps des caractéristiques de la corrélation du mouvement avec l'azimut* (The Paraconical Pendulum with Isotropic Support. Determination of the Variations over Time of the Characteristics of Correlation of the Movement with the Azimuth) (7 p.) (see below, *Chapter II*, §B.l, note 1).

I had attached these two Notes to my request for funds, addressed to the Director of CNRS on 26 February 1960.

In my new application of 19 May 1960 (which also yielded no result), I again cited them as well as the first results of the analysis of the two observational series of November-December 1959 and March-April 1960 (See *Chapter II* below, Section D).

## G.3 Summary chronology of the experiments performed under the aegis of IRSID. 1954-1960

• From February 1954 to 30 June 1960 I carried on my experiments on the paraconical pendulum with my two collaborators Jacques Bourgeot and Annie Rolland in the context of my laboratory at IRSID.

I was able to utilize this laboratory thanks to the decision of Pierre Ricard, at the time *President of the Metallurgic and Mining Industries, a man of absolutely exceptional generosity of spirit.* From the start I received the particularly helpful support of René Dugas, the author of two very remarkable works on the history of mechanics.[1]

• In order better to understand the conditions of development of my experiments, it seems indispensable to me to provide here a summary chronology from 1953 to 1960. This time span is conveniently separated into two periods: 1953-1959, and 1959-1960.

*Period October 1953 - February 1959*

*July 1953*: Decision by IRSID to provide me with the necessary finances for my researches.

*16 October 1953*: Installation of my laboratory at IRSID.

*1 February 1954*: Start of experiments with the paraconical pendulum.

*4 June - 9 July 1954*: The first month-long series of enchained experiments with the paraconical pendulum performed. Anomaly observed during the total solar eclipse of 30 June 1954.

*16 November - 22 December 1954*: The second month-long series of enchained experiments with the paraconical pendulum performed.

*3 June - 7 July 1955*: The third month-long series of enchained experiments with the paraconical pendulum performed.

*15 September 1955*: I considered *as being definitely established* the existence of the lunisolar component of 24h50m, having an amplitude *totally inexplicable* in the framework of the currently accepted theory of gravitation.

[1] See *Introduction* above, §D.1, pp. 64-65.
In July 1953 René Dugas was designated by IRSID as *Scientific Adviser*.

*From January 1956*: Very numerous visits to my laboratory by *external persons of importance*.

*6 September 1956*: Completion of my memoir, *Theory of the Paraconical Pendulum* (441 p.).

*15 February 1957*: Elaboration of the test of periodicity for autocorrelated series.

*13 May 1957*: Note to the Academy of Sciences on the generalization of the Schuster test to the case of autocorrelated series.[2]

*13 November - 23 December 1957*: Five *Notes* to the Academy of Sciences concerning my experiments with the paraconical pendulum.[2]

*22 February 1958*: Conference at the Ecole Polytechnique: "*Faut-il reconsidérer les lois de la Gravitation ? Sur une nouvelle expérience de Mécanique*" ("Should the Laws of Gravitation be Reconsidered? On a new experiment in mechanics"), organized by the Alexandre Dufour Circle.[3]

*March - April 1958*: *Installation of a second laboratory in the "Blanc Minéral" underground gallery at Bougival*, 57 meters deep and 6.5 km distant from Saint-Germain, with the support of the *Centre National de la Recherche Scientifique* (National Center for Scientific Research) and the *Comité d'Action Scientifique de la Défense Nationale* (National Defense Committee for Scientific Action).

*2 - 30 June 1958*: Preliminary experiments at Bougival and Saint-Germain.

*2 July - 1 August 1958*: *Crucial* simultaneous experiments at Bougival and Saint-Germain. Month-long enchained series with the paraconical pendulum with anisotropic support, accompanied by optical sightings at marks at Saint-Germain.

*3 November and 22 December 1958*: Two *Notes* to the Academy of Sciences concerning the experiments at Bougival and Saint-Germain.[2]

*19 January and 9 February 1959*: Two Notes to the Academy of Sciences concerning the influence of inclination of the bearing surface and the influence of anisotropy of the support.[2]

## *Period March 1959 - June 1960*

*24 - 25 September 1959*: First experiments with the paraconical pendulum with isotropic support.

---

[2] See *Introduction* above, §B.2.7, p. 51.

[3] See *Introduction* above, §B.2.9, p. 52.

*28 September - 4 October 1959*: Parallel enchained experiments with the isotropic suspension and the anisotropic suspension. Observation of the total solar eclipse of 2 October 1959.

*20 November - 15 December 1959*: Two simultaneous series of experiments at IRSID with the isotropic suspension and the anisotropic suspension.

*7 November 1959*: Conference at the French Society of Civil Engineers, "*Should the Laws of Gravitation be Reconsidered? New results, review, and perspectives*", organized by the Alexandre Dufour Circle.[3]

*23 February 1960*: Refusal by the Academy of Sciences to publish my *Note* on the results of my observations of optical deviations of sightings at marks of July 1958.[4]

*16 March - 16 April 1960*: Two simultaneous series of experiments at IRSID with the isotropic suspension and the anisotropic suspension.

*30 June 1960*: Shutdown of the IRSID laboratory.

• From 1954 to 1960, *by myself*, I tackled a crushing task. Simultaneously I was obliged: to conceive and direct the experiments; to design and calculate the apparatus used; to direct and interpret all the numerical calculations for analysis; to survey all the experimental and theoretical publications on pendulum research; to elaborate a complete theory of the movement of the paraconical pendulum, and in particular of its lunisolar components; to analyze and extend all the contributions in the literature on finding periodicities, and to elaborate an appropriate test for autocorrelated time series; and, in parallel, to obtain financing for the experiments, and to do frequent lobbying for that purpose.[5]

---

[4] See *Introduction* above, §D.3.2, pp. 70-72, and *Chapter III* below, §B.4, pp. 339-340.

[5] In my "*Note sommaire sur les recherches sur le pendule paraconique*" ("Short note on the researches with the paraconical pendulum") of 15 January 1957 (24 p.) I wrote (pp. 18-19):

> *"It would be a serious error to underestimate the great difficulty of the problems under consideration.*
>
> *"The theoretical and experimental study of the phenomena in question requires deep knowledge of very diverse disciplines (mechanics, geophysics, astronomy, statistics, and mathematics), and demands delicate experiments that need a lot of time and care; and afterwards it requires analytical calculations that are often very long, and numerical calculations some of which are impossible without the aid of electronic calculation machines.*
>
> *"Research of this type requires a lot of patience, indeed of relentlessness, and a lot of time! In such a field, nothing can be done in a hurry. If there is one lesson that I have distilled from these three years of experiments, this is it."*

In particular, I had to confront two gaps in the literature which in this case were *quite essential*. First, no author had ever calculated the influence of the Sun and the Moon upon the movement of the pendulum. Second, no test of periodicity was available for the case of autocorrelated series. These were the two *particularly difficult major* questions that I was compelled to resolve.[6, 7]

• After my Conference of 22 February 1958, and with the support of a number of members of the Academy of Sciences, in particular Albert Caquot, Pierre Tardi, et Marie-Joseph Kampé de Fériet, it was decided to perform two crucial experiments by observing the movements of two identical paraconical pendulums, one at IRSID, and the other at Bougival in a gallery 57 meters underground and 6.5 km away.[8]

These crucial experiments took place in July 1958, and *were crowned with spectacular success*.[9]

The first period 1953-1958 thus culminated with these two crucial experiments of July 1958. *Paradoxically, and in spite of the astounding success of these two crucial experiments, I was faced with growing hostility and major difficulties in obtaining financing*. At the end of 1959, due to the lack of funds, the decision had to be taken to close the two laboratories at IRSID and at Bougival in June 1960.[10]

---

[6] See *Introduction* above, §B.3.2, pp. 54-55.

[7] During all this period, and in parallel, I continued my teaching of economics at the National Superior School of Mines and at the Statistics Institute of the University of Paris, and also published numerous memoirs on monetary dynamics, on the Soviet economy, on the European Community, and on mining research (see *Autoportraits*, 1989, pp. 127, 128, 135, 138, 140, and 141 [48]).

[8] These experiments were jointly financed by the *Comité d'Action Scientifique de la Défense Nationale* (National Defense Committee for Scientific Action) and by the *Centre National de la Recherche Scientifique* (National Center for Scientific Research).

Previously, on 9 October 1956 and 5 April 1957, I had presented two requests for finance to CNRS, which were rejected.

[9] See *Section C* above, §4, pp. 160-161.

[10] The premature demise of Pierre Ricard on 4 April 1956, followed by that of René Dugas on 15 June 1957, deprived me of two extremely precious supports that were indeed *irreplaceable*.

*Without the premature loss of Pierre Ricard, I am certain that I would have been able to continue my experiments during 1960.*

## *G.4 The cost of the experiments*

From 1 January 1954 to 30 June 1960, i.e. for around six and a half years, the expenses needed for the experiments were relatively great.

In *francs of the time*,[1] the expenses were approximately as follows:[2] 1954-1957: 5 millions per year; 1958: 8 millions;[3] 1959-1960: 6 millions. In total from 1954 to 1960, the implementation of the experiments cost around 34 millions,[4], [5] of which about 50% was expenses for personnel.[6]

---

[1] *Approximately, one million 1958 francs represents a hundred thousand 1996 francs* (I remind the reader that the conversion from old francs to new francs in January 1959 was performed on the basis of one new franc for a hundred old francs).

*In the following, the expenses from January 1959 to June 1960 are evaluated in old francs.*

[2] The remuneration of my two collaborators Jacques Bourgeot and Annie Rolland represented approximately two millions per year. The overtime payments to workers at IRSID who helped with the continuous experiments represented around 300,000 francs for each month-long series.

[3] Expenses corresponding to the two laboratories at Saint-Germain and Bougival.

[4] In this total of 34 millions the personnel expenses represented around 17 millions, of which about 13 millions was for my two collaborators and about 4 millions for the remuneration of other observers during the continuous observational series and for payment for mathematical work by some of my students at the *Ecole des Mines* (School of Mines) and the Statistical Institute of the University of Paris.

[5] From 1 October 1953 to 31 December 1956, and from 1 January 1959 to 30 June 1960, all these expenses were financed by the *Institut de Recherche de la Sidérurgie* (Institute for Iron and Steel Research - IRSID).

The *Comité d'Action Scientifique de la Défense Nationale* (National Defense Committee for Scientific Action - CASDN) shouldered the expenses from 1 January 1957 to 31 December 1958.

The *Centre National de la Recherche Scientifique* (National Center for Scientific Research - CNRS) contributed 3,500,000 francs to the expenses in 1958, and 2,500,000 francs to the expenses in 1959.

[6] The above estimates of the cost of the experiments are taken from my four *Notes*: that of September 1955, *Note sommaire sur les mouvements du pendule conique* (Short note on the movements of the conical pendulum) (18 p.), p. 6; that of 10 November 1956, *Note sommaire sur les travaux expérimentaux et théoriques effectués du 1er octobre 1953 au 1er octobre 1956 dans le cadre de l'Institut de Recherche de la Sidérurgie* (Short note on the experimental and theoretical work performed from 1 October 1953 to 1 October 1956 under the aegis of the Institute for Iron and Steel Research) (12 p.), pp. 6-7; that of 2 December 1957, *Recherches sur les mouvements du pendule paraconique* (Research upon the movements of the paraconical pendulum) (6 p.), p. 4; and that of 15 April 1959, *Note sur l'état actuel des recherches et le financement des travaux* (Note upon the current state of research and the financing of the work) (3 p.), pp. 2-3.

In total, the expenditure of 34 millions from 1 January 1954 to 30 June 1960 was financed as follows:

| | | |
|---|---|---|
| IRSID | 18 million | 52.9 % |
| CASDN | 10 million | 29.4 % |
| CNRS | 6 million | 17.7 % |
| | | 100.0 % |

It proved harder and harder to finance these expenses. The premature death of Pierre Ricard on 4 April 1956 deprived me of *absolutely essential* help; from the end of 1956 pressure at IRSID for terminating support increased, and it was the *Comité d'Action Scientifique de la Défense Nationale* (National Defense Committee for Scientific Action), presided over by General Bergeron and afterwards by General Guérin, that, from 1 January 1957 to 31 December 1958, provided the bulk of the finance for my experiments.

*However high the expenses for my experiments may have been, their cost-effectiveness was very high in terms of the scientific stakes at issue.*[7]

At the beginning of 1959, *and precisely because of the total success of the crucial experiments of July 1958*, the National Defense Committee for Scientific Activity (CASDN) *considered that the effective and incontestable existence of the phenomenon previously under discussion* had been conclusively proved, that subse-

[7] In my Memoir of October 1956 (10 p.) addressed to CNRS, I wrote:

> *"A new phenomenon has been demonstrated, and it is undeniable that, from the point of view of our theoretical conceptions of the physical world, its consequences may be very considerable...*
>
> *"In the past, the systematic study of new phenomena has proved to be of extraordinarily fecund...*
>
> *"We cannot too definitely underline that pursuit of these researches does not currently pose any risk. The existence of the phenomenon being studied is effectively certain, and the order of magnitude of the difference between the observed effects and the calculated effects is no less certain...*
>
> *"Considering everything, the total cost of the finances we are asking for is relatively modest as compared with the results that can be anticipated. As in all cases in which a new phenomenon is concerned, the marginal cost-effectiveness of the expenditure that can be made is certainly very great, and at the present time, in our opinion, there is no possible comparison between the scientific profitability of 25 millions dedicated to the study of the movements of the paraconical pendulum and the scientific profitability of a similar marginal expenditure in the atomic field."*

quently their own action should be considered as completed, and that it was proper for CNRS to take charge of further pursuit of the experiments.[8]

[8] In my Memoir of 15 April 1959, I wrote to CNRS:

*"In the current state of the discussion, the general advice of all the experts consulted is that my experimental researches should be continued...*

*"Up till now my research has been financed:*

> *a) by the* Institut de Recherche de la Sidérurgie *(Institute of Iron and Steel Research), from 1 October 1953 to 1 January 1957;*
>
> *b) by the* Comité d'Action Scientifique de la Défense Nationale *(National Defense Committee for Scientific Action), from 1 January 1957 to 1 January 1959; and*
>
> *c) by CNRS in 1958 as complement, for a total sum of 3,500,000 francs.*

*"The involvement of IRSID was a start-up involvement that was only possible thanks to the breadth of vision of Mr. Ricard, now deceased. This involvement could not be continued, because it was outside the normal field of activity of IRSID.*

*"The involvement of CASDN was only on a temporary basis. In January 1956 CNRS refused me all support, and, faced with that situation, CASDN became involved on the condition that its help would be limited to financing the research needed for providing an incontestable proof of the existence of the phenomenon.*

*"This proof was provided by my experiments of June 1958 that were carried out simultaneously in two laboratories at Saint-Germain and at Bougival 6 km away, the second being located in a subterranean gallery with more than 60 meters of soil cover. These experiments indeed showed that the phenomena observed in the two laboratories had comparable periodic structures.*

*"This proof having been obtained, CASDN considers that its action has been completed.*

*"From this point in time, only CNRS has the possibility of taking charge of the financing for continuation of my experiments."*

In response to my request for support, CNRS indicated to me on 27 June 1960 that it could not take charge of financing my researches, and it directed me to a different institution.[9] With regard to the shutdown of the IRSID laboratory on 30 June 1960, it was only too evident that, without a scientific guarantee from a competent Commission chosen from both the Academy of Sciences and the *Centre National de la Recherche Scientifique* (National Center for Scientific Research), recourse to a new procedure for financing would have been totally unrealistic.[10]

---

[9] In his letter of 27 June 1960, the Director-general of CNRS wrote to me:

> *"The commission on general mechanics to which I submitted your request has determined that this problem, which would require very great expenditures, is beyond their means, and suggests that you should apply to the* Comité Interministériel de la Recherche Scientifique *(Interministerial Commitee for Scientific Research), 68 rue de Bellechasse (Paris 7)."*

[10] It is certain that a favorable scientific opinion would have made it possible for me to find the necessary finances in industry.

Certainly one can well understand that CNRS might not have had available all or even a part of the financial means needed for continuing the research on the paraconical pendulum and the optical sightings at marks, in one form or another, at IRSID or elsewhere, but it should have presented a *well-founded scientific analysis* of the interest of that research.

If it had been opposed, it should not have sheltered itself behind the pretext of the magnitude of the financial means involved and its incapacity to contemplate them; it should have *explicitly* justified its unfavorable opinion.

In any case, and as several members of the Academy of Sciences and the Mechanics Commission of CNRS expressly requested, CNRS should have set up a Commission charged with expressing an opinion after having heard all relevant submissions, including mine. In the forum provided by such a Commission, its different members would have been able to explain their respective positions *explicitly and in a properly based manner.*[11]

[11] In fact, and in view of the *intensive* discussions that never ceased to develop since the start of 1957, this would have been a relatively easy task (see §3 below, p. 228-231).

## *G.5 Dogmatic oppositions*

### Multiple objections and incessant propagation of rumors

1 - *In order to clear the path for obtaining the necessary resources for pursuit of my experiments, I was compelled incessantly to underline their very great scientific interest, which was very obvious after 1956.*[1]

I was constantly and strongly supported by certain members of the Academy of Sciences. But, equally, I was never freed from the necessity of facing multiple objections - some of them completely unfounded - and very powerful dogmatic opposition, more or less explicit, and from the necessity of struggling with the incessant propagation of rumors that cast doubt upon the validity of my experiments and of my results.[2,3]

---

[1] In my Memoir of 10 November 1956 I particularly wrote:

> *"The calling into question of the currently accepted theory of gravitation clearly presents considerable interest.*
>
> *"Without in any way prejudging the explanation which finally will be given of this phenomenon, here we are undoubtedly confronted with a new and unexplained phenomenon, and the entire history of physics is a witness to the interest that such a phenomenon can present.*
>
> *"This interest is all the greater because, up till now, the history of celestial mechanics has been only a story of stunning success, with any anomalies detected being generally only a negligible fraction of the phenomena...*
>
> *"A new phenomenon has been brought to light, and it is undeniable that, from the point of view of our theoretical conceptions of the physical world, the consequences may be very considerable..."*
>
> *"In the past, the systematic study of new phenomena has been extraordinarily fecund."*

[2] See for example §E.6 above, pp. 188-192.

My opponents gave continual proof both of passive credulity in relation to established truths and also aggressive skepticism as far as my own work was concerned.

As I wrote in my memoir of 1958, *Should the Laws of Gravitation be Reconsidered?* (p. 104) [23]:

> *"I am struck by the fact that my adversaries are only in agreement upon one point - notably that I am wrong. But this agreement is only general, and if my adversaries actually explain their points of view, they cannot fail to notice that their positions are mutually contradictory. To take only one example, one opponent considers that the existence of a lunisolar phenomenon is beyond doubt, but that this phenomenon is easily explicable within the framework of current theory. Another accepts that the observed periodic structure, if real, is absolutely inexplicable, but disputes its reality. The greatest service that my adversaries could render me would be to get together and produce a common refutation of the theories I have advanced. I strongly doubt that they could cooperate upon such a common refutation."*

[3] A recent letter sent to the editor of "*La Jaune et la Rouge*" on 24 September 1996 as a response to my article of September 1996, "*Les expériences de Dayton C. Miller 1925-1926 et la Théorie de la relativité*" ("The Experiments of Dayton C. Miller in 1925-1926 and the Theory of Relativity") [55] shows that *these rumors are still very much alive and continue to be spread.*

In fact, it was absolutely impossible for me to answer arguments *that were never put in precise writing*. It was indeed very easy for certain people to spread rumors in corridors that my results were not based upon anything real, or that the interpretation I was putting on them was pure fantasy. *But they took very good care not to formulate this point of view in a clear manner, unequivocally and publicly, because this would have given me the chance to answer.*

## My work and the theory of relativity

2 - It is undeniable that my work has been interpreted as incompatible with the theory of relativity.

• In December 1957 I sent my *Note* of 4 November 1957 on the interpretation of the constancy of the speed of light[4] to Louis de Broglie, and on 24 April 1958 I made a request to him for this *Note* to be published in the Proceedings of the Academy of Sciences.

This request was refused by the two perpetual secretaries Robert Courrier and Louis de Broglie in their letter of 5 May 1958 for the following reasons:

> *"This note is founded upon the very controversial work of M. Hély and does not seem to be compatible with the well-established principles of the theory of relativity, and does not seem to us to be suitable as the object of such a presentation."*

---

[4] See *Introduction* above, §B.3.3, note 3, p. 55, and *Chapter VI* below, §C.1, pp. 510-514.
This Note generalized Hély's equation

(1) $\Delta\varphi - \frac{1}{c^2}\frac{\partial^2\varphi}{\partial t^2} - 2\frac{k}{c}\frac{\partial\varphi}{\partial t} - k^2\varphi + 4\pi k\delta = 0$

to the case of anisotropy of space.

This text was at the very least astounding, because, as I had indicated in my letter of 24 April: "*The results described in my Note are unarguably new, and their mathematical exactness is beyond doubt.*"[5]

• In the *Summary* attached to the invitation to my Conference of 22 February 1958, I wrote:

> *"The abnormal lunisolar components in the movement of the paraconical pendulum seem to be connected with difficulties or anomalies encountered in the study of numerous phenomena in mechanics, optics, and electromagnetism, and tending to demonstrate the existence of a certain anisotropy of space (notably the experiments of Michelson, Morley, and Miller). If this connection could be confirmed, it would obviously be of considerable importance, because it would give indirect support to Miller's work and would tend to confirm its validity."*

In view of the importance of the result of Michelson's experiment - claimed to be "*negative*" - as far as the very foundations of the theory of relativity are concerned,[6] my text, as it related to the experiments of Miller, can only have reinforced the hostility towards me of all the members of the Academy of Sciences who were convinced of the total validity of the theory of relativity.[7]

**Non-scientific positions**

3 - In all the discussions that took place from 1956 onwards in the heart of the Academy of Sciences and CNRS, it was *absolutely understandable*, and *absolutely justified*, that people should have been *initially* skeptical in the face of experiments which cast doubt upon the theory of gravitation that had been ceaselessly verified for centuries.

---

[5] The position of Louis de Broglie was in fact *even more amazing*, because it was *in flagrant contradiction* to the quotation from one of his works that I have placed at the head of this volume before the *Summary* (p. 7 above).

In fact, Hély's work that I have utilized *reduces to a pure mathematical theorem whose validity is incontestable*.

[6] See *Chapters IV and VII* below. Also see my article of August-September 1996 in the magazine of the Polytechnique, La Jaune et la Rouge, "*Les Expériences de Dayton C. Miller 1925-1926 et la Théorie de la Relativité*" ("The experiences of Dayton C. Miller, 1925-1926, and the Theory of Relativity") [55]. This article, and my responses to the *Reader's letters*, are reproduced in *Appendix H* of the *Second Volume* of this work (p. 31 above).

[7] See also, notably, note 1 of §G.2.1 above, p. 215.

It was *completely understandable* and *completely legitimate* that the responsible scientific authorities should have refused to finance costly experiments, as long as some uncertainties did remain as to the reality of the observed anomalies.

But it was *utterly inadmissible* for them to deny everything en bloc and *a priori*, and for them to refuse to examine the analyses I presented by basing their attitudes on preconceived judgments, prejudices, and "*established truths*".

With regard to my experimental results, I had brought down all the discussions to three fundamental questions: - the first, *do the demonstrated diurnal lunisolar periodicities really exist?*; the second - *are these periodicities and their amplitudes explicable within the framework of the currently accepted theory of gravitation?*; - the third, *can these periodicities be explained from other known phenomena?*[8]

In fact, nobody was capable of presenting any valid objections to my three completely affirmative answers to these three fundamental questions: - the periodicities in question really exist; - they are totally inexplicable within the framework of the currently accepted theory of gravitation; - they cannot be reduced to other known phenomena.

The stronger were the peremptory *en bloc* denials of the validity of my experiments and the more they were incessantly repeated, the more did my adversaries reveal themselves incapable of presenting precise, well-founded, and public objections to these three fundamental questions.[9]

What would have been reasonable, would have been to examine the validity or invalidity of my arguments for my three answers in an objective manner and in the framework of dissenting discussions. But at no time was I given any chance to be heard.

---

[8] *Section B* above, pp. 102-141.

[9] One sole exception: the note of Jean Goguel on 21 April 1958 to the Academy of Sciences [152], which was immediately refuted by the crucial experiments of July 1958 (see §C.2.3 above, note 7, p. 148).

In 1959 the Mechanics Commission of CNRS proposed to the Directorate of CNRS to set up a Commission composed of independent and competent persons. For me, such a Commission would have had the immense advantage of being a tribunal that could finally hear me and pass a judgment as to the validity or invalidity of my experiments according to strictly scientific criteria.

Jean Coulomb, the director-general of CNRS, sent me a letter on 13 May 1959 informing me of the establishment of this Commission.[10] But finally, as far as I know and for reasons of which I am ignorant, this commission was not set up by the CNRS Directorate.[11]

---

[10] Here is the text of Jean Coulomb's letter of 13 May 1959:

> *"My Dear Colleague,*
>
> *"The Mechanics Commission has examined your requests, but has not thought it proper to take an immediate decision. They ask to be further informed by a Commission of specialists.*
>
> *"Monsieur Peres has given me the list of members of this Commission, and I will take the necessary steps to call them together.*
>
> *"Like yourself, I regret these delays, and ask you to believe, my Dear Colleague, in my devoted sentiments."*

[11] It seems best to me to reproduce here the passage of my Conference of 7 November 1959 in which I gave some suggestions regarding the contemplated establishment of this Commission for information, the only information on the subject that ever reached me:

> *"After months and months of discussions, and following the Report of one of the members of the Mechanics Commission of CNRS who is present here, and whom I thank for having been so kind as to attend, the Mechanics commission has proposed to the CNRS Directorate the establishment of a Commission formed of independent persons, and in particular of members of the Academy of Sciences who are interested in this matter.*
>
> *"The Mechanics Commission of CNRS is composed of competent people. It recognizes the great difficulty of the subject...*
>
> *"I came back from the United States a few days before the meeting of the Directorate. I met with two persons whose weight was decisive in the matter. One of them who holds a high post in CNRS said to me: "I have always been a supporter of this Commission". He sent me a friendly letter in May. The other occupies an elevated situation in the University. Both of them told me, the day before the meeting of the Directorate: "We are going to set up this Commission, and the only question for discussion is to know who will participate."*
>
> *"The next day a negative decision was taken. A credit of 2,500,000 francs was granted to me, and was understood to be the last financial assistance from CNRS."*

In fact, as I underlined in my Conference of 7 November 1959:

*"The non-establishment of this Commission is somewhat strange. It is equivalent to a refusal to inform oneself. What can one think of a Tribunal which judges without appeal, without application, and without instruction, and which refuses all inquiries?*

*"I strongly demand that this Commission should be set up, so that I can be heard and judged. The denial of justice is the hardest thing to bear, and here I protest publicly against an attitude that seems to me to be unscientific, namely the deliberate refusal to seek for information."*

In any case, it was at least abnormal that Notes *presented by members of the Academy of Sciences* could be purely and simply refused, *without any reasons being explicitly stated.*

## G.6 A scientifically incomprehensible decision. The total cessation of my experiments after the full success of the crucial experiments of July 1958

At the end of 1959, for lack of any financial support, the decision was taken to close my laboratory at IRSID on 30 June 1960.

*With the passage of time, this decision today appears scientifically incomprehensible, and really completely inadmissible.*

A new phenomenon had been brought to light. In order to test its validity, a decision was taken to perform two crucial experiments. *These crucial experiments confirmed in a stunning manner the existence of the observed anomalies, totally inexplicable in the framework of accepted theories.*[1], [2]

---

[1] *No competent person ever cast any doubt upon the impossibility of explaining the amplitude of the observed periodicities, in particular of the periodicity of 24h50m, within the framework of accepted theories.*

To take only one example, during his visit to my laboratory at IRSID in November 1956, Joseph Péres, a member of the Academy of Sciences, expressed *his total agreement on that impossibility.* As for the real existence of the detected periodicities, he was very strongly impressed by the results obtained (see the account of this visit in my *Note* of 30 November 1956, *Note sur les mouvement du pendule paraconique pour M. Caquot* (Note for Mr. Caquot on the movement of the paraconical pendulum), 3 pages).

Moreover, *at no time did any specialist in the theory of gravitation* ever controvert *the order of magnitude calculation* of the amplitudes of the periodic lunisolar components which I published in my *Note* of 16 December 1957 to the Academy of Sciences, *Théorie du pendule paraconique et influence lunisolaire* (Theory of the paraconical pendulum and the lunisolar influence) [17], respectively represented for the Moon and the Sun, in CGS units, by the coefficients

$$C_l = g\frac{M_l}{M_t}\frac{r_e^2}{d_l^3} = 0.862 \times 10^{-13} \qquad C_s = g\frac{M_s}{M_t}\frac{r_t^2}{d_s^3} = 0.396 \times 10^{-13}$$

and my adversaries certainly did not lack any desire to contradict me.

[2] In his letter of May 1959 to Wernher von Braun, director of the National Aeronautics and Space Administration, General Paul Bergeron, ex-President of the *Comité d'Action Scientifique de la Défense Nationale* (National Defense Committee for Scientific Action), wrote:

> *"Before writing to you, I considered it to be necessary to visit the two laboratories of Professor Allais (one of which was located sixty meters underground), accompanied by eminent specialists - two of whom were professors at the Polytechnic School. During a discussion that lasted several hours, no significant cause of error could be found, nor any attempt at explanation that resisted analysis.*
>
> *"I also think I should tell you that, during the last two years, more than ten members of the Academy of Sciences and more than thirty eminent persons, gravitational specialists in various capacities, have come to visit either his laboratory at Saint-Germain, or his underground laboratory at Bougival.*
>
> *"Profound discussions have taken place, not only on these occasions, but also several times in various scientific forums, notably at the Academy of Sciences and at the* Centre National de la Recherche Scientifique *(National Center for Scientific Research). None of these discussions was able to evolve any explanation in the framework of currently accepted theories."*

These crucial experiments had swept away all the objections that had been presented. *New perspectives were undeniably opened, because it was definitively established that some of the fundamental postulates of the theory of gravitation had been invalidated by the experimental data.*

However, instead of my researches being carried further, a decision was taken to stop everything. Instead of continuing with experiments that *from now on would be risk-free* because the existence of the phenomenon had been established, *the responsible scientific authorities stopped providing any further financing*.

How can *such an absurd* decision be explained, considering that, during that same year of 1959, on the recommendation of Wernher von Braun, director of the National Aeronautics and Space Administration, the *Institute of Aeronautical Sciences* had decided to translate my memoir of 1958, "*Should the Laws of Gravitation be Reconsidered?*" and to publish the English version in its journal *Aerospace Engineering* [35, 37, 36]? And considering that, in that same year, the American *Gravity Research Foundation* had awarded me a prize for my experiments with the paraconical pendulum, and that in France I had received the *Galabert Prize* of the French Society of Astronautics!

*How can such a decision be explained, in view of the fact that eminent members of the Academy of Sciences considered that my research should be continued, and that they never ceased to give me their support?*[3]

---

[3] In my *Note* to CNRS of 15 April 1959, *Note sur l'état actuel des recherches et le financement des travaux du Professeur Allais* (Note on the present state of research and the financing of the work of Professor Allais), I wrote:

> *"Messrs. Caquot, Cot, Darrieus, Kampé de Fériet, Léauté, Pérard, Roy, Tardi, and Thiry, members of the Academy of Sciences, all think that my research should be continued. The same holds for very numerous scientific persons who have visited my laboratories, of whom some will undoubtedly be members of the Academy of Sciences in the future."*

*Certainly the reason is the blind domination in all epochs of obscure and fanatic forces, always active and effective and always incompetent,*[4] *striving to ensure the domination of "established truths" and opposing themselves to the progress of science.*[5, 6]

This opposition to all progress nowadays probably continues as strong as it has been in the past,[7] and I do not have any illusions as to how the present work will be greeted today, so intolerable does any expression of dissent appear as soon as it collides with "*established truths*".[8, 9]

---

[4] In his letter of 1 June 1960 (§E.6.9, note 10, p. 192 above), already quoted, Jean Leray affirmed magisterially:

> *"By the way, modern Foucault pendulums function without friction, without any linkage; these are the artificial satellites."*

Such an assertion certainly deserves to be prominently displayed in an anthology of idiocies of the mandarins of science.

One really can ask by what aberration such an "authority" could be elected to the *Mechanics Section* of the Academy of Sciences, a field in which he obviously was utterly incompetent.

[5] See *Chapter IX* below, pp. 659-674.

On my experiments, Louis Rougier wrote in July 1959 (*Scandal at Polytechnique*) [236]:

> *"What one detects in him [Allais] are revolutionary ideas, raising questions about postulates that are considered as dogmas by certain members of the Academy of Sciences."*

Louis Rougier's memoir is reproduced in *Annex I D* of the *Second Volume* of this work.

[6] The entombment of my experiments on the paraconical pendulum with anisotropic support has been very similar to the burying of the experiments of Miller in 1925-1926 and of his memoir of 1933 (*Chapter IV* below), which I discussed in my memoir of 1958, *Should the Laws of Gravitation be Reconsidered?* (p. 102, note 38) [23]:

> *"The pure and simple burying of Miller's memoir* (of 1933) *seems to me to be one of the scandals of contemporary physics."*

[7] As far as concerns myself, rumors in 1959 about my experiments on the paraconical pendulum and tending to discredit them were again circulated in the very heart of the Academy of Sciences.

[8] On these dogmatic and blind positions, see *Chapter IX* below, pp. 659-674, and *Chapter X*, pp. 685-689.

[9] Today it is *difficult to imagine* the climate that prevailed in 1959.

Before my Conference of 7 November 1959 at the Society of Civil Engineers of France, authoritative and well-wishing persons whose judgment was sound said to me: *"If you tell the truth, you are lost; you will excite hostilities against you which will never be neutralized."*

(My Conference of 7 November 1959).

Certainly, there are *innumerable* examples of errors committed by scientific authorities in all epochs, but in the case of my experiments on the paraconical pendulum which were crowned *in a spectacular manner* by the crucial experiments of July 1958, to deny the evidence - *decisive and stunning evidence* - was and remains particularly shocking.

Today, after the roughly thirty-eight years that have elapsed from the publication in 1958 by *Perspectives X* of my memoir "*Doit-on reconsidérer les lois de la Gravitation* ?" ("Should the Laws of Gravitation be Reconsidered?") [23], I can only confirm what I wrote at the end of that memoir, *before I had performed the crucial experiments of July 1958:*[10]

> *"I advance my conclusions in the full knowledge of the objections which have been raised in an explicit and precise manner, and which I have subjected to deep examination. At this time, it has not seemed to me that any of these objections can stand.*
>
> *"I perfectly understand that the facts that I insist upon and the interpretation that I put upon them arouse,* a priori, *doubt and skepticism. I understand even better all the reservations that people have as to my results and conclusions, since I ceaselessly addressed them to myself for three years, from 1953 to 1956. During those three years I systematically avoided making any report on my results, even when they were particularly striking, like those relating to the solar eclipse of 30 June 1954.*
>
> *"I am convinced that skepticism is the only scientific position one should adopt when new results, resulting from limited experimentation, tend to cast into question the validity of principles that seem to have been constantly confirmed by innumerable previous observations.*

[10] p. 104. My comments at the start of 1958 bear witness to the hostility with which I had been confronted after the publication of my first *Notes* to the Academy of Sciences in 1957.

> *"But, while it is scientific to adopt a prudent and skeptical attitude, it is not scientific to condemn without a hearing. Neither is it scientific to condemn en bloc without saying with exactly which point one does not agree...*
>
> *"Dogmatism and sectarianism are not scientific principles. It is the facts and the facts alone that should determine theory, and not the reverse."*

In any case, I am not a man to bow before ignorance and fanaticism, and here I can only heed the admonition of Auguste Lumière:[11]

> *"No matter what be the arguments and proofs upon which something really new is based, it is never accepted until after the passage of a very long time, often twenty or thirty years."*
>
> *"Everyone who has prematurely abandoned the struggle for his ideas has seen his conceptions sink into oblivion."*

---

[11] Auguste Lumière, 1942, *Les Fossoyeurs du Progrès. Les Mandarins contre les Pionniers de la Science* (The Gravediggers of Progress. The Mandarins against the Pioneers of Science), p. 347 [188].

# Chapter II
# MY EXPERIMENTS UPON THE PARACONICAL PENDULUM WITH ISOTROPIC SUPPORT 1959-1960

## *Determination of the direction of anisotropy of space*

*It is always the concrete phenomenon that decides if a theory should be accepted or rejected. There is, and there can be, no other criterion of the truth of a theory, except its agreement more or less perfectly with concrete phenomena.*
Vilfredo Pareto *

# A THE DETERMINATION OF THE DIRECTION OF ANISOTROPY OF SPACE

## *A.1 My experiments upon the paraconical pendulum with the anisotropic support*

During five years, from 1954 to 1958, I kept using the same type of suspension in spite of its very marked anisotropy. The reason was very simple. Before anything else, I had to become certain as to the reality of the observed diurnal periodicities, and, for that, it was essential to have series of observations performed *under completely comparable conditions.*

After the two *crucial* experiments of June-July 1958 at Bougival and Saint-Germain had been performed, *the proof had been definitively provided of the existence of a new phenomenon, completely inexplicable in the framework of the classical Newtonian theory of gravitation, whether or not complemented by the corrections of the theory of relativity.*

It had thus been definitively established that the classical theory of gravitation (resulting from the application of the twin principles of inertia and of universal gravitation, in the framework of the current theory of relative movements to a Galilean space), whether or not completed by the corrections of the theory of relativity, leads to orders of magnitude for the lunisolar action *absolutely undetectable* on the experimental level, *of the order of twenty million times less than the observed effects.*

As a consequence it became possible, indeed desirable, even *necessary*, to perform experiments with a support as isotropic as possible.

---

* Vilfredo Pareto, 1909, *Manuel d'Economie Politique* (Manual of Political Economy), Giard, 1927, p. 16 [213].

## A.2 *Determination of the direction of anisotropy of space. A fundamental step*

### The limit plane

1 - *The limit plane*, the existence of which was demonstrated in my experiments on the paraconical pendulum with anisotropic support, is the result of the *combined* action of the average direction of astronomical influences and the direction of anisotropy of the support.[1]

However, while it was possible to determine the influence of the anisotropy of the support with high accuracy,[2] determination at each instant of the average astronomical influences in direction and in magnitude remained impossible.

By eliminating the influence of the anisotropy of the support, the implementation of an isotropic support *made it possible to determine the anisotropy of space in direction and in magnitude*. That was the objective of the experiments I performed in 1959-1960.

### Analysis of the deviations $\Delta$ of the plane of oscillation of the paraconical pendulum from determinate azimuths

2 - The method that I followed[3] was based upon the preliminary determination of the deviation of the plane of oscillation of the paraconical pendulum from determinate azimuths.

By itself, analysis of these deviations provided essential data on the periodic structure of the movement of the paraconical pendulum with isotropic support.[4]

---

[1] *Chapter I*, §E.7 above.

[2] *Chapter I*, §E.3 above.

[3] Section C below.

[4] Section F below.

### A fundamental step

3 - In fact, the determination of the direction of the anisotropy of space was *an absolutely fundamental step* in my experiments upon the paraconical pendulum.[5] I unfortunately was not able to continue with this line of research, due to the shutdown of my laboratory in June 1960.[6]

### The anisotropy of space

4 - The anisotropy of space is *a fact of experiment, absolutely independent of any hypothesis. It is an inescapable fact of experiment.*

It is however possible to give it an interpretation by formulating a hypothesis: *the hypothesis of an anisotropy of inertial space.*[7] This is the reason why, in the following, my discussion will examine this hypothesis and its implications for the movements of the paraconical pendulum with isotropic support.[8]

---

[5] In my Note of 15 November 1959, *Etat des recherches du professeur Allais au 15 novembre 1959* (State of the research of Professor Allais on 15 November 1959) , I wrote (pp. 2-3):

> *"An isotropic suspension for the paraconical pendulum has been constructed, and an experimental technique has been defined that makes it possible to demonstrate, at each instant, the directions of anisotropy in space and to localize them in a precise manner...*
>
> *"In the event that these directions of anisotropy prove to be the same as those already found by Miller with his Michelson-type interferometer, which is at least possible if not probable, then Miller's results will be confirmed, and by the same token the foundations of the theory of relativity must be considered as contradicted by experiment. The scientific importance of such a prospect cannot escape anyone.*
>
> *"In any case, even if this does not prove to be the case, the implementation of this isotropic support and the definition of an experimental technique that enables it to be used most profitably will constitute the greatest advance made since the start of my research on the movement of the paraconical pendulum in 1953.*
>
> *"Under any hypothesis, as far as can be judged at the moment, it will be possible to obtain significant results very rapidly if the activity of the laboratory can be maintained during the next eighteen months."*

[6] *Chapter I*, §G.2, 4 and 5 above.

In fact, I was able to mention the very first results obtained with the isotropic suspension in my *Conference of 7 November 1959* (Section B, and §1 and §2 of Section C below ).

[7] *Chapter I*, Section F.

[8] Section I below.

# B THE IMPLEMENTATION OF AN ISOTROPIC SUPPORT

## *B.1 Experimental arrangement*

### The objective

1 - During the year 1959 I was able to implement an *approximately isotropic*[1] support, *i.e.* one which had no effect of convergence to a specific direction. My objective was to eliminate, as much as possible, any anisotropy derived from the support arrangements.[2]

It was necessary, not only to determine the direction of anisotropy of space at each instant, but also to examine the influence of the suppression of the anisotropy of the support on the periodic anomalies detected previously.

### The support

2 - The support which was implemented was very massive. A square steel plate *S''* 130 cm on a side and 2 cm thick, weighing 230 kg, was squeezed against the ceiling by a supporting bridge. Three cylinders (20 cm in diameter, 30 cm high, and weighing 65 kg) supported a triangular plate weighing 150 kg, in which a cutaway circular support *S'* of 34 cm diameter and weighing 6.5 kg *was able to turn.*[3]

---

[1] Allais: - *Pendule Paraconique. Réalisation d'un Support aussi Isotrope que Possible* (Paraconical Pendulum. Implementation of a Support, being as Isotropic as Possible), IRSID, 5 November 1959, 10 p. [34]; - *Pendule Paraconique à Suspension Isotrope. Détermination des Variations au Cours du Temps des Caractéristiques de la Corrélation du Mouvement avec l'Azimut* (Paraconical Pendulum with Isotropic Support. Determination of the Variations over Time of the Characteristics of the Correlation of its Movement with its Azimuth), IRSID, 10 November 1959, 7 p. [33]

[2] My objective was also to eliminate any anisotropy deriving from the suspension arrangements of the paraconical pendulum, due to the stirrup associated with the appendix (*Chapter I*, §A.1.1 above).

[3] This cutaway support was analogous in conception to the cutaway support of the anisotropic suspension (*Chapter I*, §A.1.1 above).

*During each elementary experiment, the appendix of the cutaway support was oriented in the direction of the initial plane of oscillation.* Thus, the influence of the appendix remained the same, whatever was the azimuth of the initial plane of oscillation, at least to the first approximation, and the pendulum could oscillate in all directions.[4]

The horizontality of the bearing surface *S* which carried the ball was maintained within less than about 10 sexagesimal seconds, the effect of its inclination being utterly negligible.[5]

---

[4] Whereas beforehand the plane of oscillation of the pendulum could only shift through a total angle of 210 grads (*Chapter I*, §A.1.1 above).

[5] See *Chapter I* above, §B.3.3.

## *B.2 A support having practically no anisotropy*

The measurement of the anisotropy of the support was performed *by following the same method* as for the paraconical pendulum with anisotropic support.[1]

*In fact, the observed anisotropy results from two elements: the anisotropy of the support and the average anisotropy of space.*

### Measuring the anisotropy of the support. Five series of correlation experiments

1 - *Table I* presents the results obtained from five series of experiments upon correlation with azimuth, performed between 24 September and 16 October 1959. Each series of experiments included 20 elementary 14-minute experiments.

The correlation calculations made it possible to determine the azimuths $\Sigma_1$ and $\Sigma_2$ in the equation

(1) $$\bar{\dot{\phi}} = a_0 + a_1 \sin 2(\phi - \Sigma_1) + a_2 \sin 4(\phi - \Sigma_2)$$

representing the movement of the pendulum in each azimuth during each experiment of 14 minutes.[2]

Overall, the variations of the angles $\Sigma_1$ and $\Sigma_2$ were *of much greater magnitude* than in the previous correlation experiments with azimuth in the case of the anisotropic support, because the angles $\Sigma_1$ and $\Sigma_2$ varied over intervals of 100 and 70 grads respectively during the period considered. The angle $\Sigma$ corresponds to the azimuth for which the sum of the sinusoids corresponding to $\Sigma_1$ and $\Sigma_2$ intersects, from top to bottom, starting from the highest peak, the ordinate axis at a point equal to the average value of this sum.

---

[1] See *Chapter I* above, §E.3.

[2] By the same experiments, I endeavoured to set up the value of the angle $\phi$ of the plane of the disk with respect to the initial plane of oscillation, such that during a 14-minute experiment the Foucault effect of $-2.94$ grads (*i.e.* $-0.21$ grad per minute) should be compensated on average. This angle $\phi$ corresponds to an average nil value of the quantity $a_0$, and is of the order of 4 grads (*Table I*).

It is accordingly the case that the average of the $a_0$ of experiments 2, 3, 4, and 5 of *Table I*, *i.e.* $+0.049$ grad per minute, is much less in absolute value than the value $|a_0| = |-0.104|$ in the first experiment (quite comparable to $-0.127$, the average of the $a_0$ in *Table X* of *Chapter I*, §E.3, p. 180).

I remind the reader that the Foucault effect of $-0.550 \times 10^{-4}$ radian per second corresponds to an effect of $-0.21$ grad per minute (see *Chapter I* above, §E.3.3, p. 178).

The angle $\phi$ only appears to have an appreciable effect upon $a_0$.

*For the five series of experiments,*[3] *the averages of the azimuths* $\Sigma_1$, $\Sigma_2$ *and* $\Sigma$ *determined by the calculation are near the azimuth of the direction South-North.*

### A practically isotropic support

2 - A *very major* difference appears between the results of *Table I* for the new support and those of *Table X* of *Chapter I* which related to the anisotropic support.

Both for the angles $\Sigma$ and for the minor axes $2b$, the second harmonic coefficients $a_2$ and $2b_2$ are *of the same order of magnitude* as the first harmonic coefficients $a_1$ and $2b_1$, while in the case of the anisotropic support the coefficients $a_2$ and $2b_2$ were about *50 times smaller* than the coefficients $a_1$ and $2b_1$. This structure, *entirely different* from that for the anisotropic support, *is completely incompatible* with any effective anisotropy of the new support.

This realization makes us appreciate that the effect of anisotropy of the support can be considered as *practically nil*, at least in the first approximation, and that the measured effects correspond to the variation over time of the anisotropy of space.[4]

### The average anisotropy of space during the experiments performed

3 - If we consider the correlation

(2) $$\bar{\phi} = a_0 + a_1 \sin 2(\phi - \Sigma_1)$$

[3] All performed between 7h20m and 19h40m.

[4] This induction is *entirely confirmed* by analysis of the series of observations of November-December 1959 and March-April 1960 (§D.2 below).

we have for the average of the five series of experiments, in grads per minute (*Table I*)

(3) $$\overline{\dot{\phi}} = 0.018 - 0.051 \sin 2 (\phi - 405.8)$$

*i.e.* in radians per second (keeping $\phi$ in grads)

(4) $$\overline{\dot{\phi}} = 0.471 \times 10^{-5} - 0.133 \times 10^{-4} \sin 2 (\phi - 405.8)$$

*The corresponding anisotropy is about 13 times more feeble than the anisotropy due to the anisotropic suspension.*[5]

*Since I have been led to consider that this anisotropy cannot be attributed to anisotropy of the support, it must correspond to the average anisotropy of space when the experiments were performed.*

[5] $0.677/0.051 = 13.3$ (*Table X* of *Chapter I*, §E.3.3, and *Table I* below).

An analogous calculation can be performed by considering the estimates of the minor axis of the ellipse. We then have $0.174/0.018 = 9.67$. However, the calculation performed from the $\overline{\dot{\phi}}$ is *more precise*. The estimates of the minor axis of the ellipse are, in fact, not as good as the estimates of the azimuths.

*TABLE I*

## *ISOTROPIC SUSPENSION*

CHARACTERISTICS OF THE CORRELATION
OF THE MOVEMENT WITH THE STARTING AZIMUTH
SAINT-GERMAIN

Pendulum used: $P_1$ q = 20 grads p = 2
$\phi$ = azimuth of the disk

| PERIOD OF OBSERVATION | | | $\varphi$ | DISPLACEMENT OF AZIMUTH IN GRADS/MINUTE $\Phi' = a_0 + a_1 \sin 2(\Phi - \Sigma_1) + a_2 \sin 4(\Phi - \Sigma_2)$ | | | | | | MINOR AXIS OF ELLIPSE IN CM AT EXPERIMENT END $2b = 2b_0 + 2b_1 \sin 2(\Phi - \Sigma'_1) + 2b_2 \sin 2(\Phi - \Sigma'_2)$ | | | | | | | | | |
|---|---|---|---|---|---|---|---|---|---|---|---|---|---|---|---|---|---|---|---|
| | | | | $300 < \Sigma_1 < 500$ (400) | | | $350 < \Sigma_2 < 50$ (450) | | | $300 < \Sigma'_1 < 500$ (100) | | | $350 < \Sigma'_2 < 50$ (450) | | | | | | |
| n | DAYS | HOURS | | $a_0$ | $a_1$ | $\Sigma_1$ | $a_2$ | $\Sigma_2^{(2)}$ | $\Sigma'''$ | $2b_0$ | $2b_1$ | $\Sigma'_1$ | $2b_2$ | $\Sigma_2'^{(3)}$ | $\Sigma'''$ | $2b_1/a_1$ | $2b_2/a_2$ | $a_2/a_1$ | $b_2/b_1$ |
| 1 | 24.9.59 | 13h à 19h40 | 0 | -0.104 | -0.040 | 381.88 | -0.077 | 409.81 | 405.0 | +0.013 | -0.011 | 386.68 | -0.016 | 411.60 | 404.0 | +0.275 | +0.208 | +1.925 | +1.455 |
| 2 | 25.9.59 | 8h à 16h | 7 | +0.106 | -0.015 | 386.17 | -0.036 | 403.57 | 400.0 | +0.017 | -0.014 | 356.13 | -0.008 | 414.05 | 335.0 | +0.933 | +0.222 | +2.400 | +0.571 |
| 3 | 28.9.59 | 7h30 à 14h | 5 | +0.037 | -0.090 | 370.25 | -0.164 | 329.66 | 338.0 | +0.006 | -0.034 | 381.57 | -0.034 | 328.99 | 346.0 | +0.378 | +0.207 | +1.822 | +1.000 |
| 4 | 7.10.59 | 7h20 à 15h20 | 4 | +0.001 | -0.040 | 421.17 | -0.059 | 380.42 | 390.0 | +0.000$_5$ | -0.005 | 409.19 | -0.006 | 377.20 | 389.0 | +0.125 | +0.101 | +1.475 | +1.200 |
| 5 | 16.10.59 | 8h à 16h40 | 4 | +0.051 | -0.072 | 469.52 | -0.036 | 462.23 | 465.0 | +0.017 | -0.030 | 478.37 | -0.028 | 456.52 | 465.0 | +0.417 | +0.778 | +0.500 | +0.933 |
| AVERAGES | | | | +0.018 | -0.051 | 405.80 | -0.074 | 397.14 | 399.6 | +0.011 | -0.018 | 402.39 | -0.018 | 397.67 | 387.8 | +0.426 | +0.303 | +1.624 | +1.032 |

(1) $\Sigma$ ($\Sigma'$) = position where the sum of the two sinusoids intersects the ordinate axis at position equal to average value from top to bottom from highest peak
(2) $\Sigma_2$ = determination closest to $\Sigma_1$
(3) $\Sigma'_2$ = determination closest to $\Sigma'_1$

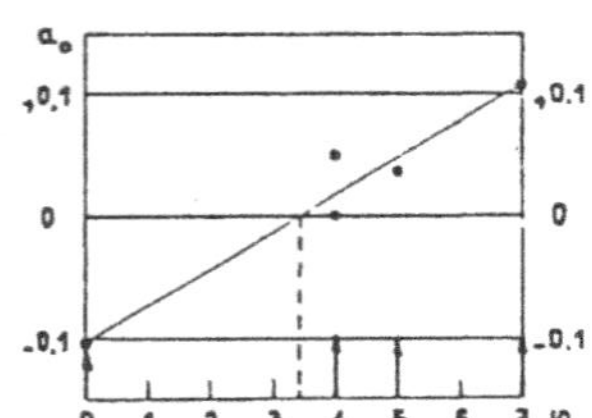


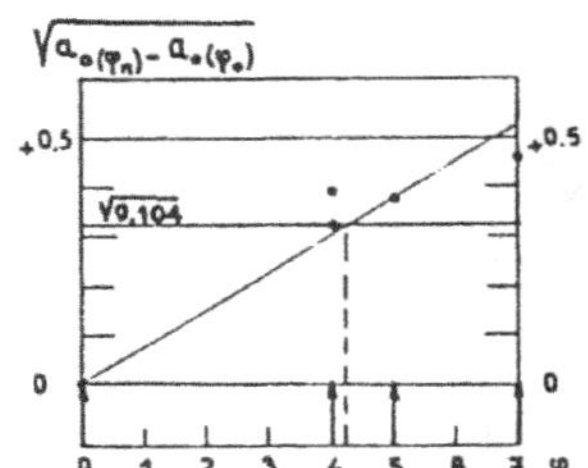


ARRANGEMENT OF SUPPORT

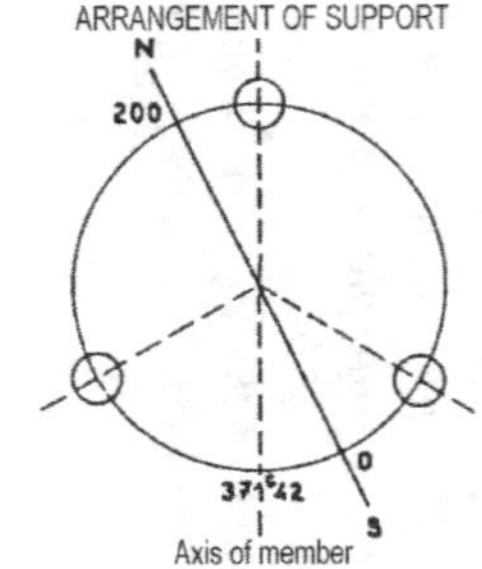


*Source:* Table X A b 1 of my Conference of 7 November 1959, and my *Note* of 25 November 1959, *Implementation of a support as isotropic as possible* (reproduction by photocopy).

# C DETERMINATION OF THE AZIMUTH *X* OF ANISOTROPY OF SPACE METHOD OF ANALYSIS

## *C.1 Two possible methods of analysis: enchained series and mobile correlations*

### Method of enchained series

1 - A series of *enchained* experiments performed with an isotropic support for determining the azimuth of the direction of the anisotropy of space over time could only be considered as being capable of giving good results if the azimuth of the plane of oscillation of the pendulum in such a series of experiments were to shift sufficiently quickly, so as to remain sufficiently close to the equilibrium position which it would have reached if this position of equilibrium remained fixed, *this position of equilibrium corresponding to the direction of anisotropy of space*.

Now, nothing *a priori* allows us to be sure that this will be the case. The result is that, with the technique of enchained series, there is a risk of only obtaining a very distorted representation of the movement of the azimuth of anisotropy of space.[1]

[1] For example, this is like the representation that one would obtain of the position of a cyclist on a circular track, if one could only observe a dog pursuing this cyclist, the dog running at one third of the cyclist's speed when he is 1/8 of a circuit (*i.e.* 50 grads) away from the cyclist, and the dog's running speed being zero when he is on the same diameter as the cyclist or at a right angle to him.

### Method of mobile correlations

2 - In order to overcome this difficulty, I had recourse to the following technique: ten directions for launch are chosen, spaced regularly 20 grads apart, and a first series of ten launches in the ten above directions is performed, one *every 20 minutes*, with the order of the directions being chosen at random; and then a second series of ten launches is performed in the same order; and so on.[2]

For each series of ten experiments, we are thus in a position to determine the azimuth $\Sigma$ of stable equilibrium by a correlation calculation with the azimuth, by considering the ten displacements of the azimuth of the pendulum during each series of ten consecutive 14-minute experiments as a function of the ten starting azimuths.[3]

By this technique, we can detect the average characteristics of the correlation of the movement of the pendulum with the azimuth during a period of 200 minutes = 3h 20m of ten consecutive 20-minute experiments *framing the instant in question*.[4]

*A priori* we may consider that the information yielded by such a technique of exploration will always be quite close to reality, and also will include much more data than that given by the method of enchained series.[5]

---

[2] In grads, these ten azimuths were the following: 240, 340, 300, 380, 280, 220, 360, 200, 320, and 260.

[3] The method is that of §B.2.1 , pp. 243-244 above, used for measuring the anisotropy due to the support.

[4] Each elementary 14-minute experiment accordingly plays a part in ten correlation calculations.

[5] In fact, the information in each series of ten consecutive 20-minute experiments is the result of measurements of the variations of azimuth from ten different azimuths distributed around the entire horizon.

## *C.2 Principle of calculation by the method of mobile correlations*

For simplification, let us consider calculating the first harmonic of equation (1) of §B.2.1 above.

The principle of the calculation is as follows. Let $\Delta_i$ be the variation of the azimuth of the plane of oscillation of the pendulum in the 20-minute experiment of rank $i$. We consider the correlation

$$(1) \qquad \begin{aligned} \Delta_i &= \Delta_i^* + \varepsilon_i \qquad (n-5 \leq i \leq n+4) \\ &= M + P\sin 2(X - A_i) + \varepsilon_i \end{aligned}$$

where $A_i$ represents the starting azimuth, and $X$ the direction of anisotropy of space at that instant.

We have

$$(2) \qquad \Delta_i^* = M + P\cos 2A_i \sin 2X - P\sin 2A_i \cos 2X$$

We calculate

$$(3) \qquad \begin{aligned} M_n &= \tfrac{1}{10}\sum_{n-5}^{n+4} \Delta_i \\ B_n &= \tfrac{1}{10}\sum_{n-5}^{n+4} \Delta_i \cos 2A_i \\ C_n &= \tfrac{1}{10}\sum_{n-5}^{n+4} \Delta_i \sin 2A_i \end{aligned}$$

We then determine the average value $X_n$ of $X$ during the ten launches considered using the equations

$$(4) \qquad \sin 2X_n = B_n/P_n \qquad \cos 2X_n = C_n/P_n$$

$$(5) \qquad P_n = \sqrt{B_n^2 + C_n^2}$$

The angle $2X_n$ is perfectly determined to 400 grads. The angle $X_n$ is therefore determined to *200 grads*. We choose the determination $X_n$ so that the difference $X_n - X_{n-1}$ should be as small as possible.

*In general we observe that $X_n - X_{n-1}$ remains sufficiently small for there to be no indeterminacy for the chosen value of $X_n$.*

The calculation of the second harmonic of the direction of anisotropy is performed in an analogous manner.

### *C.3 Illustration of the method of mobile correlations*

By way of illustration,[1] *Table II* gives the results obtained during a first series of continuous observations performed from 16h40m on 30 October to 19h on 2 November, with an interruption of 13h on 2 November.[2] *Table II* only gives the results of the correlations every two hours.[3]

Since each azimuth value is only determined up to 200 grads by the corresponding correlation, *we pass from each value to the next value by taking, for that next value, the determination of* $\Sigma$ *that is closest to the previous value.*

*Graph I* also gives a representation of the variations over time of the azimuths $\Sigma$, $\Sigma_1$ and $\Sigma_2$, with the azimuths $\Sigma_1$ and $\Sigma_2$ corresponding to the first and second harmonics of the fitting sinusoids, and the azimuth $\Sigma$ corresponding to the position where the sum of the two fitting sinusoids of the first and second order intersects the straight line representing the average value $(\Sigma_1 + \Sigma_2)/2$ of the displacement from top to bottom starting from its highest peak.[4]

One sees that, during a single period of 24h, the variation of these azimuths was greater than 100 grads.

The azimuth $X(t)$ of the anisotropy of space can be identified with the azimuth $\Sigma$ determined by the correlation corresponding to the instant $t$.

---

[1] See my *Note* of 10 November 1959, *Pendule paraconique à suspension isotrope, Détermination au cours du temps des caractéristiques de la corrélation avec le mouvement de l'azimut* (Paraconical pendulum with isotropic support. Determination over time of the characteristics of the correlation with the movement of the azimuth) [33].

[2] This interruption was necessitated by the fatigue of the operators, due to a long series of previous observations.

[3] In this paragraph, for simplification, *I have preserved the notation of my Note of 10 November 1959* (note 1 above). The direction of anisotropy $X$ is denoted by the symbol $\Sigma$. *Table II* and *Graph I* of my Note of 10 November 1959 have been reproduced by photocopying.

[4] Since the two sinusoidal components of $\phi$ have comparable amplitudes, it is not sufficient to limit oneself only to consideration of the first of them.

*TABLE II*

**ISOTROPIC SUSPENSION**

***VARIATION OVER TIME OF THE CHARACTERISTICS OF THE CORRELATION OF THE MOVEMENT WITH THE AZIMUTH***

RATE OF DISPLACEMENT IN AZIMUTH IN GRADS PER MINUTE

$$\dot{\Phi} = a_0 + a_1 \sin 2(\Phi - \Sigma_1) + a_2 \sin 4(\Phi - \Sigma_2)$$

$\Sigma$ = azimuth where the sum of the two sinusoids intersects the average axis from top to bottom, starting from the highest peak

$200 < \Sigma_1 < 400$ Determination of the $\Sigma_2$ that is closest to $\Sigma_1$

| t DAY | t TIME | $a_0$ | $a_1$ | $\Sigma_1$ | $a_2$ | $\Sigma_2$ | $\Sigma$ | $\sqrt{a_1^2+a_2^2}$ | $a_2/a_1$ |
|---|---|---|---|---|---|---|---|---|---|
| 30.10.59 | 20" | _0.087 | _0.087 | 379.24 | _0.182 | 330.87 | 339 | 0.202 | 2.092 |
| | 22 | _0.041 | _0.098 | 341.87 | _0.113 | 325,79 | 330 | 0.150 | 1.153 |
| 31.10.59 | 0" | _0.032 | _0.106 | 329.60 | _0.042 | 304.83 | 321 | 0.115 | 0.396 |
| | 2 | _0.061 | _0.142 | 327.75 | _0.057 | 302,63 | 318 | 0.153 | 0.401 |
| | 4 | _0.149 | _0.167 | 324.45 | _0.053 | 306.89 | 317 | 0.175 | 0.317 |
| | 6 | _0.124 | _0.159 | 336.74 | _0.021 | 290.63 | 335 | 0.160 | 0.132 |
| | 8 | _0.150 | _0.079 | 336.06 | _0.033 | 309.31 | 324 | 0,086 | 0.418 |
| | 10 | _0.090 | _0.053 | 346.75 | _0.067 | 323,35 | 330 | 0.085 | 1.264 |
| | 12 | _0.063 | _0.057 | 332.77 | _0.086 | 312.53 | 318 | 0.103 | 1,509 |
| | 14 | _0.018 | _0.071 | 331.38 | _0.086 | 302.33 | 311 | 0.112 | 1.211 |
| | 16 | _0.040 | _0.080 | 315.16 | _0.072 | 294.65 | 303 | 0.108 | 0.900 |
| | 18 | _0.068 | _0.051 | 385.38 | _0.044 | 383.28 | 385 | 0.067 | 0.863 |
| | 20 | _0.084 | _0.060 | 375.43 | _0.054 | 365.25 | 370 | 0.081 | 0.900 |
| | 22 | _0.085 | _0.102 | 383.14 | _0.043 | 384.55 | 383 | 0.111 | 0.422 |
| 1.11.59 | 0" | _0.074 | _0.125 | 370.91 | _0.014 | 377.08 | 370 | 0.126 | 0.112 |
| | 2 | _0.069 | _0.033 | 365.64 | _0.025 | 333.12 | 345 | 0.041 | 0.758 |
| | 4 | _0.087 | _0.073 | 376.69 | _0.066 | 403.85 | 394 | 0.098 | 0.904 |
| | 6 | _0.083 | _0.064 | 400.55 | _0.099 | 405.84 | 406 | 0.118 | 1.547 |
| | 8 | _0.065 | _0.073 | 376.72 | _0.054 | 333,82 | 353 | 0.091 | 0.740 |
| | 10 | +0.033 | _0.035 | 305.17 | _0.036 | 265.18 | 279 | 0.050 | 1.029 |
| | 12 | _0.043 | _0.091 | 303.32 | _0.071 | 260.19 | 279 | 0.115 | 0.780 |
| | 14 | _0.013 | _0.106 | 305.08 | _0.115 | 296.52 | 300 | 0.156 | 1.085 |
| | 16 | _0.093 | _0.103 | 296.26 | _0.191 | 299.08 | 298 | 0.217 | 1.854 |
| Interruption of observations | | | | | | | | | |
| 2.11.59 | 12 | +0.027 | _0.073 | 417.26 | _0.071 | 447.13 | 437 | 0.102 | 0.973 |
| | 14 | +0.007 | _0.080 | 340.36 | _0.021 | 404.63 | 335 | 0.083 | 0.263 |
| | 16" | _0.074 | _0.062 | 308.54 | _0.074 | 329.66 | 324 | 0.097 | 0.119 |
| AVERAGE OF VALUES | | _0.063 | _0.086 | 346.62 | _0.069 | 334.35 | 339 | 0.115 | 0.852 |

*Source:* My *Note* of 10 November 1959, *Paraconical pendulum with isotropic support. Determination of the variations over time of the characteristics of the correlation with the movement of the azimuth* (reproduced by photocopying).

*GRAPH I*

**ISOTROPIC SUSPENSION**

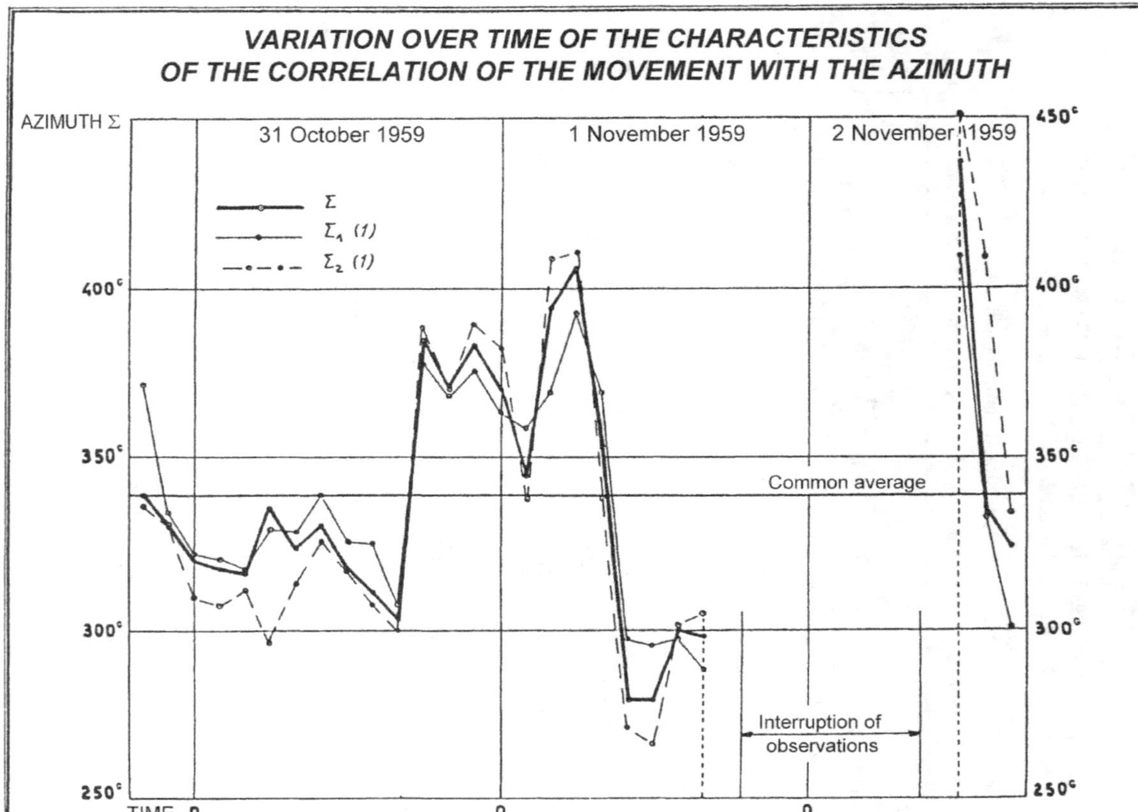


(1) *Curves representative of* $\Sigma_1$ *and* $\Sigma_2$ *obtained by bringing the averages of* $\Sigma_1$ *and* $\Sigma_2$ *onto the average of* $\Sigma$

*Source:* My *Note* of 10 November 1959 (see *Source* of *Table II*) (reproduction by photocopying).

## *C.4 The method of mobile correlations and its difficulties in practical application*

### A method that is optimal in principle

1 - In principle, the determination of the azimuth $X(t)$ of the plane of anisotropy at each instant by the method of mobile correlations might *a priori* appear to be optimal.

However it is rather laborious, because for $N$ observations it involves the calculation of $2(N-9)$ correlations.[1]

Moreover, the application of this method was found to be relatively difficult, because of great fluctuations, in certain cases of the order of 200 grads, in the angle $\Sigma$ calculated from the correlations determining $\Sigma_1$ and $\Sigma_2$, *due to the influences of defective balls*. Some uncertainty in the determination of $\Sigma$ resulted from this problem.

### The perturbing influence of the balls

2 - A *major* difficulty in the application of the method of mobile correlations in fact appeared due to the perturbing influence of the balls. In the *enchained* experiments, the seriously disturbing influence of a bad ball might invalidate the variation $\Delta\phi$ of the azimuth, *but this perturbation was limited to the experiment in question.* On the contrary, in the case of application of the method of mobile correlations, a bad ball *might completely* invalidate the determination of the angle $X$ of anisotropy, and this influence would extend itself over ten consecutive correlations.

In fact, suppose that due to the influence of a bad ball we get approximately $X_n = X_{n-1} + 100$ grads. Since $X_n$ is determined only to 200 grads, the determination $X_n = X_{n-1} - 100$ grads would be *just as good.*

[1] In 1959-1960 we had to perform all our calculations without the aid of computers.

## Smoothing with mobile averages

3 - I tried to eliminate this difficulty by considering correlations over 20 values instead of 10, and by considering the mobile averages $\overline{M}_n, \overline{B}_n, \overline{C}_n$ over 19 values of the $M_n$, $B_n$, $C_n$, and again the mobile averages of the $\overline{\overline{M}}_n, \overline{\overline{B}}_n, \overline{\overline{C}}_n$ over 19 values of the $\overline{M}_n, \overline{B}_n, \overline{C}_n$. Then $X_n$ are determined by the equations[2]

(1) $\sin 2X_n = \overline{\overline{B}}_n / \overline{\overline{P}}_n$ $\cos 2X_n = \overline{\overline{C}}_n / \overline{\overline{P}}_n$

(2) $\overline{\overline{P}}_n \sqrt{\overline{\overline{B}}_n^2 + \overline{\overline{C}}_n^2}$

which are analogous to the equations (4) and (5) of §2 above.[3]

In this way I was able to eliminate almost all of the difficulties we encountered.

---

[2] See my *Note* of 17 June 1985, *Analyse des mouvements du pendule paraconique à suspension isotrope. 20 novembre 1959, 7h. 20 mn. - 15 décembre 1959, 15 h. 20 mn.* (Analysis of the movements of the paraconical pendulum with isotropic support. 7h20m on 20 November 1959 - 15h20m on 15 December 1959), (49 p.) [46].

[3] Analogous formulas were used for calculating the second harmonic of the direction $X$ of the anisotropy of space.

# D THE VARIATIONS OF THE AZIMUTH OF THE ANISOTROPY OF SPACE NOVEMBER-DECEMBER 1959 AND MARCH-APRIL 1960

## *D.1 The series of observations of November-December 1959 and March-April 1960*

From 20 November to 15 December 1959 and from 16 March to 16 April 1960, we performed two month-long series of observations of the plane of oscillation of the paraconical pendulum in ten azimuths, *according to the method of mobile correlations* explained in *Section C* above, using the isotropic suspension and a pendulum identical to that used during the crucial experiments of July 1958.[1, 2]

[1] The series of observations of November-December 1959 had to be stopped prematurely due to the fatigue of the operators. *It accordingly only includes 25 days of observations, whereas the series of March-April 1960 includes 31 days of observations.*

The first series of experiments was carried on from 7h on 20 November to 15h on 15 December; and the second was from 9h on 16 March to 15h on 16 April.

[2] In parallel with these observations, two series of *enchained* observations of the paraconical pendulum with *anisotropic support* were performed, using a pendulum identical to that used with the isotropic support.

A detailed analysis of this set of observations will be presented in the *Second Volume* of this work, *Chapters I and II.*

See also my review Memoir of 12 June 1995, *Analyse Empirique et Théorique des Anomalies du Pendule Paraconique* (Empirical and Theoretical Analysis of the Anomalies of the Paraconical Pendulum), Revision of my Memoir of March 1992, Section VI, pp. 37-44 [54].

One of the objectives we aimed at by performing observations with the isotropic suspension and observations with the anisotropic suspension *in parallel* was to determine the exact influence of the anisotropy of the anisotropic support on the azimuth of the plane of oscillation of the paraconical pendulum.

*For the lack of time, it has never been possible to perform this analysis. It remains to be done.*

## *D.2 The average anisotropy of space*

**Overall correlations of the variations of azimuth $\Delta\phi_i$ with the azimuths $A_i$.**

1 - *Table III* gives the *average* rates of displacement in grads per minute of the azimuth of the plane of oscillation of the paraconical pendulum in the ten azimuths considered[1] during the two month-long series of observations of November-December 1959 and March-April 1960, and the two correlations *of the whole set* of variations of azimuth with the launching azimuths of the pendulum corresponding to these two month-long series.

The two correlations are *remarkably similar*. They both correspond *to the same residual anisotropy*, which is about 17 times smaller than the anisotropy corresponding to the anisotropic suspension.[2]

**Interpretation of the average anisotropies**

2 - The average direction of anisotropy for the month-long series of November-December 1959 is 323 grads, and for the month-long series of March-April 1960 is 318 grads, the angles being counted in the anticlockwise sense from the South in both cases. These two directions are *practically identical*, and they are both *near the East-West direction*.[3]

*At first sight*, these results might appear to be contradictory to the results of the correlations performed from 24 September to 16 October 1959 (*Table I* of §B.2) which yielded similar coefficients of anisotropy, but whose average direction of anisotropy was 406 grads, *i.e. virtually oriented along the meridian*.

---

[1] Note (2 ) of §C.1 above.

[2] *Chapter I* above, §E.3, *Table X*, p. 180, and *Table III* below .

I remind the reader that *Table X* of *Chapter I* and *Table III* below give the variations of the azimuths *in grads per minute*. We therefore have (*Table III*): $0.677/0.0404 = 16.75$.

[3] We have (*Table I* of §B.1): $a_1 = -0.051$ grad per minute; and (*Table III* below): $\overline{r} = -0.0404$ grad per minute. We also have $X = 318.5$ and $325.5$ grads, with $\overline{X} = 322$ grads.

However, it should be remarked that these experiments (*Table I*) were limited in number,[4] that *on average* they were performed near the meridian passage of the Sun, and that accordingly they may be considered as resulting from the dominant influence of the Sun at the times of the experiments.[5]

Moreover, *from the entirety of these experiments* we may conclude that the anisotropy due to the new support was virtually nil, so that the new support could be considered as being isotropic.[6] If, in fact, there had been any anisotropy of the support, then *it would have manifested itself in the same manner* in the experiments of *Table I* and in the experiments of *Table III.*

*In fact, the average anisotropy observed during the two month-long periods of observation may thus be considered as being the average anisotropy of space during the two periods in question, and this average anisotropy of space, detected by the movements of the paraconical pendulum, remained practically constant and oriented approximately East-West.*

---

[4] They only correspond to 102 14-minute experiments, while the month-long series of observations of November-December 1959 and of March-April 1960 respectively correspond to 1824 and 2220 14-minute experiments.

[5] The average of the average times of the experiments was around 12h30m (*Table I* of §B.2 above, p. 246).

[6] The analysis of *Table I* has already led to this same conclusion (§B.2. above).

*In fact, if there had been any anisotropy of the support, it would have appeared in all the experiments without any change*. This was evidently not the case. If therefore there was any anisotropy due to the new support, it may be considered as negligible.

*TABLE III*

## AVERAGE ANISOTROPY OF SPACE

***Paraconical pendulum with isotropic support***
***Observations of the two month-long series***
***of November-December 1959 and March-April 1960***
***Average variations Δ of the pendulum azimuths in grads per minute***
***during the 14-minute experiments***

| Azimuths $A_i$ | 200 | 220 | 240 | 260 | 280 | 300 | 320 | 340 | 360 | 380 |
|---|---|---|---|---|---|---|---|---|---|---|
| November-December 1959 | | | | | | m = - 0.140 | | | | |
| δ = Δ- m | - 0.018 | - 0.023 | 0.008 | 0.022 | 0.047 | 0.024 | 0.022 | 0.004 | - 0.039 | - 0.052 |
| March-April 1960 | | | | | | m = - 0.151 | | | | |
| δ = Δ- m | - 0.012 | 0.011 | 0.030 | 0.038 | 0.038 | 0.007 | - 0.001 | - 0.012 | - 0.068 | - 0.029 |

*Legend:* $\delta = \Delta - m$ in grads per minute during a 14-minute experiment

*Fitting for November-December 1959:* correlation coefficient : $R = 0.945$ $1-R^2 = 0.107$
$\delta_i^* = r \sin 2(A_i - X)$ $r = 0.0399$ grad $X = 318.5$ grads

*Fitting for March-April 1960:* correlation coefficient : $R = 0.930$ $1-R^2 = 0.136$
$\delta_i^* = r \sin 2(A_i - X)$ $r = 0.0410$ grad $X = 325.5$ grads

*Averages for the fittings:* $\overline{r} = 0.0404$ $\overline{X} = 322$ grads $\overline{m} = -0.145$ grad per minute

*Comparison with the anisotropy of the anisotropic support (Chapter I, §E.3, Table X)*
Anisotropic suspension: m = - 0.127 grad per minute r = 0.677 grad per minute

*Ratio of the coefficients of anisotropy:* 0.677 / 0.0404 = 16.75

*Sources:* Calculations 1105 and 1106 (10 March 1996).

## D.3 *Determination of the variation over time of the direction of anisotropy of space X during the two periods of November-December 1959 and March-April 1960*

By applying the method of mobile correlations associated with smoothing of the data,[1] the directions of anisotropy of space *X* were determined every twenty minutes for the two month-long periods of November-December 1959 and March-April 1960.

The two *Graphs II and III* show the variations of the direction of anisotropy of space *X* during these two periods.

During these two periods, the variations of the azimuth of the direction of anisotropy *X* were *considerable*: in November-December 1959, around 1800 grads in *the anticlockwise sense* in 25 days, *i.e.* about 4.5 turns; and in March-April 1960, around 900 grads in *the clockwise sense* in 31 days, *i.e.* about 2.25 turns. For the first period the average variation per day was around 70 grads, while for the second period it was only around 30 grads, *i.e.* nearly two and a half times smaller.[2]

---

[1] §C.1, §C.2, and §C.4.3 above.

[2] $70/30 = 2.33$.

The magnitude of the variations of the azimuth of the direction of anisotropy is confirmed by the *enchained* observations of the paraconical pendulum with isotropic support in September-October 1959 (Section H below, pp. 315-319).

The difference in sense of the variation between the two periods can be explained by the influence of semi-annual, annual, and long duration periodicities (see below, *Chapter V*, Section B).

This analysis will be performed in the *Second Volume* of this work, *Chapter II*, Section A (p. 28 above).

In fact, the variations of the direction of anisotropy *essentially* result from astronomical influences.

The two *Graphs II and III* thus illustrate *an essential difference between the isotropic suspension and the anisotropic suspension.* In November-December 1959 the azimuth of the plane of oscillation of the paraconical pendulum with *anisotropic* suspension effectively only oscillated between the two extreme values of 342 and 500 grads, and in March-April 1960 it only oscillated between the two extreme values of 350 and 406 grads. The corresponding total variations of azimuth were only 158 and 56 grads respectively.[3] The total amplitude of the azimuthal variations of the enchained series was thus nearly three times smaller in March-April 1960 than in November-December 1959.[4]

In fact, however great the variations of the azimuth of the direction $X$ of the anisotropy of space corresponding to the isotropic suspension may be, they are *much smaller* than the variation that would correspond to the Foucault effect *if it were only to appear by itself during one single continuous experiment.* This effect is about 300 grads per day, and it would always appear in the clockwise sense.[5]

It should be underlined that the nature of the two *Graphs II and III is entirely different* from that of the *Graphs* representing the azimuths of the plane of oscillation of the paraconical pendulum *during the enchained experiments.*[6] Here, we do not in fact have a representation *of the azimuths* $\phi$ *of the plane of oscillation* of the paraconical pendulum *during enchained experiments*, but rather a representation *of the directions X of anisotropy of space* as determined from the correlation calculations of the variations of the plane of oscillation of the paraconical pendulum as a function of the starting azimuths.[7]

---

[3] See *Table I* of §A.3 of *Chapter I* above.

Analysis of the correspondence between the azimuths of the direction of anisotropy of the paraconical pendulum for the isotropic suspension and for the anisotropic suspension will be performed in the *Second Volume* of this work, *Chapter II*, Section A.

[4] It is remarkable that these two ratios $70/30 = 2.33$ and $158/56 = 2.82$ are very comparable.

[5] The Foucault effect at Saint-Germain is $-\omega \sin L = -0.55 \times 10^{-4}$ radian per second. A rotation of -400 grads in 31.761 hours, *i.e.* a rotation of $-302.26$ grads per day, corresponds to this angular value of the rotation of the Foucault pendulum.

[6] Such as, for example, the Graphs representing the azimuths $\phi$ of the plane of oscillation of the paraconical pendulum in June-July 1955 (*Graph II*, *Chapter I* above, §A.2, p. 89), or in July 1958 at Bougival (*Graph XXIV*, *Chapter I* above, §C.2.4, p. 154).

[7] According to the method of mobile correlations explained in *Section C* above.

*GRAPH II*

**PARACONICAL PENDULUM WITH ISOTROPIC SUPPORT**
**AZIMUTHS OF THE DIRECTION OF ANISOTROPY OF SPACE**
***determined from the month-long series of observations***
***18h on 20 November - 6h on 15 December 1959***

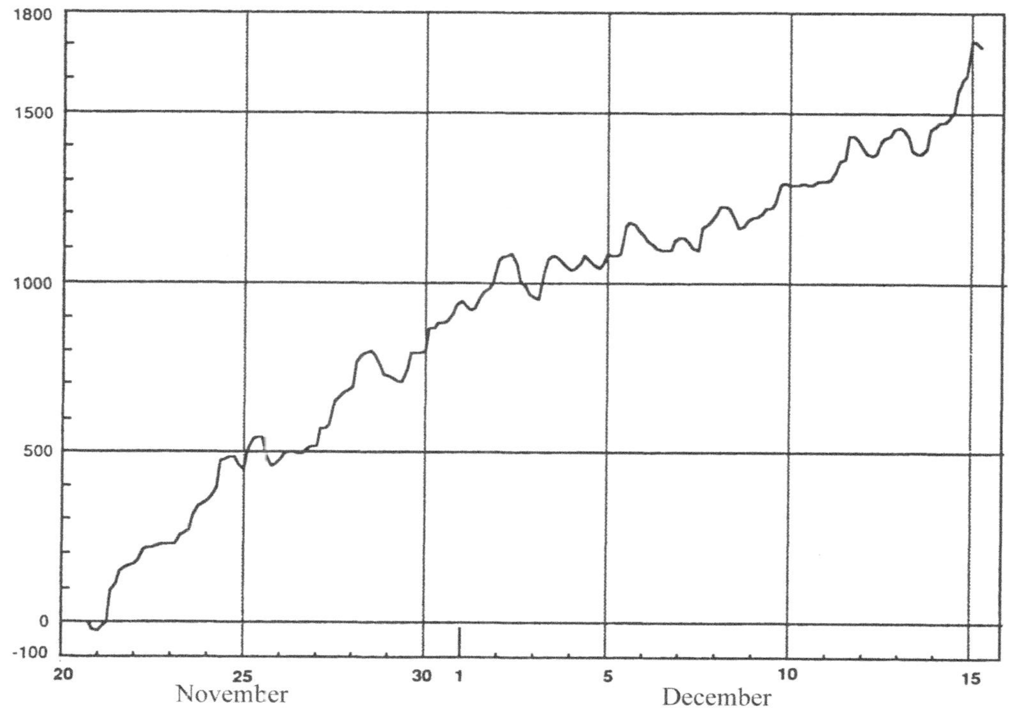


*Legend:* N=197 values of the azimuth of anisotropy, taken every 3 hours. The azimuths are reckoned in grads, positively from South in the anticlockwise sense.

*Source:* *Graph* 10842 and *Table* 12708 (18 October 1985).

*GRAPH III*

**PARACONICAL PENDULUM WITH ISOTROPIC SUPPORT**
**AZIMUTHS OF THE DIRECTION OF ANISOTROPY OF SPACE**
***determined from the month-long series of observations***
***21h on 16 March - 18h on 15 April 1960***

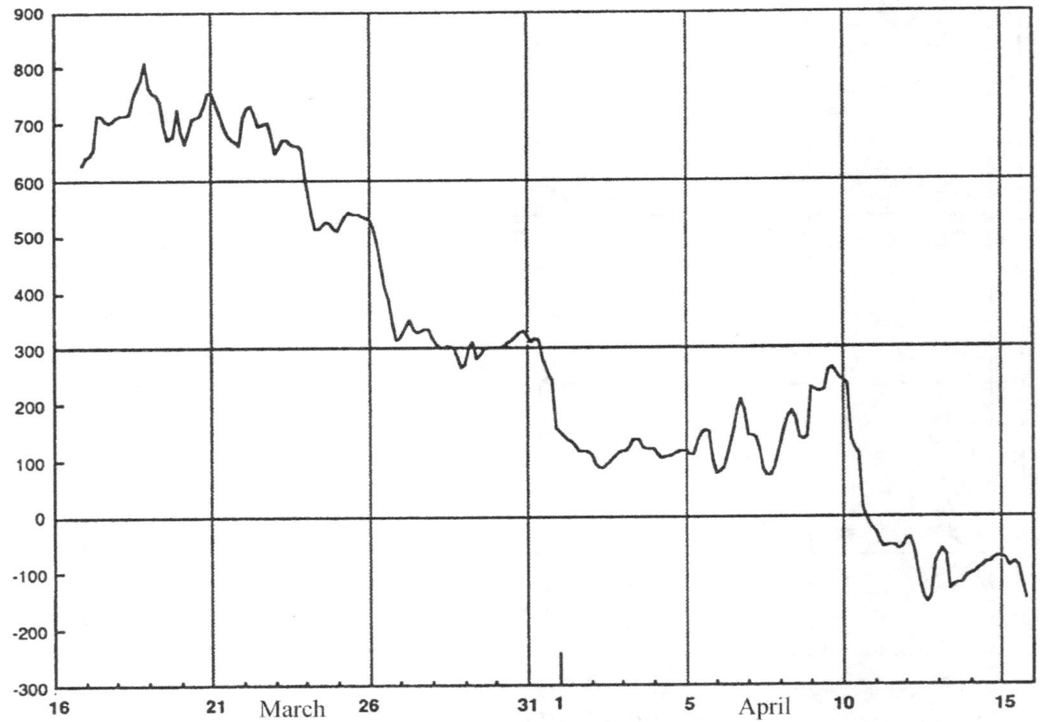


*Legend:* N=240 values of the azimuth of anisotropy, taken every 3 hours. The azimuths are reckoned in grads, positively from South in the anticlockwise sense.

*Source:* *Graph* 10977 and *Table* 12705 (19 December 1985).

## *D.4 Variations over time of* cos 2X *corresponding to the direction X of anisotropy of space in November-December 1959 and March-April 1960*

### Graphs representing cos 2X

1 - *Graphs IV and V* represent the oscillations of the azimuth $X$ of anisotropy of the paraconical pendulum with isotropic support for the two month-long series of November-December 1959 and March-April 1960, taking $\cos 2X$ along the ordinate instead of $X$.

The Graphs representing $\cos 2X$ and $\sin 2X$ have the advantage of better representing the variations over time of the effective azimuthal situation of the plane of anisotropy of the paraconical pendulum with isotropic support.[1] For simplification and for illustration, I only show the two Graphs relating to $\cos 2X$.

[1] *One value* of the direction of anisotropy, *and one only*, corresponds to each value of $2X$. Thus, if $X$ varies through 200 grads, $2X$ varies through 400 grads.

The value $\cos 2X = 1$ corresponds to the meridian, and the value $\cos 2X = -1$ corresponds to the East-West direction. For $\cos 2X = 0$, the direction of anisotropy coincides with one of the two directions angled at 50 grads with respect to the meridian. Analogous conclusions can be presented for the graphs representing $\sin 2X$.

For analysis of the oscillations of the direction of anisotropy $X$, it would also be interesting to consider the second representation corresponding to $\cos 2X$ and $\sin 2X$ in parallel with the first that corresponds to $X$ (see §H.2 below, pp. 317-319).

In fact, while the determination of the direction of anisotropy $X$ raises difficulties due to the perturbing influence of the balls (§C.4.2 above), *the values of* $\cos 2X$ *and* $\sin 2X$ *are always perfectly determined, and they involve no difficulties of interpretation.*

Since the directions of anisotropy $X$ are reckoned positively in the anticlockwise sense from the meridian, a crossing of the meridian by the direction of anisotropy corresponds to a value of $X$ that is a multiple of 200 grads, and to a value of $2X$ that is a multiple of 400 grads, *i.e.* to a value of $\cos 2X$ equal to unity.

As can be understood from *Graphs IV and V*, it is clear at very numerous points that one bad ball is enough to cause the direction of anisotropy $X$ to cross the meridian, or not, as the case may be.[2]

[2] For the averages of $\cos 2X$ and $\sin 2X$, we have:
November-December 1959 : $\overline{\cos 2X} = -0.0698$, $\overline{\sin 2X} = -0.0167$
March-April 1960 : $\overline{\cos 2X} = -0.477$, $\overline{\sin 2X} = -0.208$

*GRAPH IV*

**PARACONICAL PENDULUM WITH ISOTROPIC SUPPORT**
**AZIMUTHS X OF THE DIRECTION OF ANISOTROPY OF SPACE**
**REPRESENTATION OF COS 2X**
***18h on 20 November - 6h on 15 December 1959***

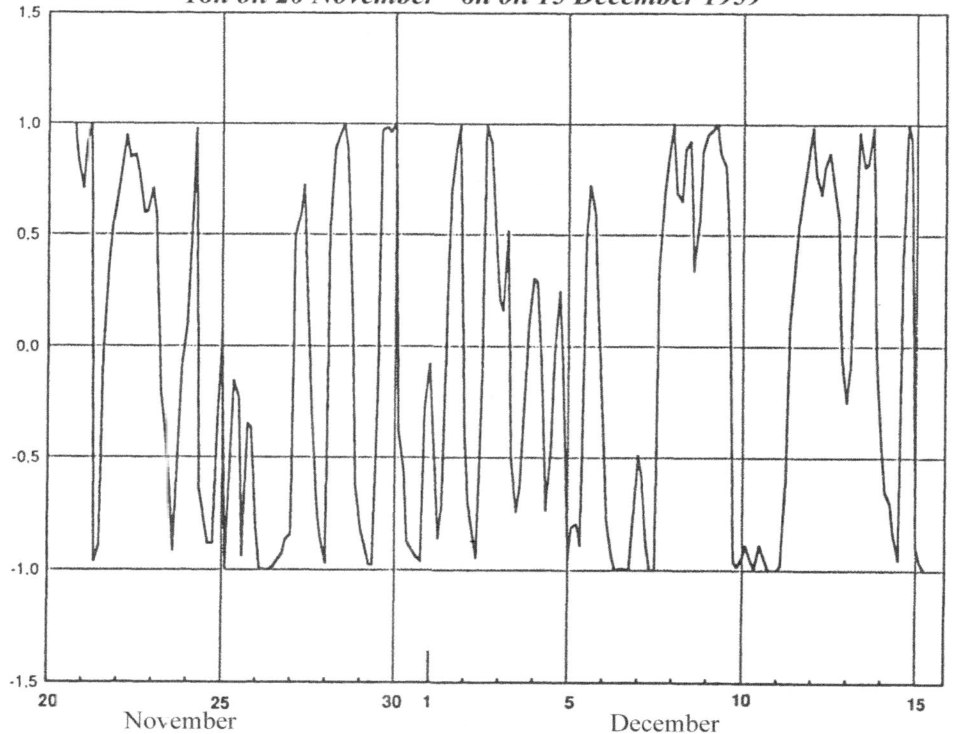


*Legend:* N=197 values of cos 2X, taken every 3 hours. Average value of cos 2X: $\overline{\cos 2X} = -0.0698$
cos 2X = +1 : direction of meridian; cos 2X = -1 : direction East-West

*Sources:* *Graph II - Table* 12708 (22 May 1996)
Calculation 1114 - *Graph* 13805 (23 May 1996)

*GRAPH V*

**PARACONICAL PENDULUM WITH ISOTROPIC SUPPORT**
**AZIMUTHS X OF THE DIRECTION OF ANISOTROPY OF SPACE**
**REPRESENTATION OF COS 2X**
***21h on 16 March - 18h on 15 April 1960***

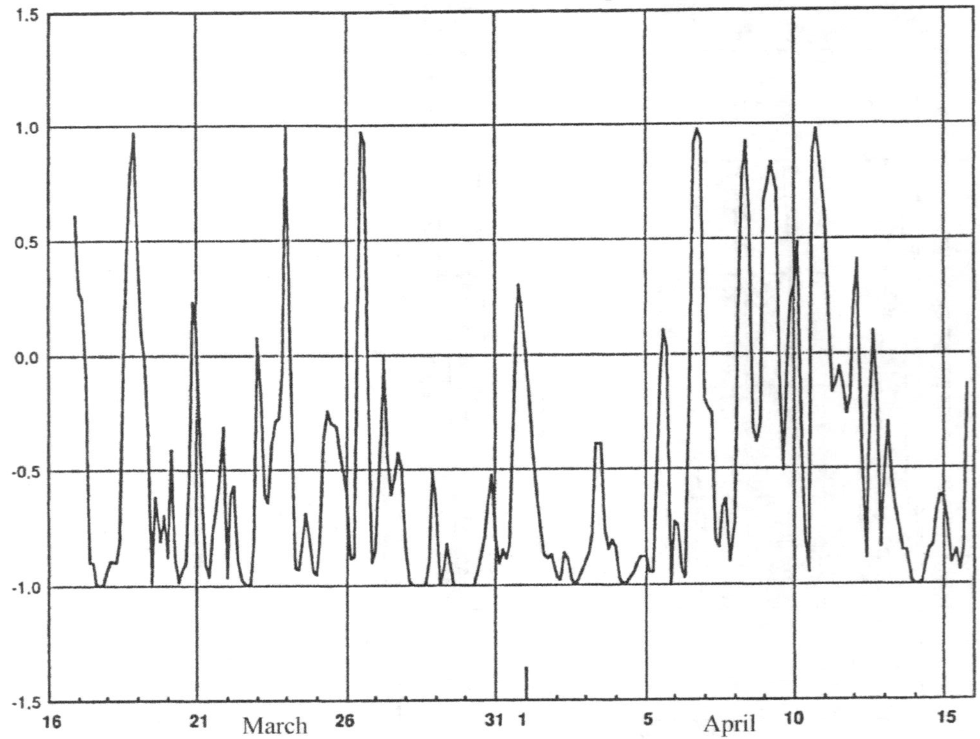


*Legend:* N=240 values of cos 2X, taken every 3 hours. Average value of cos 2X: $\overline{\cos 2X} = -0.477$
cos 2X = +1 : direction of meridian; cos 2X = -1 : direction East-West

*Sources:* *Graph III - Table* 12707 (21 May 1996)
Calculation 1113 - *Graph* 13804 (23 May 1996)

**Cumulative sums of the differences** $(\cos \mathbf{2X} - \overline{\cos \mathbf{2X}})$

2 - While *Graphs IV and V* show no clear regularities, the case is quite different for the graphs representing the cumulative values

$$\sum_{1}^{N} (\cos 2X - \overline{\cos 2X})$$

of the differences between the values of $\cos 2X$ and their average $\overline{\cos 2X}$ for the two periods of November-December 1959 and March-April 1960.[3] Furthermore, a very significant similarity appears between these two Graphs.

*Graph VI* in fact represents the correlation between the cumulative values of $(\cos 2X - \overline{\cos 2X})$ for the two series of observations of November-December 1959 and March-April 1960. The similarity of the curves *demonstrates the existence of the same underlying periodic structure during the two periods.*

[3] The summation has a double effect: on the one hand, it attenuates the random effect of the balls; and on the other hand, it relatively reinforces the amplitudes of those periodicities whose periods are greater.

*GRAPH VI*

**PARACONICAL PENDULUM WITH ISOTROPIC SUPPORT**
**AZIMUTHS X OF THE DIRECTION OF ANISOTROPY OF SPACE**
**CUMULATIVE VALUES Σ (cos 2X - $\overline{\text{cos 2X}}$)**
***November-December 1959 and March-April 1960***

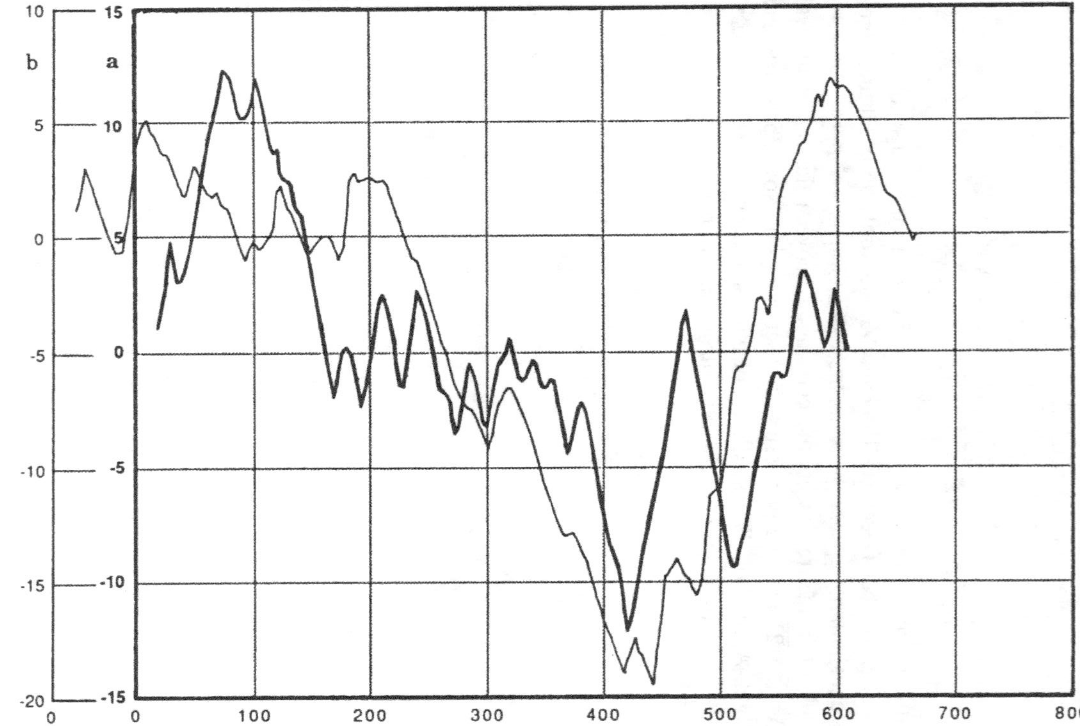


*Legend:* a ▬▬ November-December 1959. Σ (cos 2X - $\overline{\text{cos2X}}$)
b —— March-April 1960. Σ (cos 2X - $\overline{\text{cos2X}}$)

*Sources:* a ▬▬ *Graph IV* above; and calculation 1232 and *Graph* 13863 (11 octobre 1996)
b —— *Graph V* above; and calculation 1231 and *Graph* 13862 (11 octobre 1996)

# E THE PERIODIC DIURNAL AND MONTHLY STRUCTURE OF THE DIRECTION OF ANISOTROPY OF SPACE NOVEMBER-DECEMBER 1959 AND MARCH-APRIL 1960

## *E.1 Diurnal periodicities of the azimuth X of anisotropy of space*

*Table IV* shows the amplitudes for eight periods, according to the current theory of lunisolar periodicities, of the periodic components of the direction *X* of anisotropy of space, and comparison thereof with the coefficients corresponding to this theory.

Both for the period of March-April 1960 and for the period of November-December 1959, we see very marked lunisolar periodicities of 24h, 24h50m, 12h, and 12h25m. We also see other significant periods.[1]

Relatively to the diurnal periodicities of the enchained series of the paraconical pendulum with *anisotropic* suspension, the diurnal periodicities of the direction of anisotropy of space deduced from the observations of the paraconical pendulum with *isotropic* suspension present certain essential overall characteristics.

[1] The shown amplitudes result from fittings of the values of the azimuth *X* of the direction of anisotropy to sinusoids having the indicated periods, together with a linear trend.

I remind the reader that the series of November-December 1959 only includes 25 days of observations, against 30 days of observations for the series of March-April 1960.

The calculations in *Table IV* were performed in 1996.

**Amplitudes of the periodic diurnal components**

1 - First of all, the amplitudes of the periodic diurnal components of 24h and 24h50m in the direction of anisotropy are, on average, *five times* greater than the amplitudes of the corresponding components of the series of the paraconical pendulum with anisotropic support.[2]

The relative amplitudes of the two components of 24h50m and 24h are *of comparable order* of magnitude for the two month-long series of November-December 1959 and March-April 1960.[3]

---

[2] For the double-amplitudes *2R* in grads, we have, from *Table IV* and *Table I* of *Chapter I*, §A.3.1

| | | Direction of anisotropy (isotropic suspension) 2R (1) | Azimuths of enchained series (anisotropic suspension) 2R (2) | Ratio of amplitudes (3) = (1) / (2) |
|---|---|---|---|---|
| 24h | Nov-Dec 1959 | 9.9 | 2.5 | 4.0 |
| | Mar-Apr 1960 | 2.1 | 1.8 | 1.2 |
| | averages | 6 | 2.15 | 2.6 |
| 24h50m | Nov-Dec 1959 | 20.3 | 1.3 | 15.6 |
| | Mar-Apr 1960 | 4.2 | 1.5 | 2.8 |
| | averages | 12.25 | 1.4 | 9.2 |
| | General averages | 9.1 | 1.8 | 5.1 |

We have: $9.1/1.8 = 5.1$ and $(4.0+1.2+15.6+2.8)/4 = 5.9$
$6/2.15 = 2.79$ and $12.25/1.4 = 8.75$

It should be underlined that, since the observational series of November-December 1959 only included 25 days, the waves of 24h and 24h50m in this series are not perfectly separated.

[3] Thus, for the *isotropic* suspension, we have $9.9/2.1 = 4.4$ for the period of 24h; and for the period of 24h50m, we have $20.3/4.2 = 4.8$.

Moreover, for the observations with the *isotropic* suspension of November-December 1959 and March-April 1960, we also have $20.3/9.9 = 2.1$ and $4.2/2.1 = 2$ respectively.

For the enchained experiments with the *anisotropic* suspension, we have $2.5/1.8 = 1.39$ and $1.3/1.5 = 0.87$.

*The outstanding feature is thus a considerable average amplification, of the order of five to one, of the amplitudes of the diurnal periodic components of the direction of anisotropy corresponding to the isotropic suspension, while the relative periodic structures remain of comparable orders of magnitude.*[4]

## Diurnal periodic structures of the direction $X$ of anisotropy of space and lunisolar coefficients

2 - Yet again, as previously for the enchained series,[5] we see that the periodic structure of the direction of anisotropy of space is *totally* different from the periodic structure of the coefficients of the lunisolar periodicities as deduced from the current theory of tides (*Table IV* below).

The outstanding periodicities of the direction $X$ of anisotropy of space in fact correspond to the waves $M_1$ and $O_1$ of 24.84h and of 25.82h, while for the lunisolar coefficients of the theory of tides the outstanding periodicities are the periodicities $K_1$ and $M_2$ of 24h and of 12.42 hours. *In both cases, the two outstanding periodicities represent more than 40% of the total amplitudes by themselves.*[6, 7]

---

[4] A detailed analysis of the diurnal periodicities of the paraconical pendulum with isotropic support in November-December 1959 and in March-April 1960 will be given in the *Second volume* of this work, *Chapter II*, Section A (p. 28 above).

[5] See in particular *Table XI* of *Chapter I*, §E.3 (p. 187).

For the diurnal periods, the ratios $a/b$ of this *Table XI* are: 0.552, 3.92, 0.465, and 4.16. They are of the same order of magnitude as the ratios $a/b$ of *Table IV* below, which are 0.376, 5.56, 1.87, and 5.92.

[6] The average double-amplitudes of the eight periodicities for November-December 1959 and March-April 1960 are respectively 9.94 and 4.85 grads (79.5/8 and 38.8/8), whose ratio is very close to 2 ($9.94/4.85 = 2.01$).

The ratio of the trends of *Graphs VII and VIII* is also of the order of 2 ($67.4/30.05 = 2.21$), and so is the ratio of the amplitudes ($144.3/68.2 = 2.12$).

[7] A *very remarkable* circumstance should be underlined. The fittings to a period of 31.761 hours, corresponding to the Foucault effect, of the two series of November-December 1959 and March-April 1960 for the direction of anisotropy associated with linear trends gives respective double-amplitudes of 22.2 and 11.8 grads, which are quite comparable with the double-amplitudes of 20.3 and 4.2 grads for the wave of 24h50m.

I remind the reader that at Saint-Germain the Foucault effect $-\omega \sin L$ corresponds to a rotation in the clockwise sense of $0.55 \times 10^{-4}$ radians per second, *i.e.* to a rotation of 400 grads in 31.761 hours (§D.3 above, note 5).

*TABLE IV*

**AZIMUTHS OF THE DIRECTION OF ANISOTROPY OF SPACE**
**DIURNAL AND SEMI-DIURNAL LUNISOLAR EFFECTS**
***Observed double-amplitudes and coefficients of lunisolar periodicities***

| Periodicities | $K_1$ 23 h 93 | $M_1$ 24 h 84 | $O_1$ 25 h 82 | $Q_1$ 26 h 87 | $S_2$ 12 h | $M_2$ 12 h 42 | $N_2$ 12 h 66 | $2N_2$ 12 h 92 | Total of double amplitudes |
|---|---|---|---|---|---|---|---|---|---|
| Double-amplitude of the wave: 2R in grads | | | | | | | | | |
| November-December 1959 [1] | 9.89 | 20.3 | 17.0 | 11.6 | 5.13 | 1.74 | 5.86 | 8.03 | 79.5 |
| March-April 1960 [1] | 2.11 | 4.21 | 9.93 | 5.55 | 1.04 | 5.17 | 3.77 | 6.98 | 38.8 |
| November-December 1959 [2] | 12.4 | 25.5 | 21.4 | 14.6 | 6.45 | 2.19 | 7.37 | 10.1 | 100 |
| March-April 1960 [2] | 5.44 | 10.9 | 25.6 | 14.3 | 2.68 | 13.3 | 9.72 | 18.0 | 100 |
| Average of relative values: a | 8.92 | 18.2 | 23.5 | 14.5 | 4.56 | 7.74 | 8.55 | 14.1 | 100 |
| Theory of lunisolar periodicities | | | | | | | | | |
| Coefficients [1] | 0.706 | 0.0977 | 0.377 | 0.0730 | 0.563 | 0.908 | 0.232 | 0.0235 | 2.98 |
| Relative values: b | 23.7 | 3.28 | 12.6 | 2.45 | 18.9 | 30.5 | 7.78 | 0.789 | 100 |
| Ratio of relative values | | | | | | | | | |
| Ratios a/b | 0.376 | 5.55 | 1,87 | 5,92 | 0.241 | 0.254 | 1.10 | 17.9 | 1 |

*Legend:* (1) Absolute values (2) Relative values=absolute values / total of double-amplitudes (or of coefficients)

*Sources:* 1) Azimuths of the directions of anisotropy, in grads: Graphs II and III of §II D
2) Coefficients of the current theory of lunisolar periodicities: Schureman, 1941, Manual of Harmonic Analysis and Prediction of Tides, pp. 164-165. Coefficients of periods that are very close together are added.
3) Harmonic analysis calculations: C-1116 and 1117; C-1119 to 1121; C-1126 to 1131; C-1136; and C-1153 and 1154 (24 March - 18 June 1996)

## E.2 Sidereal monthly lunar periodicity of the azimuth X of anisotropy of space

*Graphs VII and VIII* show the fittings of the month-long series of the direction $X$ of anisotropy of space for November-December 1959 and March-April 1960 to a sinusoid of period equal to the sidereal period of the Moon, *i.e.* 27.322 days, together with a linear trend.

The two coefficients of multiple correlation are respectively 0.994 and 0.965. The double-amplitudes $2R$ of the two sinusoids are respectively 288 grads and 136 grads. They are of comparable order of magnitude, but the first is approximately double the second.[1]

The relative importance of the amplitudes of the sidereal lunar periodic component of 27.322 days is much greater for the azimuth $X$ of anisotropy of space than for the lunisolar coefficients of the current theory.[2]

---

[1] We have $288.6/136.4 = 2.116$.

The two double-amplitudes of 288 and 136 grads are around three and a half times greater than the sums of the corresponding amplitudes of *Table IV*. In fact we have: $288.6/79.5 = 3.63$ and $136.4/38.8 = 3.52$.

We also have: $79.5/38.8 = 2.049$ (*Table IV*), a ratio which is very close to the above ratio of 2.116.

[2] If the double-amplitude of the periodic component of 27.322 days is denoted by $2R_S$, we have (*Table IV* and *Graphs VII and VIII*): $2R_S/2R_{24} = 288.6/9.89 = 29.18$ in November-December 1959 and $2R_S/2R_{24} = 136.4/2.11 = 64.64$ in March-April 1960.

Now, for the lunisolar coefficients (Schureman 1941, id., p. 164 [244]), we have $2R_S/2R_{24} = 0.1269/0.706 = 0.179$ (the sum of the coefficients of the neighboring waves of Schureman of numbers 141, 73, and 74, *i.e.* 27.322, 27.554, and 27.092 days, is in fact $0.0399 + 0.0827 + 0.0043 = 0.1269$).

We thus see that the relative importance represented by the ratio $2R_S/2R_{24}$ *is around 360 times greater* ($64.64/0.179 = 361$) *for the azimuths X than for the lunisolar coefficients of the current theory*. For November-December 1959, this ratio is around 160 ($29.18/0.179 = 163$).

It is *very remarkable* that, taking account of their average respective dates, the two sinusoids corresponding to the period of 27.322 days are *almost exactly in phase*, which would not be the case if we were to consider the fittings corresponding to the synodic period of the Moon, which is 29.53 days.[3] This agreement in phase *demonstrates the existence of a very strong coherence* between the data for the two periods of November-December 1959 and March-April 1960.[4]

---

[3] The distance in days of the peak of the fitting sinusoid of November-December 1959 from 0h on 1 January 1960 (*Graph VII*) is:

$$365.25 - 332.50 = 32.75 \text{ days}$$

And the distance between 0h on 1 January 1960 and the peak of the fitting sinusoid of March-April 1960 is 76.21 days. From the above, the distance in days between the peaks of the two fitting sinusoids is (*Graph VIII*):

$$32.75 + 76.21 = 108.96 \text{ days}$$

This distance is *almost exactly equal to four times the sidereal period of the Moon, i.e. 27.321 days*. In fact we have

$$108.96/27.321 = 3.988$$

[4] In fact, for these two fittings, the linear trends *can be interpreted as being elements of longer periodicities* (see note 2 of §D.3 above, p. 259, and *Chapter V* below , §B.4).

*GRAPH VII*

**DIRECTION X OF ANISOTROPY OF SPACE**

***20 November - 15 December 1959***

*Fitting of the azimuths to the sidereal period of the Moon of 27.322 days associated with a linear trend*

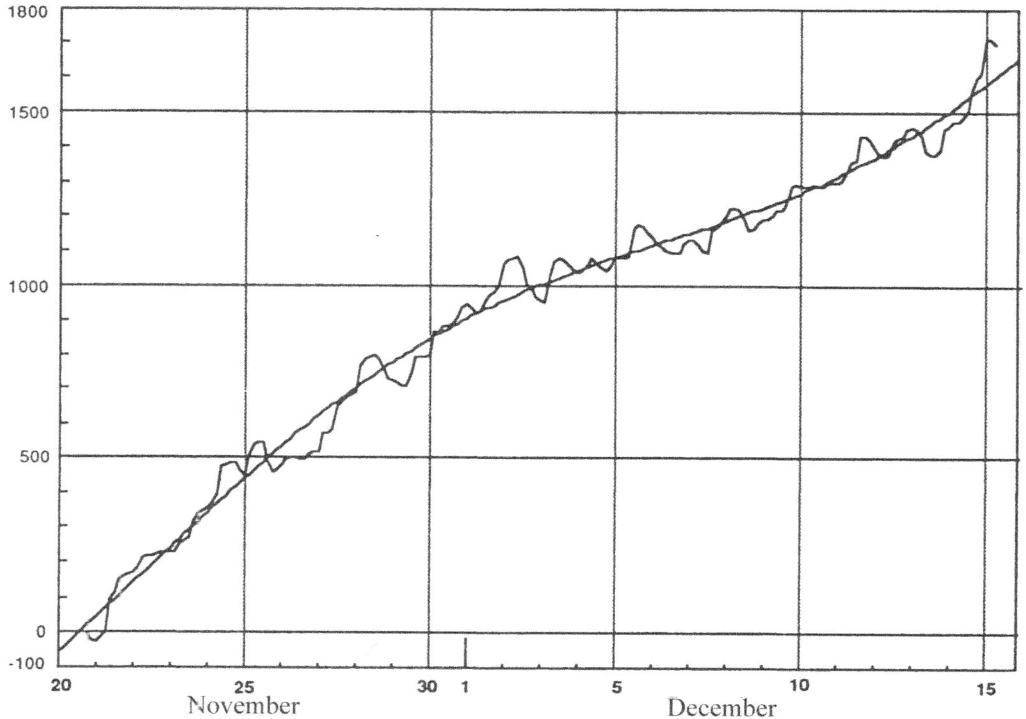


*Legend:* $\Sigma = 421$ grads; $R = 0.994$; $1\text{-}R^2 = 0.013$; $m = 904.37$ grads; $N = 197$
trend = + 67.4 grads per day; $r = 144.3$ grads; date of peak of sinusoid: 12h on 29 November 1959.
($s_0$=332.5 days from 0h on 1 January 1959). See the *Legend* of *Graph XXVII* of *Chapter I.*

*Sources: Graph II, Calculation* 1112 (23 May 1996), and *Graph* 13803.

*GRAPH VIII*

**DIRECTION X OF ANISOTROPY OF SPACE**

***16 March - 16 April 1960***

*Fitting of the azimuths to the sidereal period of the Moon of 27.322 days associated with a linear trend*

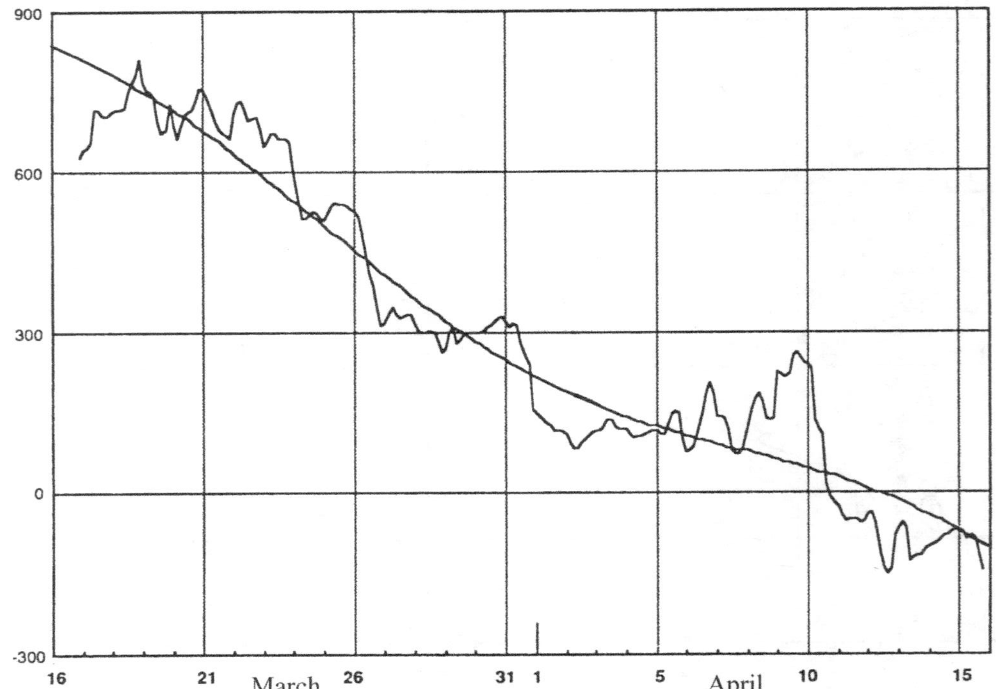


*Legend:* $\Sigma$ = 279 grads; R = 0.965; 1-$R^2$ = 0.068; m = 290.7 grads; N = 240
trend = - 30.5 grads per day; r = 68.2 grads; date of peak of sinusoid: 5h on 18 March 1960.
($s_0$=76.21 days from 0h on 1 January 1960). See the *Legend* of *Graph XXVII* of *Chapter I.*

*Sources: Graph III, Calculation* 1111 (21 May 1996), and *Graph* 13801.

## *E.3 Sidereal monthly lunar periodicity of the cumulative values of the differences* $(\cos \mathbf{2X} - \overline{\cos \mathbf{2X}})$

Graphs IX and X represent the fittings of the cumulative values of the differences $\cos 2X - \overline{\cos 2X}$ for the series of azimuths of the direction $X$ of anisotropy of space for November-December 1959 and March-April 1960.[1, 2]

[1] $\overline{\cos 2X}$ represents the average of the $\cos 2X$.

[2] For comparison of the phases, see *Table VIII* of §G.3 below , p. 313.

*GRAPH IX*

**DIRECTIONS X OF ANISOTROPY OF SPACE**

**CUMULATIVE VALUES OF $\cos 2X - \overline{\cos 2X}$**

***20 November - 15 December 1959***

*Fitting of the azimuths to the sidereal period of the Moon of 27.322 days associated with a linear trend*

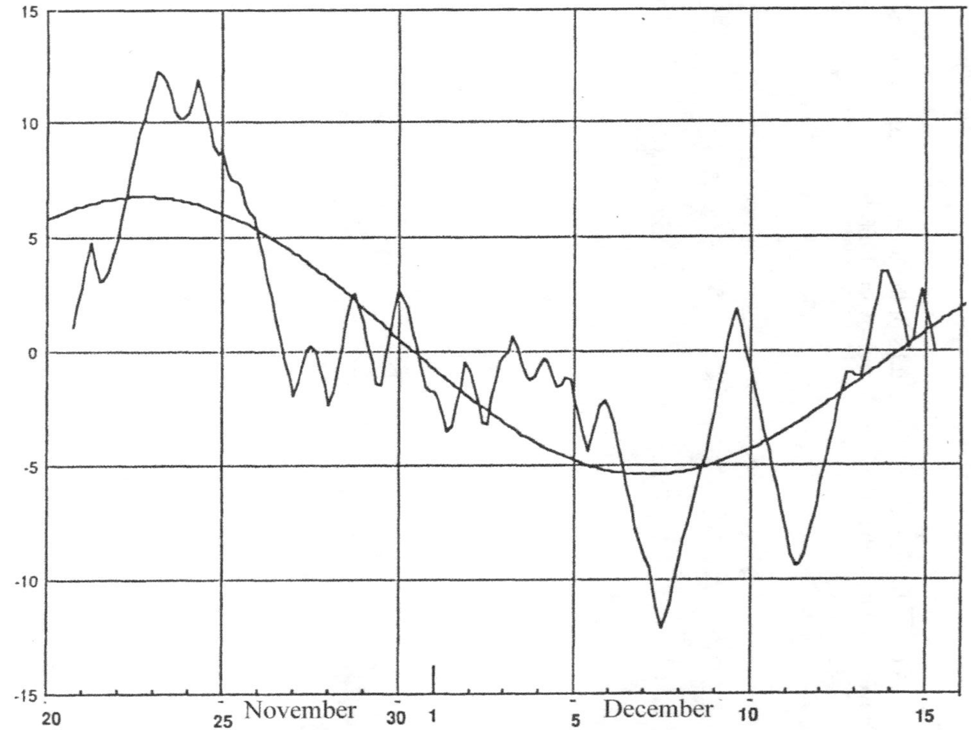


*Legend:* $\Sigma$ = 5.374 grads; R = 0.810; 1-$R^2$ = 0.344; m = 1.698; N = 197
trend = - 0.00428; r = 5.378; date of peak of sinusoid: 1h on 23 November 1959.
($s_0$=326.09 days from 0h on 1 January 1959). See the *Legend* of *Graph XXVII* of *Chapter I.*

*Sources: Graph IV* above, *Calculation* 1233 (14 October 1996), and *Graph* 13864.

*GRAPH X*

**DIRECTIONS X OF ANISOTROPY OF SPACE**
**CUMULATIVE VALUES OF cos 2X - $\overline{\text{cos 2X}}$**
***16 March - 16 April 1960***
*Fitting of the azimuths to the sidereal period of the Moon of 27.322 days associated with a linear trend*

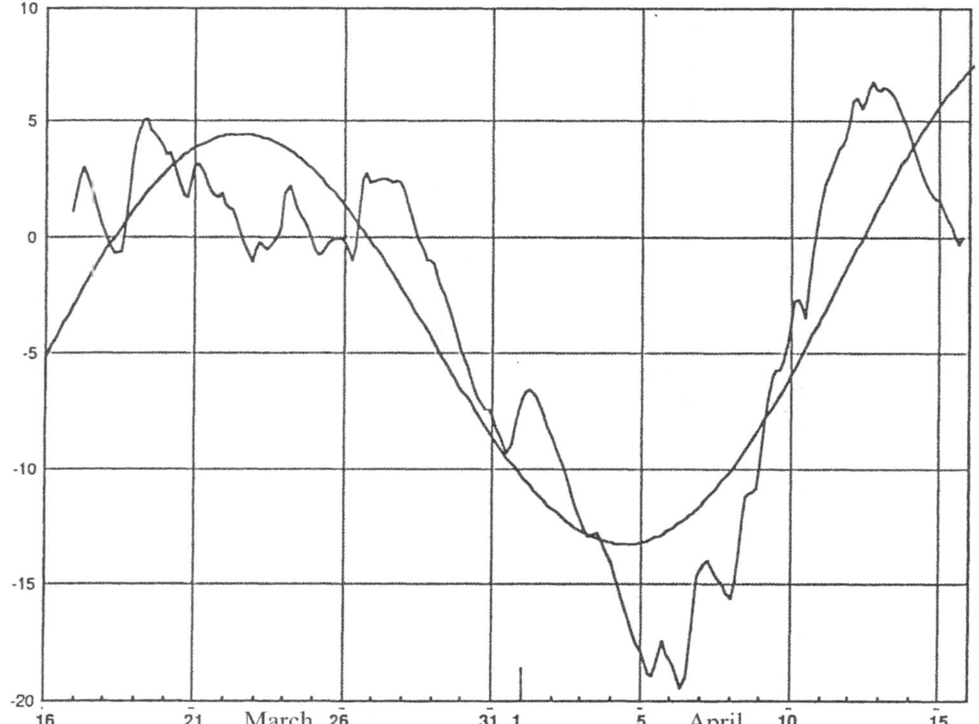


*Legend:* $\Sigma = 7.255$ grads; $R = 0.879$; $1\text{-}R^2 = 0.227$; $m = -6.755$; $N = 240$
trend = 0.00748; r = 10.043; date of peak of sinusoid: 3h on 22 March 1960.
($\varepsilon_0 = 80.4$ days from 0h on 1 January 1960). See the *Legend* of *Graph XXVII* of *Chapter I.*

*Sources: Graph V* above, *Calculation* 1234 (14 October 1996), and *Graph* 13865.

# F PERIODIC DIURNAL AND MONTHLY STRUCTURE OF THE DISPLACEMENTS OF THE PLANE OF OSCILLATION OF THE PARACONICAL PENDULUM WITH ISOTROPIC SUPPORT FROM THE DIRECTIONS NORTH-SOUTH AND EAST-WEST

## *F.1 Deviations $\Delta$ of the plane of oscillation of the paraconical pendulum from the meridian*

The direction of anisotropy of space was deduced from the observations of November-December 1959 and March-April 1960 by the method of mobile correlations[1] from the displacements $\Delta$ of the plane of oscillation of the paraconical pendulum with isotropic support *in ten specific azimuths*.

It is naturally very instructive to analyze the displacements of the plane of oscillation of the paraconical pendulum, and their periodic structures, *in each of these azimuths*.

*By way of illustration*, and for simplification, I shall at first limit myself here to examining the diurnal and monthly periodic structure of the displacements $\Delta$ of the plane of oscillation of the paraconical pendulum with isotropic support from the meridian.[2]

---

[1] *Section C* above.

[2] No particular significance should be attributed to the completely arbitrary choice of this particular azimuth for this illustration. In fact, one would find analogous results to the following for all the other azimuths.

During each of the two month-long periods of observations, the plane of anisotropy of space is in fact determined by starting from ten different equally spaced directions in succession, and as a result, *on average*, the periodic structure of the deviations $\Delta$ is, to a large part, independent of the azimuths in question (§II.C.1 above, p. 248).

In §5 and §6 below I provide some discussion upon the deviations of the plane of oscillation of the paraconical pendulum from the East-West direction, and upon their relationship with the North-South deviations.

*Graphs XI and XII* represent the displacements $\Delta$ in grads per minute from the meridian as a function of time from the start of the experiments, during the two month-long periods of observations of November-December 1959 and March-April 1960.[3]

The fluctuations of the $\Delta$ are *very marked.* They result both from their diurnal periodic components and from the influence of the balls, which was significant.

In November-December 1959 and March-April 1960, the averages of the $\Delta$ were respectively $-0.159$ and $-0.162$ grad per minute, and their standard deviations were respectively 0.201 and 0.203 grad per minute.[4] *These are practically identical values.*

---

[3] For the meridian, we naturally only have a single value for each series of experiments of 3h20m (20m $\times$ 10 = 3h20m). We thus have $N = 182$ values for the series of observations of November-December 1959 and $N = 222$ values for the series of observations of March-April 1960.

I remind the reader that the series of November-December 1959 only included 25 days, so that the separation of the waves of 24h50m and 24h could only be very partially guaranteed.

Separation of these two waves really requires a duration of 30 days.

[4] The average values $-0.159$ and $-0.162$ for the azimuth $A = 200$ grads are naturally identical to the values shown in *Table III* of *Section D* above (p. 258). In fact, for $A = 200$ grads, we have

$\overline{\Delta} = m + \overline{\delta} = -0.140 - 0.018 = -0.158$ and $\overline{\Delta} = m + \overline{\delta} = -0.151 - 0.012 = -0.163$

We have

$-0.158 \times 14 = -2.21$ grads and $-0.163. \times 14 = -2.27$ grads

while the Foucault effect corresponds to $-2.94$ grads in 14 minutes (§I.A.4, note 1, p.93 above).

*GRAPH XI*

**ISOTROPIC SUPPORT**
**DEVIATIONS Δ FROM THE MERIDIAN IN GRADS PER MINUTE OF THE PLANE OF OSCILLATION OF THE PARACONICAL PENDULUM**
***9h40m on 20 November - 13h on 15 December 1959***

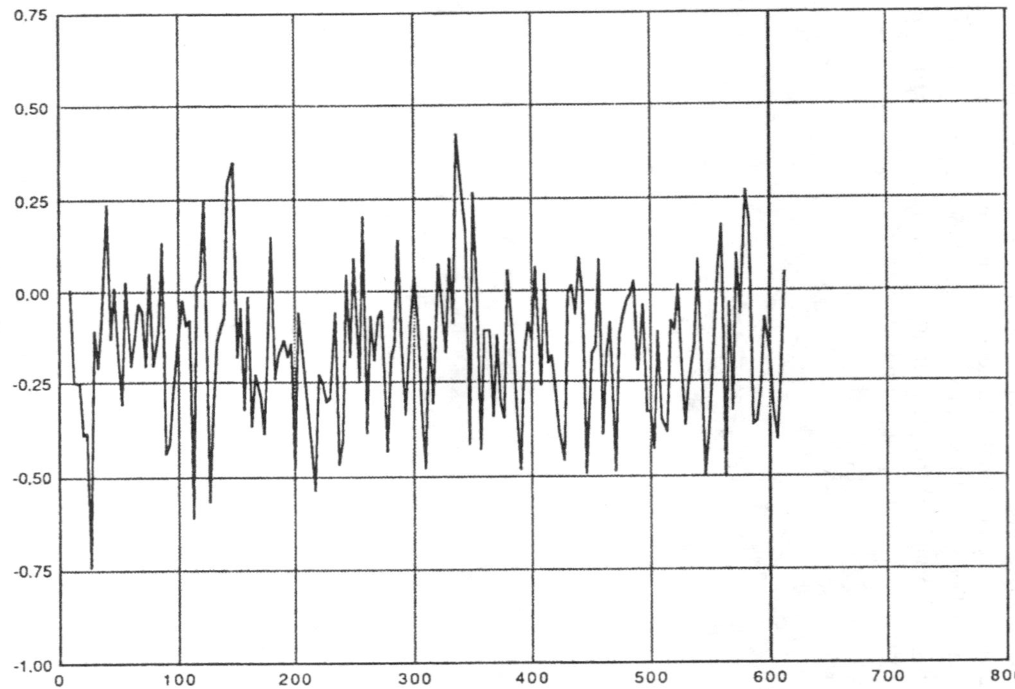


*Legend:* The Δ (every 3h20m) are shown in grads per minute and are reckoned positively in the anticlockwise sense.
The abscissas s are shown *in hours* from 0h on 20 November: $9.66 \leq s \leq 613$
Number of values of Δ: N = 182 [(613 - 9.66) / 3.33 = 181 = N - 1]
Average of the Δ: $\overline{m}$ = -0.159 grad per minute; standard deviation = 0.201 grad per minute.

*Source:* *Graph* 13812 (5 June 1996).

*GRAPH XII*

**ISOTROPIC SUPPORT**
**DEVIATIONS Δ FROM THE MERIDIAN IN GRADS PER MINUTE**
**OF THE PLANE OF OSCILLATION OF THE PARACONICAL PENDULUM**
***9h20m on 19 March - 2h on 16 April 1960***

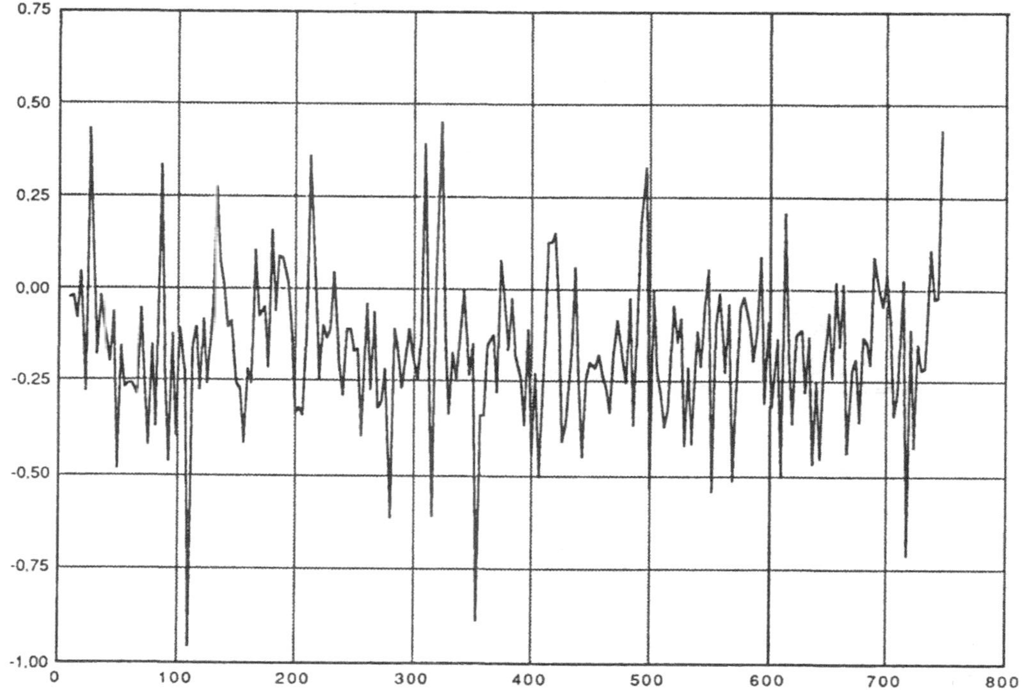


*Legend:* The Δ (every 3h20m) are shown in grads per minute and are reckoned positively in the anticlockwise sense.
The abscissas s are shown *in hours* from 0h on 16 March: $9.33 \leq s \leq 746$
Number of values of Δ: N = 222 [(746 - 9.33) / 3.33 = 221 = N - 1]
Average of the Δ: $\overline{m}$ = -0.162 grad per minute; standard deviation = 0.208 grad per minute.

*Source:* *Graph* 13810 (5 June 1996).

## *F.2 Periodic diurnal components of the deviations of the plane of oscillation of the paraconical pendulum from the meridian*

**Structure of the periodic diurnal components of the deviations $\Delta$.**

1 - *Table V*, whose presentation is analogous to that of *Table IV* of §E.1, shows the double-amplitudes of the periodic diurnal components of the displacements $\Delta$ of the paraconical pendulum from the meridian for the two periods of November-December 1959 and March-April 1960.

In this Table, the double-amplitudes of the diurnal components of $\Delta$ are given in hundredths of a grad per minute; in other words, the values of these double-amplitudes in grads per minute are multiplied by 100.

**Double-amplitudes of the periodic components of the deviations $\Delta$**

2 - The total of the double-amplitudes of the eight periodic components of the deviations $\Delta$ considered is virtually the same in November-December 1959 and in April 1960, and is of the order of 0.40 grad per minute (*Table V*).

For the wave of 24h50m, the double-amplitudes of the variations of the azimuth per minute are respectively, for the two month-long series, 0.0445 and 0.0651 grad per minute, *i.e.* values that respectively correspond to variations of $1.165 \times 10^{-5}$ and $1.704 \times 10^{-5}$ radian per second, corresponding to an average of $1.434 \times 10^{-5}$ rad/sec on average.[1]

[1] In fact we have

$$0.045 \frac{1}{60} \frac{\pi}{200} = 1.165 \times 10^{-5} \text{ rad/sec} \qquad 0.0651 \frac{1}{60} \frac{\pi}{200} = 1.704 \times 10^{-5} \text{ rad/sec.}$$

Their average is

$$\left(1.165 \times 10^{-5} + 1.704 \times 10^{-5}\right)/2 = 1.434 \times 10^{-5} \text{ rad/sec.}$$

The corresponding variations for the wave of 24h50m and for the two series of *enchained* values of November-December 1959 and March-April 1960 for the paraconical pendulum with anisotropic support are respectively $0.0454 \times 10^{-5}$ and $0.0524 \times 10^{-5}$ radian per second, corresponding to an average of $0.0489 \times 10^{-5}$ rad/sec, *i.e.* about *thirty times smaller*.[2] This difference is explained by the suppression, in the case of the isotropic support, of the restoring torque effect of the anisotropic support.

We have seen that, for the Moon, the calculated value of the wave of 24h50m is at most of the order of $10^{-13}$ radian per second.[3] The observed variations $\Delta$ are here *more than a hundred million times greater than the calculated values*.[4]

The direct measurements of the variations $\Delta$ of the angle of oscillation of the pendulum from the meridian thus confirm the conclusions of *Chapter I*.[5] *The amplitude of the periodic component of 24h50m of the observed movements of the paraconical pendulum is totally inexplicable in the framework of the current theory of gravitation.*

---

[2] From *Table I* of §A.3.1. of *Chapter I* above, we in fact have $2R_{25} = 1.3$ and 1.5 grads for the series of November-December 1959 and of March-April 1960. We thus have

$$\frac{1.3}{12.5}\frac{1}{3600}\frac{\pi}{200} = 0.0454 \times 10^{-5} \text{ rad/sec} \qquad \frac{1.5}{12.5}\frac{1}{3600}\frac{\pi}{200} = 0.0524 \times 10^{-5} \text{ rad/sec.}$$

and we have

$$1.165 \times 10^{-5}/0.0454 \times 10^{-5} = 25.7 \qquad 1.704 \times 10^{-5}/0.0524 \times 10^{-5} = 32.5$$

$$(25.7 + 32.5)/2 = 29.1$$

[3] *Table VII* of §B.2.5 of *Chapter I* above, p. 129.

[4] We have in fact

$$1.165 \times 10^{-5}/1.018 \times 10^{-13} = 114 \times 10^6 \qquad 1.704 \times 10^{-5}/1.018 \times 10^{-13} = 167 \times 10^6$$

$$\left(114 \times 10^6 + 167 \times 10^6\right)/2 = 140.5 \times 10^6$$

[5] *Chapter I* above, §B.2.6, p. 124.

**Relative amplitudes of the periodic components of the deviations Δ**

3 - In *relative* values, the average values compared to the total of the amplitudes of the variations Δ for the periods of 24h and 24h50m are more or less the same as for the azimuths of the direction of anisotropy.[6]

**Periodic structure of the deviations Δ from the meridian of the plane of oscillation of the paraconical pendulum, and the coefficients of the lunisolar forces**

4 - As previously for the direction $X$ of anisotropy (*Table IV* of §E.1.2), we see that the periodic structure of the deviations Δ (*Table V* below) is *totally* different from the periodic structure of the lunisolar periodicities corresponding to the current theory of gravitation.

For the lunisolar periodicities of the theory of tides, the amplitude of the wave of 24h in fact represents around 24% of the total of the amplitudes, while for the deviations Δ it only represents around 7% of the total of the amplitudes.

In the same way, for the deviations Δ, the amplitude of the wave of 24.84h is more or less double that of the wave of 24h, while for the lunisolar periodicities of the theory of tides the amplitude of the wave of 24h is around seven times as great as that of the wave of 24h50m.[7, 8]

Comparison of *Tables IV and V* demonstrates *the existence* of a great coherence between the periodic structures of the directions $X$ of anisotropy and the deviations Δ from the meridian of the plane of oscillation of the paraconical pendulum with isotropic support.

---

[6] In fact we have (*Table V* below): $13.5/6.96 = 1.94$, while we have (*Table IV*, p. 272): $18.2/8.92 = 2.04$.

Here as well, we see that the values corresponding to the ratios $a/b$ of the two *Tables IV and V* on pages 272 and 287 are of the same order of magnitude.

See note 5 on page 271 above.

[7] In fact we have (*Table V*): $13.5/6.96 = 1.94$ and $23.7/3.28 = 7.22$.

[8] The calculations of *Table V* were performed on 18 June 1996 (Calculations 1137 to 1152).

*TABLE V*

**DISPLACEMENTS FROM THE MERIDIAN OF THE PLANE OF OSCILLATION OF THE PARACONICAL PENDULUM IN HUNDREDTHS OF A GRAD PER MINUTE DIURNAL AND SEMI-DIURNAL LUNISOLAR EFFECTS**

***Observed double-amplitudes and coefficients of lunisolar periodicities***

| Periodicities | $K_1$ 23 h 93 | $M_1$ 24 h 84 | $O_1$ 25 h 82 | $Q_1$ 26 h 87 | $S_2$ 12 h | $M_2$ 12 h 42 | $N_2$ 12 h 66 | $2N_2$ 12 h 92 | Total of double amplitudes |
|---|---|---|---|---|---|---|---|---|---|
| Double-amplitude of wave: 2R in hundredths of a grad per minute | | | | | | | | | |
| November-December 1959 [1] | 4.44 | 4.45 | 6.41 | 4.15 | 4.95 | 6.95 | 2.23 | 8.39 | 42.0 |
| March-April 1960 [1] | 1.32 | 6.51 | 6.60 | 1.67 | 4.41 | 7.49 | 6.59 | 5.09 | 39.7 |
| November-December 1959 [2] | 10.6 | 10.6 | 15.3 | 9.88 | 11.8 | 16.5 | 5.31 | 20.0 | 100 |
| March-April 1960 [2] | 3.32 | 16.4 | 16.6 | 4.21 | 11.1 | 18.9 | 16.6 | 12.8 | 100 |
| Average of relative values: a | 6.96 | 13.5 | 16.0 | 7.04 | 11.5 | 17.7 | 11.0 | 16.4 | 100 |
| Current theory of lunisolar periodicities | | | | | | | | | |
| Coefficients [1] | 0.706 | 0.0977 | 0.377 | 0.0730 | 0.563 | 0.908 | 0.232 | 0.0235 | 2.98 |
| Relative values [2]: b | 23.7 | 3.28 | 12.6 | 2.45 | 18.9 | 30.5 | 7.78 | 0.789 | 100 |
| Ratio of relative values | | | | | | | | | |
| Ratios a/b | 0.294 | 4.12 | 1.27 | 2.87 | 0.608 | 0.580 | 1.41 | 20.8 | 1 |

*Legend:* (1): Absolute values. The amplitudes in grads per minute corresponding to the paraconical pendulum are multiplied by 100.
(2): Relative values=absolute values / total of double-amplitudes (or of coefficients)

*Sources:* 1) *Displacements,* from the plane of the meridian, *of the azimuth* of the plane of oscillation of the paraconical pendulum (Tables 6629 and 6630 of 1960).
2) Coefficients of the current theory of lunisolar periodicities: See *Table IV* of §E.1.

## F.3 Sidereal monthly lunar periodicity of the deviations $\Delta$ from the meridian of the plane of oscillation of the paraconical pendulum

In view of the results obtained in analysis of the azimuths $X$ of anisotropy of space, we are led to investigate whether the two month-long series of the deviations $\Delta$ present a sidereal monthly periodicity having period equal to 27.322 days. In fact, the amplitudes of the fittings are quite small, and by themselves are relatively insignificant.[1]

On the other hand it is *very remarkable* that, considering the dates of their respective peaks, the two fitting sinusoids to the deviations $\Delta$ are *almost exactly in phase*,[2] which would not be the case if one were to consider the fittings corresponding to the Moon's synodic period of 29.53 days. This agreement in phase, *very similar* to that detected for the azimuths $X$ of the direction of anisotropy,[3] *demonstrates the existence of a very strong coherence* between the observational data for the two periods of November-December 1959 and March-April 1960.

---

[1] For the sidereal lunar period of 27.322 days, the double-amplitudes $2r$ of the fittings of November-December 1959 and March-April 1960 are respectively equal to 0.0225 and 0.0388 grad per minute. Their ratios to the standard deviations of the $\Delta$ are thus respectively: $0.0225/0.201 = 0.112 = 1/8.92$, and $0.0388/0.2076 = 0.187 = 1/5.35$.

For the fittings corresponding to the azimuths $X$ of anisotropy (§E.2 above, pp. 275 and 276), we have $288.68/421.21 = 0.685 = 1/1.46$ and $136.37/278.93 = 0.489 = 1/2.045$.

The amplitudes of the periodic components of 27.322 days of the deviations $\Delta$ are thus relatively less marked than in the case of the azimuths of anisotropy $X$.

[2] The distance in days from 0h on 1 January 1960 of the peak of the fitting sinusoid for November-December 1959 is (*Graph XIII*):

$$365.25 - 335.95 = 29.30.$$

The distance between 0h on 1 January 1960 and the peak of the fitting sinusoid for March-April 1960 is 80.33 days (*Graph XIV*).

From the above, the distance in days between the peaks of the two fitting sinusoids is:

$$29.30 + 80.33 = 109.33 \text{ days}$$

This distance is *almost exactly equal* to four times the sidereal period of the Moon which is 27.322 days:

$$109.63/27.32 = 4.013$$

(See, in note 3 of §E.2 above, p. 274, the very similar calculation for the azimuths $X$ of the direction of anisotropy).

[3] Note 3 of §E.2 above, p. 274.

The observed agreement in phases appears *yet more striking* because the variations from one moment to the next of the deviations $\Delta$ are very great, and because the correlation coefficients are very small (*Graphs XI and XII* above, and *XIII and XIV* below).

*Graphs XIII and XIV* show the moving averages over 9 values of the deviations $\Delta$ of *Graphs XI and XII*, as well as fittings of the deviations $\Delta$ to sinusoids of period equal to 27.322 days, *i.e.* the sidereal period of the Moon.

The moving averages over 9 values[4] present all the characteristics of almost periodic functions.[5]

---

[4] Each series of 9 values corresponds to a time interval of 30 hours (3h20m $\times$ 9 = 30h).

[5] For almost periodic functions, see *Chapter I*, §A.5.4, p. 101.

It is very remarkable that certain parts of *Graph XIV* are almost identical to certain parts of *Graph XIII*, which indicates the existence of the same underlying periodic structure.

*GRAPH XIII*

**ISOTROPIC SUPPORT**
**DEVIATIONS Δ FROM THE MERIDIAN IN GRADS PER MINUTE OF THE PLANE OF OSCILLATION OF THE PARACONICAL PENDULUM**
***9h40m on 20 November - 13h on 15 December 1959***
***Moving averages over 9 values, and fittings of the Δ to a sinusoid of 27.322 days***

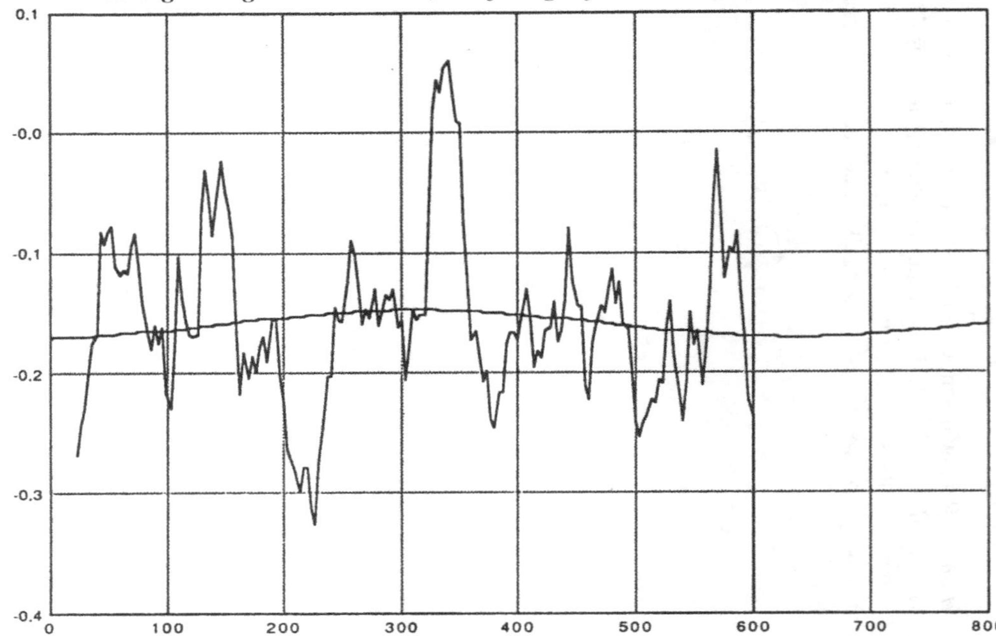


*Legend:* See *Graph XI* above. The abscissas are graduated in hours.
—— fitting sinusoid of the Δ to a period of 27.322 days (equal to 27.322 x 24 = 655.728h)
maximum of the sinusoid $s_0$ = 311.44 hours (s = 0 corresponds to 0h on 20 November), *i.e.* 23h on 2 December (corresponding to s = 335.95 days from 0h on 1 January 1959).
$\Sigma$ = 0.201 grad/minute; 2r = 0.0225 grad/minute; m = - 0.159 grad / minute; R = 0.0378; $1-R^2$ = 0.9986

*Sources: Graph* 13819 and *Calculation* 1166 (25 June 1996).

*GRAPH XIV*

**ISOTROPIC SUPPORT**
**DEVIATIONS Δ FROM THE MERIDIAN IN GRADS PER MINUTE OF THE PLANE OF OSCILLATION OF THE PARACONICAL PENDULUM**
***9h20m on 16 March - 2h on 16 April 1960***
***Moving averages over 9 values, and fittings of the Δ to a sinusoid of 27.322 days***

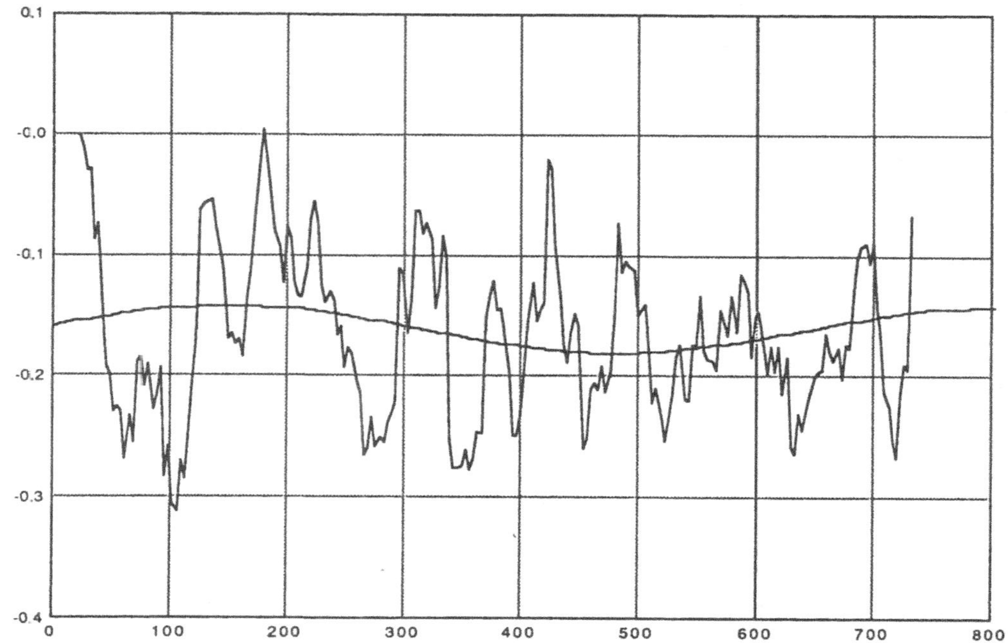


*Legend:* See *Graph XI* above. The abscissas are graduated in hours.

—— fitting sinusoid of the Δ to a period of 27.322 days (equal to 27.322 x 24 = 655.728h)
maximum of the sinusoid $s_0$ = 152.23 hours (s = 0 corresponds to 0h on 16 March), *i.e.* 8h on 22 March (corresponding to s = 80.33 days from 0h on 1 January 1960).
Σ = 0.208 grad/minute; 2r = 0.0388 grad/minute; m = - 0.162 grad / minute; R = 0.0645; $1\text{-}R^2$ = 0.9958

*Sources: Graph* 13818 and *Calculation* 1167.

## *F.4 Structural correspondence between the directions X of anisotropy of space and the deviations Δ from the meridian of the plane of oscillation of the paraconical pendulum*

A very strong underlying coherence exists between the directions *X* of anisotropy of space and the deviations *Δ* from the meridian of the plane of oscillation of the paraconical pendulum.

*By way of illustration*, we consider the period of November-December 1959, and consider the cumulative sums of the residues of the two following correlations.

*First correlation*: the correlation of the azimuths *X* of anisotropy of space with a sinusoid whose period is the sidereal monthly period of 27.322 days, along with a linear trend. This is the correlation shown in *Graph VII* above.

*Second correlation*: correlation of the moving average over 9 values of the deviations *Δ* with respect to the meridian of the plane of oscillation of the paraconical pendulum with a sinusoid whose period is the sidereal monthly period of the Moon, *along* with a linear trend.[1]

*Graph XV* shows that the two cumulative sums of the residues of these two correlations are virtually identical, up to a translation.

This agreement implies *a very strong underlying coherence* between the azimuths *X* of anisotropy and the deviations *Δ* of the displacements from the meridian of the plane of oscillation of the paraconical pendulum.

---

[1] This correlation is different from the correlation corresponding to *Graph XIII* above, which was performed against the periodicity of 27.322 days without including any trend.

Now, precisely, while this trend is completely negligible in the case of *Graph XIII*, it is no longer negligible if we consider the cumulative sums of the residues.

This agreement is *yet more remarkable* because the deviations $\Delta$ only correspond to the meridian, whereas the azimuths $X$ are determined based upon consideration of the deviations of the plane of oscillation of the pendulum in the ten azimuths in question.[2]

[2] The agreement stems from the fact that *the two curves of Graph XV derive from the same periodic structure that underlies all the observations*.

*GRAPH XV* **COMPARATIVE CUMULATIVE SUMS OF RESIDUES OF THE CORRELATIONS OF AZIMUTHS X OF ANISOTROPY AND OF MOBILE AVERAGES OVER 9 VALUES OF DEVIATIONS Δ WITH A SIDEREAL LUNAR PERIODICITY OF 27.322 DAYS, INCLUDING A LINEAR TREND**

***November - December 1959***

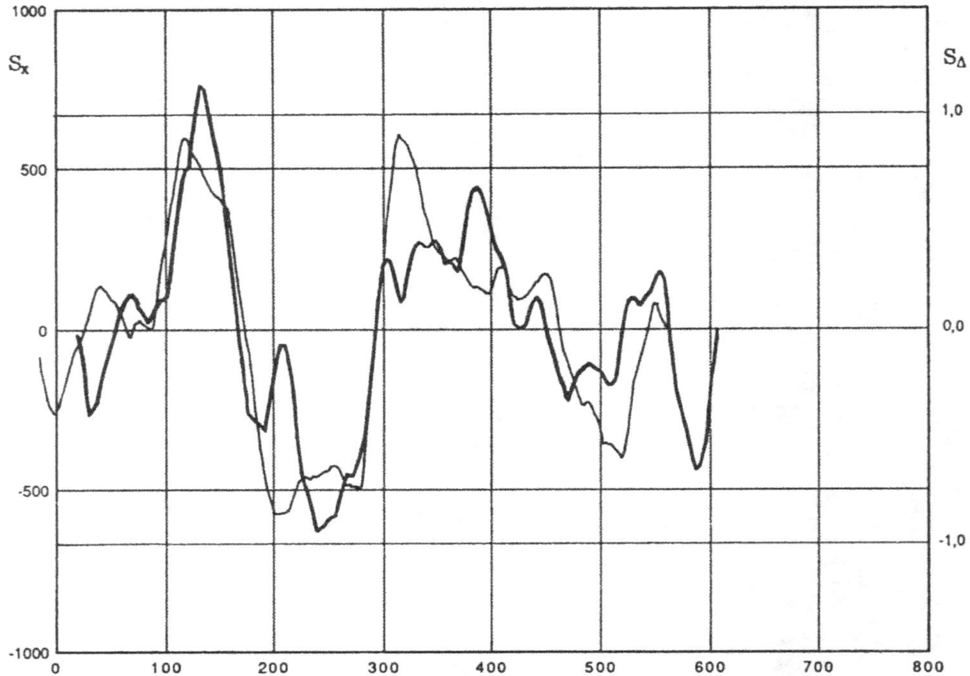


*Legend:* —— Cumulative sums $S_x$ of the residues of X; —— Cumulative sums $S_\Delta$ of the residues of Δ;
*Scale of s in hours*: corresponds to the $S_x$; s = 0 at 0h on 20 November 1959
*Scale of ordinates*: corresponds to $S_x$ on left and to $S_\Delta$ on right
*Time shift*: the curve for $S_\Delta$ is around Δs = 41.3h ahead of the curve for $S_x$

*Sources:* curve $S_x$ : *Calculation* 1112 and *Graph* 13834 (24 July 1996); curve $S_\Delta$: *Calculation* 1171 and *Graph* 13832 (23 July 1996)..

## *F.5 Cumulative sums of the differences $\Delta - \overline{\Delta}$ between the deviations $\Delta$ and their averages $\overline{\Delta}$ in the azimuths North-South and East-West*

The previous discussions show the difficulties of approaches based upon consideration of the deviations $\Delta$ from the meridian of the plane of oscillation of the paraconical pendulum, due to the influence of the balls.

This realization leads one to consider the cumulative values of the differences $\Delta - \overline{\Delta}$, where $\overline{\Delta}$ represents the average value of the $\Delta$. These cumulative values in fact make it possible to reduce the perturbing influence of the balls.

I have extended this analysis to the deviations $\Delta$ from the East-West direction of the plane of oscillation of the paraconical pendulum. The curves representative of these deviations have an appearance completely analogous to that of *Graphs IX and X* above (pp. 278 and 279).

### Comparative periodic structure of the cumulative values of the $\Delta - \overline{\Delta}$

1 - Comparison of the four Graphs representing the cumulative values of the differences $\Delta - \overline{\Delta}$ for the two azimuths North-South and East-West and for the two periods of November-December 1959 and March-April 1960 brings out very significant regularities and the same underlying periodic structure in the different series. I limit myself to two of these comparisons, by way of illustration.[1]

[1] For four representative Graphs, there are six combinations of two. *They are all significant.*

*Graph XVI* represents the cumulative values of the differences $\Delta - \overline{\Delta}$, on the one hand for the series of November-December 1959 and the East-West direction, and on the other hand for the series of March-April 1960 and the North-South direction. The correspondence of the two representations is *all the more remarkable because this is a case of two different month-long series of observations and two different azimuths*.

*Graph XVII* represents the cumulative values of the differences $\Delta - \overline{\Delta}$, on the one hand for the series of November-December 1959 and the North-South direction, and on the other hand for the series of March-April 1960 and the East-West direction. Here as well, the correspondence of the two representations is all the more remarkable because this is a case of two different month-long series of observations and two different fittings.

The two *Graphs XVI and XVII* demonstrate the existence in the deviations $\Delta$ *of the same underlying almost periodic structure, very evident in spite of the perturbing influence of the balls*.

*GRAPH XVI*

**COMPARATIVE CUMULATIVE VALUES $\Sigma(\Delta - \overline{\Delta})$**

***November-December 1959 - Values of $\Delta$ in the East-West direction***
***March-April 1960 - Values of $\Delta$ in the North-South direction***

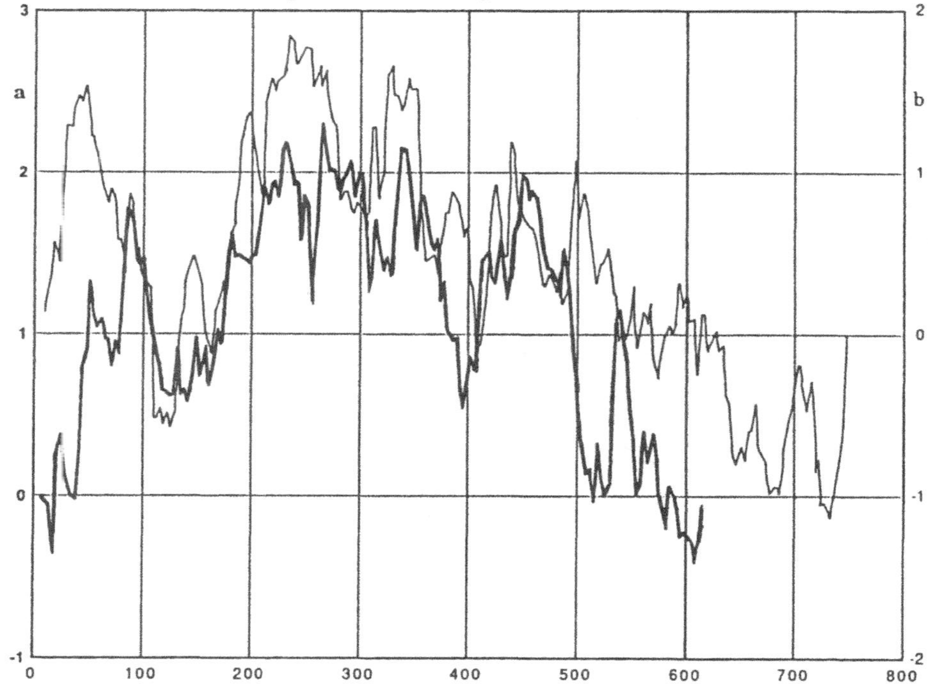


*Legend:* time unit=1 hour; values every 3h20m

a —— November-December 1959. $\Sigma(\Delta - \overline{\Delta})$ corresponding to the East-West direction.
8h on 20 November - 14h40m on 15 November. $8 \leq s \leq 614.666$; N=183 (s=0 at 0h on 20 November)

b —— March-April 1960. $\Sigma(\Delta - \overline{\Delta})$ corresponding to the North-South direction.
9h20m on 16 March - 2h on 16 April. $9.333 \leq s \leq 746$; N=222 (s=0 at 0h on 16 March)

*Sources:* a —— *Graph* 13848 (9 September 1996); b —— *Graph* 13854 (12 September 1996).

*GRAPH XVII*

**COMPARATIVE CUMULATIVE VALUES $\Sigma(\Delta - \overline{\Delta})$**

***November-December 1959 - Values of Δ in the North-South direction***

***March-April 1960 - Values of Δ in the East-West direction***

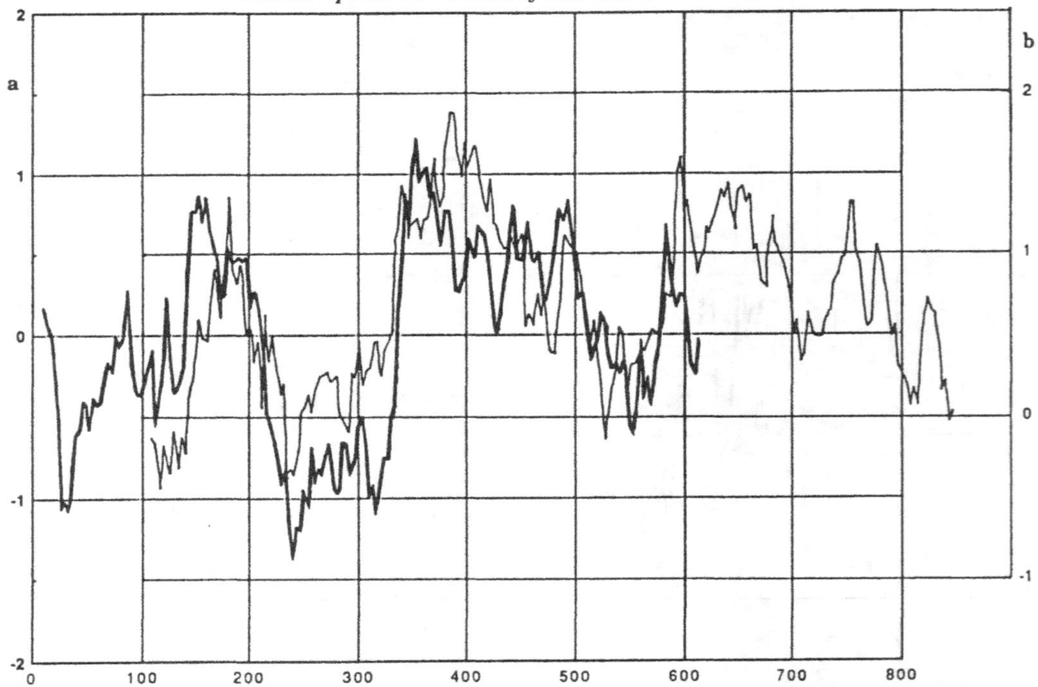


*Legend:* time unit=1 hour; values every 3h20m

a —— November-December 1959. $\Sigma(\Delta - \overline{\Delta})$ corresponding to the North-South direction. 9h40m on 20 November - 13h on 15 December. $9.666 \leq s \leq 613$; N=182 (s=0 at 0h on 20 November)

b —— March-April 1960. $\Sigma(\Delta - \overline{\Delta})$ corresponding to the East-West direction. 10h on 16 March - 2h40m on 16 April. $10 \leq s \leq 746.666$; N=222 (s=0 at 0h on 16 March)

*Sources:* a —— *Graph* 13852 (11 September 1996); b —— *Graph* 13850* (12 September 1996).

**Sidereal monthly lunar periodicity of 27.322 days of the cumulative values of the differences $\Delta - \overline{\Delta}$**

2 - *Graphs XVIII, XIX, XX and XXI* represent fittings to the sidereal period of 27.322 days of the cumulative values of the differences of the deviations $\Delta$ with their averages during the two periods of November-December 1959 and March-April 1960, and for the directions North-South and East-West.

The two *Graphs XVIII and XIX* show a difference in phase of 6 days for the directions North-South and East-West during the period of November-December 1959.

On the other hand, the two *Graphs XX and XXI* show a remarkable agreement in phase for the directions North-South and East-West during the period of March-April 1960.[2, 3]

[2] Since the derivative $\dot{y}(t)$ of a periodic function $y(t)$ is a quarter of a period ahead of $y(t)$, comparison of *Graphs XIII and XIV*, pp. 290 and 291, and of *Graphs XVIII and XIX* allows us to verify that, in the first approximation, this is indeed so.

For *Graphs XIII and XVIII* corresponding to the North-South direction and to the period of November-December 1959, we have respectively

$$s_0 = 335.95 \qquad s_0 - (27.322/4) = 342.73 - 6.83 = 335.90$$

and for *Graphs XIV and XX* corresponding to the North-South direction and to the period of March-April 1960 we have respectively

$$s_0 = 80.33 \qquad s_0 - (27.322/4) = 89.24 - 6.83 = 82.41$$

[3] On the agreements of phase between the different series, see *Section G* below.

*GRAPH XVIII*

**CUMULATIVE VALUES $\Sigma(\Delta - \overline{\Delta})$**

***20 November - 15 December 1959: North-South direction***

*Fitting to the Moon's sidereal period of 27.322 days with a linear trend*

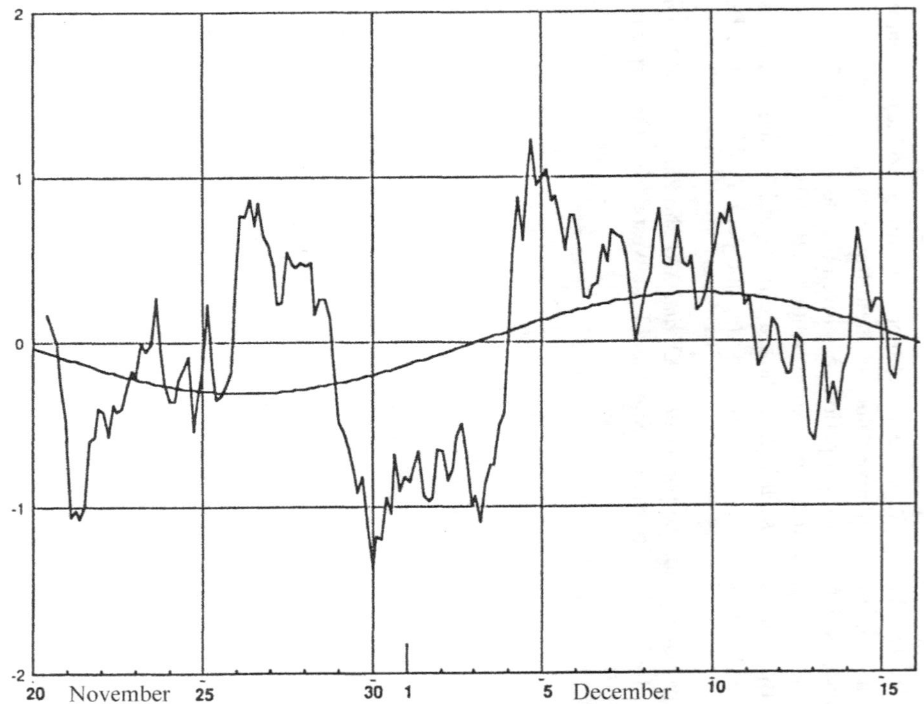


*Legend:* $\Sigma$ = 0.589 grad; R = 0.371; 1-$R^2$ = 0.862; $\sigma$ = 0.547; m = 0.028; r = 0.324; N = 182
trend = - 0.000132 per day; date of peak of sinusoid: 17h32m on 9 December 1959.
($s_0$=342.73 days from 0h on 1 January 1959; $s_0$ - 6.83 = 335.90).
See the *Legend* of *Graph XXVII* of *Chapter I.*

*Sources: Calculation* 1215, *Graph* 13853 (12 September 1996).

*GRAPH XIX*

**CUMULATIVE VALUES $\Sigma(\Delta - \bar{\Delta})$**

***20 November - 15 December 1959: East-West direction***

*Fitting to the Moon's sidereal period of 27.322 days with a linear trend*

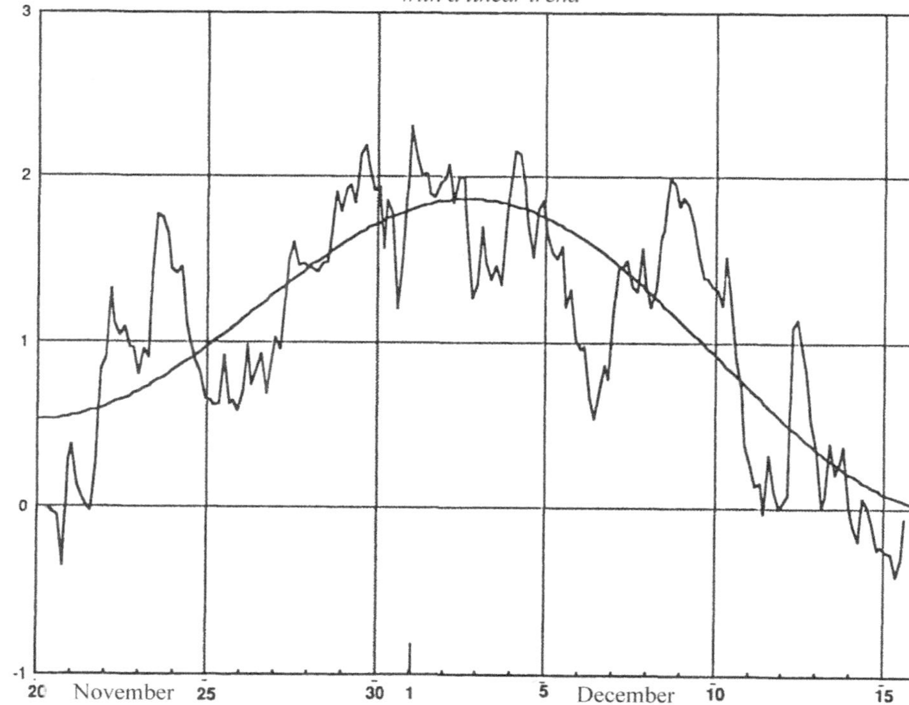


*Legend:* $\Sigma = 0.692$ grad; $R = 0.803$; $1\text{-}R^2 = 0.355$; $\sigma = 0.413$; $m = 1.326$; $r = 0.801$; $N = 183$
trend = - 0.000844 per day; date of peak of sinusoid: 4h46m on 3 December 1959.
($s_0$=336.20 days from 0h on 1 January 1959; $s_0 - 6.83 = 329.37$).
See the *Legend* of *Graph XXVII* of *Chapter I.*

*Sources: Calculation* 1213, *Graph* 13849 (9 September 1996).

*GRAPH XX*

**CUMULATIVE VALUES $\Sigma(\Delta - \overline{\Delta})$**
***16 March - 16 April 1960: North-South direction***
*Fitting to the Moon's sidereal period of 27.322 days with a linear trend*

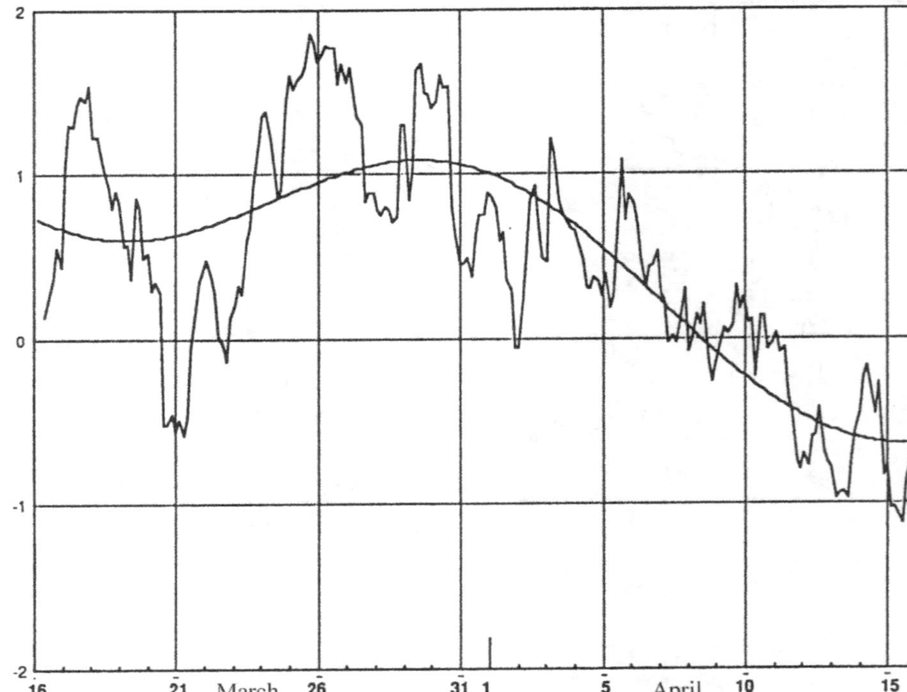


*Legend:* $\Sigma = 0.726$ grad; $R = 0.781$; $1\text{-}R^2 = 0.390$; $\sigma = 0.453$; $m = 1.220$; $r = 0.512$; $N = 222$
trend = - 0.00188 per day; date of peak of sinusoid: 5h41m on 31 March 1960.
($s_0 = 89.24$ days from 0h on 1 January 1960; $s_0 - 6.83 = 82.41$).
See the *Legend* of *Graph XXVII* of *Chapter I.*

*Sources: Calculation* 1216, *Graph* 13855 (6 September 1996).

*GRAPH XXI*

**CUMULATIVE VALUES $\Sigma(\Delta - \bar{\Delta})$**

***16 March - 16 April 1960: East-West direction***

*Fitting to the Moon's sidereal period of 27.322 days with a linear trend*

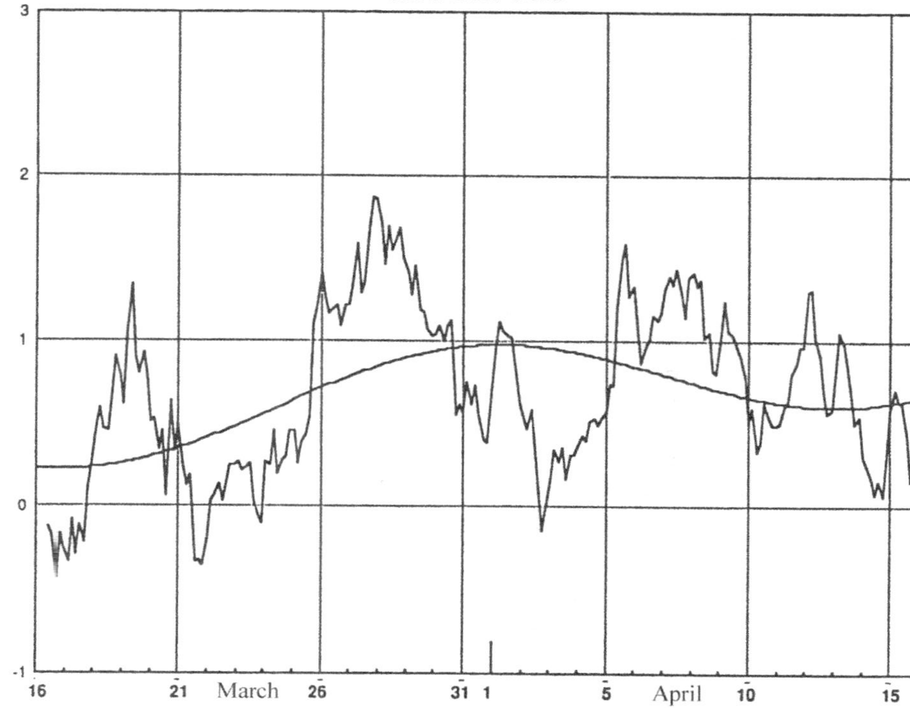


*Legend:* $\Sigma = 0.507$ grad; $R = 0.457$; $1\text{-}R^2 = 0.791$; $\sigma = 0.451$; $m = 0.484$; $r = 0.277$; $N = 222$
trend = 0.000573 per day; date of peak of sinusoid: 6h18m on 31 March 1960.
($s_0 = 89.26$ days from 0h on 1 January 1960; $s_0 - 6.83 = 82.43$).
See the *Legend* of *Graph XXVII* of *Chapter I*.

*Sources:* *Calculation* 1214, *Graph* 13851 (6 September 1996).

### *F.6 Structural similarities between the cumulative values of the* $\Delta - \overline{\Delta}$ *and the cumulative values of the* $\cos \mathbf{2X} - \overline{\cos \mathbf{2X}}$

Between the cumulative values

$$\sum \left(\Delta - \overline{\Delta}\right) \quad \text{and} \quad \sum \left(\cos 2X - \overline{\cos 2X}\right)$$

we can detect *significant structural similarities* for the two periods of November-December 1959 and March-April 1960, and for the directions North-South and East-West.

Thus, and for example, *Graph XXII* shows a significant structural correspondence between the two series

(a) $\sum \left(\cos 2X - \overline{\cos 2X}\right)$ in November-December 1959; and

(b) $-\sum \left(\Delta - \overline{\Delta}\right)$ in November-December 1959, direction East-West one being shifted with respect to the other.

This naturally does not mean that the series (b) is a linear function of the series (a). It merely means that *both of them correspond to one and the same almost periodic underlying structure.*

*GRAPH XXII*

**STRUCTURAL SIMILARITY OF THE CUMULATIVE VALUES**

**$-\Sigma(\Delta - \overline{\Delta})$ and $\Sigma(\cos 2X - \overline{\cos 2X})$**

***Period 20 November - 15 December 1959, direction East-West***

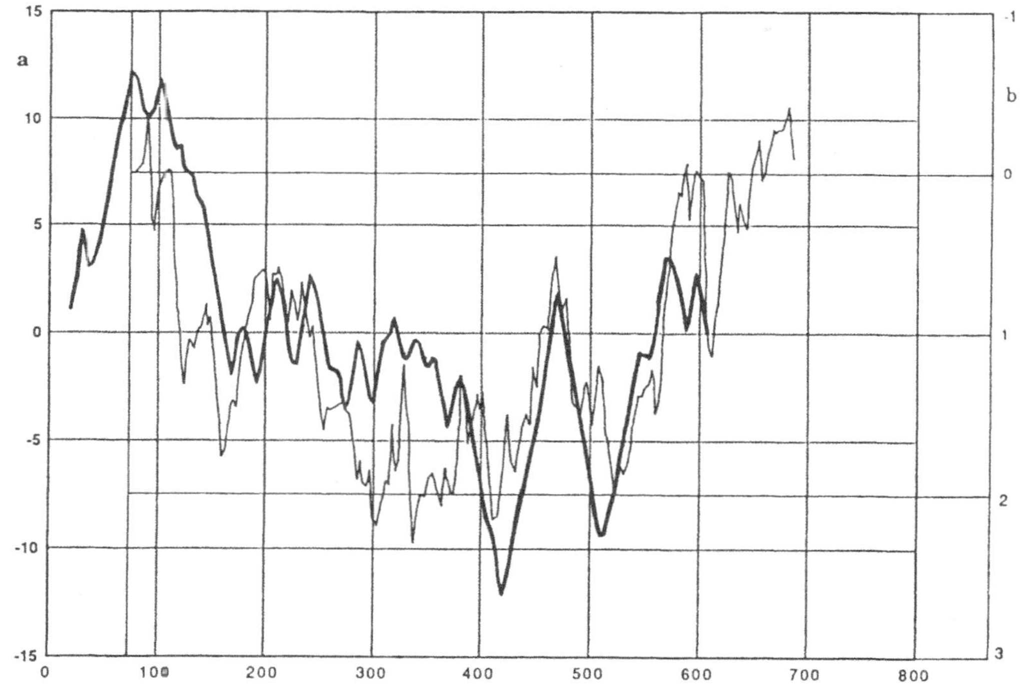


*Legend:* time unit=1 hour. s=0 at 0h on 20 November 1959.

a —— $\Sigma(\cos 2X - \overline{\cos 2X})$

b —— $-\Sigma(\Delta - \overline{\Delta})$, East-West direction.

10h on 16 March - 2h40m on 16 April. $10 \leq s \leq 746.666$; N=222 (s=0 at 0h on 16 March)

*Sources:* a —— *Calculation* 1232, *Graph* 13863 (11 October 1996); b —— *Table* 12836, *Graph* 13848 (9 September 1996).

# G PHASES OF THE SIDEREAL LUNAR MONTHLY PERIODIC COMPONENTS OF 27.322 DAYS OF THE CHARACTERISTICS OF THE MOVEMENT OF THE PARACONICAL PENDULUM AND THE DECLINATION OF THE MOON

## *G.1 Paraconical pendulum with anisotropic support, directions X of anisotropy of space, and deviations Δ from the directions North-South and East-West of the plane of oscillation of the paraconical pendulum with isotropic support. Phases of the sidereal lunar periodicities of 27.322 days compared together*

The fittings of a period of 27.322 days to the azimuths $X$ of the directions of anisotropy of space and the deviations $\Delta$ of the plane of oscillation of the paraconical pendulum with isotropic support from the directions North-South and East-West have demonstrated remarkable agreements in phase between the month-long observational periods of November-December 1959 and March-April 1960.

Agreements in phase have also been observed for the enchained series of the paraconical pendulum with anisotropic support at Bougival and at Saint-Germain in July 1958.[1]

*Table VI* below demonstrates the agreements in phase of the various fittings to the period of 27.322 days, *taking as reference* the period of November-December 1959 to which corresponds a very good fitting of the azimuths $X$ of anisotropy of space.[2]

[1] *Graphs XXVII and XXVIII* (*Chapter I*, §C.3, pp. 158-159) representative of the average $(S+B)/2$ and the half-difference $(S-B)/2$. The two fitting sinusoids have their respective peaks at 20h on 24 July 1958 and 0h on 24 July 1958.

[2] *Graph VII* of §E.2 above, p. 275.

The series 1, 2, 3, and 5 of *Table VI* exhibit very remarkable agreements in phase, the deviations of the phases $s_0^*$ being different from their average 331.66 by one day at the most. Similarly, the series 4 and 6 have practically identical phases.

It should also be pointed out that the average of the phases of the six series, *i.e.* $\overline{s_0^*} = 333.15$, only differs by a day and a half from the value of $s_0^*$ corresponding to the minimum declination of the Moon at 21h on 1 December 1959, which is 334.87.

Since most correlations include values of $1 - R^2$ that are relatively high, the quality of the fittings *would remain substantially the same if, in the fittings, the values of $s_0^*$ in Table VI were replaced by the value of $s_0^*$ corresponding to the minimum declination of the Moon.*

*TABLE VI*

## PARACONICAL PENDULUM
## FITTINGS TO THE SIDEREAL LUNAR PERIOD OF 27.322 DAYS
## COMPARISON OF THE PHASES OF THE FITTING SINUSOIDS

| Periods | q | Series | R | $1-R^2$ | r | Time of maximum | $s_0$ | p | $s_0^*$ |
|---|---|---|---|---|---|---|---|---|---|
| July 1958 | 1 | (S + B) / 2 | 0.629 | 0.604 | 1.02 | 24 July 20h | 204.85 | 18 | 331.40 |
| | 2 | (S - B) / 2 | 0.680 | 0.537 | 1.70 | 24 July 0h | 204.03 | 18 | 330.57 |
| November-December 1959 | 3 | X | 0.994 | 0.013 | 144.3 | 29 November 12h | 332.50 | 0 | 332.50 |
| | 4 | Δ | 0.0378 | 0.9986 | 0.0113 | 2 December 23h | 335.95 | 0 | 335.95 |
| March-April 1960 | 5 | X | 0.965 | 0.068 | 68.2 | 18 March 5h | 76.21 | 4 | 332.17 |
| | 6 | Δ | 0.0645 | 0.9958 | 0.0194 | 22 March 8h | 80.34 | 4 | 336.30 |
| Time of minimum declination of Moon in Nov-Dec 1959: 21h on 1 December, $s^*_0 = 334.87$ | | | | | | | | | |

*Legend: Value of r:* r is expressed in grads for the series 1, 2, 3, and 5, and in grads/minute for the series 4 and 6.

*Calculation of $s^*_0$* (date of peak of periodic component nearest to 29 November 1959)

July 1958: $s_0 - 365.25 + 27.322\ p$ (p is an integer)

March-April 1960: $s_0 + 365.25 - 27.322\ p$

Averages of the $s^*_0$: (1, 2, 3, 4, 5, 6) = 333.15 (1, 2, 3, 5) = 331.66

(1, 2) = 331.0 (3, 5) = 332.33 (4, 6) = 336.12

*Sources: Graphs XXV and XXVI, Chapter I*, §C.3; *Graphs IV and V*, and *VIII and IX* of §E.2 and §E.3 above.

*Calculations*: 1 (Calculation 944*); 2 (Calculation 948*); 3 (Calculation 1112); 4 (Calculation 1166); 5 (Calculation 1111); 6 (Calculation 1167); (January-February 1996).

## G.2 *Deviations $\Delta$ of the plane of oscillation of the paraconical pendulum with isotropic support from the directions North-South and East-West and the cumulative values $\sum\left(\Delta-\overline{\Delta}\right)$. Phases of the sidereal monthly lunar periodicities of 27.322 days compared*

The Graphs representing the deviations $\Delta$ of the plane of oscillation of the paraconical pendulum with isotropic support from the East-West direction for the two periods of November-December 1959 and March-April 1960 have appearances completely comparable to *Graphs XI and XII* of §F.1 above. The influence of the balls also seems to be very important here.

Consideration of the cumulative values $\sum\left(\Delta-\overline{\Delta}\right)$ of the differences between the values of $\Delta$ and their averages $\overline{\Delta}$ makes it possible both to reduce the influence of the balls and also to amplify the amplitudes of the periodicities that have relatively longer periods.[1]

*Table VII* enables us to compare the phases $s_0^*$[2] of the sidereal monthly lunar periodicities of 27.322 days for the deviations $\Delta$ with the cumulative values $\sum\left(\Delta-\overline{\Delta}\right)$ for the directions North-South and East-West, for the two periods of November-December 1959 and March-April 1960, while taking the period of November-December 1959 as reference period, as for *Table VI.*

In order to be able to compare the phases of $\Delta$ and $\sum\left(\Delta-\overline{\Delta}\right)$, it should be pointed out that, for any sinusoidal function $y(t)$, the derivative $\dot{y}(t)$ is a quarter of a period $T/4$ ahead of $y(t)$. *Table VII* takes account of this fact, as shown in its *Legend.*

---

[1] See §F.5 above, p. 295.

[2] The dates of the peaks during November-December 1959 of the 27.322 day periodicities have been made comparable, as indicated in *Table VII.*

It is very remarkable that the averages of the $s_0^*$ for the $\Delta$ and the $\sum\left(\Delta-\overline{\Delta}\right)$ are substantially identical, 332.63 and 332.04 in November-December 1959 and 338.53 and 338.47 in March-April 1960. This equality is even more remarkable because the correlation coefficients corresponding to the deviations $\Delta$ are very small.

In the same way, the general average 335.46 of the $s_0^*$ is, to less than a day, almost the same as the value of 334.87 of the $s_0^*$ corresponding to the minimum declination of the Moon in November-December 1959.

*TABLE VII*

**PARACONICAL PENDULUM WITH ISOTROPIC SUPPORT**
**DEVIATIONS Δ OF OSCILLATION PLANE FROM DIRECTIONS N-S AND E-W**
**AND CUMULATIVE SUMS $\Sigma (\Delta - \bar{\Delta})$**

***Comparison of phases of fitting sinusoids***

*November-December 1959 and March-April 1960*

| Series | | $\Sigma (\Delta - \bar{\Delta})$ | | | | | Δ | | | | Averages (3+5) / 2 |
|---|---|---|---|---|---|---|---|---|---|---|---|
| | | R | $1-R^2$ | 1 $s_0$ | 2 $s_0 - 6.83$ | 3 $s^*_0$ | R | $1-R^2$ | 4 $s_0$ | 5 $s^*_0$ | |
| 1959 | N - S | 0.371 | 0.862 | 342.73 | 335.90 | 335.90 | 0.0378 | 0.9986 | 335.95 | 335.95 | 335.92 |
| | E - W | 0.803 | 0.355 | 336.20 | 329.37 | 329.37 | 0.0925 | 0.991 | 328.44 | 328.44 | 328.90 |
| | Averages | | | 339.46 | 332.63 | 332.63 | | | 332.04 | 332.20 | 332.41 |
| 1960 | N - S | 0.781 | 0.390 | 89.24 | 82.41 | 338.47 | 0.0645 | 0.9958 | 80.34 | 336.30 | 337.38 |
| | E - W | 0.457 | 0.791 | 89.26 | 82.49 | 338.59 | 0.0213 | 0.9995 | 84.68 | 340.64 | 339.71 |
| | Averages | | | 89.25 | 82.42 | 338.53 | | | 82.51 | 338.47 | 338.55 |
| General average | | | | | | 335.58 | | | | 335.33 | 335.46 |
| Time of minimum declination of Moon in November-December 1959: 21h on 1 December, $s^*_0 = 334.87$ | | | | | | | | | | | |

*Legend:* Calculation of $s^*_0$ (date of peak of periodic component nearest to 29 November 1959)

1) The phase of $\Sigma (\Delta - \bar{\Delta})$ is a quarter of a period (27.322/4 = 6.830 days) behind that of Δ.
2) For March-April 1960: $s^*_0 = s_0 + 365.25 - 4 \times 27.322 = s_0 + 255.96$ (*Table VI*)

*Sources:* $\Sigma (\Delta - \bar{\Delta})$ : Calculations 1215, 1213, 1216, and 1214
Δ : Calculations 1166, 1208, 1167, and 1209 (June-September 1996)

### G.3 Directions X of anisotropy of space and cumulative values $\sum\left(\cos 2\mathbf{X}-\overline{\cos 2\mathbf{X}}\right)$. Phases of sidereal monthly lunar periodicities of 27.322 days compared

*Table VIII* shows a comparison of the phases of the fittings to the sidereal monthly lunar period of 27.322 days of the azimuths $X$ of anisotropy of space, and the cumulative values of the differences $\cos 2X-\overline{\cos 2X}$.

As has been already indicated in *Table VI*, the agreement of the phases of $X$ for the two periods of November-December 1959 and March-April 1960 is very remarkable (332.50 and 332.17). This is all the more the case in view of the fact that the calculations that have led to the two *Graphs II and III* representing the directions $X$ of anisotropy are relatively complex.

It is also very remarkable that the average of the phases of the cumulative values $\sum\left(\cos 2X-\overline{\cos 2X}\right)$ is substantially the same as the average of the phases of the azimuths $X$ (331.09 days against 332.33 days). This equality seems even more remarkable when we consider *Graphs II, III, IV, and V* that represent the azimuths $X$ of anisotropy and the $\cos 2X$.[1]

[1] The relation of the fittings of the cumulative values $\sum\left(\cos 2X-\overline{\cos 2X}\right)$ to the sidereal lunar period of 27.322 days with the fittings of the azimuths $X$ is a *very complicated* dependence, a deep analysis of which would require many calculations for simulation.

*TABLE VIII*

**PARACONICAL PENDULUM WITH ISOTROPIC SUPPORT**

**DIRECTIONS OF ANISOTROPY X AND CUMULATIVE VALUES $\Sigma$ (cos 2X - $\overline{\cos 2X}$)**

***Comparison of phases of fitting sinusoids***

*November-December 1959 and March-April 1960*

| Periods | q | Series | R | $1-R^2$ | r | Time of maximum | $s_o$ | p | $s_o^*$ |
|---|---|---|---|---|---|---|---|---|---|
| November-December 1959 | 1 | X | 0.994 | 0.013 | 144.3 | 29 November 12h | 332.5 | 0 | 332.50 |
| | 2 | $\Sigma(\cos 2X - \overline{\cos 2X})$ | 0.810 | 0.344 | 5.38 | 23 November 1h | 326.09 | 0 | 326.09 |
| March-April 1960 | 3 | X | 0.965 | 0.068 | 68.2 | 18 March 5h | 76.21 | 4 | 332.17 |
| | 4 | $\Sigma(\cos 2X - \overline{\cos 2X})$ | 0.879 | 0.227 | 10.04 | 22 March 3h | 80.14 | 4 | 336.10 |
| Time of minimum declination of Moon in November-December 1959: 21h on 1 December, $s^*_0 = 334.87$ | | | | | | | | | |

*Legend:* See the Legend of *Table VI*

Averages of the $s^*_0$ : (1, 3) = 332.33 (2, 4) = 331.09
: (1, 2, 3, 4) = 331.71

*Sources:* *Graphs* : *VII, IX, VIII, X* (May-October 1996)
*Calculations* : 1 (Calculation 1112); 2 (Calculation 1233); 3 (Calculation 1111); 4 (Calculation 234) (21 May-14 October 1996)

## *G.4 Overall view*

Taken together, the three Tables give evidence of agreements of phase that are extremely remarkable, in view of the *very different* natures of the series in question: $X, \Delta, \sum(\Delta - \overline{\Delta})$, and $\sum(\cos 2X - \overline{\cos 2X})$.[1]

Whatever theoretical interpretation may be given to these different series, their coherence constitutes *an undeniable experimental fact* whose validity is *incontestable*.[2]

[1] *Tables VI, VII and VIII*, pp. 308, 311 and 313.

[2] This coherence also gives proof of the quality of the observers: on the one hand of Jacques Bourgeot and Annie Rolland, my collaborators at IRSID, and on the other hand of the employees at IRSID who participated in the continuous observations of November-December 1959 and March-April 1960.

I remind the reader that, during his turn for observations, each observer had to perform every twenty minutes, not only physical operations relating to the paraconical pendulum with isotropic support, but also operations relating to the paraconical pendulum with anisotropic support.

These observations were extremely tiring, and required very great professional dedication and very great skill from the observers.

# H ENCHAINED EXPERIMENTS FROM 28 SEPTEMBER TO 4 OCTOBER 1959 AND TOTAL SOLAR ECLIPSE OF 2 OCTOBER 1959

## *H.1 Variations of azimuth $\phi$ of plane of oscillation of paraconical pendulum with isotropic support during the enchained experiments of September-October 1959*

### Amplitudes of the azimuth variations

1 - *Graph XXIII* shows the variations of the hourly values of the azimuth $\phi$ of the plane of oscillation of the paraconical pendulum with *isotropic support* during the *enchained* experiments of September-October 1959.[1]

During this period of five days, the amplitude of the azimuth variations was considerable, around 350 grads. Its order of magnitude was quite comparable to the amplitudes of the variations of the direction *X* of anisotropy of space for November-December 1959 and March-April 1960 (*Graphs II and III* above).[2]

[1] For the principle of enchained experiments, see §A. 1.2. of *Chapter I* above.

In fact, from 16h20m on 28 September to 16h on 30 September 1959, the experiments were *doubly enchained*. A comparative analysis of these two series of observations is presented in the *Second volume* of this work, *Chapter II*, Section A (on the principle of doubly enchained experiments, see §B.1.1. of *Chapter I* above).

In *Graph XXIII*, the hourly values correspond to the even observations from 28 September to 30 September.

[2] For *Graph II*, the average variation *for a period of five days* was around 350 grads; for *Graph III*, it was around 150 grads (§D.3 above, pp. 261 et 262).

These enchained experiments of September-October 1959 were performed *in parallel* with *enchained* experiments with the paraconical pendulum with anisotropic support.[3]

### The total solar eclipse of 2 October 1959

2 - The enchained experiments with the isotropic support preceded and followed *the total solar eclipse of 2 October 1959*, which took place from 10h50m to 14h UT.[4]

We saw that, from 16h on 28 September to 20h on 2 October, the azimuth of the plane of oscillation of the *isotropic* pendulum remained between azimuth 0 grad (the direction of the meridian) and azimuth 100 grads (the East-West direction). After the end of the eclipse, the azimuth of the plane of oscillation of the *isotropic* pendulum abruptly returned, and this plane turned through 350 grads in the anticlockwise sense in 34 hours, while in parallel the plane of oscillation of the pendulum with *anisotropic* suspension continually oscillated near the meridian within the interval (-10 grads, -40 grads).

In the 12 hours before the eclipse, the plane of oscillation of the isotropic pendulum progressively approached the meridian.

The observations performed with the paraconical pendulum *with isotropic support* thus confirm the comparative analysis given above of the two eclipses of 1954 and 1959 in the case of the anisotropic suspension.[5] *While taking place, both the eclipses had the effect of bringing the plane of oscillation of the paraconical pendulum towards the common direction of the Moon and the Sun on the meridian.*

---

[3] See *Chapter I* above, §A.3.I., *Table I*, p. 92.

[4] See *Chapter I*, §D.2, pp. 166-167.

[5] *Chapter I* above, Section D.3, pp. 158-159.

## *H.2 Variations during the enchained experiments of September-October 1959 of* $\cos 2\phi$ *and* $\sin 2\phi$*, corresponding to the azimuth* $\phi$ *of the paraconical pendulum with isotropic support.*

The two *Graphs XXIV* show the oscillations of the paraconical pendulum with isotropic support from 28 September to 4 October 1959, with $\cos 2\phi$ in one case, and $\sin 2\phi$ in the other case, being shown along the ordinates instead of $\phi$.

These *Graphs* have the advantage of better representing the variations over time of the *effective* azimuthal situation of the plane of oscillation of the paraconical pendulum with isotropic suspension.

In fact, *one single direction and one only* of the plane of oscillation of the paraconical pendulum corresponds to each value of $2\phi$.[1] The value $\cos 2\phi = 1$ corresponds to the meridian, and the value $\cos 2\phi = -1$ corresponds to the East-West direction. For $\cos 2\phi = 0$, the plane of oscillation of the pendulum can be either one of the two directions inclined at 50 grads with respect to the meridian.

Analogous considerations hold for $\sin 2\phi$.

In general, for analysis of the movements of the paraconical pendulum with isotropic support and for analysis of the direction of anisotropy $X$, it is helpful to use this second representation (*Graphs XXIV*) in parallel with the first (*Graph XXIII*).[2]

[1] In fact, if $\phi$ varies through 200 grads, $2\phi$ varies through 400 grads. In the calculations for the mobile correlations, the azimuth $X$ is only determined to 200 grads (§C.4 above).

[2] See §D.4 above, pp. 263-266.

*GRAPH XXIII*

**PARACONICAL PENDULUM WITH ISOTROPIC SUPPORT**

***Enchained series***

***16h20m on 28 September 1959 - 3h40m on 4 October 1959***

***Hourly azimuths $\phi$ in grads***

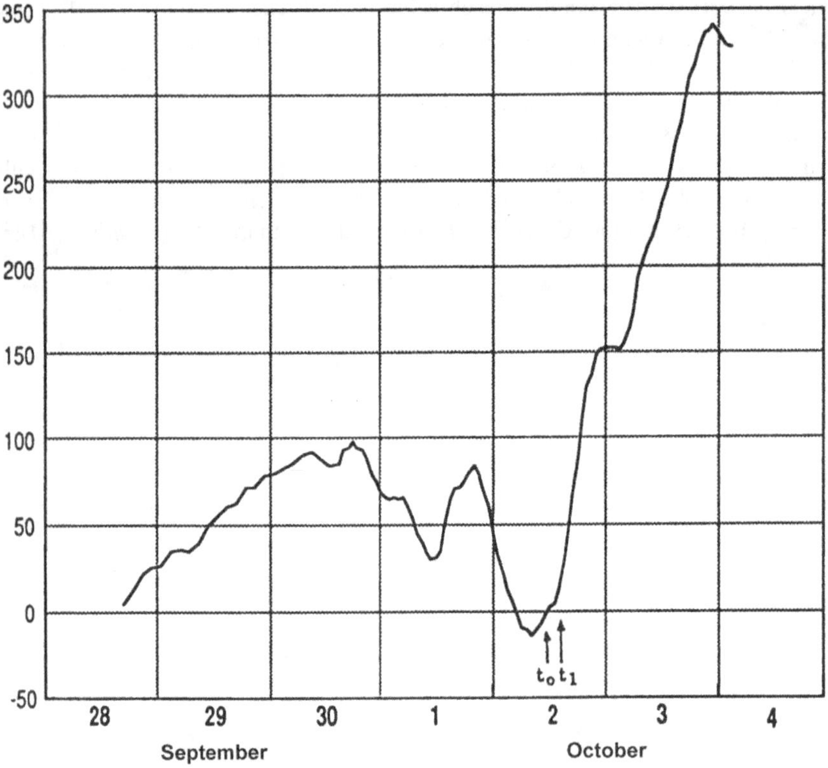


*Legend:* $t_0$ and $t_1$, start and end of the total solar eclipse of 2 October 1959.
Azimuths 0 and 200 grads correspond to the meridian.
Azimuths 100 and 300 grads correspond to the East-West direction.

*Sources:* *Graphs 8617* (20 May 1996) and *Table 7461* (4 November 1982).

*GRAPH XXIV*

**PARACONICAL PENDULUM WITH ISOTROPIC SUPPORT**

***Enchained series***

***16h20m on 28 September 1959 - 3h40m on 4 October 1959***

**cos 2ϕ and sin 2ϕ**

ϕ =hourly values of the azimuth of the plane of oscillation

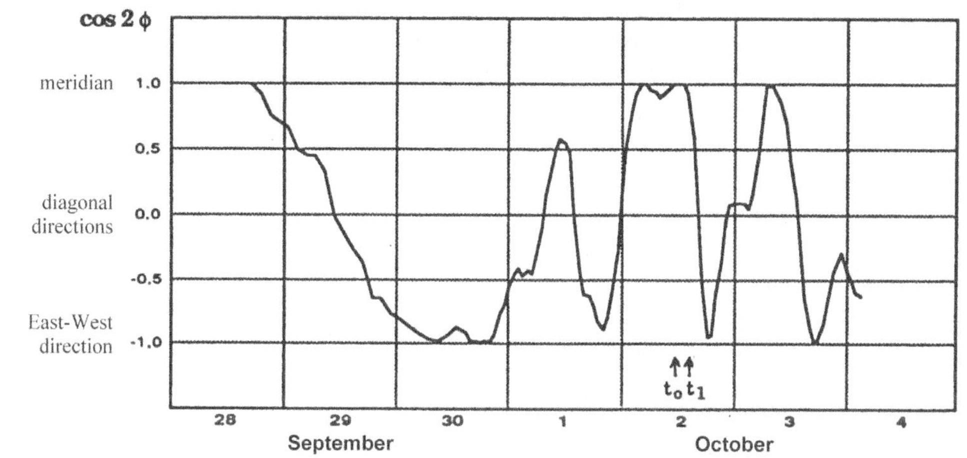


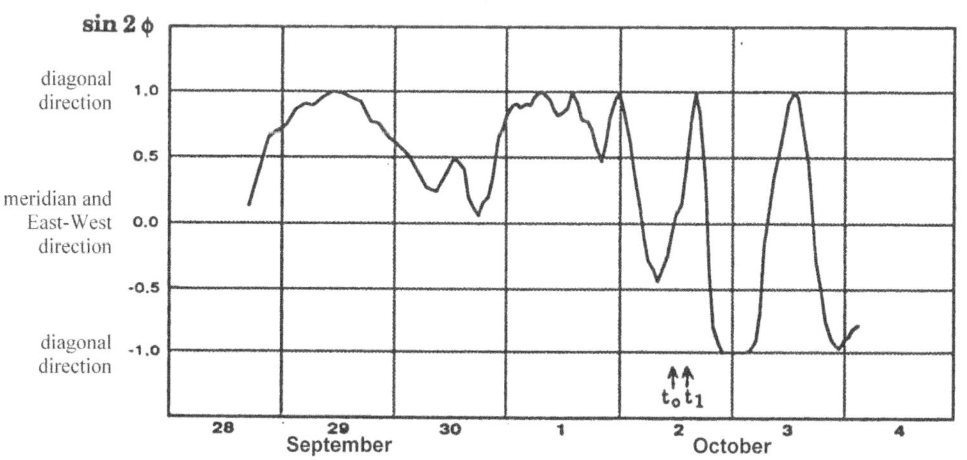


*Legend:* cos 2 ϕ = +1 : meridian (ϕ = 0 and 200 grads)
cos 2 ϕ = -1 : direction East-West (ϕ = 100 and 300 grads)
sin 2 ϕ = 0 : meridian (ϕ = 0 and 200 grads)
and direction East-West (ϕ = 100 and 300 grads)
sin 2 ϕ = +1 : ϕ = 50 and 250 grads
sin 2 ϕ = -1 : ϕ = -50 and 150 grads

*Source: Graph XXIII*

# I MOVEMENT OF THE PLANE OF OSCILLATION OF THE PARACONICAL PENDULUM WITH ISOTROPIC SUPPORT UNDER THE HYPOTHESIS OF ANISOTROPY OF INERTIAL SPACE

## *I.1 Equations of movement and calculation of the coefficients of anisotropy*

From the data given for the paraconical pendulum *with anisotropic support*,[1] it follows that, to the first approximation, under the hypothesis of anisotropy of inertial space, the movement of the paraconical pendulum with *isotropic support* can be represented by the two equations:

(1) $\dot{\phi} = -\omega \sin L + \frac{3}{8} p \alpha \beta \qquad p = \sqrt{g/l}$

(2) $\dot{\beta} = \sum_i \frac{p}{2} \alpha \varepsilon_i \sin 2\,(X_i - \phi)$

Here, $-\omega \sin L$ represents the Foucault effect, $\alpha$ and $\beta$ represent the major and the minor axis of the ellipse of oscillation of the paraconical pendulum in radians, $X_i$ represents the direction of the celestial body $i$, and $\varepsilon_i$ represents the coefficient of the anisotropy that it induces.

*Table IX* presents the principle of the calculation of $\varepsilon_i$ in the case of the Moon and the periodicity of 24h50m.

[1] *Chapter I*, §F.3, *Table XIII*, p. 212.

From the equality of the theoretical values and the observed values of $\dot{\phi}$, we can deduce the equality

(3) $(3/32)\Delta t\, p^2\alpha^2\varepsilon_l \overline{\sin 2\,(X_l-\phi)} = 1.434\times 10^{-5}$

(4) Order of magnitude of $\varepsilon_l = 1.434\times 10^{-5}/9.32 = 1.54\times 10^{-6}$

The influence of $\varepsilon_l$ is thus of the order of magnitude of a millionth of gravity.[2] Here of course it is only possible to discuss the order of magnitude.

---

[2] We have found (*Chapter I*, §F.2.3, note 7, p. 204) that

(1) Foucault force/gravity $= 3.20\times 10^{-6}$

and we have

(2) Amplitude of the periodicity of 24h50m/Foucault effect
$= 1.434\times 10^{-5}/0.55\times 10^{-4} = 0.260$

Accepting that the forces are proportional to their effects, we thus have

(3) Periodic force/gravity $= 0.260\times 3.20\times 10^{-6} = 0.832\times 10^{-6}$

The periodic effect of 24h50m thus corresponds to a force of the order of a millionth that of gravity.

Thus we find the same estimate for the order of magnitude of $\varepsilon$ by a calculation *of an entirely different nature*.

In fact, the estimation $1.54\times 10^{-6}$ of the order of magnitude of $\varepsilon_l$ is the more precise one.

*TABLE IX*

## EQUATIONS OF MOVEMENT OF THE PLANE OF OSCILLATION IN THE CASE OF THE ISOTROPIC SUPPORT UNDER THE HYPOTHESIS OF ANISOTROPY OF INERTIAL SPACE

*Calculation of the coefficients of anisotropy* $\varepsilon_i$

*Variations* $\dot{\phi}$ *and* $\dot{\beta}$ *of* $\phi$ *and* $\beta$

(1) $\dot{\phi} = -\omega \sin L + \frac{3}{8} p \alpha \beta$ $\qquad p = \sqrt{g/l} = = \sqrt{981/83} = 3.44$

(2) $\dot{\beta} = \sum_i \frac{p}{2} \alpha \varepsilon_i \sin 2(X_i - \phi)$ $\qquad p\ \alpha/4 = 0.086$

*Orders of magnitude of the* $\varepsilon_i$ *- Illustration in the context of the Moon - Periodicity of 24h50m:*

*Theoretical value of the average component* $\dot{\phi}_{tl}$ *of* $\dot{\phi}$ during a single experiment of duration $\Delta t = 14$ min = 840 sec.[1, 2]

(3) $\dot{\phi}_{tl} = \frac{3}{8} p \bar{\alpha} \bar{\beta}$ $\qquad \bar{\alpha} = 0.1$

$$= \frac{3}{8} p \bar{\alpha} [\frac{1}{2} \Delta t \frac{p}{2} \bar{\alpha} \varepsilon_1 \overline{\sin 2(X_1 - \phi)}]$$

$$= \frac{3}{32} \Delta t\ p^2 \bar{\alpha}^2 \varepsilon_1 \overline{\sin 2(X_1 - \phi)}$$

$$= 9.32\ \varepsilon_1 \overline{\sin 2(X_1 - \phi)}$$

(4) Order of magnitude of $\dot{\phi}_{tl} = 9.32\ \varepsilon_l$ rad/sec.

*Observed value* (§F.2.2, p. 284)

(5) $\dot{\phi}_{ol} = 1.434\ 10^{-5}$ rad. / sec.

*Order of magnitude corresponding to* $\varepsilon_l$ *:*

(6) $1.434\ 10^{-5} / 9.32 = 1.54\ 10^{-6}$

-----

(1) The lines ——— mean that these are average values over the duration $\Delta t$ of the experiment, =14 minutes.

(2) From equation (3), the order of magnitude of the effect of anisotropy of inertial space upon the $\dot{\phi}_l$ is equal to $(3/32)\ p^2\ \bar{\alpha}^2\ \Delta t\ \varepsilon_l$.

From the current theory of gravitation, the effect of universal attraction upon $\dot{\phi}_{tl}$ is proportional to $(3/32)\ \bar{\alpha}^2\ \Delta t\ K_l$ (*Table VII*, §B.2, of *Chapter I* above, p. 129).

We thus have :

$$(3/32)\, p^2 \bar{\alpha}^2\, \Delta t\ \varepsilon_1 / (3/32)\, \bar{\alpha}^2\, \Delta t\ K_1 = (g/l)(\varepsilon_1 / K_1)$$

## *I.2 The second order differential equation in* $\mathbf{u} = \mathbf{X} - \phi$

In order to simplify the notation, we can write the two equations (1) and (2) above as equations (1) and (2) of *Table X*.

We can define the average direction $X$ of anisotropy and the average coefficient $\varepsilon$ of anisotropy by Equations (5) and (6) of *Table X*. Naturally the $X_i$ and $\varepsilon_i$ are functions of time, and the same is true for $X$ and for $\varepsilon$.

Finally, equations (1) and (2) above can be written (equations 7 and 8 of *Table X*)

(1) $\dot{\phi} = -\omega \sin L + \lambda\beta \qquad \lambda = (3/8)\, p\alpha$

(2) $\dot{\beta} = \mu \sin 2\,(X - \phi) \qquad \mu = (1/2)\, p\alpha\varepsilon$

where $X$ denotes the direction of anisotropy of inertial space, and $\varepsilon$ is the *average* coefficient of the anisotropy of inertial space.

By differentiating equation (1) with respect to time, and replacing $\dot{\beta}$ by expression (2), we obtain the second order differential equation (equation 11 of *Table X*)

(3) $\ddot{u} + \lambda\mu \sin 2u = \ddot{X} \qquad u = X - \phi$

where $\ddot{X}$ represents the second differential of $X$.[1]

This equation allows us to discuss the modes of movement of the paraconical pendulum with isotropic support easily, in the case in which $u = X - \phi$ remains relatively small and the variations of $X$ are small with respect to the variations of $u$.

[1] A detailed study of equation (3) will be given in the *Second volume* of this work (*Chapter II*, Section A, p. 28 above).

Equation (3) without its second term is formally identical to the equation for oscillations of a one-dimensional pendulum with arbitrary amplitudes, whose analysis is classical (see for example Jules Haag, 1952, *Les mouvements vibratoires* (Vibratory Movements), P.U.F., Vol. I, no. 210, pp. 237-239 [154]).

In this case, equation (3) is written (equation 12 of *Table X*)

(4) $\ddot{\phi} + \Omega^2 \phi = \Omega^2 X \qquad \Omega^2 = 2\lambda\mu$

Discussion of this equation shows that the direction $\phi$ of the plane of oscillation of the pendulum tends to oscillate around the direction $X$ of anisotropy (equation 15 of *Table X*) with its own period (equation 12 of *Table X*).[2]

(5) $\Theta = 2\pi/\sqrt{2\lambda\mu}$

*Overall, equation (4) allows us to determine, in the first approximation, the movement of the paraconical pendulum with isotropic support during each 14-minute experiment.*

---

[2] Since we have (equation 2 of *Table X*)

(1) $\mu_l = (1/2)\,p\alpha\varepsilon_l = 0.172\varepsilon_l$

and (equation 6 of *Table IX*)

(2) $\varepsilon_l = 1.54 \times 10^{-6}$

we have

(3) $\mu_l = 0.172 \times 1.54 \times 10^{-6} = 0.265 \times 10^{-6}$

From equation (5) of *Table X*, the coefficient $\mu$ depends in a relatively complex manner on the $\mu_i$. Since we still do not know all the periodic components of the movement of the paraconical pendulum, we can only formulate *very approximate* hypotheses.

*By way of illustration*, and accepting the value $\mu = 2 \times 10^{-6}$ for example, we would have, for $\lambda = 0.129$ (*Table X*)

(4) $\Omega = \sqrt{2\lambda\mu} = 0.718 \times 10^{-3} \qquad \Theta = 2\pi/\Omega = 8747$ seconds $= 2.43$ hours

It goes without saying that a calculation like this can only give an order of magnitude.

What is certain is that the period $\Theta$ is relatively small with respect to the periods of the periodicities determining the movement of the paraconical pendulum, and that accordingly the direction $X(t)$ can be considered as being constant during a single oscillation ($\Omega$). For a period $\Theta = 2.43$ hours, a 14-minute experiment corresponds to a fraction of around a tenth of the period $\Theta(14/145.8 = 0.096)$. From equations (15) and (16) of *Table X*, it follows that during a 14-minute experiment the deviation between the azimuth $\phi$ and the direction $X$ of anisotropy diminishes, and that accordingly the plane of oscillation shifts towards the direction of anisotropy.

An analogous conclusion holds for the paraconical pendulum with anisotropic support. We thus see that my choice in 1954 to perform experiments with a duration of 14 minutes is justified *a posteriori*. This is also confirmed by observation (see *Graph VI* of *Chapter I* above, §B.1.1, p. 104).

See note 15 of §F.3.3 of *Chapter I* above, pp. 209-210.

## *TABLE X*

### EQUATIONS OF MOVEMENT OF THE PARACONICAL PENDULUM IN THE CASE OF THE ISOTROPIC SUPPORT UNDER THE HYPOTHESIS OF ANISOTROPY OF INERTIAL SPACE

*Second-order equation in $u = X - \phi$*

*Simplified expressions of the equations of movement (Table IX, equations 1 and 2)*

(1) $\dot{\phi} = -\omega \sin L + \lambda \beta$ $\quad\quad$ $\lambda = (3/8)\, p\, \alpha = 0.129$

(2) $\dot{\beta} = \sum_i \mu_i \sin 2 (X_i - \phi)$ $\quad\quad$ $\mu_i = (1/2)\, p\, \alpha\, \varepsilon_i = 0.172\, \varepsilon_i$

$X_i$ and $\varepsilon_i$ represent the azimuth and the coefficient of anisotropy due to celestial body i.

*Average direction of anisotropy of inertial space*

(3) $\mu \sin 2 (X - \phi) = \sum_i \mu_i \sin 2 (X_i - \phi)$

(4) $\mu \sin 2X = \Sigma \mu_i \sin 2X_i$ $\quad\quad$ $\mu \cos 2X = \Sigma \mu_i \cos 2X_i$

(5) $\mu = \sqrt{\Sigma \mu_i^2 + 2 \Sigma \mu_i \mu_j \cos 2 (X_i - X_j)}$ $\quad\quad$ $\mu = (1/2)\, p\, \alpha\, \varepsilon = 0.172\, \varepsilon$

(6) $\mathrm{tg}\, 2X = \Sigma \mu_i \sin 2X_i \,/\, \Sigma \mu_i \cos 2X_i$

*Expressions for $\dot{\phi}$ and $\dot{\beta}$*

(7) $\dot{\phi} = -\omega \sin L + \lambda \beta$ $\quad\quad$ $\lambda = (3/8)\, p\, \alpha = 0.129$

(8) $\dot{\beta} = \mu \sin 2 (X - \phi)$ $\quad\quad$ $\mu = (1/2)\, p\, \alpha\, \varepsilon = 0.172\, \varepsilon$

*Differential equation in $\phi$*

(9) $\ddot{\phi} - \lambda \mu \sin 2 (X - \phi) = 0$

(10) $u = X - \phi$ $\quad\quad$ $\ddot{u} = \ddot{X} - \ddot{\phi}$

(11) $\ddot{u} + \lambda \mu \sin 2u = \ddot{X}$

*Linearization* for small values of $u = X - \phi$ and for $X(t)$ *varying slowly*

(12) $\ddot{\phi} + \Omega^2 \phi = \Omega^2 X$ $\quad\quad$ $\Omega^2 = 2 \lambda \mu$ $\quad\quad$ $\Theta = 2\pi / \sqrt{2 \lambda \mu}$

(13) $\phi(t) = A \cos \Omega t + B \sin \Omega t + X(t)$

(14) Initial conditions for t=0 $\quad\quad$ $\phi = \phi_o$ $\quad\quad$ $\dot{\phi}_o = -\omega \sin L$

***Solution***

(15) $\phi(t) = (\phi_o - X_o) \cos \Omega t - \dot{X}_o \dfrac{\sin \Omega t}{\Omega} - \omega \sin L \dfrac{\sin \Omega t}{\Omega} + X(t)$

(16) $\dot{\phi}(t) = \Omega\, (X_o - \phi_o) \sin \Omega t - \dot{X}_o \cos \Omega t - \omega \sin L \cos \Omega t + \dot{X}(t)$

***Illustration***

(17) $\mu = 2\; 10^{-6}$ $\quad\quad$ $\lambda = 0.129$ $\quad\quad$ $\Omega = \sqrt{2 \lambda \mu} = 0.718\; 10^{-3}$

$\Theta = 2\pi / \Omega = 8747$ seconds $= 2.43$ hours $\quad\quad$ $\Omega\, \Delta t = 0.603$ radian

$\sin \Omega\, \Delta t = 0.567$ $\quad\quad$ $\cos \Omega\, \Delta t = 0.824$ $\quad\quad$ ($\Delta t = 840$ sec.)

# J OVERALL VIEW

Overall, the implementation of an isotropic support in 1959 made it possible to obtain *very significant and very coherent* results, which marked *a decisive step* forward in my experiments with the paraconical pendulum.

## *J.1 Determination of the direction* X *of anisotropy of space (Sections A, B, and C)*

1 - First of all, a method of analysis - *the method of mobile correlations* - was defined that made it possible to determine at each instant *the direction X of anisotropy of space*, and its variations over time.

2 - This anisotropy of space is *a fact of observation*, but it can be interpreted as corresponding to an anisotropy of inertia according to the direction in question of the anisotropy of space, and as resulting from astronomical influences.

This interpretation *is a hypothesis*, but *everything happens as though it is justified.*

3 - Application of the method of mobile correlations to the observational series of November-December 1959 and March-April 1960 demonstrates the existence *of the same average anisotropy of space*, about one-seventeenth as strong as that which characterized the anisotropic suspension, with *the anisotropy due to the new support being considered as practically non-existent.*

*This average anisotropy of inertial space is oriented approximately East-West.*

## *J.2 Azimuth X of anisotropy of space in November-December 1959 and March-April 1960 (Sections D and E)*

1 - The method of mobile correlations enables us to determine the direction *X* of anisotropy of space at each instant during the experiments of November-December 1959 and March-April 1960.

The direction *X* of anisotropy of space generally *varies considerably over time:* over a single month, the direction *X* varied through several hundreds of grads - in fact, about 1800 grads in November-December 1959 and about 900 grads in March-April 1960. These rotations respectively correspond to about four turns anticlockwise and to about two turns clockwise.

2 - The two month-long series of the azimuths *X* of the direction of anisotropy of space are characterized by diurnal periodicities whose amplitudes are around *five times greater*, on average, than in the case of the enchained series for the paraconical pendulum with anisotropic support.

As in the case of the anisotropic support, the lunar periodicity of 24h50m is particularly prominent.

In terms of its amplitudes, this periodic structure is *very different* from the structure that characterizes the lunisolar periodicities resulting from the current theory of gravitation.

3 - Analysis of the two month-long series of the azimuths *X* of the direction of anisotropy of November-December 1959 and March-April 1960 also demonstrates the existence of a *sidereal* (*i.e.* non-synodic) *lunar periodicity* of 27.322 days of relatively significant amplitude.

The two periodic components of 27.322 days were *very remarkably in phase.* Within two days, these periodic components attained their maxima when the declination of the Moon reached its minimum.

4 - The cumulative values of the differences $\cos 2X - \overline{\cos 2X}$ between the values $\cos 2X$ and their average $\overline{\cos 2X}$ also present remarkable sidereal monthly lunar periodicities.

## *J.3 Displacements Δ of the plane of oscillation of the paraconical pendulum from the directions North-South and East-West (Section F)*

1 - The displacements $\Delta$ of the plane of oscillation of the paraconical pendulum *from the meridian* during the two observational periods November-December 1959 and March-April 1960 were similarly characterized by the existence of a *very significant* periodic diurnal structure, roughly analogous to the periodic diurnal structure of the azimuths $X$ of anisotropy as far as the relative values of the amplitudes are concerned.

Similarly, in November-December 1959 and March-April 1960, the displacements $\Delta$ were characterized by sidereal lunar periodicities of 27.322 days that were *remarkably in phase*. As with the azimuths $X$ of anisotropy, to less than two days, these periodic lunar components attained their maxima when the declination of the Moon reached its minimum.

2 - A remarkable structural interdependence is seen between the azimuths $X$ of anisotropy and the deviations $\Delta$ from the meridian of the plane of oscillation of the paraconical pendulum with isotropic support.

3 - The observed periodic components of 24h50m of the plane of oscillation of the paraconical pendulum with isotropic support from the meridian have amplitudes *more than a hundred million times greater* than the amplitudes calculated from the current theory of gravitation, and by this very fact are *totally inexplicable* in the framework of that theory.

4 - The displacements $\Delta$ of the plane of oscillation of the paraconical pendulum with isotropic support from the East-West direction have periodic structures very similar to those corresponding to the direction North-South, in particular as to the existence of a sidereal monthly lunar periodicity of 27.322 days.

5 - The same is true for the cumulative values of the differences $\Delta - \overline{\Delta}$ between the deviations $\Delta$ and their average $\overline{\Delta}$ in the directions North-South and East-West.

## *J.4 Remarkable agreements in phase (Section G)*

1 - Not only do the azimuths $X$ of anisotropy and the displacements $\Delta$ from the meridian of the plane of oscillation in 1959 and 1960 exhibit *a remarkable agreement in phase with the declination of the Moon*, but also the same holds for the azimuths of the *enchained* series for the paraconical pendulums with anisotropic support *at Bougival and at Saint-Germain in 1958*.

2 - The same is true for the cumulative values of the differences $\Delta - \overline{\Delta}$ and for the cumulative values of the differences $\cos 2X - \overline{\cos 2X}$.

## *J.5 Enchained series for the paraconical pendulum with isotropic support (Section H)*

1 - The enchained and doubly enchained series of experiments performed from 28 September to 4 October 1959 with the paraconical pendulum and the isotropic support demonstrated *considerable* variations in azimuth.

In particular, a variation of 350 grads in two days was observed, while in parallel the azimuth of the anisotropic suspension stayed virtually stationary within an interval of the order of 50 grads.

2 - These same experiments allowed for a *very significant* comparison of the effects of the total solar eclipse of 2 October 1959 upon the simply enchained series performed, on the one hand with the isotropic suspension, and on the other hand with the anisotropic suspension.

## J.6 Anisotropy of inertial space (Section I)

The observed movements of the paraconical pendulum with isotropic support can be easily explained by *supposing an anisotropy of inertial space of the order of* $10^{-6}$.

## J.7 Very significant results

Overall, the construction of an isotropic support in 1959 and the corresponding experiments made it possible to obtain results *of absolutely exceptional importance*.

*Not only did they confirm the existence of periodic structures that are completely inexplicable in the framework of current theories, having amplitudes a hundred million times greater than the amplitudes calculated with current theory, but also they made it possible to demonstrate the existence of a direction of anisotropy that was variable over time, and to specify its azimuth at each moment.*[1]

[1] Unfortunately it was not possible to continue these *very promising* experiments, due to the shutting down of my laboratory at Saint-Germain in June 1960.

If the experiments could have been continued, undoubtedly the best path to pursue would have been to employ two isotropic suspensions, *using one with enchained experiments and the other with the method of mobile correlations*.

Even with only one suspension, it would also have been possible to perform enchained observations with even-numbered experiments and to apply the method of mobile correlations for the odd-numbered experiments.

# Chapter III
# MY EXPERIMENTS OF JULY 1958 ON OPTICAL DEVIATIONS OF SIGHTINGS AT MARKS, AND THEIR CONTINUATION DURING FEBRUARY-MARCH 1959

*The most essential characteristic of scientific technique is that it proceeds from experiment, not from tradition. The experimental habit of mind is a difficult one for most people to maintain; indeed, the science of one generation has already become the tradition of the next.*

Bertrand Russell *

# A EXPERIMENTS ON OPTICAL DEVIATIONS OF SIGHTINGS AT MARKS AND AT COLLIMATORS, IRSID, 1958, AND IGN, 1959

## My experiments on optical sightings at marks at IRSID in 1958

Comparison of known anomalies in precision leveling and triangulation operations[1] compared to the anomalies that I detected in the movement of the paraconical pendulum with anisotropic support[2] compelled me to perform *a series of optical sightings North-South and South-North at fixed marks*, in parallel with the crucial experiments of Saint-Germain and Bougival in July 1958.[3]

---

* Bertrand Russell, 1931, *The Scientific Outlook*, Chapter 6 [239].

[1] J. Vignal, *Nivellement de précision. Publications Techniques de l'Institut Géographique National* (Accuracy Leveling, Technical Publications of the Institut Géographique National), Paris, 1948 [274]. Also see P. Tardi, *Traité de Géodésie*, Vol. 1, Part 1 (Treatise on Geodesy), Gauthier-Villars, Paris, 1951, pp. 370-376 [258].

M. Allais, 21 May 1958, *Anomalies of triangulation and leveling operations. Possible explanations, and comparison with experiment*, IRSID, 5p [18]. In this *Note*, I wrote:

"*The operations of triangulation and leveling give evidence of systematic anomalies (systematic errors of Lallemand, those related to the direction of operation and those related to lateral refraction in the operations of triangulation). The relative order of magnitude of these anomalies is the same as that determined in study of various phenomena (Miller's experiments, falling bodies, and the paraconical pendulum). It seems highly probable to me that they derive from one and the same cause...*

"*This analysis suggests a very simple experiment. It is only necessary to take sightings, simultaneously in azimuth and elevation, from one station S at two marks S' and S", one towards the South and one towards the East, in a continuous manner at the rate of one observation every half-hour, over the duration of a month. It seems probable to me that the observed variations will show periodic perturbations of the same characteristics as the ones that I found, and that Miller appears to me to have detected. It would also be desirable that, as far as possible, the one mark at S should be sighted from S' and S", simultaneously in azimuth and elevation.*"

[2] *Chapter I*, above.

[3] *Chapter I, Section C*, above.

These observations fully met my expectations, and yielded *very striking* results. The variations of the readings made with the aid of fixed telescopes at fixed marks were *equal and in the same sense* for the North-South and South-North sightings. Moreover, a very remarkable correspondence was noted between the periodic components of the sightings at marks and the periodic components of the azimuths of the paraconical pendulum with anisotropic support.

### Experiments on optical sightings at marks and at collimators at IGN in 1959

- The 1958 experiments at IRSID were followed in 1959 at the Institut Géographique National (IGN) by experiments that gave evidence for *a very remarkable structure of deviations of sightings at marks and sightings at collimators.* Overall, the 1959 experiments at IGN *completely* confirmed the essential results of the experiments at IRSID in July 1958 relating to deviations of sightings at marks. Moreover, they showed evidence of monthly lunar periodicities of sightings at marks and at collimators that are *completely inexplicable in the framework of currently accepted theories.*[4]

[4] *Trials of continuous automatic registration* of optical deviations of sightings at marks were performed at Palaiseau and at IGN by Michel Kasser in 1992-1993. I give a brief analysis of these trials in *Chapter III* of the second volume of this work (p. 29 above). Also see note 1 on page 371 below and *Chapter VIII* below, Sections A.1, note 2 on page 647, and B.3, note 1 on page 656.

# B MY EXPERIMENTS ON OPTICAL DEVIATIONS OF SIGHTINGS AT MARKS, IRSID, JULY 1958

## *B.1 Experimental arrangements for the optical observations at IRSID in July 1958*

Two pedestals were installed in the basement below my laboratory in St-Germain. A fixed mark and a telescope with an azimuthal circle (model 40) of the Institut Géographique National[1] were installed on each of these pedestals, about three meters below ground level. The front of each telescope was spaced from its corresponding mark by about 8.30 meters. The directions of the sightings were substantially North-South and South-North. Ten readings were performed with a micrometer every twenty minutes.

Due to certain faults in the mounting of the telescopes which were not remedied until 15 July 1958, only the observations of the second half of July can be considered as worthy of consideration.

In this period, the personal equations of the observers appear to have been relatively unimportant. They were corrected for as appeared appropriate, *but the following results in connection with the raw readings are independent of these corrections.*

---

[1] These telescopes were graduated in grads, *positively in the clockwise direction*, with the readings being made *in centesimal seconds*.

A positive variation of the reading corresponded to a displacement of the image of the mark towards the right of the observer.

## B.2 *Harmonic analysis of the diurnal deviations of the sightings at marks*

The Buys-Ballot filter method[1] was applied for cycles of 24h, 25h, and 12h.[2] The results obtained are given in *Table I*. There is very little difference between the results for the *raw* readings and those for the readings *that were corrected* as much as possible for the personal equations of the observers. It is seen that the variations are in the *same sense* for both of the telescopes, and that the amplitudes of the waves of 24h and 25h are *of the same order of magnitude*. The cycles of 24h and 25h obtained for the North and South telescopes are presented in *Graphs I and II.*

The waves of 12h and 12h30m were completely separated over a period of fourteen days, but the waves of 24h and 25h could not be. However, calculation shows that a sinusoid of 24h analyzed with a Buys-Ballot filter of 25h over a period of 14 days suffers an amplitude reduction of 47%. As a result, the cycles of 25h obtained over the fortnight in question cannot be considered as a non-eliminated residue of the wave of 24h.[3]

This result is confirmed by examination of the periodogram of the South telescope *(Graph III)*.[4] It is seen that the ordinate corresponding to 24h50m has an amplitude comparable to that of the ordinate corresponding to 24h.

One can therefore conclude that a *very remarkable* periodic structure is present in the variations of sightings with a fixed telescope at a fixed mark, *these variations having previously been attributed to accidental causes.*

---

[1] See above, *Chapter I*, § A. 5. 1, note 1 on page 96.

[2] Due to lack of time, the cycles of 12h30m were not calculated in 1960.

[3] See *Volume Two* of this work, *Chapter VI* (p. 30 above).

[4] This periodogram was prepared with the aid of an IBM 704 computer, operating upon values spaced apart by ten minutes.

TABLE I

## TELESCOPES
## RESULTS OF BUYS-BALLOT FILTER
### FITTING SINUSOIDS
### JULY 1958 – SECOND HALF (3+4)

| RAW READINGS | | | | | | |
|---|---|---|---|---|---|---|
| | 24ᴴ | | 12ᴴ | | 25ᴴ | |
| | 2 R | Θ | 2 R | Θ | 2 R | $\Theta - \Theta_{12L}$ |
| North Telescope | 1.174 | 10ʰ58 | 0.348 | 7ʰ53 | 1.106 | _8ʰ49 |
| South Telescope | 1.364 | 9ʰ17 | 0.250 | 9ʰ02 | 1.172 | _7ʰ25 |

| CORRECTED READINGS | | | | | | |
|---|---|---|---|---|---|---|
| | 24ᴴ | | 12ᴴ | | 25ᴴ | |
| | 2 R | Θ | 2 R | Θ | 2 R | $\Theta - \Theta_{12L}$ |
| North Telescope | 1.234 | 10ʰ17 | 0.348 | 7ʰ35 | 1.340 | _8ʰ57 |
| South Telescope | 1.198 | 9ʰ55 | 0.356 | 7ʰ38 | 1.050 | _7ʰ37 |

2R is expressed in centesimal seconds

Θ = UT time of the summit

$\Theta_{12L}$ = time of lunar meridian passage on an average day

GRAPH I

## TELESCOPES – $24^H$ CYCLE
### SAINT-GERMAIN – JULY 1958 3+4

NORTH TELESCOPE

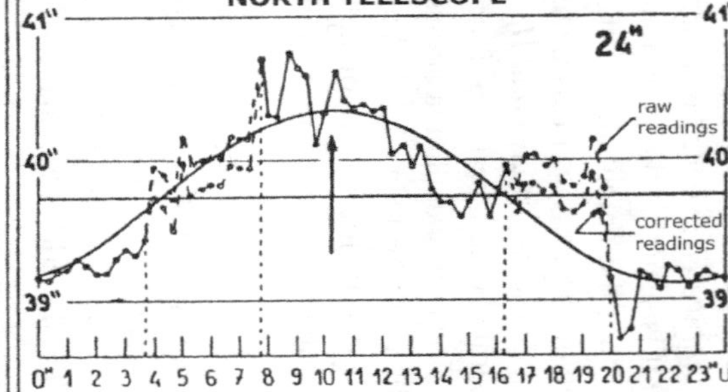


SOUTH TELESCOPE

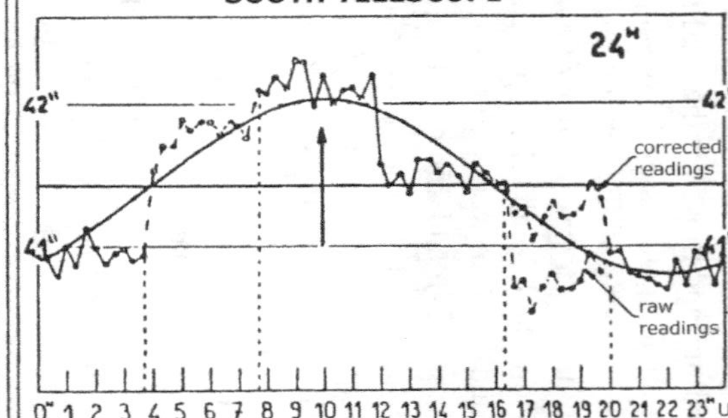


*Sinusoids fitted to corrected values*

$$y = \bar{y} + R \cos 2\pi \frac{t - \Theta}{T}$$

| | q | $\bar{y}$ | R | Θ |
|---|---|---|---|---|
| North Telescope | 15 | 39.74 | 0.617 | 10ʰ17 |
| South Telescope | 15 | 41.42 | 0.599 | 9ʰ55 |

T = 24

q = number of cycles

Θ = UT time of the summit

dotted verticals : times of operator change

GRAPH II

## TELESCOPES – 25H CYCLE

SAINT-GERMAIN – JULY 1958 3+4

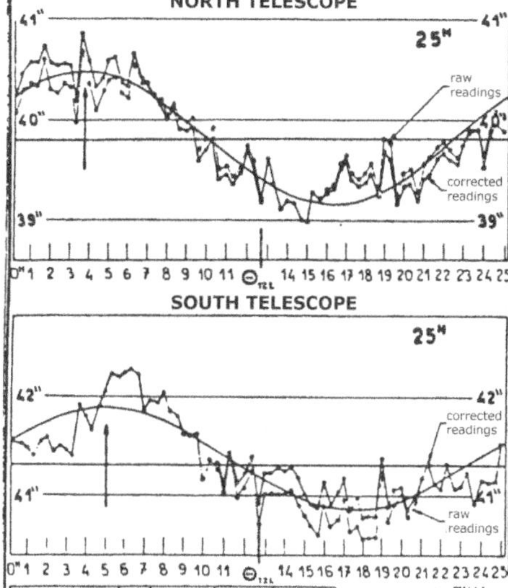


GRAPH IV

## COMPARISON OF THE AVERAGE OF THE TWO TELESCOPES – 25H CYCLE, AND OF THE AVERAGE OF THE TWO PENDULUMS

BOUGIVAL AND SAINT-GERMAIN – JULY 1958 3+4

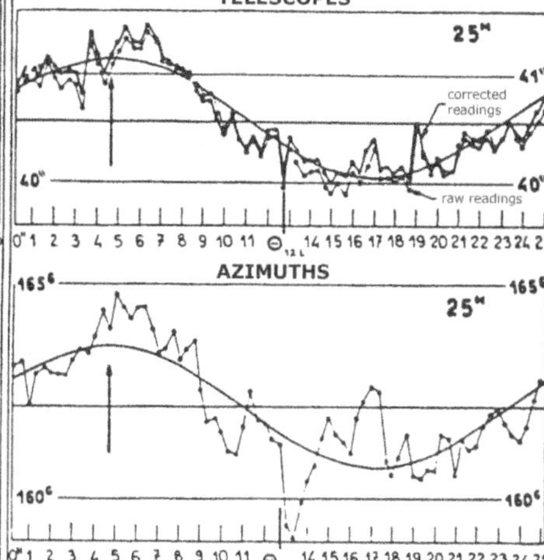


| | q | $\bar{y}$ | R | $\Theta\text{-}\Theta_{12L}$ |
|---|---|---|---|---|
| North Telescope | 14 | 39.83 | 0.670 | – 8H57 |
| South Telescope | 14 | 41.36 | 0.525 | – 7H37 |

*Fitting sinusoids*
*(telescopes : corrected values)*

$$y = \bar{y} + R \cos 2\pi \frac{t - \Theta}{T}$$

T = 25

| | q | $\bar{y}$ | R | $\Theta\text{-}\Theta_{12L}$ |
|---|---|---|---|---|
| Telescopes | 14 | 40.60 | 0.557 | – 8H06 |
| Azimuths | 14 | 162.15 | 1.440 | – 8H01 |

q = number of cycles

$\Theta\text{-}\Theta_{12L}$ = time of the summit against time of lunar meridian passage of mean day (considered at mean time)

GRAPH III

## PERIODOGRAM OF THE RAW READINGS AT THE SOUTH TELESCOPE

SAINT-GERMAIN – JULY 1958 3+4

N = 2n–1 = 361. 359. 357. 355. 353. 351. 349. 347. 345. 343. 341. 339

$N = 361 \left\{ \begin{array}{l} \Sigma = 1.57 \\ \Sigma^2 = 2.46 \end{array} \right.$  $\qquad r_p^2 = \frac{R_p^2}{2\Sigma^2}$

| μ | 0.69 | 1.60 | 2.30 | **3.00** | **3.90** | 4.60 | 7.00 | 9.00 | 11.00 |
|---|---|---|---|---|---|---|---|---|---|
| $e^{-\mu}$ | 0.50 | 0.20 | 0.10 | **0.05** | **0.02** | 0.01 | 0.001 | 0.0001 | 0.00001 |

for μ = n $\left\{ \begin{array}{l} r'^2_p = 2n \frac{\ell_p}{N} \\ \text{with } a \; 0.96 \end{array} \right.$ $\qquad \ell_p = \frac{1 - a^2}{1 - 2a\cos\frac{2\pi}{p} + a^2}$

*Legend*

the probability that a point of the periodogram is above the line μ=n is equal to $e^{-n}$.(1)

(1) CRAS, 244, 1957, p.2469, and 245, 1957, p.2467

Estimator of correlogram implied by this test : ρ = 0.96

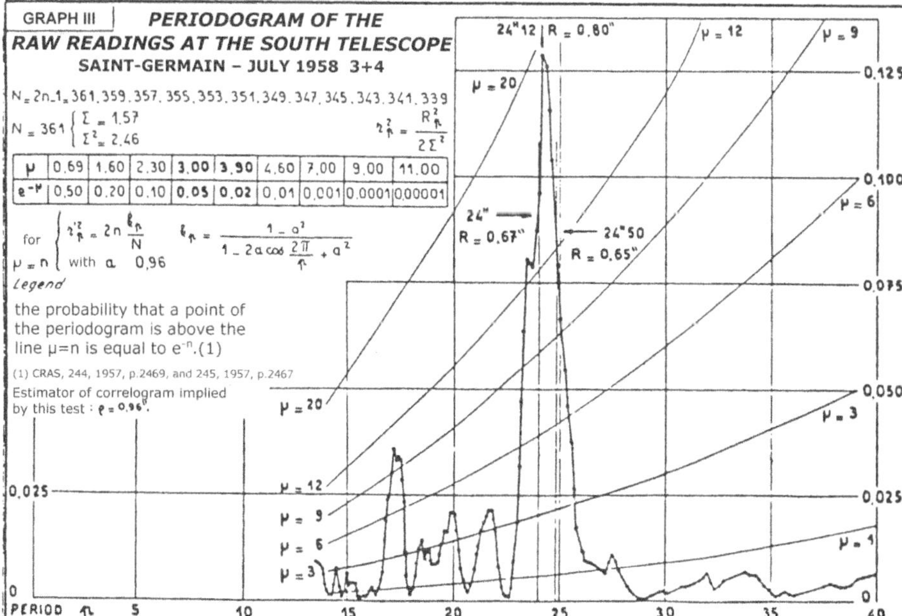

## B.3 Correspondence between the optical deviations of sightings at marks and the azimuths of the paraconical pendulum in July 1958

Finally, we wish to emphasize a situation that is *also extremely remarkable*: if the cycles of 25h for the half-sum of the azimuths of the two pendulums installed at Bougival and at Saint-Germain and for the half-sum of the readings of the two telescopes[1] are considered for the second fifteen days of July 1958, *these cycles are substantially in phase. The agreement of the phases is accurate to five minutes.* This correspondence is apparent for both the two weeks into which the considered fortnight can be divided.[2]

[1] Which eliminates the wave of 24h.
See above, *Chapter I*, § C.2.2, note 4 on page 147.

[2] See above, *Chapter I*, § B.1.2, pp. 105-112

## *B.4 The refusal by the Academy of Sciences to publish my Note of 23 February 1960 on the optical observations at IRSID in July 1958*

The preceding text and graphs are virtually a *complete* reproduction[1] of my Note of 23 February 1960 presented by Pierre Tardi to the Academy of Sciences under the title "*Existence of periodic components in the variation of readings corresponding to sightings at a fixed mark performed with a fixed telescope, as correlated with the movements of the paraconical pendulum*" [38].[2]

Due to the explicit opposition *of a single member* of the Academy of Sciences, Jean Leray,[3] *whose incompetence was only matched by his fanaticism*, the publication of this Note, accepted in plenary session, *which had already been printed by the services of the Academy of Sciences* and whose proofs had been sent to me by the secretariat of the Academy, *was finally refused without my receiving a hearing*.[4]

---

[1] With the exception of the correction of an error in the legend of *Table I*. *Centesimal* seconds were used, not *sexagesimal* seconds as was inadvertently wrongly indicated in the 1960 publication.

At the end of that Note I stated "*These conclusions appear to be confirmed by the analysis we have made of the results of observations taken in similar circumstances by the Institut Géographique National over the period 23 February to 26 March, 1959, which we will report later.*"

The text of my *Note of 22 February 1960* will be reproduced in *Annex II* of the *second volume* of this work (p. 33 above).

[2] This Note was presented late, because during the first semester of 1959 I was invited by the University of Virginia to lead a research seminar concerning the European Economic Union.

At the time I naively thought that this invitation would help my election as successor to the Chair of Economy in the Ecole Polytechnique, which had been declared vacant. But, on the contrary, my absence allowed the development *of a powerful cabal* that caused my candidature to fail.

On this point, see the third edition of my *Traité d'Economie Pure* (Treatise of Pure Economics, Clément Juglar, 1994), *Third Introduction*, pp. 124-125 [53].

This cabal spread its influence to my work in physics, and finally its influence culminated in the closure of my laboratory on 30 June 1960.

For more details of this cabal, see Louis Rougier, July 1959, *Scandale à Polytechnique* (Scandal at Polytechnique, 14 p.) [236]. This memoir is published in the volume *Un Savant Méconnu - Portraits d'un Autodidacte* (An Underrated Scientist - Portraits of an Autodidact), Editions Clément Juglar, 2002. It is also reproduced in *Annex I.D* of the *second volume* of this work (p. 31 above).

[3] At the time, Jean Leray was a professor of differential and functional equations at the Collège de France, and apparently had never performed one single experiment of any kind. He had obstinately refused to come and visit my two laboratories at IRSID and at Bougival.

[4] See above, *Chapter I*, § G.5 and G.6; and *Chapter X*, § B.2 below in this work. Also see *Chapter VIII* and *Annex I.A* in the *second volume* of this work (pp. 30 and 31 above).

This refusal was *fundamentally completely unjustified* and did no honor whatever to the Academy of Sciences, and it greatly discouraged me.

The motivation for this fierce opposition by Jean Leray was not very clear. However, after I had given him every reasonable explanation both verbally and in writing, Jean Leray had no hesitation in writing to me on 18 December 1960:[5]

> "*The publication of your Notes, wherever it may take place, will cast doubt not only upon the methods that you use in the physical sciences but also in the economic sciences; maybe, in that sense, this publication will be useful.*"

I could only answer him on 11 January 1961:

> "*Your mental attitude in scientific matters seems to me to betray itself unambiguously in the* a priori *judgment that you do not mind to make - quite gratuitously - of my work in economy, about which you know nothing whatever, and in which subject you have no competence, at least as far as I know.*
>
> "*This judgment shows that, as far as relates to myself, the concerns that inspire you are quite other than scientific. This explains a lot to me, and I must admit that I derive some comfort from the fact.*
>
> "*It is impossible to bring sight to a blind man or hearing to a deaf one, and even more impossible to make a person who wishes neither to see nor to hear do so; and accordingly I prefer to abandon arguing the cause of scientific impartiality before your personal court.*
>
> "*If the scientific method is really the one that you appear to espouse, then I am happy to be described as not being a scientist.*"

[5] I do not know what Jean Leray's reaction may have been when I received the *1988 Nobel Prize in Economic Sciences*!

## *B.5 Some supplementary comments on the optical observations of July 1958 at IRSID*

I feel it necessary to supplement my 1960 text on the anomalies of optical sightings at marks with a few observations.

### Extreme minuteness of the observed effects

1 - First of all, the observed optical effects are *extremely small.* The amplitudes of the observed cycles of 25h, 24h, and 12h are of the order of one centesimal second, *i.e.* around $1.6 \times 10^{-6}$ radians.[1]

### Global deviations

2 - As I have indicated,[2] the observations made during the previous four weeks of observations from 19 June to 15 July 1958 *cannot be employed in calculation, due to difficulties in the adjustment of the telescopes, due to the magnitude of observational errors, and due to the minute amplitude of the periodic components* as brought into evidence during the last two weeks from 17 to 31 July 1958. *But this is not true for the overall deviations, whose amplitude considerably exceeds the possible order of magnitude of observational errors.* Accordingly *the totality of the observations at IRSID, from 19 June to 31 July 1958, can certainly be used for determination of the global deviations (improperly described as "drifts") of these sightings at marks.*

[1] One centesimal second represents $1.571 \times 10^{-6}$ radians. At ten meters, a deviation of 1" centesimal second accordingly corresponds to a displacement of 0.01571 mm. An amplitude of the order of a centesimal second for the 25 hour cycle corresponds to an azimuthal variation of a centesimal second in 12h30m, *i.e.* of the order of $3 \times 10^{-11}$ radians/sec $(1.571 \times 10^{-6}/(12.5 \times 3600) = 3.49 \times 10^{-11})$, which is around fifty thousand times smaller than the variation per second of the wave of 25h for the azimuth of the paraconical pendulum with anisotropic support, which is around $10^{-5}$ radians per second (*Chapter I,* § B.2.6, pp. 123-124) $(0.186 \times 10^{-5}/3.49 \times 10^{-11} = 5.33 \times 10^{4})$.

[2] § 1 above.

In fact it is seen that, from 19 June to 31 July 1958, the global deviations of the sightings at marks, *counted positively* in the clockwise sense, were respectively -30" and -44", so that they were *in the same sense and of the same order of magnitude* for the North and South telescopes, corresponding respectively to *average deviations per day of around -0.7" and -1"*.

## Monthly lunar periodicity

3 - *Graph V* represents fitting of the average deviation $(N+S)/2$ of the North and South *telescope sightings at marks* from 1 to 31 July 1958[3] to a linear trend and to a sinusoid of period equal to the sidereal period of the Moon, *i.e.* 27.322 days.[4]

It is seen that, upon the linear tendency, a periodic lunar component is superimposed whose total amplitude is of the order of 5", in other words of the order of four times the amplitude of the diurnal components.

The linear tendency is certainly only local, and in fact *can be interpreted as an element having a periodic fluctuation whose period is six months.*[5]

[3] For simplification of the calculations, the azimuths at 0h on each day are considered, *not the daily averages.*

In contrast to the calculations in *Graphs I through IV* above which were made in 1959, the calculation in *Graph V* (calculation 943) was performed on 18 January 1996.

[4] For consideration of the lunar sidereal period, see *Chapter I*, § C.3, note 3, p. 157.

I remind the reader that the notations used in the *legends* of the graphs are as follows: $\Sigma$ = standard deviation of the series; $R$ = correlation coefficient; $m$ = average of the fitting; $r$ = semi-amplitude of the fitting sinusoid; $\sigma$ = standard deviation of the correlation residues (*Chapter I*, § A.5.5, p. 101, and § C.3, *Graph XXVII*, p. 158 above).

[5] See *Chapter V*, § C.1, pp. 447-449, below.

The periodic monthly lunar component of the average deviation $(N+S)/2$ reached its maximum on 6 July 1958 at 22h, and its minimum on 20 July at 14h.[6]

## The almost periodic structure of the observations

4 - As for the paraconical pendulum with anisotropic support, an *almost periodic* structure was observed in the observations performed with the North and South telescopes.[7]

In view of the observed amplitudes of the periodicities (in *absolute* and *relative* values), this almost periodic structure can *in no way be explained* by luni-solar deformation of the soil or by thermal effects.[8]

---

[6] In order to compare the fitting of *Graph V* for the sightings at marks with the fittings of *Graphs XXVII and XXVIII* of *Chapter I* (pp. 158 and 159 above) that correspond to the paraconical pendulum with anisotropic support, it is necessary first to consider that the deviations from marks are counted *positively in the clockwise sense*, while the azimuths of the paraconical pendulum are counted *positively in the anticlockwise sense*.

*Counting all the angles in the anticlockwise sense*, it is seen that the maximum of the sinusoid of *Graph V*, $(N+S)/2$, corresponds to 14h on 20 July ($s_o = 200.6$ days), while, for the azimuths $(S+B)/2$ of *Graph XXVII* of *Chapter I*, the maximum corresponds to 20h on 24 July ($s_o = 204.85$). Therefore, *to the first approximation*, there is agreement in phase.

But this agreement in phase only appears with a spacing apart of 4 days 6 hours, with the deviations of the sightings at marks being 4 days 6 hours ahead of the half-sum $(B+S)/2$ of the azimuths of the pendulum.

However, all these calculations are only rough. *Essentially they assume the linearity of local trends, which is a very simplistic hypothesis* that certainly is only *an approximation*, because the trends can be interpreted as corresponding to segments of sinusoids *of semi-annual period*.

As a consequence it may be considered that, at least in the first approximation, the lunar component of the optical deviation $(N+S)/2$ is *in phase agreement* with the monthly lunar components of the azimuths $(S+B)/2$ and $(S-B)/2$ corresponding to the paraconical pendulum that are practically in phase ($s_o = 204.85$ and $s_o = 204.03$) (*Graphs XXVII and XXVIII of Chapter I*, pp. 158-159 above).

[7] *Chapter I*, § A.5.4, above, p. 101.

[8] This question will be examined in the *second volume* of this work, *Chapter III*, Section A (p. 29 above).

However, considering the results of the experiments performed at IGN in February-March 1959 (§ C.4.2 below, pp. 364-365), it may be asserted that any explanation based upon relative displacement of the pedestals, whatever the cause might be, *must be excluded*.

*GRAPH V*

**IRSID, JULY 1958**
**HALF-SUM (N+S)/2**
**OF THE AZIMUTHS OF THE SIGHTINGS AT MARKS**
**OF THE NORTH AND SOUTH TELESCOPES**
**1st to 30th July 1958 (30 days)**

*Values at 0h in centesimal seconds, counted positively in the clockwise sense*

***Fitting to the lunar sidereal period of 27.322 days, taking account of a linear trend***

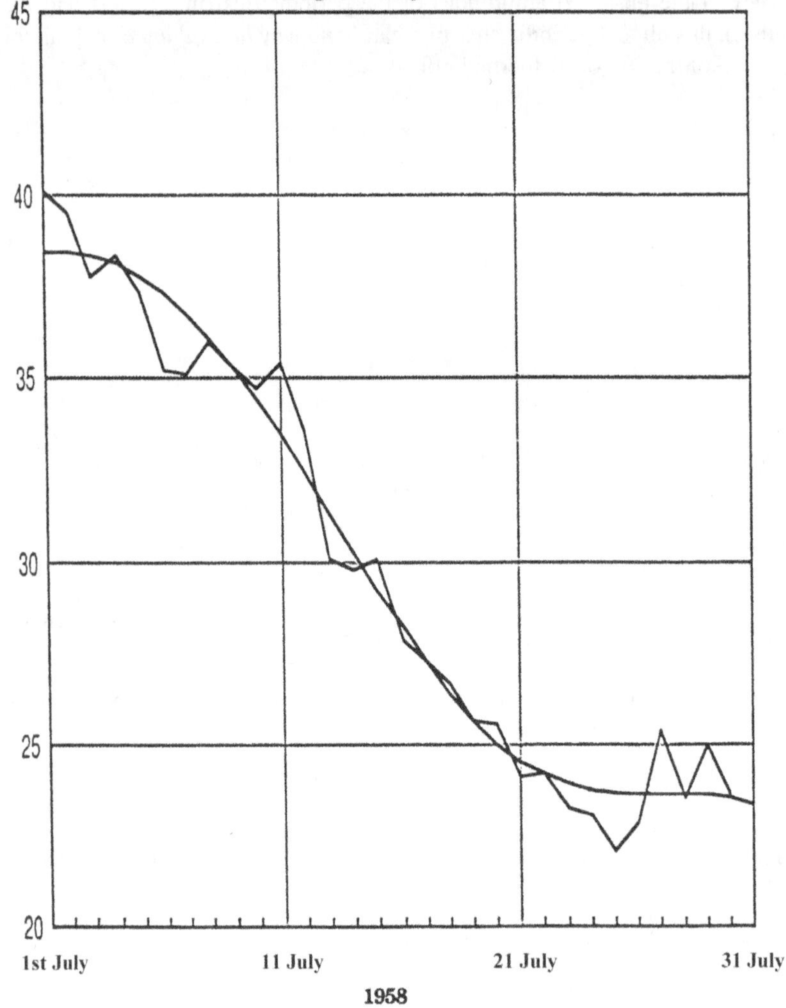


*Legend* : $\Sigma$ = 5.86" ; R = 0.986 , 1-$R^2$ = 0.0285 , m = 29.95" , trend = - 0.543" per day, r = - 2.38" , $\sigma$ = 0.989" (see the *legend* of *Graph XXV* of *Chapter I*).
Maximum of the sinusoid : 22 h on 6 July ($s_0$ = 186.9) ; minimum : 14h on 20 July ($s_0'$ = 200.6)
The dates are counted in days from 0h on 1 January 1958.

*Sources* : Calculation 943 (18 January 1996), Graph 13663*

## *B.6 Scientific implications of the experiments performed at IRSID in July 1958*

In fact, the scientific implications of the observed optical deviations of the sightings at marks were *considerable*, indeed *absolutely extraordinary*, when they were compared with the results obtained in the same period for the two pendulums at Saint-Germain and Bougival, which had *similar periodic structure.*

Not only were the anomalies of the paraconical pendulum *completely confirmed* by the continuous experiments performed *simultaneously* at Bougival and Saint-Germain, but *a new phenomenon, never previously imagined by anyone,*[1] was brought to light: the existence of deviations in sightings at marks over time, and their *absolutely astounding* agreement in diurnal and monthly phase with the azimuths of the paraconical pendulum. This was *a completely authentic major discovery.* It demonstrated the existence of a connection between phenomena *of completely different nature*: one of mechanical type, and one of optical type.

In the light of the *absolutely crucial* character of the experiments performed simultaneously at Bougival and Saint-Germain in similar conditions, which demonstrated *in an irrefutable manner* the existence of diurnal periodic components in the movement of the paraconical pendulum with anisotropic support having amplitudes completely inexplicable in the framework of the current theory of gravitation, the refusal to publish such remarkable results as those of the optical anomalies which established a connection between mechanical phenomena and optical phenomena having completely different natures, *sprang from an utter inconsistency and a partial and obscurantist attitude, scientifically completely inadmissible, and quite unworthy of such a great scientific institution as the French Academy of Sciences.*[2]

---

[1] In fact, one can only be astonished that, in view of the anomalies that had been observed in leveling operations (§ A.1 above), nobody had previously ever thought of performing these experiments, *which are nevertheless so simple and so easy to implement.*

[2] See above, *Chapter I*, § G.5 and G.6, pp. 226-236.

# C EXPERIMENTS ON OPTICAL DEVIATIONS OF SIGHTINGS AT MARKS AND AT COLLIMATORS, INSTITUT GEOGRAPHIQUE NATIONAL, FEBRUARY - MARCH 1959

## *C.1 The optical experiments on sightings at marks and at collimators performed during February-March 1959 at IGN*

From 23 February to 26 March 1959, experiments on optical sightings were performed by the *Institut Géographique National* (IGN), as suggested by me and at the initiative of Pierre Tardi, under the control of a very young geographic engineer Claude Palvadeau (a former polytechnic student who graduated in 1953) and under the aegis of the *astronomical and geodesic section* of IGN, which was led by the chief geographic engineer Jean Segons (a former polytechnic student who graduated in 1930).[1]

[1] In *August 1959*, these experiments were the subject of a report by Claude Palvadeau entitled "*Rapport sur des Expériences de Contrôle de la Stabilité des Visées Optiques*" (Report on Experiments in Controlling the Stability of Optical Sights, 14 pages with 36 annexes, of which 28 were graphs) [210]. (This report is archived at IGN.)

In four Tables (annexes 5, 6, 7, and 8), this Report particularly contains the results of the *raw* observations of sightings at marks and at collimators in centesimal seconds performed every 20 minutes, both North-South and South-North.

Unfortunately the sightings at collimators could only be performed over *27 days* instead of 30 days, which introduced a bias into the harmonic analysis calculations.

These Tables (and also the Report) do not explicitly show the moments when significant discontinuities took place in the readings due to adjustments or accidental shocks.

In the months of February and March 1959, I was at the University of Virginia as a "*Distinguished Visiting Scholar*" for five months (see note 2 of § B.4 above, p. 339). Therefore I was not able to participate in the implementation of these experiments and in the procedures for analyzing the observations.

Due to lack of space, I unfortunately cannot give any description here of the *very elaborate* analyses I made of these observations in 1960, 1981, and 1991.

**Experimental Arrangements**

1. - Sightings *at marks and at collimators* were performed simultaneously every 20 minutes. All the measuring instruments and marks were mounted on two massive pedestals 25.6 meters apart in an underground corridor below IGN, substantially in a North-South orientation.[2]

The experiments were carried out day and night without interruption, from 16h on 23 February to 16h on 26 March.[3]

The observations were made *in centesimal seconds and counted positively in the clockwise sense.*

---

[2] From Palvadeau's Report:

*"On each pedestal there were installed: a complete azimuth circle, with its support and its yaw axis telescope; an azimuth circle telescope; a mark with a vertical black line...*

*"Two groups of aimings were established: aimings at marks, focusing at a finite distance, and aimings at collimators, focusing at infinity.*

a) The sightings at collimators

*"The upper telescope of the azimuthal circle, focused at infinity, was used as a collimator: its reticle was lit by a bulb mounted in the place of, and instead of, the eyepiece.*

*"The yaw axis telescope, also focused at infinity, was used to make the sightings with the movable wires only.*

*"The yaw axis telescope on the South pedestal was sighted upon the collimator on the North pedestal, and vice versa.*

b) The sightings at marks

*"The telescopes were focused at a finite distance (25.6 m) for aiming at the marks (with the movable wires). Here as well, the "South telescope" viewed the "North mark", and vice versa."*

[3] From the Palvadeau Report:

*"Two teams of eight observers operated in relays each week... On each telescope, a series of 10 sightings using the movable wire was performed every 20 minutes."*

## C.2 *Diurnal variations of the optical sightings at marks and at collimators at IGN during February-March 1959*

Overall, the Report of Claude Palvadeau comes to the conclusion that there were no significant cycles of 24h, 25h, 12h, or 12h30m.[1] But in reality this conclusion is *largely* due to the *very great* number of observers – twenty-six – *and to the method by which compensation was made* with the personal equations of these observers.[2]

In fact, I had the cycles of 24h, 25h, 12h, and 12h30m calculated from Palvadeau's *raw data* in 1960 by my collaborator Jacques Bourgeot.[3] *Notably*, Bourgeot found a *very significant* cycle of 25h for the sightings from the North telescope at its collimator, having a total amplitude 2R=1.1" (*Graph VI*), and *very marked* cycles of 12h30m for the sightings at marks North-South and South-North, having amplitude of 0.70" (*Graphs VII and VIII*).[4]

---

[1] At the end of his Report, Palvadeau wrote (p. 12):

"*As far as diurnal variations are concerned, harmonic analysis does not extract anything definite... The important phenomenon seems to be a period of the order of a month, which appears for all four telescopes with amplitudes of the same order of magnitude (5" to 10").*

[2] It follows from Palvadeau's comments that he performed the calculations for diurnal cycles *from values that had been corrected*, after "*elimination*" of the personal equations of the observers, *deduced from the "closing errors"*.

In fact, as viewed by Palvadeau, the "*personal equations*" of the observers include a systematic element $\Delta_a$ which should be determined, and a personal error $\Delta_b$ made by the observer in his reading.

It is not possible to take as an objective "*elimination of the personal equations of the observers*" while considering the "*closing errors*", precisely because these personal equations and these "*closing errors*" include two elements, a systematic element $\Delta_a$ *which must be determined* and an error $\Delta_b$ *which cannot be eliminated except by averaging* when one calculates the cycles by the Buys-Ballot method.

From this, it follows that the cycles of 24h and 25h calculated by Palvadeau are fundamentally biased, and accordingly that his conclusion (p. 12) that no demonstrated periodicities of 24h and 25h are present *is essentially founded upon an erroneous method of calculation.*

[3] See my Memoir of 8 April 1991, "*Analyse des Observations de l'Institut Géographique National. Février-Mars 1959*" (Analysis of the Observations of the Institut Géographique National, February-March 1959)[51] which groups together the 24 graphs established by Bourgeot in 1960.

[4] Overall, and even if one considers the adjustments made by Jacques Bourgeot, the results *are incontestably poorer* than those obtained in 1958 at IRSID.

The *major* reason is undoubtedly that the observations performed at IGN were made *by 26 different operators*, while at IRSID, *over 16 hours of every day*, the observations were made *only by four different operators*, and there were only *seven operators* in all.

These cycles *calculated from the raw data* with simple correction of trends, which Palvadeau did not extract, demonstrate the *totally inappropriate* character of the corrections that he made to the raw data.

From my calculations of 1991, the standard deviation of the sightings by a single operator at a given instant is around 1.54" for sightings at a mark and around 3.60" for sightings at a collimator.[5] From this it follows that, for the cycles of 24h and 25h, the amplitudes are *of the same order of magnitude* as the standard deviations of the errors of the observations, which helps to explain the poor quality of some of the fittings.[6]

*Graph VI* corresponds to the *North telescope, to the sightings at a collimator*, and to a cycle of 25h. Its total amplitude is around 1.08",[7] and its maximum on 28 February 1959 is at 23h56m. At the middle of the period considered, it corresponds to a difference $\Theta - \Theta_{12L} = -7h8m$ between the time of the maximum of the cycle and the time of meridian passage of the Moon.[8]

---

[5] Palvadeau (p. 6) gives the value of 1.50" for the standard deviation of the sightings of a single operator, but does not distinguish between the sightings at marks and the sightings at collimators. He writes (p. 7):

*"For the 24h period, the error to be feared at each point of the graph is around 0.5". For the period of 25h, it is around 1.2". These figures show the difficulty of demonstrating a phenomenon of amplitude near 1" by means of observations participated in by a large number of observers."*

[6] For the sightings at marks and the cycles of 24h, the averages over 30 days thus suffer from errors whose standard deviation is around $1.54/\sqrt{30} = 0.28$". For the sightings at collimators and the cycles of 24h, the averages over 25 days suffer from errors whose standard deviation is around $3.6/\sqrt{25} = 0.72$". And, for the cycles of 25h, the standard deviations are of the same order of magnitude.

[7] The reduction of the amplitude of a wave of 24h50m analyzed with a 25h filter over 24 oscillations is around 0.96 (see *Chapter VI* of the *second volume* of this work, p. 30 above).

In connection with the consideration of a cycle of 25h instead of 24h50m, see Note 2 of § 3 of *Chapter I*, p. 90 above.

[8] On 28 February 1959, the maximum of the first sinusoid of 25h is located at 23h56m. But, since the real cycle is 24h50m, the cycle of 24h50m advances ten minutes a day on the cycle of 25h. The result is that, at the center of the period analyzed, the total advance of the average cycle is around 10 minutes $\times$ 12 = 2 hours.

From this, it follows that on 1 March 1959 the maximum of the 24h50m cycle is located at $\Theta = -2h4m$. Now, on 1 March 1959, the time of meridian passage of the Moon at the meridian $\Theta_{12L}$ was 5h4m. We therefore have:

$$\Theta - \Theta_{12L} = -2h4m - 5h4m = -7h8m$$

which is a figure entirely comparable with the average time difference for Bougival and Saint-Germain of -7h39m in *Table IX* of § C.2.2 of *Chapter I*, pp. 147 and 149 above (*equal to (- 7h23m - 7h53m)/2 = - 7h39m*).

*This coincidence confirms the reality of the cycle of Graph VI above, calculated in 1960 by Bourgeot.*

*Graph VII* corresponds to the *North telescope for sightings at a mark,* and to a cycle of 12h30m (25h/2) for a total duration of 700 hours. The amplitude of the cycle is around 0.72", and the maximum of the first cycle at 12h30m corresponds to 18h58m on 24 February 1959.

*Graph VIII* corresponds to the *South telescope for sightings at a mark,* and to a cycle of 12h30m (25h/2). The amplitude of the cycle is around 0.68", which is almost identical to that of 0.72" for the *North telescope.* The maximum of the first cycle at 12h30m corresponds to 20h2m on 24 February. To about one hour, this is in phase with the corresponding cycle for the *North telescope (Graph VII).*[9]

*Taken together, these results confirm the reality of the cycles of 24h50m and 12h25m.*[10]

[9] The times of the two maxima of the first cycles of *Graphs VII and VIII* are indeed 18h58m and 20h2m.

Thus, it may be considered that, in practice, the 12h30m cycle of the sightings at a mark of the North telescope is in phase with the 12h30m cycle of the sightings at a mark of the South telescope.

[10] Because of the time-slippage of 50 minutes a day of the 24h50m cycle with respect to the 24h cycle, the influence of the personal equations of the observers is attenuated for the cycle of 24h50m while being reinforced for the cycle of 24h, *due to the changing of operators at fixed times.*

*GRAPH VI*

**I.G.N.**

**SIGHTINGS AT COLLIMATORS - NORTH TELESCOPE**

**25h CYCLE.**

**Raw values after trend correction**

*in centesimal seconds, counted positively in the clockwise sense*

***28 February - 25 March 1959***

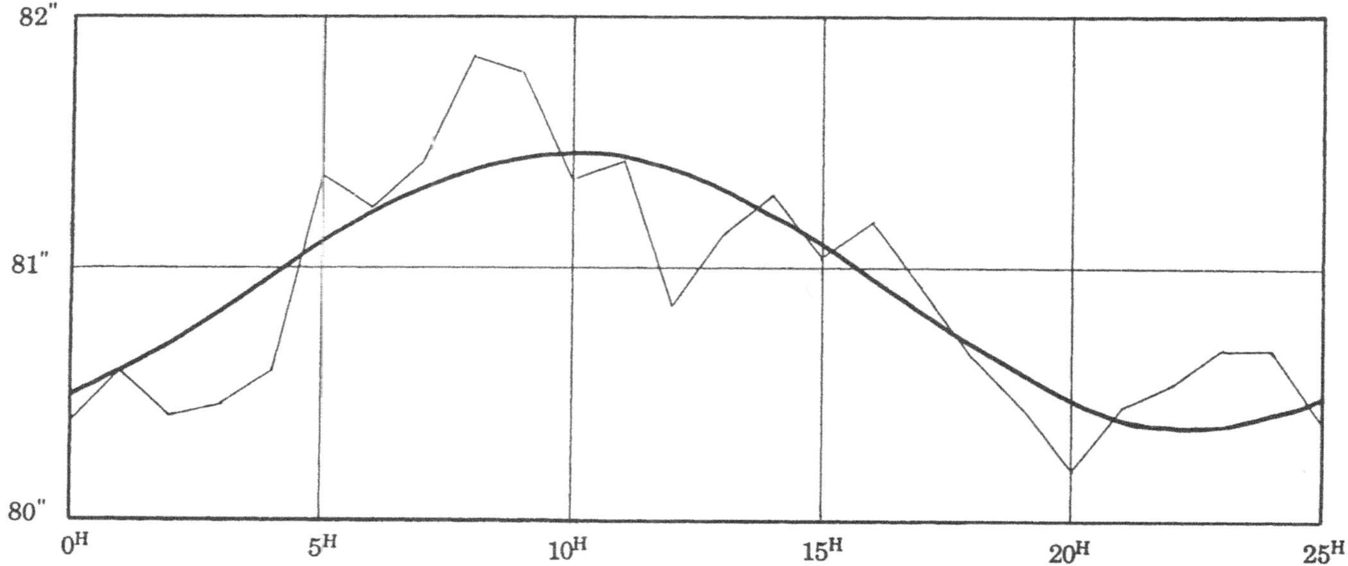


*Legend* : **Fitting sinusoid: half-amplitude: r=0.54"; time of maximum: 9h56m UT.**
**The cycle was calculated by the Buys-Ballot method, from 14h UT on 28 February 1959 to 13h UT on 25 March 1959.**
**The calculation thus encompassed** ***24 cycles of 25h*** **(i.e. 600 hours).**
**The maximum of the first cycle corresponds to 14h on 28 February + 9h56m = 23h56m.**

*Sources* : ***Analysis of the observations of the Institut National Géographique*****, 8 April 1991 (C-5040, p. 7) (Calculation by Bourgeot, 1960)**

*GRAPH VII*

**I.G.N.**

**SIGHTINGS AT MARK - NORTH TELESCOPE**
**CYCLE OF 12h.30m.**
**Raw values after trend correction**

*in centesimal seconds,*
*counted positively in the clockwise sense*

***24 February - 25 March 1959***

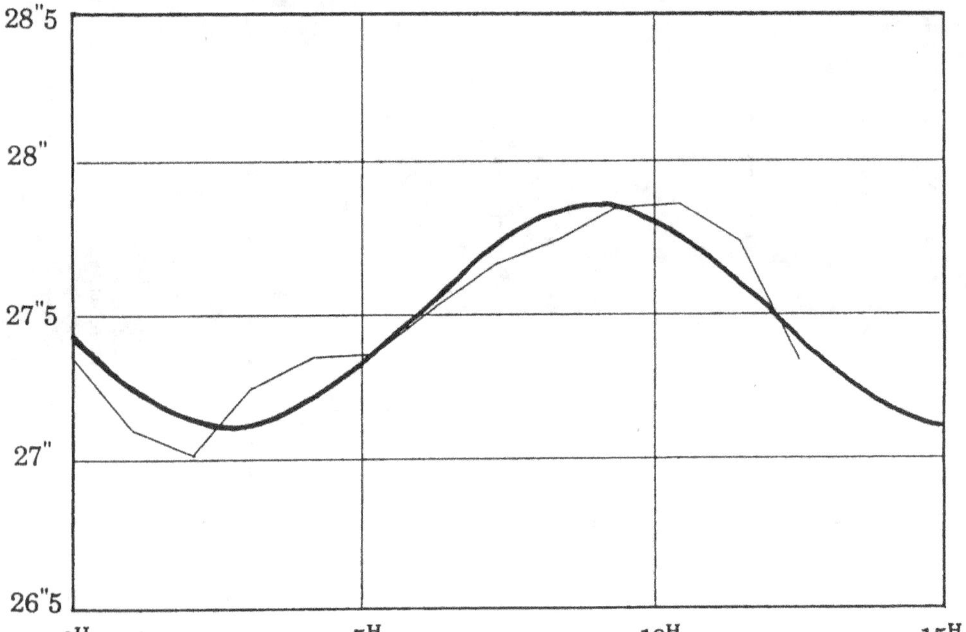


***Legend*** **: Fitting sinusoid: half-amplitude: r=0.36"; time of maximum: 8h58m.**
**The cycle was calculated by the Buys-Ballot method, from 10h on 24 February 1959 to 13h on 25 March 1959. The calculation thus encompassed *56 cycles of 12h30m* (i.e. 700 hours).**
**The maximum of the first cycle corresponds to 10h on 24 February + 8h58m = 18h58m UT.**

***Sources :Analysis of the observations of the Institut National Géographique*, 8 April 1991 (C-5040, p. 19) (Calculation by Bourgeot, 1960)**

*GRAPH VIII*

**I.G.N.**
**SIGHTINGS AT MARK - SOUTH TELESCOPE**
**CYCLE OF 12h.30m.**
**Raw values after trend correction**
*in centesimal seconds,*
*counted positively in the clockwise sense*
***24 February - 25 March 1959***

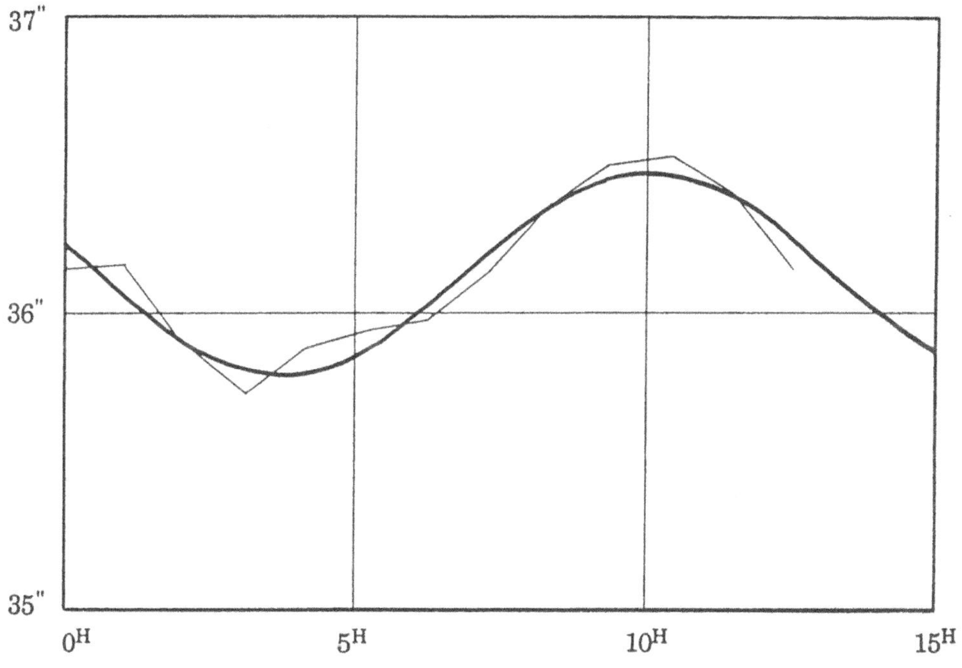


***Legend*** : **Fitting sinusoid: half-amplitude: r=0.34"; time of maximum: 10h2m.**
**The cycle was calculated by the Buys-Ballot method, from 0h on 24 February 1959 to 13h on 25 March 1959. The calculation thus encompassed *56 cycles of 12h30m* (i.e. 700 hours).**
**The maximum of the first cycle corresponds to 10h on 24 February + 10h2m = 20h2m UT.**

*Sources* : ***Analysis of the observations of the Institut National Géographique*, 8 April 1991 (C-5040, p. 20) (Calculation by Bourgeot, 1960)**

## *C.3 Monthly sidereal lunar periodicities of sightings at marks and at collimators at IGN in February-March 1959*

### Global deviations over the month

1 - While the personal equations of the observers had an impact of considerable significance upon the uncertainty of the diurnal and semi-diurnal cycles,[1] their effect upon the *global* deviations observed over the course of a month was relatively *much more feeble.*

The Palvadeau Report sets out results that are *absolutely essential* as far as concerns the observed global deviations, which appear to be *linear and of clearly significant amplitude.* These deviations are as follows *in centesimal seconds counted positively in the clockwise sense*:

*Sightings at marks* (31 days)

| | | | | |
|---|---|---|---|---|
| North telescope : | -35" | *i.e.* | -1.1" | per day |
| South telescope : | -27" | " | -0.9" | " " |
| Average : | -31" | " | -1" | " " |

*Sightings at collimators* (27 days)[2]

| | | | | |
|---|---|---|---|---|
| North telescope : | +58" | *i.e.* | 2.1" | per day |
| South telescope : | -78" | " | -2.9" | " " |
| Average of absolute values : | 68" | " | 2.5" | " " |

The average daily deviations of the sightings at collimators, which are of the same order of magnitude, are thus *two and a half times greater in absolute value* than those of the sightings at marks, which similarly are of the same order of magnitude. They are *in the same sense for the sightings at marks, and in the opposite sense for the sightings at collimators.*

---

[1] Particularly for the cycles of 24h and 12h, due to the observers being changed over *at the same times* of day and night.

[2] The observations corresponding to the sightings at collimators could only be started from 27 February 1959 (see Note 1 on p. 346 above).

*The average deviations of the sightings at marks are of the same order of magnitude as those observed at IRSID.*[3]

*Considerable importance must be attached to the average global deviations*, which are of the order of 2.5 centesimal seconds per day for the sightings at collimators and of the order of 1 centesimal second per day for the sightings at marks, and which are seen steadily for four weeks.

*Graphs IX and X*[4] represent the deviations of the sightings at marks and of the sightings at collimators. And *Graph XI* represents the correspondence of the averages $(N+S)/2$ and $(S-N)/2$ of the sightings at marks and at collimators.

[3] § B.5.2 above, pp. 341-342.

[4] The curves of *Graphs IX and X* were deduced from measurements made upon photographic enlargements of Graphs 9, 11, 10, and 12 of the *Palvadeau Report of August 1959.*

*GRAPH IX*

**I.G.N.**

**SIGHTINGS AT MARKS**

**NORTH AND SOUTH TELESCOPES**

**Daily Averages**

*in centesimal seconds counted positively in the clockwise sense*

**23 February - 25 March 1959 (31 days)**

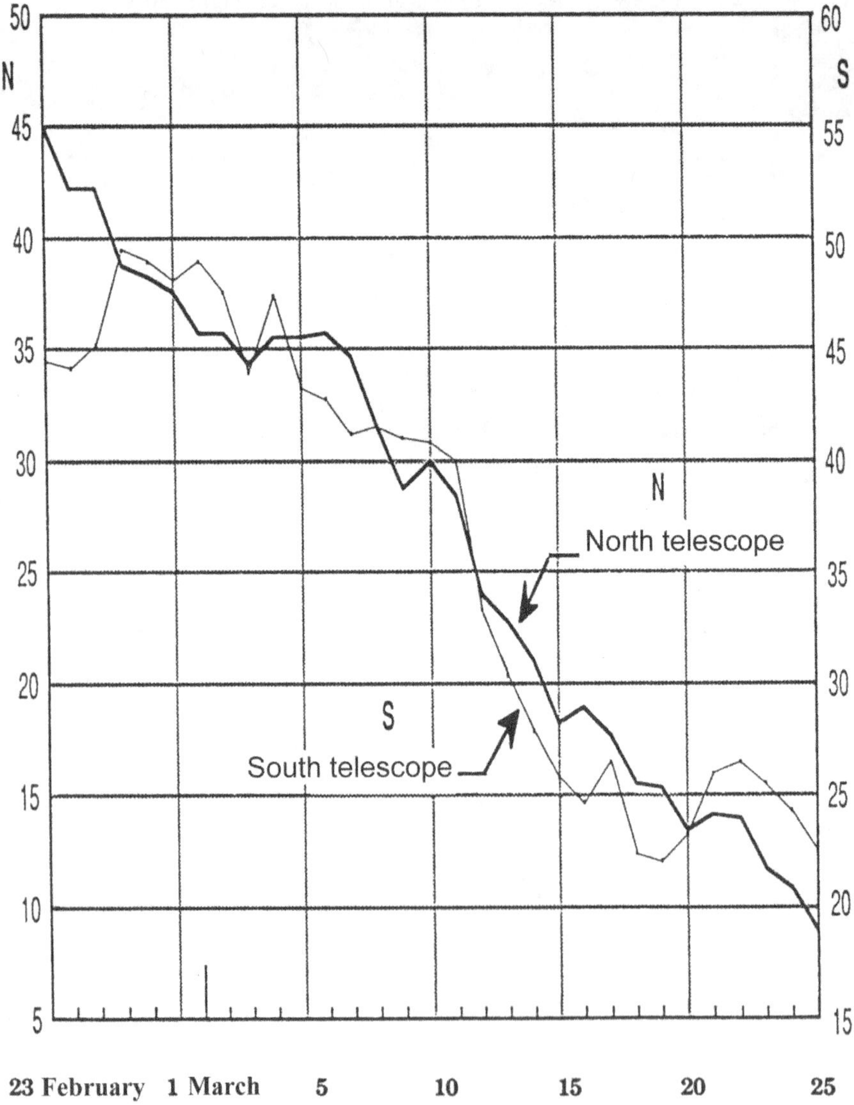


*Legend* : The total variation of the azimuth scales of the N and S sightings is the same, i.e. 45 centesimal seconds.

*Sources* : Graphs 13587 and 13588 (Graphs 9 and 11 of the Palvadeau Report of August 1959)

*GRAPH X*

**I.G.N.**

**SIGHTINGS AT COLLIMATORS**

**SOUTH AND NORTH TELESCOPES +S AND -N**

**Daily Averages**

*in centesimal seconds counted positively in the clockwise sense*

**27 February - 25 March 1959 (27 days)**

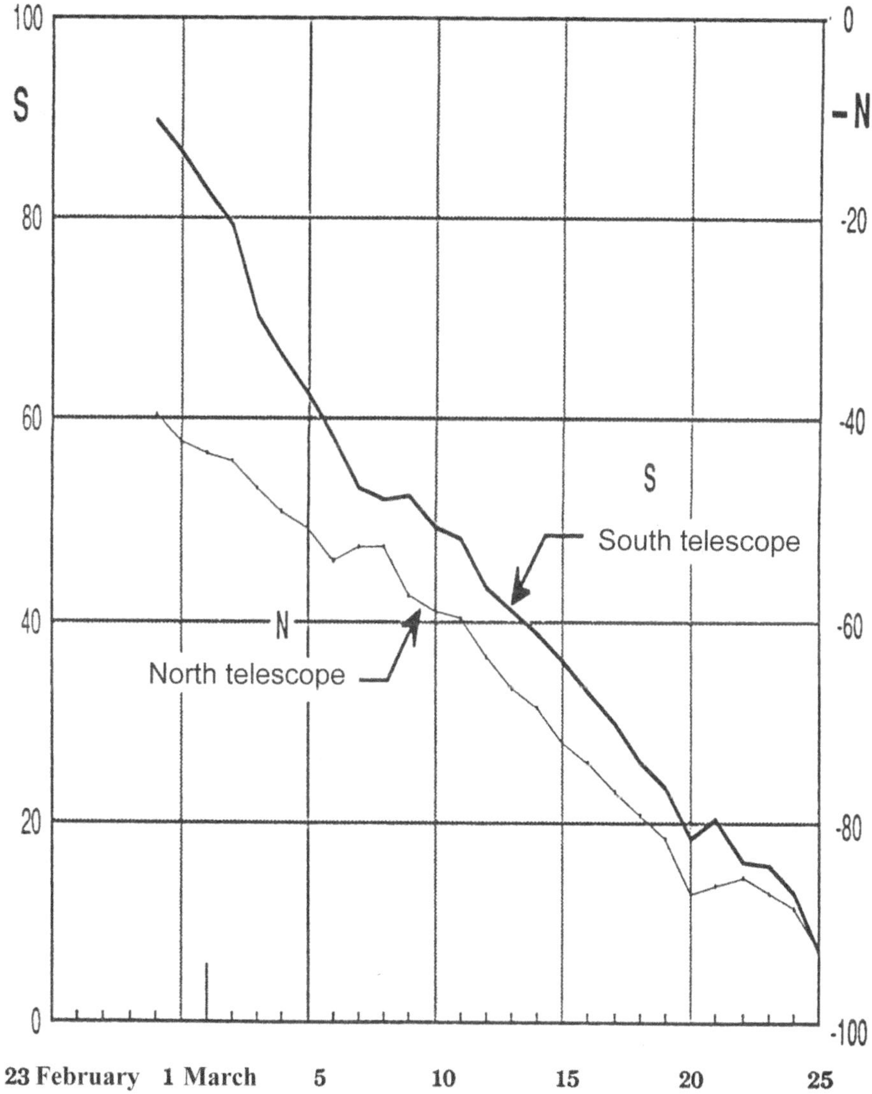


*Legend* : The total variation of the azimuth scales of the N and S sightings is the same, i.e. 100 centesimal seconds.

*Sources* : Graphs 13590 and 13589 (Graphs 10 and 12 of the Palvadeau Report of August 1959)

*GRAPH XI*

**I.G.N.**
**COMPARISON OF GRAPHS**
**OF SIGHTINGS AT MARKS (N+S)/2**
**AND SIGHTINGS AT COLLIMATORS (S-N)/2**
*in centesimal seconds counted positively in the clockwise sense*
**February - March 1959**

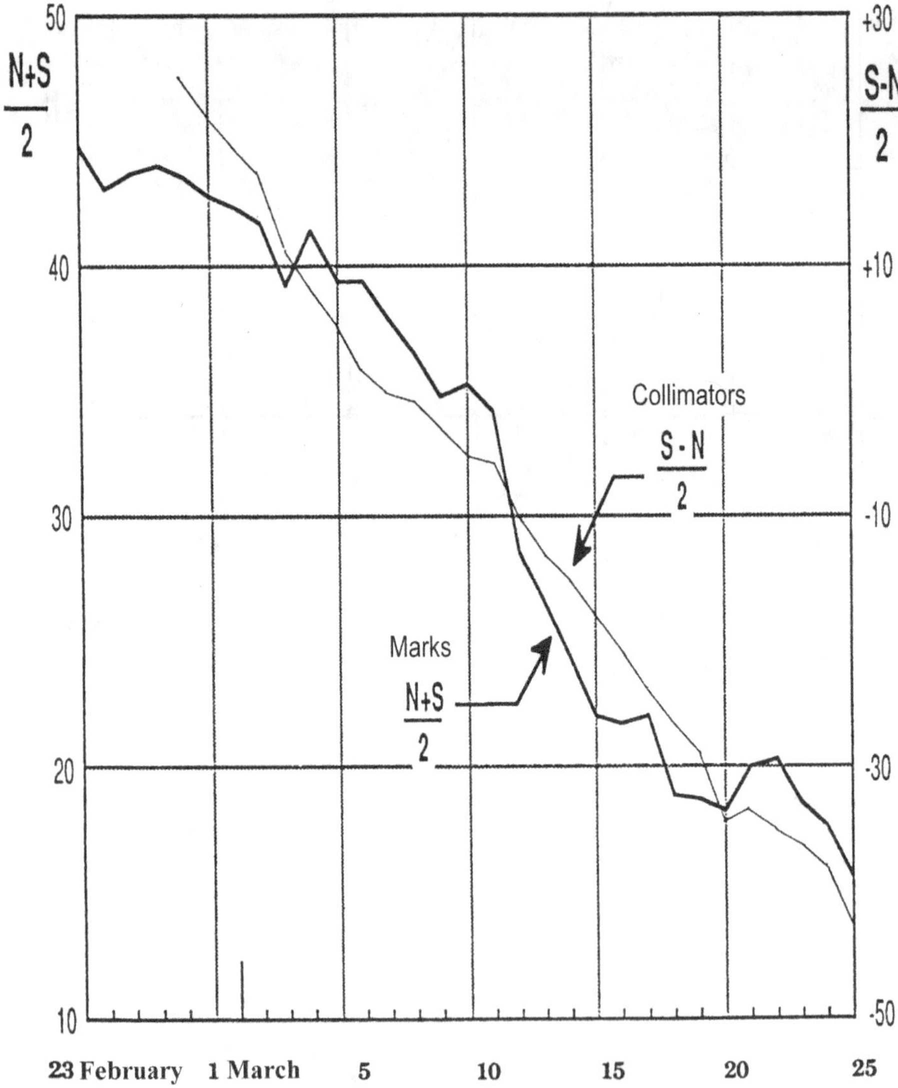


***Legend*** : The total variation of scale of the averages (N+S)/2 of the azimuths of the sightings at marks is around 40", and the corresponding variation of scale of the averages (S-N)/2 of the sightings at collimators is around 80", i .e. double the variation 40" of the scale (N+S)/2 of the sightings at marks.

***Sources*** : Graphs 13597 and 13598 (January 1996)

## Monthly lunar periodicities

2. - When the residuals of the linear fits to the global deviations are considered, analysis of the daily averages of the sightings at marks and at collimators of Palvadeau shows periodicities of the order of a month, in other words of the order of the sidereal period of the Moon, whose existence appears to be *inarguable*. Their amplitude is from 5" to 10".[5] They are *in the same sense* for the sightings at marks and appear to be *in opposite senses* for the sightings at collimators. The global deviation for the South telescope in the collimator case is *in the same sense* as the global deviations of the sightings at marks of the North and South telescopes.

*Graphs XII and XIII* represent fitting of the averages $(N+S)/2$ and $(N-S)/2$ corresponding to the *sightings at marks*, to linear trends associated with sinusoids of period equal to the sidereal period of the Moon of 27.322 days.

In fact, the linear trends can be considered as corresponding to periodic fluctuations having period equal to six months.[6] To about one day, the two cycles of *Graphs XII and XIII* are in opposite phase.[7] Their amplitudes are respectively 7.4 and 4.0 centesimal seconds.[8]

---

[5] See Note 1, page 348 above. Palvadeau also writes (p. 10):

"*In spite of the approximations by which they were obtained, the overall graphs clearly show the average drifts, and a possibility of a period near one month whose amplitude, of the order of 5 to 10 centesimal seconds, is significantly higher than the level of operator error to be feared.*"

[6] See below, *Chapter V*, § C.1.2, pp. 447-449.

[7] $s = 62.51$ for the maximum of the sinusoid of *Graph XII*, and $s = 61.19$ for the minimum of the sinusoid of *Graph XIII*. We thus have $62.51 - 61.19 = 1.32$ days ($s$ is counted in days from 0h on 1 January 1959).

[8] It is interesting to compare the phases of the 27.322 day cycles for the averages $(N+S)/2$ *of the sightings at marks* in 1958 at IRSID (*Graph V*, pp. 342 and 344) and at IGN in 1959 (*Graph XII*, p. 361).

Taking 0h on 1 January 1958 as origin, the maximum of the sinusoid of *Graph V* has abscissa $s = 186.9$. We therefore have

$$186.9 + 9 \times 27.322 - 365.25 = 67.5$$

for the abscissa in March 1959 of the maximum of the component corresponding to the 27.322 day cycle, with 0h on 1 January 1958 being taken as origin.

Now, the maximum of the 27.322 day cycle in 1959 (*Graph XII*) had abscissa $s = 62.54$. Thus there is a correspondence of phase to only five days $(67.55 - 62.54 = 5.01)$.

However, the method for calculation of the fitting sinusoids and of the *supposedly linear* trends (although actually they are not linear) *inevitably introduces biases* in the estimations of the phases. (See Note 6 of § B.5.3 above, p. 343).

Still, it may be asked whether consideration of the *synodic* period of the Moon might yield a better result. In fact, with the period of 29.53 days, the fitting to $(N+S)/2$ in July 1958 at IRSID gives $s = 182.85$ for the maximum, while the fitting for March 1959 at IGN gives $s = 61.91$. Thus we have

$$182.85 + 8 \times 29.53 - 365.25 = 53.84$$

Here, the difference is $53.84 - 61.91 = -8.07$, while it was only five days under the hypothesis of the *sidereal* period of the Moon. *This calculation therefore supports the hypothesis of a sidereal period of the Moon, as opposed to that of a synodic period.*

*Graph XIV* represents fitting of the half-sum $(S+N)/2$ of the azimuths of the *sightings at collimators*, to a linear trend associated with a sinusoid of period equal to the sidereal period of the Moon of 27.322 days. Within five days, this cycle is in phase with the cycle of *Graph XIII* that represents the half-difference $(N-S)/2$ of the sightings at marks. And, within two days, it is in opposite phase with the cycle of *Graph XII* that represents the average $(N+S)/2$ of the sightings at marks. Its amplitude is around 7 centesimal seconds.[9]

[9] I do not present the fitting corresponding to the difference $(S-N)/2$ for the sightings at collimators (calculation 950, 19 January 1996). In fact, the amplitude of the 27.322 day cycle, which is of the order of 2r = 2", is too small for the calculated curve to be appreciably different from the observed curve. Accordingly the corresponding Graph is not very readable.

The fitting sinusoid corresponding to the difference $(S-N)/2$ has its maximum at 12h on 23 March ($s_0 = 81.5$) and its minimum at 19h on 9 March ($s'_0 = 67.8$). Thus, to about two days, this sinusoid is in phase with the average $(S+N)/2$ of the sightings at marks (*Graph XIV* below).

*GRAPH XII*

**I.G.N.**
**HALF-SUM (N+S)/2**
**OF THE AZIMUTHS OF THE SIGHTINGS AT MARKS**
**Daily Averages**
**23 February - 25 March 1959 (31 days)**
*in centesimal seconds counted positively in the clockwise sense*
***Fitting to the sidereal period of the Moon of 27.322 days, with a linear trend***

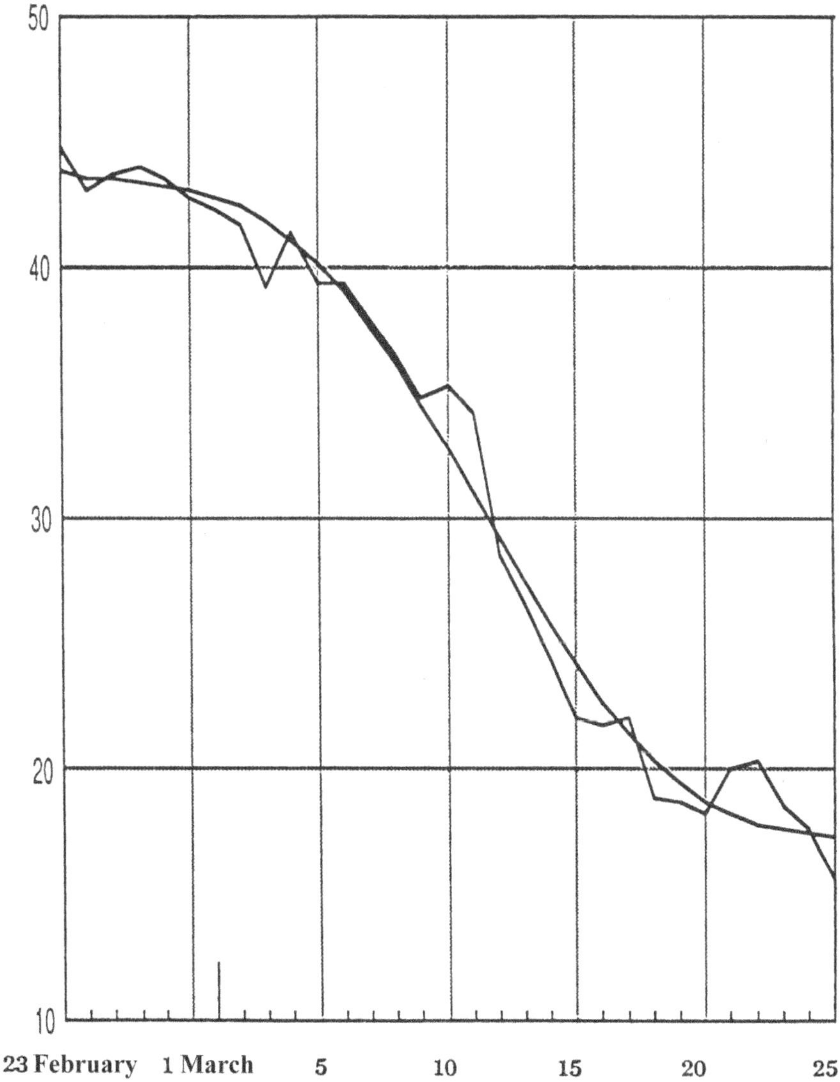


***Legend*** : $\Sigma$=10.2", R=0.992, 1-$R^2$=0.016, m=31.6", trend=-0.955" per day, r=3.70", $\sigma$=1.23"
See the *Legend* of *Graph XXV*, *Chapter I.*
The dates of the maximum and the minimum of the fitting sinusoid are respectively: 12h on 4 March ($s_0$=62.54) and 5h on 18 March ($s'_0$=76.2). The dates s are counted in days from 0h on 1 January 1959.

***Sources :*** Calculation 949 (19 January 1996) and Graph 13657.

*GRAPH XIII* **I.G.N.**

**HALF-DIFFERENCE (N-S)/2**
**OF THE AZIMUTHS OF THE SIGHTINGS AT MARKS**
**Daily Averages**
*in centesimal seconds counted positively in the clockwise sense*
**23 February - 25 March 1959 (31 days)**
***Fitting to the sidereal period of the Moon of 27.322 days,***
***with a linear trend***

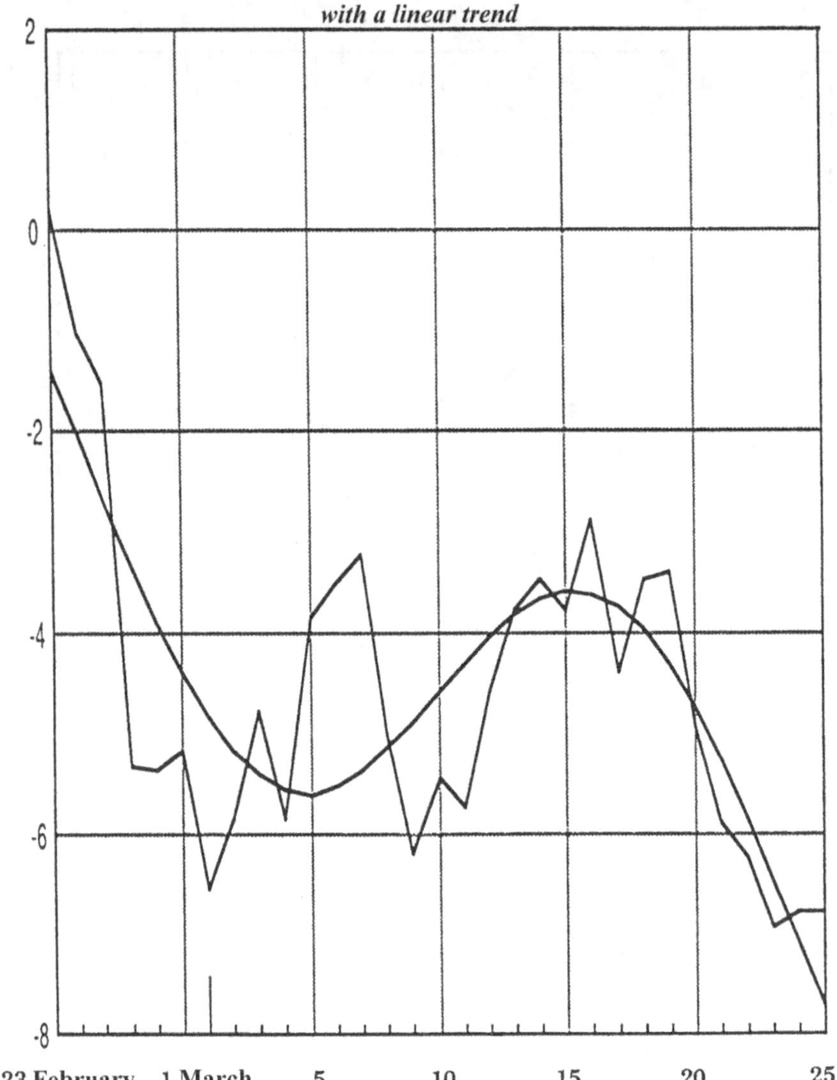


***Legend*** : Σ=1.71", R=0.773, $1\text{-}R^2$=0.403, m=-4.59", trend=-0.170" per day, r=-2.03", σ=1.09"
See the *Legend* of *Graph XXV, Chapter I.*
The dates of the maximum and the minimum of the fitting sinusoid are respectively:
20h on 16 March ($s_0$=74.35) and 5h on 3 March ($s'_0$=61.19). The dates s are counted in days from 0h on 1 January 1959.

***Sources*** : Calculation 951 (23 January 1996) and Graph 13661.

*GRAPH XIV*

**I.G.N.**

**HALF-SUM (S+N)/2**
**OF THE AZIMUTHS OF THE SIGHTINGS AT COLLIMATORS**
**Daily Averages**
*in centesimal seconds counted positively in the clockwise sense*
**27 February - 25 March 1959 (27 days)**
***Fitting to the sidereal period of the Moon of 27.322 days,***
***with a linear trend***

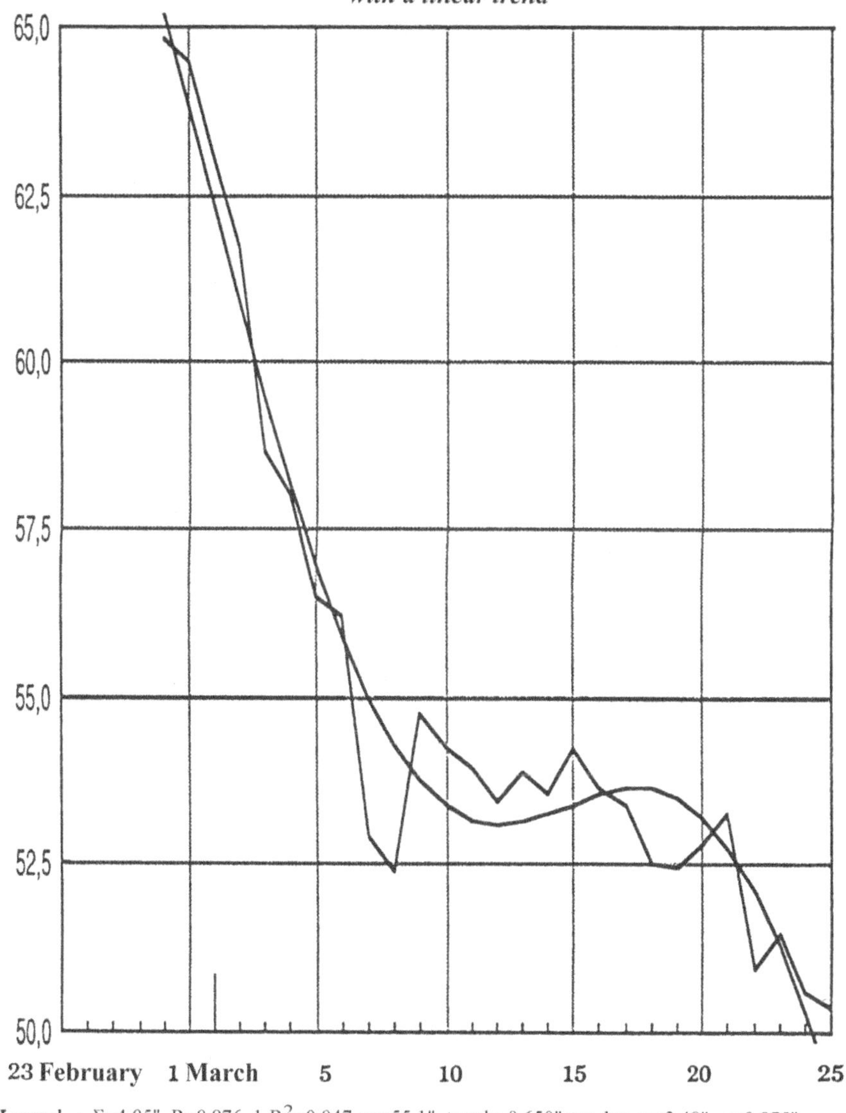


***Legend*** : $\Sigma$=4.05", R=0.976, 1-$R^2$=0.047, m=55.1", trend=-0.650" per day, r=-3.48", $\sigma$=0.875"
See the *Legend* of *Graph XXV*, *Chapter I*.
The dates of the maximum and the minimum of the fitting sinusoid are respectively:
12h on 21 March ($s_0$=79.56) and 22h on 7 March ($s'_0$=65.9). The dates s are counted in days from 0h on 1 January 1959.

***Sources :*** Calculation 952 (23 January 1996), and Graph 13662

## *C.4 Scientific implications of the experiments performed at IGN in February-March 1959*

### Almost periodic structure of the series of observations at IGN

1. In fact, all the graphs that can be drawn up for the series of observations at IGN in 1959 present *very numerous* morphological similarities both to the observations with the paraconical pendulum and also to the optical observations at IRSID in 1958: symmetries and double-symmetries, and pronounced local periodicities, are seen. These series thus present *all the characteristics of almost periodic functions.*[1]

Furthermore, the optical observations at IGN in 1959 also present *the same general characteristics* as the optical observations at IRSID in 1958.

Thus, all these observations have *great mutual coherence.*

### Significance of the results obtained at IGN

2. - The existence of *very clear* deviations *in the opposite sense* for the sightings at collimators, while on the other hand the deviations of the sightings at marks were *in the same sense*, with all the North units being mounted *upon one single pedestal* and all the South units *also* being mounted *upon another single pedestal*, shows that *none of these deviations observed for the sightings at marks and the sightings at collimators can be attributed to relative displacement of the pedestals, whatever the cause of such displacement might be.*

*Due to this fact, it is necessary to invoke factors acting upon the trajectories of the light rays.*[2]

As far as can be judged, the images of the marks corresponded to waves emitted by the lines of the marks, while on the other hand the light rays coming from the collimators essentially involved photons, and *this could explain the differences in characteristics between the sightings at collimators and the sightings at marks.*

---

[1] See above, *Chapter I*, § A.5.4, p. 101.

[2] This result is an observed fact, *totally independent of any hypothesis.*

The above fact constitutes the essential interest - *which may be termed fundamental* - of the *different* results obtained for the sightings at collimators and the sightings at marks. These results *completely* eliminate any influence of relative movement of the pedestals that support the apparatus.

It must be considered quite astonishing that no person in charge at IGN at the time (in particular Pierre Tardi) considered that here was an *absolutely fundamental* phenomenon that required explanation,[3] and that, at least, study thereof should be continued.[4]

*In summary, the experiments at IGN were extremely positive, in spite of the considerable importance of the personal equations arising from the quite excessive number of observers, and in spite of the completely inappropriate method used by Palvadeau for correcting for the personal equations of the observers. In fact, these experiments confirmed the existence of optical anomalies, and brought out a very significant structural difference between sightings at marks and sightings at collimators.*

*These experiments also confirmed the existence of structural monthly lunar periodicities, both in the sightings at marks and in the sightings at collimators.*

---

[3] *The same observation holds for myself* It is a fact that at the time, and in connection with the results obtained by comparison of the diurnal periodic components of the movements of the two pendulums at Saint-Germain and at Bougival and the diurnal periodic components of the optical sightings at Saint-Germain in July 1958, *my reflections and my analysis concentrated upon the diurnal fluctuations, and totally neglected the global deviations of the sightings at marks.*

[4] However in his *Conclusion* (p. 12), Palvadeau writes: "*It would be interesting to do a further series of experiments extending over several months, but only including a few points every day.*

*"But one cannot ignore the importance of variations, unknown but certainly present, between the personal equations of the observers, and it would be desirable to plan for an eventual repetition of similar experiments using a set-up that performs automatic registration of data points, without any human intervention.*

*"This does not diminish the fact that the actual experiments that were performed are very interesting...*

*"In any case, in view of the measurements performed, the exploration described in the present report certainly cannot claim to have exhausted the subject.*"

# D THE PHASES OF THE PERIODIC COMPONENTS ACCORDING TO THE LUNAR SIDEREAL MONTH OF 27.322 DAYS OF THE DEVIATIONS OF OPTICAL SIGHTINGS AT MARKS AND AT COLLIMATORS OF JULY 1958 AT IRSID AND OF FEBRUARY-MARCH 1959 AT IGN

## Phases of the paraconical pendulum

1. - In *Chapter II, Table VI*[1] compared the phases of the periodic sidereal monthly lunar components of period 27.322 days for the *enchained* series of the paraconical pendulum with anisotropic support of July 1958, for the azimuths X of the *directions of anisotropy* corresponding to the isotropic suspension, and *for the deviations Δ from the meridian* of the azimuths of the paraconical pendulum with isotropic support.

The month-long series of observations taken as reference is the month-long series of azimuths of the direction of anisotropy X of the paraconical pendulum with isotropic suspension in November-December 1959, for which the date of the maximum of the sinusoid was $s_0^* = 332.5$ days, the dates being counted from 0h on 1 January.

For the paraconical pendulum, all the azimuths are counted *positively in the anticlockwise sense*.

---

[1] p. 308.

## Phases of the optical sightings at marks and at collimators

2. -With the agreement to count the optical deviations similarly as *positive in the anticlockwise sense*, *Table II* below presents the phases of the different fittings to the optical observations.[2]

Since the measures of the deviations in centesimal seconds are counted *positively in the clockwise sense*, the maxima of the sinusoids of the deviations counted *positively in the anticlockwise sense* correspond to the minima of the sinusoidal fittings, for which the deviations are counted *positively in the clockwise sense*.

The same convention for orientation of the azimuths is valid in *Table II* below for the measurements of the optical deviations as for the azimuths of the paraconical pendulum, and *the same reference* is made to the month-long series of November-December 1959 as for *Table VI* of *Chapter II*, Section G.[3] This permits a comparison of the phases of the different periodic components.

## Phases of the optical deviations compared

3. - In *Table II* below, it can be seen that the series 2 and 3 of the sightings at marks at IGN, *i.e.* $(S+N)/2$ and $(S-N)/2$, are practically in phase, the difference in phase being about a day.

In the same manner, the series 4 and 5 of the sightings at collimators at IGN, *i.e.* $(S+N)/2$ and $(S-N)/2$, are practically in phase, the difference in phase being about two days.

[2] *Table II* presents in the same way the series $(S-N)/2$ for the sightings at marks and the sightings at collimators. For the sightings at marks, there is thus a change of sign due to the change from $(N-S)/2$ to $(S-N)/2$.

[3] p. 308.

The series 1 and 2, *i.e.* $(S+N)/2$, of the sightings at marks in July 1958 at IRSID and in February-March 1959 at IGN are in phase to about five days. But, in view of the biases introduced by the hypothesis of linear trends, it may be considered, *at least in the first approximation*, that these two series are in phase.

### Marks and collimators

4. - The average of the $s_0^*$ for the series 1, 2, and 3 corresponding to the sightings at marks is around 323 days, while the average of the series 4 and 5 corresponding to the sightings at collimators is around 340 days. This difference leads one to suppose that, *at least in the first approximation, the sightings at collimators are in opposite phase to the sightings at marks, provided that the South sightings at collimators are made to correspond to the North and South sightings at marks.*[4]

### The optical sightings and the azimuths of the paraconical pendulum

5. - The phase of the average deviations $(S+N)/2$ of the sightings at marks in July 1958 at IRSID (series 1) is around 327 days. This is only four days different from the phase of around 331 days of the average of azimuths $(S+B)/2$ of the paraconical pendulums at Saint-Germain and at Bougival in July 1958 (series 1 in *Table VI* of *Chapter II*, p. 308).

Thus, in *the first approximation*, it may be considered that the average $(S+N)/2$ of the sightings at marks S and N at Saint-Germain is in phase with the average $(S+B)/2$ of azimuths at Saint-Germain and at Bougival.

[4] § 3.1 above, p. 354.

**Relative values of the phases**

6. - In comparison of the phases of the different series, it is appropriate to consider two factors in particular.

a - First of all, when the correlation coefficients are not very high (of the order of 0.8, for example), the quality of the fittings may remain effectively the same with phase differences of four to five days. Even with high correlation coefficients (of the order of 0.985), the visual quality of the fittings is almost the same for differences in phase of two days.[5]

b - In the second place, the hypothesis of linearity of the trends introduces a serious bias in the calculation of the phases. In fact, and precisely to the extent that the quantities considered have (for example) periodicities of six months, the local tendency can be very different from a linear trend.

With regard to these observations it may be considered that, *overall*, the agreements in phase in July 1958 demonstrated by *Table VI* of *Chapter II* (Section G, p. 308) for the paraconical pendulum with anisotropic support and by *Table II* below for the optical deviations may be considered as *very remarkable*.

Nevertheless, there are obviously too few observations for definite conclusions to be reached.

[5] Thus if, for series 1 of *Table II* below (R=0.986), we take $s_0^*$=327.15+2=329.15, then the correlation coefficient stays relatively high (R=0.977), and visually the two fittings appear equivalent.

*TABLE II*

**OPTICAL DEVIATIONS OF SIGHTINGS AT MARKS AND AT COLLIMATORS**
**FITTINGS TO THE LUNAR SIDEREAL PERIOD OF 27.322 DAYS**
**COMPARISON OF THE PHASES OF THE FITTING SINUSOIDS**

***Optical deviations in centesimal seconds counted positively in the anticlockwise sense***

| Periods | q | Series | R | $1-R^2$ | r | Time of maximum | $s_o$ | p | $s_o^*$ |
|---|---|---|---|---|---|---|---|---|---|
| July 1958<br>Sightings at marks - IRSID | 1 | (S + N) / 2 | 0.986 | 0.029 | 2.38 | 14h on 20 July | 200.6 | 18 | 327.15 |
| February-March 1959<br>Sightings at marks - IGN | 2<br>3 | (S + N) / 2<br>(S - N) / 2 | 0.992<br>0.773 | 0.016<br>0.402 | 3.70<br>2.04 | 5h on 18 March<br>20h on 16 March | 76.2<br>74.85 | 9<br>9 | 322.1<br>320.75 |
| February-March 1959<br>Sightings at collimators - IGN | 4<br>5 | (S + N) / 2<br>(S - N) / 2 | 0.976<br>0.997 | 0.047<br>0.0061 | 3.48<br>0.995 | 22h on 7 March<br>22h on 9 March | 65.9<br>67.8 | 10<br>10 | 339.12<br>341.02 |
| November-December 1959<br>Azimuth of anisotropy X<br>Paraconical pendulum | Time of maximum: 12h on 29 November ; $s_o^* = 332.5$ | | | | | | | | |
| Time of minimum declination of the Moon in Nov.-Dec. 1959: 21h on 1 December , $s_o^* = 334.87$ | | | | | | | | | |

***Legend*** : *Value of r*: r is expressed in centesimal seconds
*Calculation of* $s_o^*$ (date of the maximum of the periodic component closest to 29 November 1959, *Table VI* of *Chapter II*)

July 1958: $s_o - 365.25 + 27.322\ p$
February-March 1959: $s_o + 27.322\ p$
Averages of the $s_o^*$ : (1, 2, 3, 4, 5,) = 330.02 (1, 2, 3) = 323.33 (4, 5) = 340.07
27.322 / 2 = 13.661 days 340.07 - 13.66 = 326.41 340.07 - 323.33 = 16.74

*Sources* : Calculations 943, 949, 951, 952, and 950.
*Graphs XII, XIII, and XIV* of § B.5.3 and § C.3.2.

# E OVERALL VIEW OF THE EXPERIMENTS ON OPTICAL SIGHTINGS IN JULY 1958 AT IRSID AND IN FEBRUARY-MARCH 1959 AT IGN

## Major anomalies

1 - My experiments on optical sightings of July 1958, associated with the *crucial* experiments of Bougival and Saint-Germain on the anomalies of the paraconical pendulum with anisotropic support, established the existence of *major optical anomalies not previously perceived*, and they also established *an extremely significant linkage between two entirely different domains - that of mechanics and that of optics.*

Despite their imperfections, the experiments at IGN in 1959 *not only confirmed the existence of these anomalies, but also extended their scope.* Particularly, these experiments showed the existence of *very significant monthly deviations of sightings at marks and sightings at collimators, having very different natures.* They also demonstrated the periodic structure of these anomalies.

In all these experiments, of which I can only give a very abbreviated description here, the difficulties that were encountered essentially arose from two circumstances that were very difficult to overcome: *the extreme minuteness of the observed effects*; *and the necessity for continuous observations over a period of a month at least*, which were very laborious to implement.[1]

[1] *Trials of continuous automatic registration* of optical deviations of sightings at marks were performed from April 1992 to January 1993 at Palaiseau and at IGN by Michel Kasser.

Although these trials were performed in very unfavorable conditions, using an apparatus chosen by Michel Kasser that, finally, proved to be very sensitive to temperature variations, and with very insufficient material means and personnel, they led to positive results and provided precious lessons.

These trials showed the *possibility of continuous automatic registration, even at very close range, all the apparatus being disposed upon the same support.*

These trials also showed, in January 1993, the existence of *similar trends*, on the one hand in a subterranean gallery at Palaiseau (with instruments fixed to the walls), and on the other hand at IGN (where all the apparatus was mounted upon the same table of granite), where the *conditions of temperature* were *totally different*. These trials thus provided evidence of the existence of *identical* optical anomalies in the two laboratories (which were far apart) and in quite different conditions.

A summary analysis of these trials is presented in *Chapter III*, Section B of the *second volume* of this work (p. 29 above).

**An Anisotropy of Space**

2 - The *major* conclusion from all these experiments is that the observed anomalies *really exist*, and that *they cannot be attributed to perturbations resulting from other phenomena.*

This conclusion is inescapable for three reasons:

- *the periodic structure of the optical anomalies observed in July 1958, similar to that of the anomalies of the paraconical pendulum* with anisotropic support and to that of the anomalies of the paraconical pendulum with isotropic support;
- the periodic lunar sidereal structure of the components of the deviations of sightings at marks on the one hand, and of those at collimators on the other hand;
- the presence of *very significant and completely different* monthly deviations *in the sightings at marks and the sightings at collimators* at IGN in 1959, *although all the apparatus was set up on the same pedestals.*

*The observed effects can only be attributed to modifications of the space between the telescopes and the marks (or the collimators), in other words to anisotropy of optical space.*

The result of my optical experiments of July 1958, in association with the crucial experiments of July 1958 in Bougival and Saint-Germain on the paraconical pendulum, is that *anisotropy of optical space is closely linked to anisotropy of the space* corresponding to the paraconical pendulum, *which may be considered as an anisotropy of inertial space.*

Both these two cases that I brought to light - the anomalies of the paraconical pendulum and the optical anomalies (anomalies that are closely mutually linked) - constituted *major discoveries* that were suffocated at the time *by the blind and fanatical obscurantism of the mandarins of pseudo-science.*

# Chapter IV
# TWO VERY SIGNIFICANT PREVIOUS EXPERIMENTS

*True physical reality is only present in the totality of experimental results.*

Louis de Broglie *

# A THE OBSERVATIONS OF ESCLANGON AND MILLER

1 - The anomalies of the paraconical pendulum and the optical anomalies of sightings at marks that I have brought to light *are not isolated anomalies*. In the last century, experimenters have demonstrated numerous anomalies.

In my memoir of 1958, "*Should the laws of gravitation be reconsidered?*" [23], I gave numerous examples of these. These are basically the anomalies observed in the classical experiment of Foucault, the anomalies observed in mechanics, and the anomalies observed in certain optical and electromagnetic phenomena. I accompanied this relatively detailed presentation with an analysis of the precision to which the laws of gravitation have been verified, as well as a general commentary on the significance of the anomalies detected.

With regard to the general conception of the present work, I can do no better than revisit my analysis of 1958.[1] I shall limit myself simply to recapitulating the essential conclusions:

> *"From examination of the observed anomalies and from the discussion of the precision with which the principles of mechanics have been verified, it results that these principles, according to all the available evidence, do not at all have the absolute value that a far too common opinion appears to attach to them. These principles have nowadays acquired a sort of metaphysical quality that is superior to all discussion. In reality mechanics is not at all a perfect science, a pure science where there is nothing further to discover. It is, and remains, an experimental science that can and should be improved.*

---

* Louis de Broglie, 1953, *La Physique quantique restera-t-elle indéterministe?* (Will quantum physics remain indeterministic?), Gauthier-Villars , p. 96 [86].

[1] p. 99-103 [23].

> *"It is in fact time to rethink all these phenomena...*
>
> *"Deeper examination of all the observed anomalies on the experimental and theoretical level seems to me to be of the highest interest, because it seems to me that it might involve the revision of certain postulates whose rigorous validity has been accepted without effective input from experiment.*
>
> *"Only the facts should guide us, not mummified principles, even if they may be useful in the first approximation. The only source for our knowledge is experiment, and any thought that definitively imprisons itself in abstract principles condemns itself thereby to sclerosis."*

2 - In fact, the literature contains two series of experiments of fundamental importance, analysis of which on at least a brief level seems to me to be absolutely necessary here, in view of the general conception of the present work: the experiments of Esclangon on the anisotropy of space, 1927-1928; and the experiments of Miller, 1925-1926. In particular, the observations of Miller seem to me to be of *major and absolutely exceptional interest.*

*These two series of experiments performed by Esclangon and Miller have remained poorly known or been rejected, as a consequence of preconceived ideas and because they have been considered to be in contradiction with "established truths".*[2]

[2] However, all those who, in the name of their convictions and of *"established truths"*, have ignored or rejected the observations of Esclangon and the observations of Miller as useless, should meditate upon the quotation placed at the head of this Chapter, on p. 373.

# B THE OPTICAL OBSERVATIONS OF ERNEST ESCLANGON, 1927-1928

## *B.1 The researches of Esclangon on the anisotropy of space, 1926-1928*

From 1926 to 1928, Ernest Esclangon published various analyses of researches that he had performed at the Observatory of Strasbourg *with the goal of demonstrating the absolute movement of the Earth from apparent mechanical and optical asymmetries of space.*[1]

### Asymmetry of sidereal space and tidal phenomena

1 - In his *Note of 12 July 1926*, Esclangon gave the results of his analysis of 166,500 hourly tidal observations taken at Pola in the Adriatic from 1 January 1898 to 31 December 1916, thus spanning 19 years. This analysis demonstrated *an asymmetry of sidereal space*, not due to luni-solar action, corresponding to a diurnal *sidereal wave* of amplitude approximately $14.4 \times 10^{-6}$ radians.

[1] Ernest Esclangon: 12 April 1926, *Sur la dissymétrie mécanique et optique de l'espace en rapport avec le mouvement absolu de la Terre* (On the mechanical and optical asymmetry of space with respect to the absolute movement of the Earth), CRAS, 1926, 1st. semester, vol. 182, pp. 921-923 [138]; 12 July 1926, *La dissymétrie de l'espace sidéral et le phénomène des marées* (The asymmetry of sidereal space and the phenomenon of the tides), CRAS, 1926, 2nd. semester, vol. 183, no. 2, pp. 116-118 [137]; 27 December 1927, *Sur la dissymétrie optique de l'espace et les lois de la réflexion* (On the optical asymmetry of space and the laws of reflection), CRAS, 1927, 2nd. semester, vol. 185, no. 26, pp. 1593-1595 [139]; 15 April 1928, *Sur l'existence d'une dissymétrie optique de l'espace* (On the existence of an optical asymmetry of space), Journal des Observateurs, vol. XI, no. 4, pp. 49-63 [140].

## On the existence of an optical asymmetry of space

In his *Note of 27 December 1927*, Esclangon demonstrated the existence of an "*optical asymmetry of space*". And on 15 April 1928 he published a detailed explanation, adding a Table of his observations.[2]

*The experimental arrangement* depended upon sighting *by autocollimation at the horizontal filament* of an astronomical telescope (an alt-azimuth of the Observatory of Strasbourg) *and its image reflected by mirrors.*[3] Series of observations were performed alternately in the directions north-west and north-east.

From 25 February 1927 to 4 January 1928, 40,000 readings were taken during 150 series of observations, *during both day and night.*

---

[2] pp. 54-57 [140]. The observed differences (denoted by $c - c'$) are given in ten-thousandths of a minute of arc, as a function of sidereal time and Greenwich time, with indications of the numbers of the readings and of the temperatures.

A ten-thousandth of a minute of arc represents $60/10,000 = 0.006$ sexagesimal seconds, *i.e.* $0.0291 \times 10^{-6}$ radians. A sexagesimal second represents 166.7 ten-thousandths of a minute of arc, *i.e.* $4.848 \times 10^{-6}$ radians $[\pi/(180 \times 3600)]$.

[3] There is *an evident connection* between this experimental procedure and the deviations of the sightings at marks that I demonstrated (*Chapter III* above). But the deviations observed by Esclangon were *vertical*, while those that I observed were *azimuthal*.

## *B.2 Analysis of the optical observations of Esclangon in 1927-1928*

A *systematic difference* is apparent between the north-west and the north-east readings - a *difference that depends only upon the average sidereal time of the observational session,* and that can be represented by a sinusoidal fluctuation *whose period is 24h sidereal.*[1] By contrast, arranged according to the average solar time, the observations are scattered without any apparent order, and without corresponding to any curve away from the axis representing their average.[2]

*Graphs I* below reproduce the two representations of the observed differences, in sidereal time and in average solar time, *just as they were published by Esclangon* in the *Journal des Observateurs* of 1928.[3] In sexagesimal seconds, the observed differences varied between −0.08" and +0.08" at around 2h30m and 14h30m sidereal. They became zero at around 8h30m and 20h30m.

[1] I remind the reader that the sidereal year is shorter than the solar year by one solar day. The sidereal day is shorter than the solar day by around four minutes, and the sidereal month is about two hours shorter than the solar month.

During a single month, the diurnal sidereal periodicity *cannot be distinguished* from the diurnal solar periodicity. These two periodicities can only be distinguished over the course of a year.

At any instant, the fixed stars are at the same azimuths at the place of observation, as they were 24 sidereal hours before. After 24 civil hours, it is the Sun that is at the same azimuth.

[2] This is exactly what Miller observed in the analysis of his four series of observations of 1925-1926 (see below, §D.1, pp. 392-393).

[3] p. 52 [140]. The graphs of Esclangon have been reproduced by photocopying on *Graphs I* below.

*Graph II* represents the averages of the observed differences $c - c'$ *for each sidereal hour* during the year, as well as the fitting sinusoid.[4] The amplitude $2r$ of the fitting sinusoid is around 12 ten-thousandths of a minute of arc, *i.e.* around 0.072 sexagesimal seconds (or $0.35 \times 10^{-6}$ radians). The ordinate of the fitting sinusoid becomes zero at around 8h30m and 20h30m. The correlation coefficient is about 0.925. *Esclangon did not perform this fitting.*

The observations of Esclangon thus demonstrate *an optical anisotropy of space in sidereal time.*[5, 6]

---

[4] These averages were calculated by me from Esclangon's *Table* of differences $c - c'$ (*Note 2* above, p. 377).

[5] Esclangon's observations also present a *semi-annual periodicity* whose minimum corresponds to 25 March, near the spring equinox (see below, *Chapter V*, §C.2, pp. 450-451).

[6] While the observations at Strasbourg gave deviations of the order of three hundredths of a sexagesimal second, new experiments performed by Esclangon at the Observatory of Paris over 18 months starting in 1932 did not give any variation of the deviations, *"at least to an order of 0"001"*.

(Esclangon, *Recherches expérimentales sur la dissymétrie optique de l'espace* (Experimental researches on the optical asymmetry of space), C.R.A.S., 1 April 1935, pp. 1165-1168 [141]).

In fact, the variations of the deviations that were found were about a hundred times smaller than those found at Strasbourg. In any case, the experimental arrangements employed at Paris were *entirely different* from those used at Strasbourg, and the deviations that were measured were *in azimuth, rather than vertically* as at Strasbourg.

## *GRAPHS I*

### ESCLANGON'S EXPERIMENTS ON THE OPTICAL ASYMMETRY OF SPACE

*Values of c-c'*
*in sexagesimal seconds*
*Daily observations from 0h to 24h*
*25 February 1927 - 4 January 1928*

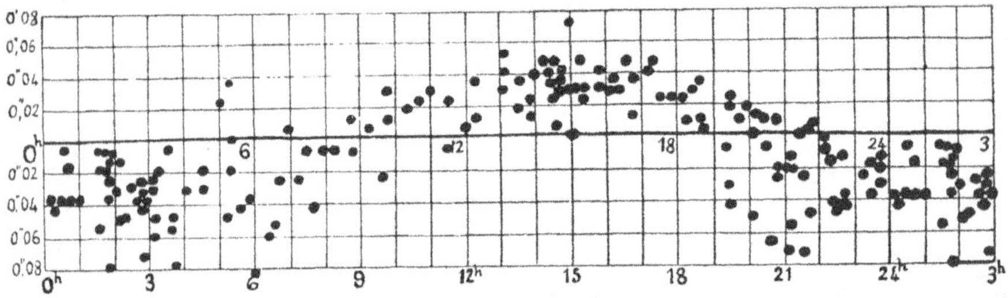


*Sidereal time*

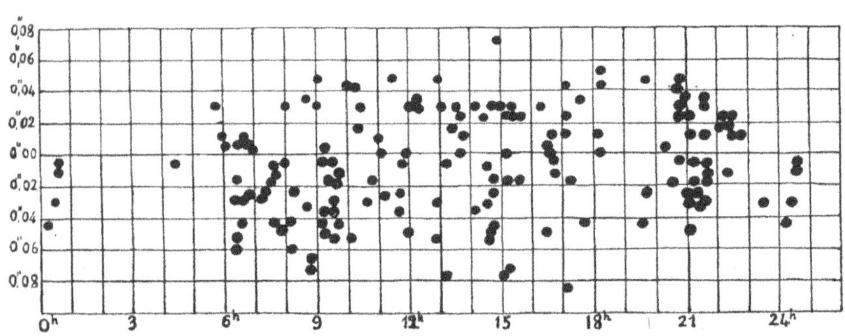


*Mean solar time*

***Legend :*** The values of c-c' indicated by Esclangon in the Table of the *Journal des Observateurs*, pp. 54-57, are in ten-thousandths of a minute of arc (*Graph II* below). A ten-thousandth of a minute of arc is equal to 0.006 sexagesimal seconds.

The order of magnitude in absolute value of the deviations c-c' is approximately three hundredths of a sexagesimal second.

***Source :*** Ernest Esclangon, 15 April 1928, *Sur l'Existence d'une Dissymétrie Optique de l'Espace* (On the Existence of an Optical Asymmetry of Space), *Journal des Observateurs*, Vol. XI, no. 4, p. 52.

*GRAPH II* **ESCLANGON'S EXPERIMENTS ON THE ASYMMETRY OF SPACE**

**Average daily observations from 0h to 24h in sidereal time**
**25 February 1927 - 4 January 1928**
*Values of c-c' in ten-thousandths of a minute of arc*

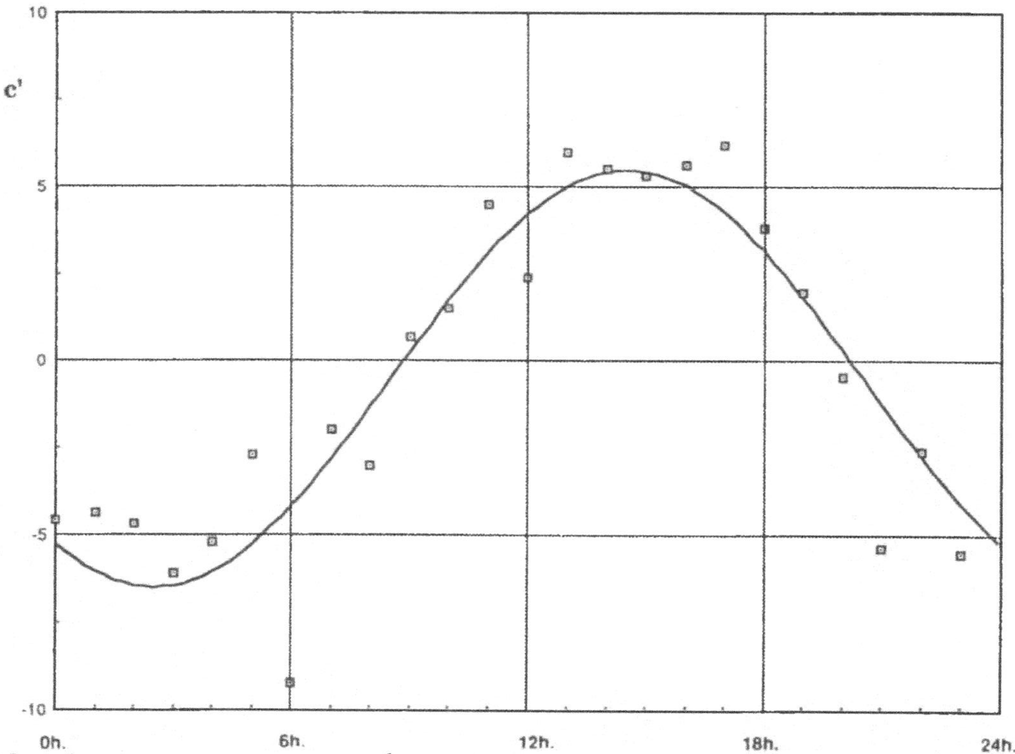


***Legend :*** fitting sinusoid: $\Sigma$=4.57; R=0.925; 1-$R^2$=0.145; m=-0.51; r=5.98; $\sigma$=1.74; times of minimum and maximum: 2h30m and 14h30m; times at which the value of the sinusoid attains its average value: 8h30m and 20h30m. (see the legend of *Graphs I* above, and the *Legend of Graph XXVII* of §C.3 of *Chapter 1*, p. 158)

***Sources :*** Calculation 718 (6 November 1995) and Graph 13575; and Esclangon, 1928, id., pp. 54-57.

# C THE INTERFEROMETRIC OBSERVATIONS OF DAYTON C. MILLER, 1925-1926

## *C.1 The experiments and analyses of Dayton C. Miller, 1925-1933*

Following the interferometric experiments of Michelson at Berlin, 1878-1881, of Michelson and Morley at Cleveland, 1887[1], of Morley and Miller at Cleveland, 1902-1906, of Miller at Mount Wilson, 1921, and of Miller at Cleveland, 1922-1924, Miller undertook at Mount Wilson, in 1925-1926, a set of *four series of continuous observations*, each over about a week, *during four different months.*[2]

[1] Michelson: *1881, The relative motion of the earth and the luminiferous aether*, The American Journal of Science: Third Series, Vol. XXII, 1881, Art. XXI, pp. 120-129 [196].

Michelson and Morley: *1887, On the relative motion of the earth and the luminiferous aether*, The American Journal of Science: Third Series, Vol. XXIV, no. 203, 1887, Art. XXXVI, pp. 333-345 [199].

On these experiments, and on those that preceded them and followed them, see particularly:

- M. Mascart, 1872-1874, *Sur les modifications qu'éprouve la lumière par suite du mouvement de la source lumineuse et du mouvement de l'observateur* (On the modifications experienced by light due to movement of the luminous source and movement of the observer), Annales de l'Ecole Normale Supérieure, 1872, pp. 157-214; and 1874, pp. 363-420 [191, 192].

- Oliver Lodge, 1893, *Aberration Problems. A Discussion concerning the Motion of the Ether near the Earth, and concerning the connection between Ether and Gross Matter; with some new Experiments.* Philosophical Transactions of the Royal Society of London, 1894, Vol. 184, pp. 727-806 [182];

- Augustin Sesmat, 1937, *L'Optique des Corps en Mouvement* (The Optics of Moving Bodies), Hermann, Paris [246].

[2] Miller: 28 April 1925, *The Ether-Drift Experiments at Mount Wilson*, Proceedings, National Academy of Sciences, Vol. II, 1925, pp. 306-314 [201]; Miller: *1926, Significance of the Ether-Drift Experiments of 1925 at Mount Wilson*, Science, Vol. LXIII, April 1926, No. 1635, pp. 433-443 [202]; Conference on the Michelson-Morley Experiment: *4-5 February 1927*, The Astrophysical Journal, Vol. LXVIII, December 1928, pp. 341-402, (the proceedings of a conference held at Mount Wilson in 1927), in which Miller's presentation is given on pp. 352-367 [198]; Miller: 1933, *The Ether-Drift Experiment and the Determination of the Absolute Motion of the Earth*, Reviews of Modem Physics, Vol. 5, July 1933, Number 3, pp. 203-242 [203]. This last memoir is *absolutely fundamental*.

In view of the previous observations, *the goal of these experiments was to determine if it was effectively possible to determine the absolute movement of the Earth with respect to the ether by measuring its speed with the interferometer.*[3]

*Even though they are of capital importance*, the validity of Miller's observations has been *denied or neglected* by almost all commentators, *or has even left people totally unaware*, basically because of the absolute ascendancy that the theory of relativity has assumed in the world of the physicists.[4]

In fact *it is indubitable that, if the validity of Miller's observations is recognized, the theory of relativity in its current form would need to be at least profoundly revised, because its initial foundation was aimed precisely at giving an explanation of the claimed absence of any positive result in the experiments of Michelson* .[5, 6]

---

[3] On his motivation, Miller wrote (Miller, 1933, id., p. 217) [203]:

> *"The Theory of Relativity had its inception at this time when Einstein published his paper entitled* "Zur Electrodynamik bewegter Körper" *in November 1905, and was elaborately developed in succeeding years ... Since the Theory of Relativity postulates an exact null effect from the ether-drift experiment which had never been obtained in fact, the writer felt impelled to repeat the experiment in order to secure a definitive result".*

[4] The scientific credentials of Miller (1866-1941) were however *indisputable*. Miller was a scientific personality of the top grade. He became a member of the Academy of Sciences of the USA in 1921, and was president of the American Physical Society in 1925.

[5] In his work of 1922 on *"The Axioms of Mechanics"* (*Les Axiomes de la Mécanique*, Gauthier-Villars, 1922, p. 102) [208], Painlevé wrote:

> *"That the formulas (of the theory of relativity deduced from the Lorentz-Einstein transformation) and the hypotheses that they reflect should account for Michelson's experiment is certain in advance, because it is this experiment that elicited them."*

On the Lorentz-Einstein transformation, see Painlevé (1922, id., pp. 100-102), and Albert Einstein, 1905, *Zur Elektrodynamik bewegter Körper* (On the Electrodynamics of Moving Bodies), Annalen der Physik, 10, p. 891 [123]; French translation: *Sur l'Electrodynamique des Corps en Mouvement*, Gauthier-Villars, 1925, pp. 12-26 [126]; notes (1), pp. 2 and 17 of the French translation, which relate to the work of Lorentz, *do not appear* in the original German publication. See also *Chapter VII* below, §A.6, pp. 569-572.

[6] On the theory of relativity, see *Chapter VII* below, pp. 547-616.

The following explanation will carefully distinguish the experimental arrangement of Miller, the *indisputable* facts that he demonstrated, and the theory that he applied in order to analyze his observations.[7]

*In essential matters, this explanation will limit itself to analysis of the observations of Miller.*[8]

[7] Miller's theory and his application of it inevitably provoke very strong reservations (see below, §F.1, pp. 417-419).

[8] In *Chapter IV* of the *second volume* of this work, I shall present an analysis of the determination of the horizontal projection of the velocity of the Earth (*Section B*), a critical analysis of Miller's theory and of his interpretation of his observations (*Section C*), and finally an analysis of the implications of two hypotheses: an anisotropy of space, and a cosmic velocity of the Earth towards the constellation of Hercules (*Section D*) (see the *summary* above, p. 29).

## C.2 Miller's experimental procedure

The Mount Wilson experiments were performed in a room of the Mount Wilson Observatory, at 1750 meters altitude and $34^\circ 13'$ latitude.

The principle of Miller's experiments is the same as that of the celebrated experiments of Michelson. From the displacements of the fringes observed during the rotation of the arms of the interferometer, it is possible to deduce the square $v^2/c^2$ of the ratio of the speed $v$ of the Earth with respect to the ether to the speed $c$ of light.[1] *In fact, what the interferometer measures is the speed difference v of light for two perpendicular directions.*

By comparison with preceding experiments, Miller's experimental procedure presented three essential novelties:

1. The path of each light ray was extended to 64 meters, so that the displacement of the fringes was made to be 27 times greater than in the first experiment of Michelson in 1881, and to be around three times greater than in the Michelson and Morley experiment in 1887;[2]
2. The observations were performed *in a continuous manner over all azimuths at all times, day and night*;
3. *Four series of continuous observations were performed over six to eight days*[3] at four different epochs.

In total, Miller effected 6,400 complete turns of the interferometer, which included 200,000 readings.[4]

---

[1] We have (Miller, 1933, id., p. 227 [203])

$$d = 2D(v^2/c^2)$$

whence

$$v = \sqrt{dc^2/2D}$$

where $d$ is the maximum displacement of the fringes observed in azimuth $A$ during a rotation of the interferometer through 180°, and $D$ is the length of the arms of the interferometer.

For an overall explanation of the experiment of Michelson, see particularly: Augustin Sesmat, 1937, id., pp. 77-129 [246].

[2] Note (2) of §C. 1 above, p. 382.

[3] Miller, 1933, id., p. 213 [203].

[4] Miller, 1933, id., p. 228 [203].

## *C.3 The observations of Miller*

### The presentation of the observations

1 - Miller did not publish the details of his observations, but *the essence* of all his original observations is presented in the form of eight very clear graphs, *four for the azimuths as functions of sidereal time, and four for the speeds.*[1] These are shown opposite (*Graphs III and IV*). These graphs represent both the observations and their moving averages.

These eight curves showing the speeds and the azimuths represent *the synthesis of all the observations taken*. These are the data from which all Miller's estimations are deduced.

Furthermore, and to facilitate comparison, I also reproduce the graphs representing the azimuths *as a function of civil time (Graphs V).*[2]

### The average deviations $\overline{A}$

Miller stated that, during an observational period of around a week, the average direction corresponding to the maximum displacement of the fringes *oscillated over 24 hours about an average position* $\overline{A}$ *which was not null, and which was different for each of the four periods considered (Graphs V).*

---

[1] Miller, 1933, id., p. 229 [203].

These graphs can be photographically enlarged and easily exploited. I reproduce these enlargements in *Chapter IV (Section A)* of the *second volume* of this work (p. 29 above).

*Miller indicated the moving averages of the observations on these graphs, thus determining twenty average observations per day (1933, id., p. 228* [203]*).*

[2] Miller, 1933, id., p. 234 [203].

The average directions $\overline{A}$ observed in the four periods considered (of 6 or 8 days) are the following according to Miller, taking North as the origin for the angles and the sense from West through North to East as positive[3]: 1 April 1925, $\overline{A} = +40°$; 1 August 1925, $\overline{A} = +10°$; 15 September 1925, $\overline{A} = +55°$; 8 February 1926, $\overline{A} = -10°$.[4]

---

[3] Miller, 1933, id., p. 235 [203], and *Graphs V* below.

*The angles* $\overline{A}$ *are thus reckoned in degrees, positively from the North in the clockwise sense.*

[4] *My own estimations* deduced from the averages $(A_M + A_m)/2$ of the maximal values $A_M$ and the minimal values $A_m$ of the azimuth $A$, and plotted on the Graphs of Miller (*Graphs III and IV* below), are the following: 40°, 12.5°, 55°, and −12.5° (*Chapter V*, §D.2, *Table II*, p. 454).

*GRAPHS III*

## INTERFEROMETRIC OBSERVATIONS OF MILLER AVERAGES OF DAILY OBSERVATIONS

***Speeds in km per second***
***Azimuths in degrees as a function of sidereal time θ measured positively from the North in the clockwise sense***
***Averages over 6 or 10 days of observations***

*1 April 1925*

***Speed***

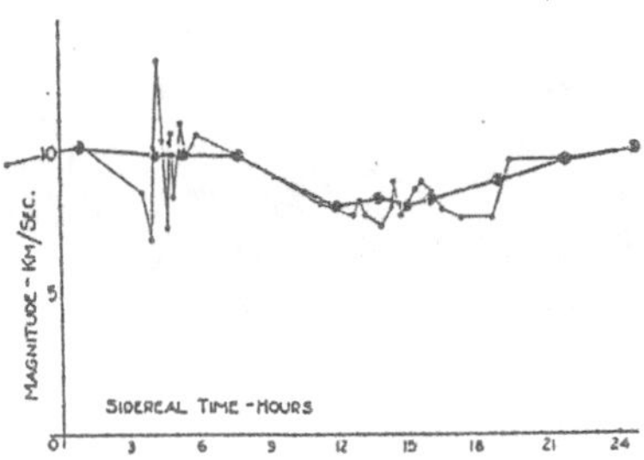


***Azimuth***

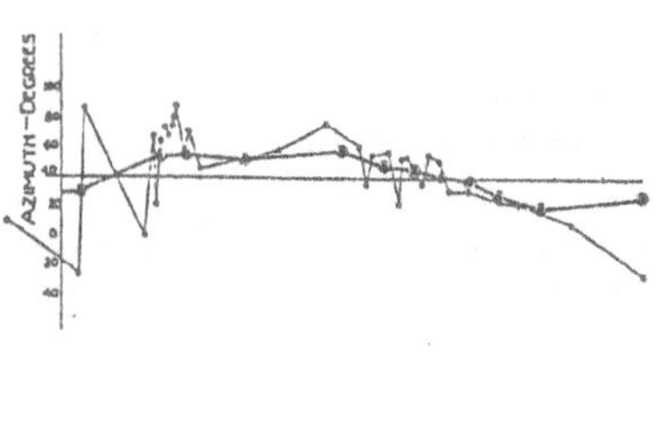


*1 August 1925*

***Speed***

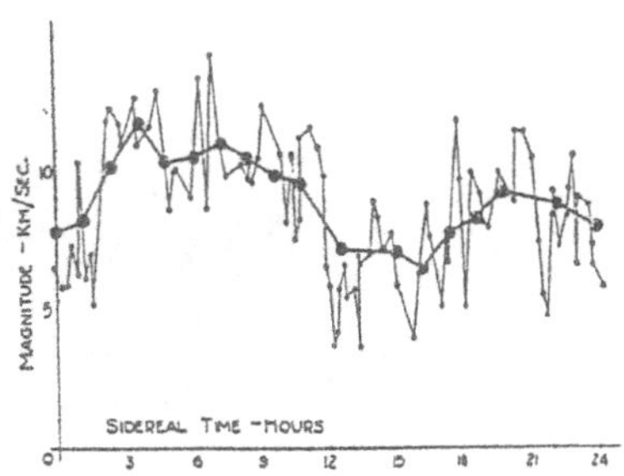


***Azimuth***

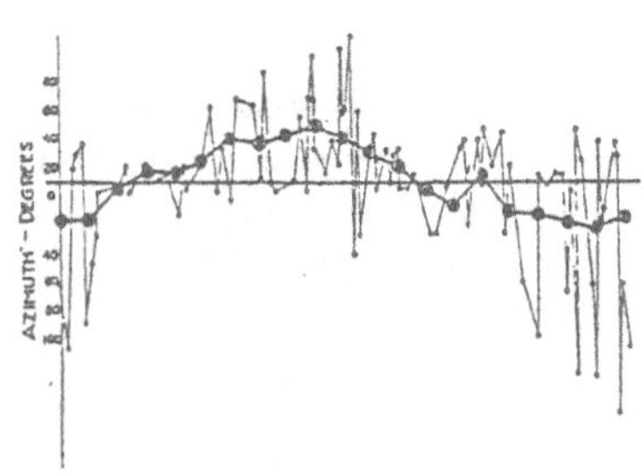


*Source :* Miller, 1933, id., p. 229.

*GRAPHS IV*

| *15 September 1925* | *8 February 1926* |
|---|---|

*Speed*

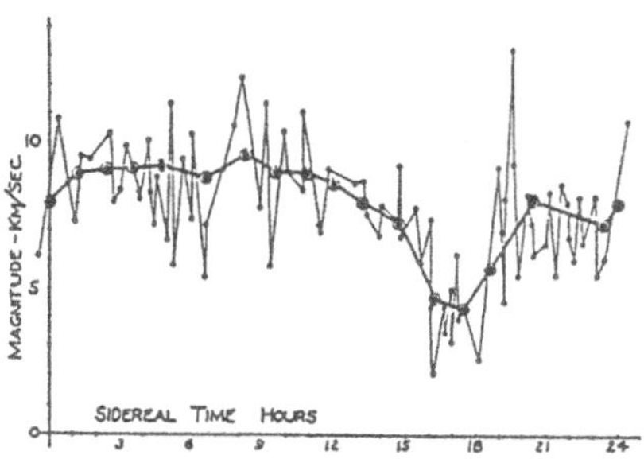


*Speed*

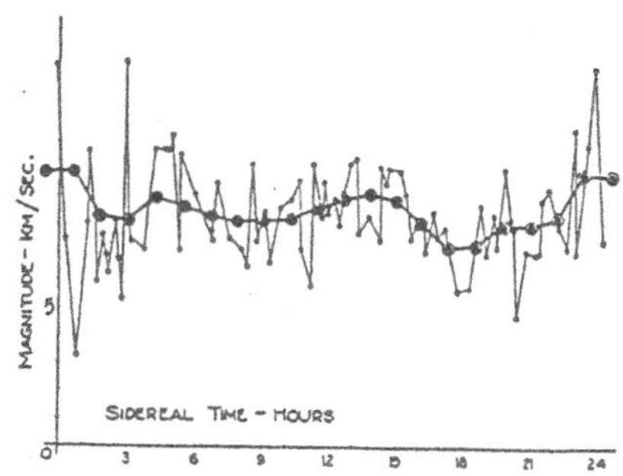


*Azimuth*

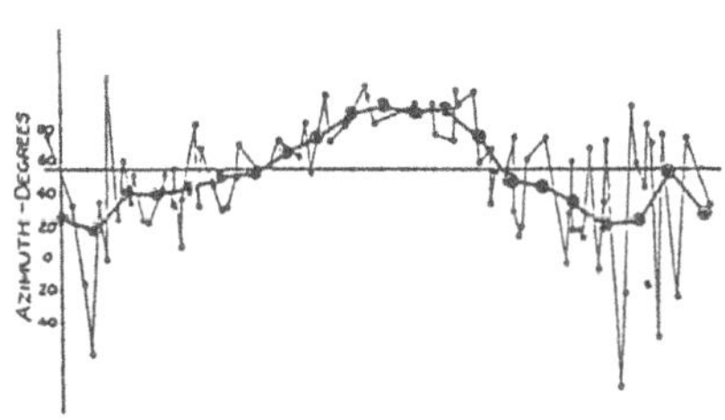


*Azimuth*

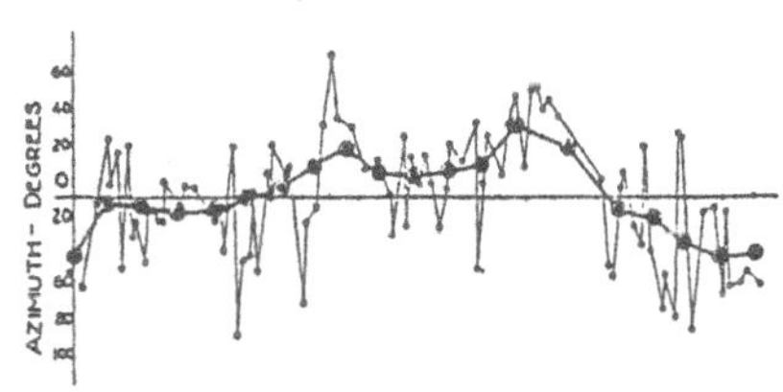


*Source :* Miller, 1933, id., p. 229.

*GRAPHS V*

## INTERFEROMETRIC OBSERVATIONS OF MILLER REPRESENTATION OF THE MOVING AVERAGES OF THE AZIMUTHS IN CIVIL TIME

*1 April 1925*

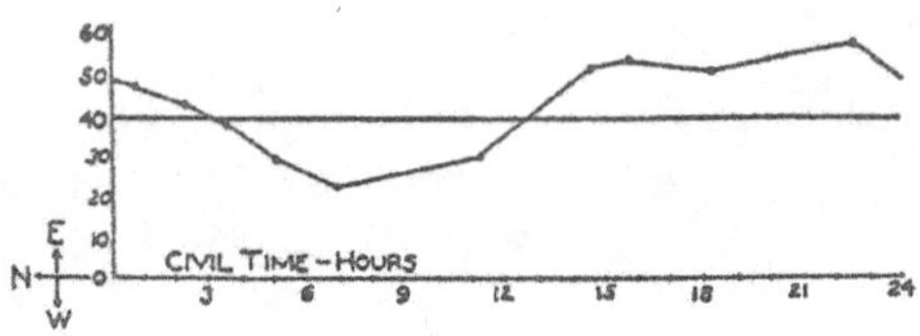


*average azimuth: 40° east*

*1 August 1925*

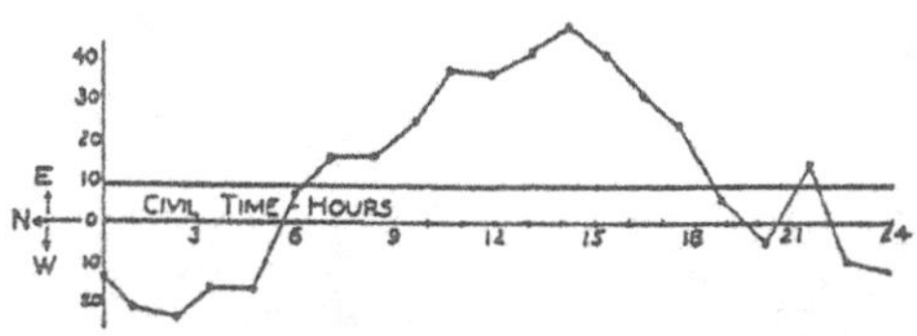


*average azimuth: 10° east*

*15 September 1925*

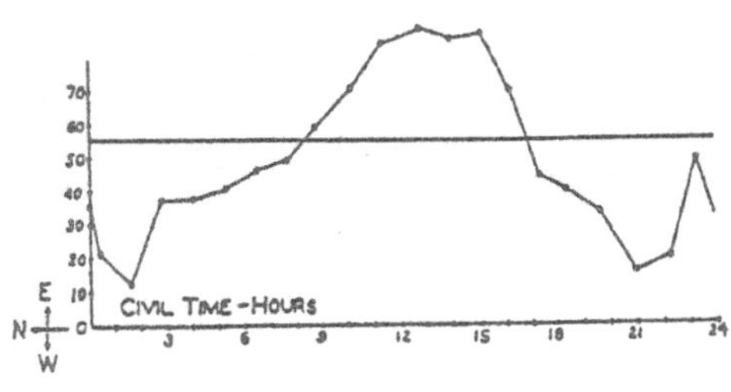


*average azimuth: 55° east*

*8 February 1926*

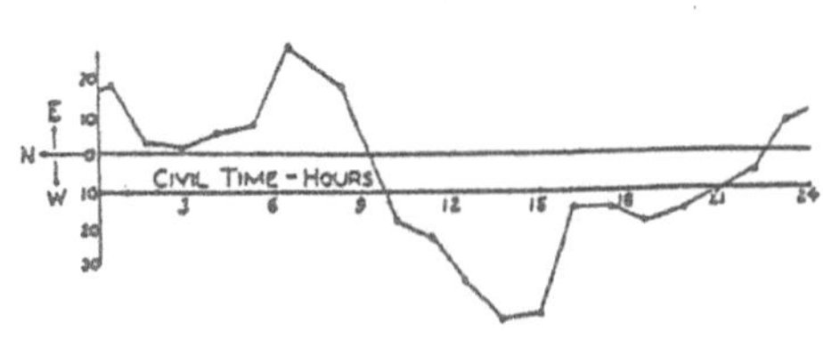


*average azimuth: 10° west*

*Source :* Miller, 1933, id., p. 234.

## C.4 Three fundamental questions

*Any appreciation of the meaning and the implications of Miller's observations comes down to three fundamental questions:*

First Question: *Did Miller's observations*[1] *result from simple perturbations like the temperature, or do they present a very real internal coherence, independent of any perverse effect?*

Second Question: *Do Miller's observations make it possible to determine variations of the speed of light according to its direction?*

Third Question: *Is it possible to deduce, from Miller's observations, the position of the Earth on its orbit?*

[1] Graphs III and IV *above.*

# D THE GREAT UNDERLYING COHERENCE OF MILLER'S OBSERVATIONS

## *D.1 The coherence of the observations in sidereal time*

If the azimuths A are plotted *as functions of civil time*, the four curves that correspond to the four periods of observations do not present any coherence *(Graphs V)*.

By contrast, when these observations for each day are plotted *as functions of sidereal time,*[1] *a coherence appears when the four curves of the azimuths are compared (Graphs III and IV). Their average shows as a periodic curve, having period equal to 24 sidereal hours (Graph VI* of the azimuths).[2]

*The same regularity in sidereal time is found for the speeds*, which correspond to the displacements of the interference fringes (*Graph VI* of the speeds).[3, 4]

These are *extremely significant* results. *They obviously correspond to an undeniable underlying reality*, and, *without any further support*, they make it possible to reject *all* the objections that have been presented with the view of completely devaluing Miller's observations.[5]

---

[1] See note (1) of §B.2 above, p. 378.

Representing two daily curves in sidereal time amounts to displacing these two curves in solar time relative to one another by an interval *in minutes* approximately equal to four times *the number of days* that separate their average dates ($24 \times 60/365.25 = 3.94$).

[2] Miller, 4 and 5 February 1927, *Conference on the Michelson-Morley Experiment, The Astrophysical Journal*, December 1928, p. 362 [198]; and 1933, id., p. 231 [203].

[3] These two regularities relating to the speeds and the azimuths were emphasized by Miller (particularly, 1933, id., p. 231 [203]). But he did not calculate the two average curves and their diurnal fittings shown in *Graphs VI*.

In relation to the *legend of Graphs VI*, see the legend of *Graph XXVII* of *Chapter I*, §C.3, p. 158.

[4] The average curves of *Graphs VI* have been calculated from *values of the moving averages of Miller taken every two hours*, as shown in *Graphs III and IV*.

[5] These regularities *in sidereal time* are *absolutely analogous* to those demonstrated by Esclangon in 1927 (§B.2 above, p. 381).

*GRAPHS VI*

## AVERAGES OF MILLER'S DAILY CURVES OF SPEEDS AND AZIMUTHS

***in sidereal time***

Averages : —▫—▫— Fitting sinusoid : ——

***Speeds*** (in km/sec)

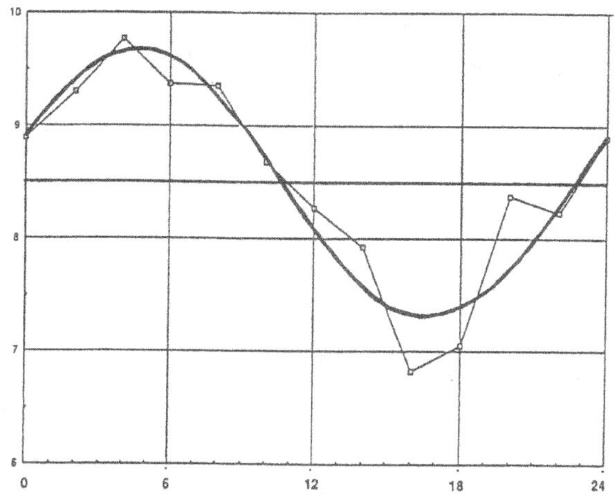


***Fitting sinusoid :*** (calculated on 4 August 1995); R = 0.945; average = 8.504; semi-amplitude = 1.177; times of intersections with the average ordinate: 10h40m and 22h40m; times of maximum and minimum: 4h40m and 16h40m.

***Sources :*** Calculation 446; and *Graph* 13526. Σ = 0.853; σ = 0.279.

***Azimuths*** (in degrees, counted positively from the North in the clockwise sense)

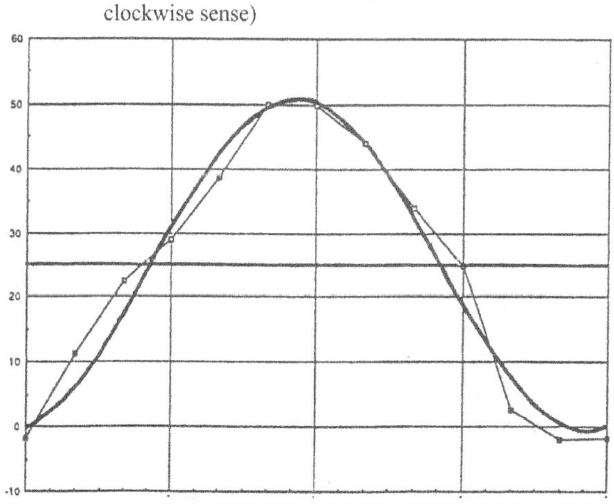


***Fitting sinusoid :*** (calculated on 4 August 1995): R = 0.985; average = 25.08°; semi-amplitude = 25.77°; times of intersections with the average ordinate: 5h7m and 17h7m; times of maximum and minimum: 11h15m and 23h15m.

***Sources :*** Calculation 447; and *Graph* 13527. Σ = 19.0; σ = 3.25.

## *D.2 The correspondence between the graphs of the speeds and of the azimuths*

On *Graphs III and IV*, we can see an approximate correspondence between the times of the maxima and the minima of the speeds and the times of intersection of the azimuth curves with the lines representing their average ordinates.[1]

This correspondence is *very marked* if we consider the *average* curves of the speeds and the azimuths for the four periods considered *(Graphs VI)*.[2] The fitting sinusoid for the speeds is maximum at 4h40m and minimum at 16h40m. The sinusoid of adjustment of the azimuths passes its average ordinate at 5h7m and at 17h7m. *These times correspond, with a relatively small gap of 27m.*[3]

It should be underlined that the curves of speeds and azimuths in Miller's *Graphs III and IV* correspond to three very different types of structure.[4] The curves of 1 August and 15 September are very similar, while the curves of 1 April and 8 February have a different appearance, and are different from one another.[5] These differences certainly correspond to some underlying reality.

---

[1] This correspondence was emphasized by Miller (notably in 1933, id., pp. 225-226 [203]), but he neither calculated the average values for the four epochs nor the corresponding diurnal fittings.

[2] These average curves, *not calculated by Miller*, were established by myself in August 1995, and they correspond to values *taken every two hours* from the moving averages of Miller, reproduced in *Graphs III and IV*.

Naturally these average curves only correspond to the averages of the observations at Miller's four dates. *They do not correspond to the averages that would be obtained if data was available for each month of the year.*

[3] The correlation coefficients of the Miller data with the fitting sinusoids *(Graphs VI)* are respectively 0.945 and 0.985. In fact these two fittings are *very remarkable*.

[4] Notably with regard to the numbers of maxima and minima during the same period of observations.

[5] Miller's four periods of observation, 8 February, 1 April, 1 August, and 15 September, are *very unevenly* spread over the course of the year. The period from 15 September to 8 February (nearly six months) does not include any observations.

The two periods of 1 April and 15 September are near to the equinoxes.

## *D.3 The diurnal fittings of the daily values of speeds and azimuths*

While, in its essentials, the body of this work was conceived and written during June to September 1995, finally, in January 1996, it seemed to me to be very desirable to complete this writing with a complementary analysis of Miller's eight fundamental Graphs of the speeds and azimuths. In fact, *this analysis can strongly contribute to the appreciation of the validity of Miller's observations, and to their interpretation.*

*Graphs VII and VIII* opposite show fittings with sinusoids of period 24h to the curves *in sidereal time* of Miller's speeds and azimuths (*Graphs III and IV* above). These fittings have been calculated from values taken *hour by hour* from the mobile averages calculated by Miller and shown in *Graphs III and IV* above.[1]

*Overall these fittings are extremely remarkable*, with correlation coefficients of 0.361, 0.981, 0.882, and 0.854 for the speeds and 0.856, 0.839, 0.970, and 0.927 for the azimuths, for the four periods centered on 8 February 1926 and 1 April, 1 August, and 15 September 1925. Only the fitting of the speeds for 8 February 1926 is rather mediocre, indeed poor.[2] Generally speaking, the correlations are better for the azimuths than for the speeds.

The overall results are shown in *Tables I* below.

The average values $(\overline{A})$ of the azimuths are $-6.87°$, $41.7°$, $12.1°$, and $53.1°$ for 8 February 1926 and 1 April, 1 August, and 15 September 1925 respectively.[3, 4]

[1] While in *Graph VI*, produced in August 1995, the values of the speeds and the azimuths are only extracted once every two hours.

[2] The average quality of the fittings for each month can be represented by the coefficients R* of *Table I*. For the four periods the coefficients are respectively 0.657, 0.960, 0.927, and 0.891, and their average is 0.859.

[3] The values corresponding to the average azimuths *indicated by Miller* (*Graphs VI* above) are respectively: $-10°$; $40°$; $10°$; and $55°$.

[4] I show the average values calculated by the daily fittings to the azimuths by $(\overline{A})$, so as to distinguish them from the averages $\overline{A}$ of the maximum values $A_M$ and the minimum values $A_m$ estimated graphically from the azimuths $A$ (*Chapter V*, §D.2, *Table II*, p. 454). These averages $\overline{A}$ are respectively $-12.5°$, $40°$, $12.5°$, and $55°$.

If the sidereal time at which the speed is minimal is designated as $\theta^*$ and the sidereal time at which the azimuth $A$ is equal to its average value $(\overline{A})$ and for which $dA/dt < 0$ is designated as $\theta^{**}$, then one realizes that these values are *very close together* for the four epochs, as shown in the third table of *Tables I*. Their averages are *practically identical.*

All these results are *very significant*, and they translate into *very remarkable underlying properties of Miller's observations.*[5]

[5] *None of these results was brought out by Miller.*

## *TABLES I*

### OBSERVATIONS OF MILLER

***Sinusoidal fittings with a period of 24h of the speeds and the azimuths***

*Hourly values in sidereal time*

***Speeds*** (in km/sec)

| | Σ | R | $1-R^2$ | m | r | σ | r/m | r/Σ | r/σ |
|---|---|---|---|---|---|---|---|---|---|
| 8 February | 0.723 | 0.361 | 0.869 | 8.25 | 0.370 | 0.674 | 0.0448 | 0.512 | 0.548 |
| 1 April | 0.736 | 0.981 | 0.0377 | 9.10 | 1.02 | 0.143 | 0.112 | 1.39 | 7.14 |
| 1 August | 1.43 | 0.882 | 0.223 | 8.86 | 1.78 | 0.672 | 0.201 | 1.25 | 2.64 |
| 15 September | 1.46 | 0.854 | 0.271 | 7.90 | 1.77 | 0.762 | 0.223 | 1.21 | 2.32 |

***References :*** Calculations 977, 979, 975, and 981 (6 to 8 February 1996). (See the *legend of Graph XXVII*, §C.3, *Chapter I*, p. 158).

***Azimuths*** (in degrees)

| | Σ | R | $1-R^2$ | m | r | σ | r/m | r/Σ | r/σ |
|---|---|---|---|---|---|---|---|---|---|
| 8 February | 20.0 | 0.856 | 0.267 | -6.87 | -24.3 | 10.4 | 3.53 | -1.21 | -2.34 |
| 1 April | 13.2 | 0.939 | 0.118 | 41.7 | 17.5 | 4.52 | 0.419 | 1.33 | 3.87 |
| 1 August | 27.7 | 0.970 | 0.0593 | 12.1 | 31.2 | 5.54 | 2.58 | 1.12 | 5.63 |
| 15 September | 23.4 | 0.927 | 0.141 | 53.1 | 30.7 | 8.80 | 0.578 | 1.31 | 3.49 |

***References :*** Calculations 976, 978, 974, and 980 (6 to 8 February 1996); $m = (\bar{A})$.

***Estimations of θ* et θ***** (in sidereal hours)

| | θ* | θ** | θ** - θ* |
|---|---|---|---|
| 8 February | 17.65 | 18.56 | 0.91 |
| 1 April | 14.55 | 15.48 | 0.93 |
| 1 August | 16.50 | 15.83 | -0.67 |
| 15 September | 17.59 | 17.78 | 0.19 |
| Averages | 16.57 | 16.91 | 0.34 |

***Legend :*** θ* = hour of speed minimum
θ** = hour of equality $A = m = (\bar{A})$
with $dA/dt < 0$

***Sources :*** the above calculations of speeds and azimuths.

***Average quality of the fittings***

| | R* | $1-R^{*2}$ |
|---|---|---|
| 8 February | 0.657 | 0.568 |
| 1 April | 0.960 | 0.078 |
| 1 August | 0.927 | 0.141 |
| 15 September | 0.891 | 0.206 |
| Averages | 0.859 | 0.248 |

***Legend :*** $R^* = \sqrt{(R_V^2 + R_A^2)/2}$

***Sources :*** $R_V$ and $R_A$, values of R for the speeds and azimuths.

*GRAPHS VII*

## HOURLY OBSERVATIONS OF MILLER
## DAILY CURVES OF SPEEDS AND AZIMUTHS

***Diurnal fittings***
***in sidereal time***

***8 February 1926*** ***1 April 1925***

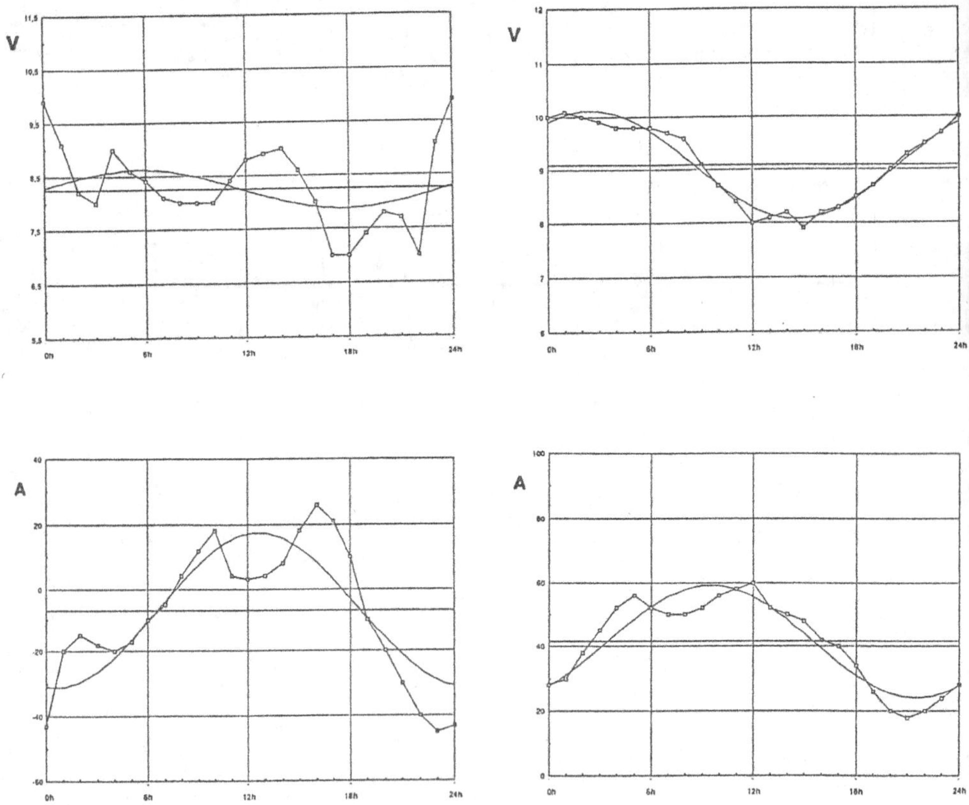


***Legend :***

observed values : fitting sinusoid with 24h period :

v = speed in km/sec A = azimuth in degrees

***Sources :***

Calculations 977, 976, 979, and 978 (7-8 February 1996)

Graphs 13711*, 13710, 13715, and 13714*

*GRAPHS VIII*

## HOURLY OBSERVATIONS OF MILLER
## DAILY CURVES OF SPEEDS AND AZIMUTHS

***Diurnal fittings***
***in sidereal time***

***1 August 1925*** ***15 September 1925***

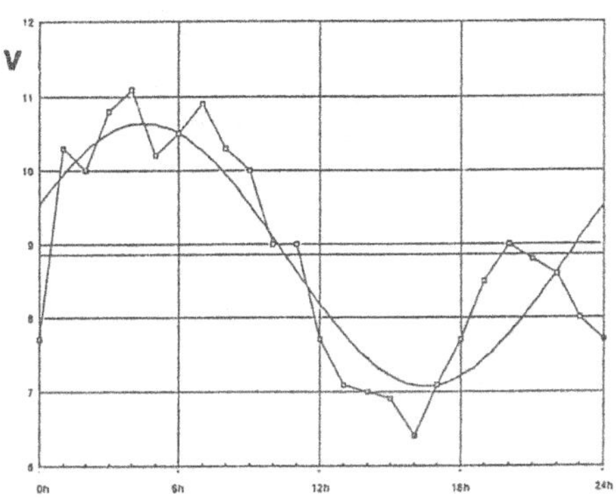


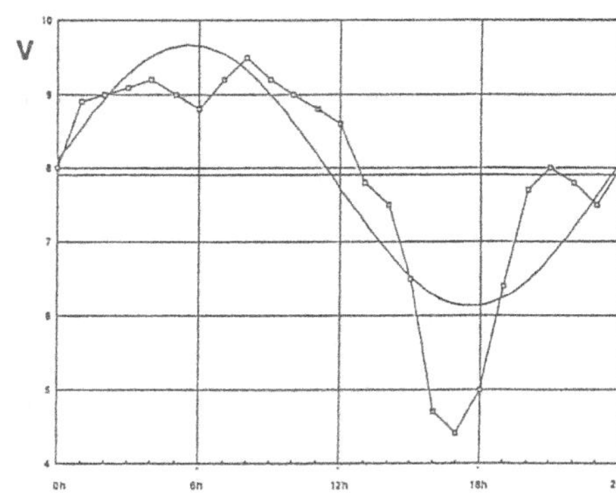


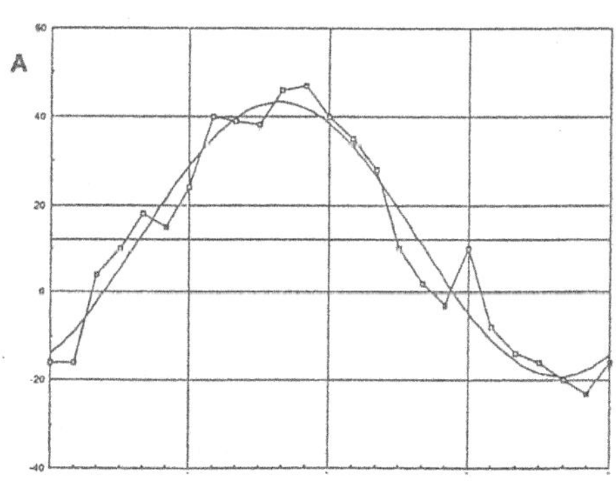


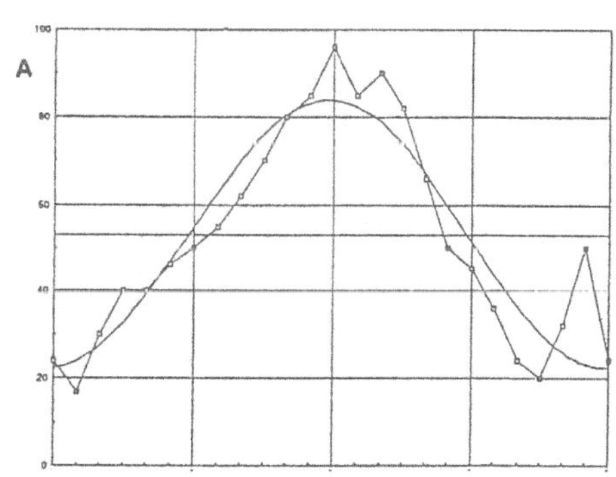


***Legend :***

observed values : -■-■-
**V** = speed in km/sec
fitting sinusoid with 24h period : ——
**A** = azimuth in degrees

***Sources :***

Calculations 975 , 974 , 981, and 980 (6-7 February 1996)
Graphs 13707, 13706, 13719, and 13718

## *D.4 The hodographs of the velocities*

### Observed hodographs

1 - In their upper portions, *Graphs IX and X* show the hodographs of the *observed* velocities, as deduced from the interferometric observations for the four periods, taken hourly from the speeds and the azimuths in *sidereal time* considered in §3 above.[1] The average azimuth ($\overline{A}$) is shown on each graph, as calculated from the extracted values of the azimuths (*Table I*).

In the simple views shown in these four graphs, a *remarkable* circumstance appears. This is that the figures representing the hodographs are, overall, *approximately perpendicular* to the directions of the average azimuths ($\overline{A}$).

### Hodographs estimated from the diurnal fittings of the velocities and the azimuths

2 - In the lower portions of *Graphs IX and X*, I have shown the *calculated* hodographs as taken from the fittings reproduced in *Graphs VII and VIII*.[2]

It is seen that, for the four epochs, the hodographs estimated in this way are *almost exactly perpendicular* to the average directions ($\overline{A}$) of the azimuths, and are symmetrical about these directions. This is an *even more remarkable circumstance*.

---

[1] For constructing a hodograph, the absolute value of the velocity vector is drawn in the direction corresponding to the azimuth, counted positively from the North in the clockwise sense. The hodograph corresponds to the representation *in polar coordinates v and A* of the positions of the ends of the velocity vectors, when drawn from the origin.

On the graphs, the indicated scales for the abscissas and the ordinates are in km/sec.

[2] The *fittings* for the hodographs are determined by replacing the observed values of the speeds and azimuths by their values estimated from the sinusoidal fittings of *Graphs VII and VIII*.

Moreover, for 8 February 1926 and 1 April, 1 August, and 15 September 1925, *the azimuths A* corresponding to the calculated hodographs assume *their minimum values*, in sidereal time, respectively at approximately 0h, 22h, 22h, and 0h, *times which are very close to one another.*[3]

Finally, the figures become progressively modified from one epoch to another. They attain their maximum dimensions around 21 September, which corresponds to the autumn equinox, and their minimum dimensions around 21 March, which corresponds to the spring equinox. *They thus depend upon the position of the Earth in its orbital trajectory.*

*All these properties are very remarkable, and they undeniably correspond to the very great underlying coherence of Miller's observations.*[4]

[3] It should be underlined that the two dates of 1 April and 15 September are *very close* to the dates of the equinoxes, 21 March and 21 September.

[4] *None of these absolutely essential properties were noticed by Miller.*

*GRAPHS IX*

## OBSERVATIONS OF MILLER

## OBSERVED HODOGRAPHS OF HOURLY VALUES AND HODOGRAPHS DRAWN FROM DAILY FITTINGS OF SPEEDS AND AZIMUTHS

**(from Graphs VII)**

***8 February 1926*** — **(Ā) = - 6,87 °**

***1 April 1925*** — **(Ā) = 41,71 °**

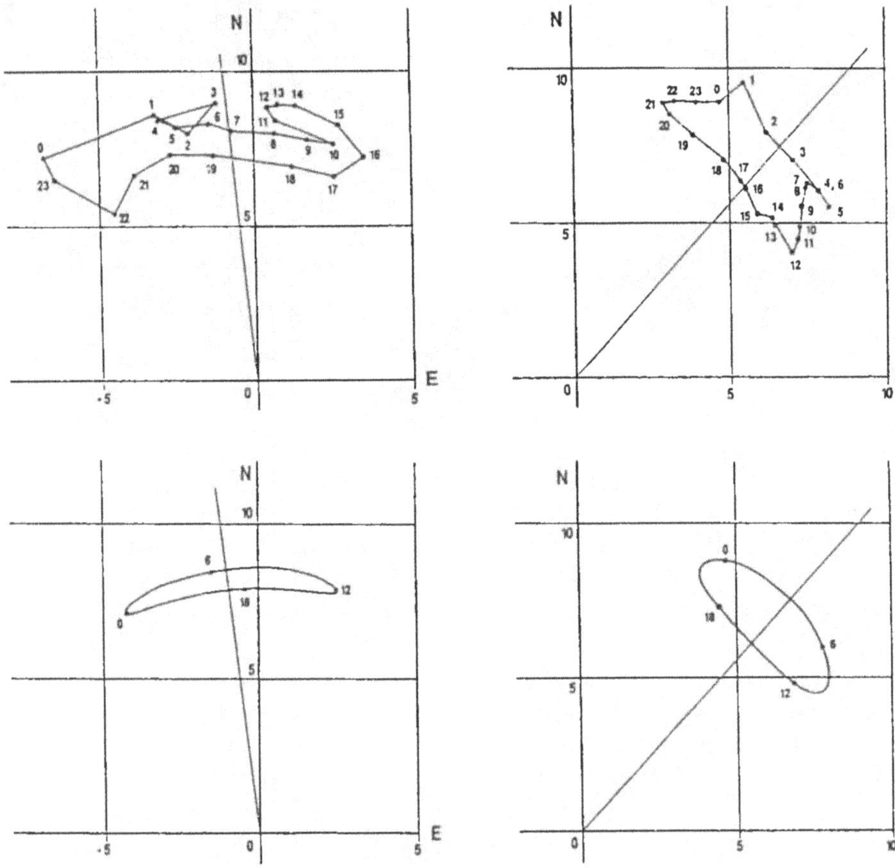


***Legend :*** ●—●—● observed values ——— calculated values

***Sources :*** Graphs 13709*, 13712, 13713*, and 13716 (7-8 February 1996)

*GRAPHS X*

## OBSERVATIONS OF MILLER

## OBSERVED HODOGRAPHS OF HOURLY VALUES AND HODOGRAPHS DRAWN FROM DAILY FITTINGS OF SPEEDS AND AZIMUTHS (from Graphs VIII)

*1 August 1925* — *15 September 1925*

**($\bar{A}$) = 12,08 °** — **($\bar{A}$) = 53,12 °**

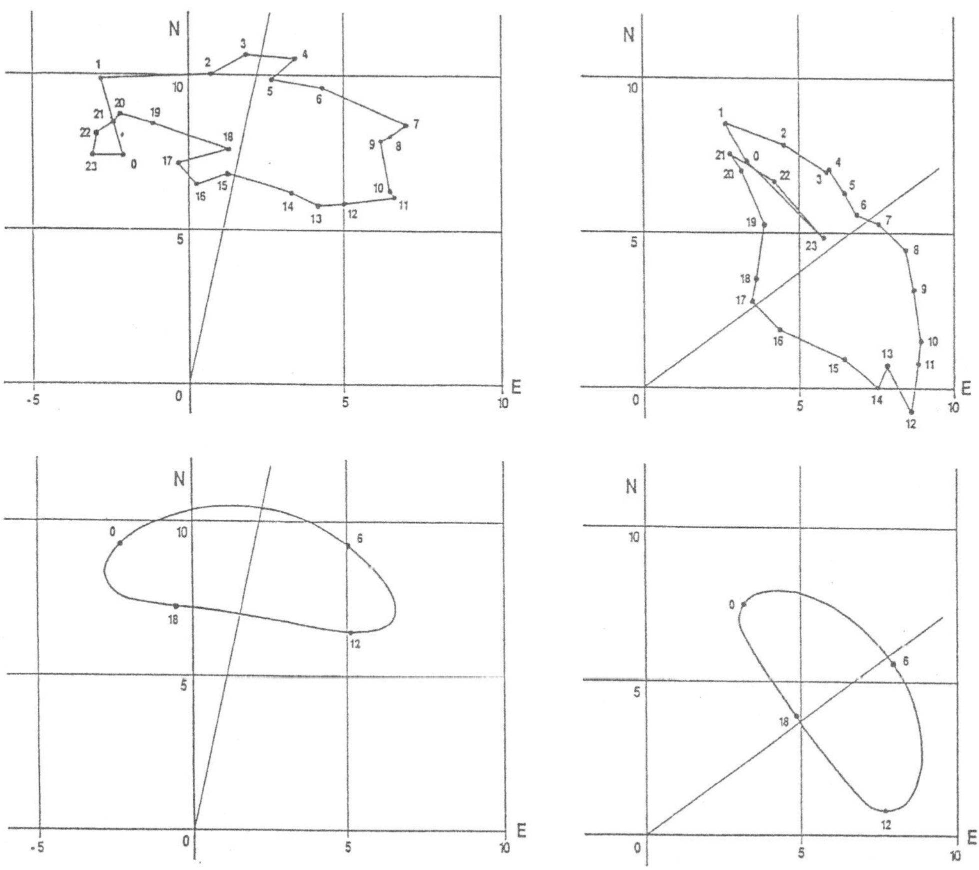


***Legend :*** ●—●—● observed values ——— calculated values

***Sources :*** Graphs 13705*, 13708, 13717, and 13720 (6-9 February 1996)

**Hodographs estimated from the diurnal fittings of the Cartesian coordinates of the velocity vector $(v, A)$**

3 - The expressions for the Cartesian coordinates $x$ and $y$ of the velocity vector $\vec{v}$ having polar coordinates $v$ and $A$ determined by the interferometer are:

(1) $X = v\cos A \qquad Y = v\sin A$

For each observational period, it is possible to calculate the daily fittings to sinusoids of period 24h of the hourly values of $X$ and $Y$ *as functions of sidereal time*, and to obtain a new estimation of the hodograph with the estimated values $X^*$ and $Y^*$ of $X$ and $Y$. This estimated hodograph is *an ellipse, the coordinates of whose center are the averages* $\overline{X}$ *and* $\overline{Y}$ *of* $X$ *and* $Y$.[5]

[5] We consider the 24 series of values of $X$ and $Y$ corresponding to the 24 hourly values of sidereal time $\theta$: 0, 1, ... 23.

The two correlations in question may be written:

(1) $X = \alpha\cos\theta + \beta\sin\theta + \gamma + \varepsilon_X$

(2) $Y = \alpha'\cos\theta + \beta'\sin\theta + \gamma' + \varepsilon_Y$

$\theta$ represents the sidereal time, *expressed in* °, equal to the value in hours multiplied by 15. $\alpha$, $\beta$, $\alpha'$, $\beta'$ are the regression coefficients.

Since the average values of $\cos\theta$ and $\sin\theta$ are zero, we have

(3) $\gamma = \overline{X} \qquad \gamma' = \overline{Y}$

where $\overline{X}$ and $\overline{Y}$ are the average values of $X$ and $Y$. The quantities $\varepsilon_X$ and $\varepsilon_Y$ are the residues of the two correlations.

We naturally have:

(4) $X^* = \alpha\cos\theta + \beta\sin\theta + \gamma$

(5) $Y^* = \alpha'\cos\theta + \beta'\sin\theta + \gamma'$

where $X^*$ and $Y^*$ represent the *estimated* values of $X$ and $Y$.

The linear equations (4) and (5) allow us to determine $\cos\theta$ and $\sin\theta$ as linear functions of $X^*$ and $Y^*$, and, substituting these values for $\cos\theta$ and $\sin\theta$ into the identity equation

(6) $\cos^2\theta + \sin^2\theta = 1$

we obtain a second-degree equation in $X^*$ and $Y^*$. This means that the hodograph estimated in this manner is *an ellipse, the coordinates of whose center are the average values* $\overline{X}$ *and* $\overline{Y}$ *of* $X$ *and* $Y$. The angle $(\bar{\bar{A}})$ connecting the origin to this center may be expressed as

(7) $(\bar{\bar{A}}) = \arctan(\overline{Y}/\overline{X})$

The results of these correlations are shown in *Table II*. Considering the linearity approximations that were made,[6] the multiple correlation coefficients are remarkably high. Their overall average is equal to $\overline{R} = 0.891$, with $1 - \overline{R}^2 = 0.206$.[7]

Ellipses representing the hodographs H* corresponding to these fittings are shown in *Graphs XI and XII*. For each of Miller's dates, the estimated hodograph H* is given in the upper figure, and the observed hodograph H overlaid upon the estimated hodograph H* is given in the lower figure.

*Overall, the representation of the observed hodographs is quite remarkable.*[8] The same is true for the *agreements in phase* between the calculated hodographs and the observed hodographs. Finally, we note that the estimated hodographs are *almost exactly* perpendicular to the directions $(\bar{\bar{A}})$.[9, 10]

*None of these regularities was noticed by Miller, but they are absolutely fundamental.*

---

[6] Relations (4) and (5) of footnote 5 above.

[7] The regression coefficients of the eight correlations of *Table II* are as follows:

$\alpha$ or $\alpha'$: −0.615; −3.40; 1.94; −1.13; 1.48; −3.60; −3.61; −2.21

$\beta$ or $\beta'$: 0.624; −0.630; −0.617; 1.75; 0.968; 2.89; 0.950; 1.66

[8] This representation is better than that of *Graphs IX and X*. It corresponds to the *diurnally periodic structure* of the coordinates of the velocity vectors.

[9] Regularities such as these *completely and definitively* exclude the contention, often put forward, that these interferometric observations represent nothing more than meaningless effects, such as effects resulting from temperature, for example. (See §E.1 below, pp. 412-413).

[10] The average value of the velocity is 8 km/sec. *(Table II)*. This corresponds to an average relative difference of around $2.7 \times 10^{-5}$ $(8/300,000)$.

*TABLE II*

## OBSERVATIONS OF MILLER

***Sinusoidal fittings of the Cartesian coordinates of the observed velocity vectors to a period of 24h***

$$X = v \cos A \qquad Y = v \sin A$$

*Hourly values in sidereal time*

$$X = \alpha \cos \theta + \beta \sin \theta + \gamma + \varepsilon_X$$
$$Y = \alpha' \cos \theta + \beta' \sin \theta + \gamma' + \varepsilon_Y$$

| Dates | | R | $1\text{-}R^2$ | $R^*$ / $1\text{-}R^{*2}$ | $\gamma$ or $\gamma'$ | $(\bar{v})$ | $(\bar{\bar{A}})$ |
|---|---|---|---|---|---|---|---|
| 8 February | X | 0.702 | 0.507 | 0.791 | 7.72 | 7.78 | - 7.43 |
| | Y | 0.871 | 0.241 | 0.374 | - 1.01 | | |
| 1 April | X | 0.943 | 0.111 | 0.939 | 6.64 | 8.86 | 41.47 |
| | Y | 0.935 | 0.126 | 0.119 | 5.87 | | |
| 1 August | X | 0.844 | 0.288 | 0.907 | 7.97 | 8.18 | 12.81 |
| | Y | 0.965 | 0.069 | 0.178 | 1.81 | | |
| 15 September | X | 0.941 | 0.113 | 0.932 | 4.40 | 7.26 | 52.67 |
| | Y | 0.923 | 0.141 | 0.131 | 5.77 | | |
| Averages | | 0.891 | 0.200 | 0.892<br>0.201 | | 8,02 | |

***Legend :***

$$R^* = \sqrt{(R_X^2 + R_Y^2)/2}$$

$\gamma = \bar{X}$ $\qquad \gamma' = \bar{Y}$ $\qquad (\bar{v}) = \sqrt{\gamma^2 + \gamma'^2}$ $\qquad (\bar{\bar{A}}) = \text{arc tg}(\gamma'/\gamma)$

$(\bar{v})$ : in km/sec $\qquad (\bar{\bar{A}})$ : in degrees

*Corresponding calculations* : 1081 and 1082 (11 April 1996),
and 1240 and 1245 (22 January 1997)

***Sources :***

*Graphs III and IV* of the speeds and azimuths, above (pp. 388 and 389).

*GRAPHS XI*

## OBSERVATIONS OF MILLER

## HODOGRAPHS OF THE COORDINATES X,Y OF THE VECTOR (v,A) DEDUCED FROM THE DAILY FITTINGS (from the correlations of Table II)

***8 February 1926*** — $(\bar{\bar{A}}) = -7,43\,°$

***1 April 1925*** — $(\bar{\bar{A}}) = 41,47\,°$

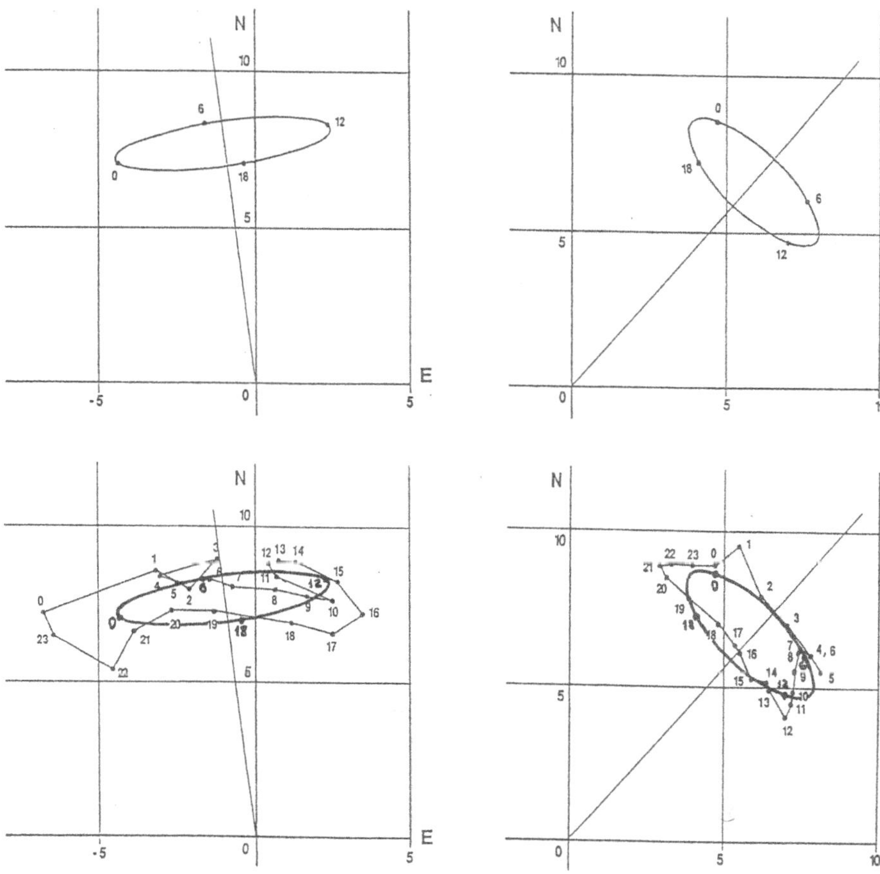


***Legend :*** observed values — calculated values
speed scale: km/sec

***Sources :*** Graphs 13772*, 13709*, 13773, and 13713* (7 February - 25 March 1996)

*GRAPHS XII*

## OBSERVATIONS OF MILLER

## HODOGRAPHS OF THE COORDINATES X,Y OF THE VECTOR (v,A) DEDUCED FROM THE DAILY FITTINGS (from the correlations of Table II)

*1 August 1925* — **$\overline{\overline{(A)}}$ = 12,81 °**

*15 September 1925* — **$\overline{\overline{(A)}}$ = 52,67 °**

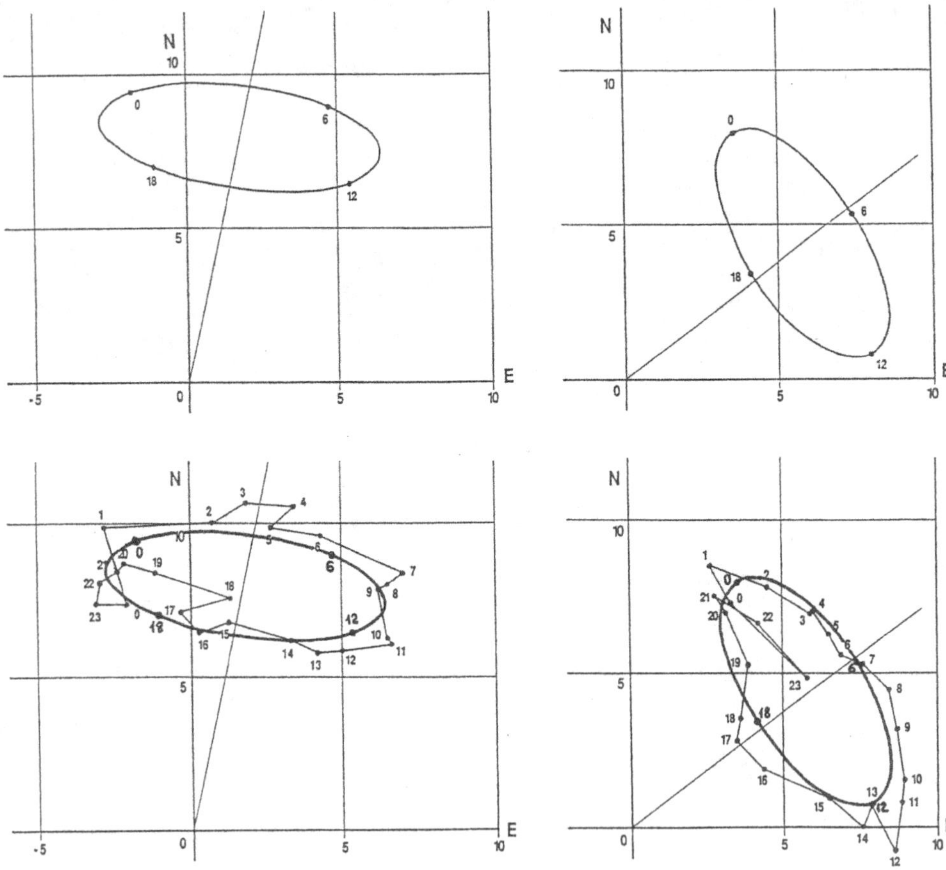


*Legend :* ●—●—● observated values ——— calculated values

speed scale: km/sec

***Sources :*** Graphs 13774, 13705*, 13775, and 13717$^{bis}$ (6 February - 27 March 1996)

### *D.5 The semi-annual and annual periodic structure of Miller's observations*

As I shall demonstrate in *Chapter V*,[1] the characteristic parameters of the eight fundamental Graphs of Miller[2] *are characterized by remarkable semi-annual and annual periodicities, which also were not noticed by Miller.*

Moreover, they also exhibit numerous *structural regularities of interdependence* between the observed speeds and azimuths, *likewise not noticed by Miller.*[3]

Not only do these regularities tend to reinforce the hypothesis of the existence of a *very real* underlying coherence in Miller's observations, but they *show a tight dependence between Miller's observations and the position of the Earth on its orbit.*

---

[1] *Chapter V*, Sections D and E, pp. 452-468 and 469-483.

[2] It is really astonishing that Miller never thought of correlating the average azimuths $\overline{A}$ with the position of the Earth on its orbit. If he had done this, he would have noticed that the azimuths have *a very remarkable semi-annual period whose phase coincides with the spring equinox* (see below, *Chapter V*, §D.3.1, pp. 455-456 and 458).

[3] *Chapter V*, §D.5, pp. 465-466.

## D.6 *An undeniable and very great underlying coherence in Miller's observations*

### Very remarkable properties of the eight fundamental Graphs of Miller

1 - It is certain that Miller's observations have *very great internal coherence*.

a - A very prominent coherence appears when the variations of the azimuths and speeds are considered, *not in civil time, but in sidereal time.*[1]

b - We see that, *effectively and at least approximately*, the maxima and the minima of the speeds occur when the angles

$$(1) \qquad A^* = A - \overline{A}$$

are nil, in other words when the direction of the arm of the interferometer is the same as that of the average azimuth $\overline{A}$.[2]

c - The diurnal fittings to the speeds and the azimuths bring out the *very marked* internal coherence of Miller's observations.[3]

d - Consideration of the hodographs of the velocities similarly brings out *an extremely remarkable underlying coherence* in Miller's observations,[4] *which also was not noticed by Miller.*

e - The characteristic parameters of Miller's eight fundamental Graphs are characterized *by remarkable annual and semi-annual periodicities.*[5] *Miller also did not notice these.*

---

[1] §D.1 above. This coherence in *sidereal time* is very similar to that which can be seen in the observations of Esclangon (§B.2 above, pp. 378-381).

[2] §D.2 above.

[3] §D.3 above.

[4] §D.4 above.

[5] §D.5 above.

## An affirmative answer to the three fundamental questions

2 - *All these regularities make it possible to answer the three fundamental questions in the affirmative with absolute certainty.*[6]

- First, they *undeniably* show that Miller's observations did not result from *fortuitous perturbations* or perverse influences (due to temperature, for example). *Absolutely certainly*, they correspond to a *very real* phenomenon.
- In the second place, they show that *the speed of light varies with its direction.*
- Finally, they show *a tight dependence between Miller's observations and the position of the Earth on its orbit.*

---

[6] §C.4 above, p. 391.

The totality of all these properties shows that the criticisms of Miller's experiments (Section E below, pp. 412-416) are *utterly unjustified.*

# E CRITICISMS AIMED AT MILLER'S EXPERIMENTS

Miller's experiments have been the target of objections of two types.

## *E.1 Miller's results were due to the combined effect of random perturbations and temperature perturbations*

In April 1955, in the American publication *Reviews of Modern Physics*, Robert S. Shankland, an old collaborator of Miller's,[1] with S. W. McCuskey, F. C. Leone, and G. Kuerti, published a study "*New Analysis of the Interferometer Observations of Dayton C. Miller*" [249], attributing the effects obtained by Miller to random perturbations and particularly to effects resulting from *perturbations of the temperature*,[2] and concluding:[3]

> *"Miller's extensive Mount Wilson data contain no effect of the kind predicted by the aether theory and, within the limitations imposed by local disturbances, are entirely consistent with a null result at all epochs during a year".*

---

[1] Miller, 1933, id., p. 242 [203].

[2] For details of the precautions taken by Miller in connection with temperature effects, to which Shankland and his associates did not give sufficient attention, particularly see: Miller, 25 April 1925, *Ether-Drift Experiments at Mount Wilson* (§C.1, note 2 above, p. 382), pp. 310-311 [201]; Miller, 1928, *Conference on the Michelson-Morley Experiment*, p. 359 [198]; and Miller, 1933, id., pp. 211, 212, 218 and 220 [203].

Miller was a *very experienced* experimenter who completely analyzed the possible effects of temperature with very different experimental conditions, and he was quite justified in writing (1933, id., p. 311 [203]):

> *"The experiments proved that under the conditions of actual observation, the periodic displacement (of the interference fringes) could not possibly be produced by temperature effects".*

On the qualities of Miller as an experimentalist, see below, *Chapter VII*, §B.3, note 7, p. 586.

[3] Shankland *et al.*, 1955, Summary, p. 167 [249].

However, in the light of the *deep and extensive* study I have devoted to Shankland's paper, this conclusion seems *completely biased and totally unfounded*, because of the *very great coherence* that appears in analysis of Miller's observations, *which on all the evidence could never have resulted from local perturbations* - neither the *diurnal sidereal periodicity* in Miller's eight fundamental Graphs; nor the *remarkable correspondences* between the Graphs of the speeds and the azimuths;[4] nor the *periodic structure* of the diurnal fittings to the speeds and the azimuths; nor the *extremely remarkable* properties of the hodographs; nor *the structural correspondences* observed between the azimuths and the speeds; nor *the semi-annual and annual periodicities* of the parameters characterizing Miller's fundamental eight Graphs, and particularly the semi-annual periodicity of the observed average deviations $\overline{A}$ of the azimuths.[5]

[4] It is necessary to underline that the determinations of the azimuth for which the displacement of the fringes is maximum, and the magnitude of this displacement which determines the velocity, are two *independent* estimations (Miller, 1933, id., pp. 211-213 and pp. 225-226 [203]).

As Miller indicated (Conference on the Michelson-Morley Experiment of 4-5 February 1927 [198]):

> *"The determination of the direction of the earth's motion is dependent only upon the direction in which the telescope points when the observed displacement of the fringes is a maximum; it is in no way dependent upon the amount of this displacement or upon the adjustment of the fringes to any particular zero position."*

[5] See above, *Section D*, pp. 392-411.

See also *Chapter V* below, §E.2, note 10, p. 479.

## *E.2 The negative results of other interferometric experiments demonstrate that Miller's experiments were invalid*

It is also claimed that *all* other interferometric experiments ever performed *have yielded negative results.*

### The previous interferometric experiments of Michelson and Morley, and of Morley and Miller

1 - In fact, there is *no contradiction* between Miller's results and the previous results of Michelson and Morley.[1]

In his memoir of 1933,[2] Miller illustrates in a chart the compatibility of the results of the Michelson and Morley experiments in 1887, and of those of Morley and Miller in 1902, 1904 and 1905, with the results that he (Miller) obtained in 1925-1926. In all cases, the speeds are of the order of 8 or 9 km/sec at comparable sidereal times.

In particular, the Michelson and Morley experiment in 1887 gave an estimate of 8 km/sec for the velocity, and at the time this result was interpreted as being an error of measurement.[2]

---

[1] Three names appear prominently in connection with the fundamental interferometer experiments of Michelson: Albert A. Michelson (1852-1931); Edward W. Morley (1838-1923); and Dayton C. Miller (1866-1941).

The first interferometric experiment was performed in 1881 by Michelson, and the second in 1887 by Michelson and Morley. These were followed by three other experiments in 1902, 1904, and 1905 performed by Morley and Miller. After Morley's death in 1923, Miller carried on alone and performed the experiments of 1925-1926.

Each time, the experimental arrangements employed were considerably improved.

[2] Miller, 1933, id., p. 207.

## Miller's previous interferometric experiments

2 - In his address of 28 April 1925 to the *National Academy of Sciences*,[3] Miller comments upon the results obtained in three series of experiments: (I) 15 April 1921; (II) 8 December 1921; (III) 5 September 1924. These results are entirely consistent with those of Mount Wilson in April 1925.

In particular, Miller presented two very suggestive Figures[4] that showed the very *significant* consistency between the observations of 1921 and of 1925.[5]

## Interferometric experiments subsequent to Miller's experiments of 1925-1926

3 - Miller's experiments of 1925-1926 were followed by a whole series of experiments: those of Kennedy in 1925; those of Piccard and Stahel in a balloon in June 1926, in Brussels in November 1926, and on Mt. Rigi in Switzerland at 1,800 meters altitude in September 1927; those of Michelson, Pease and Pearson in 1926-1927; and those of Joss in Jena in 1930.[6]

All these experiments yielded results considered to be *"negative"*.[7]

---

[3] Miller, 1925, pp. 308-314 [201] (Note 2 of §E.1 above, p. 412).

[4] id., pp. 309 et 312 [201].

[5] Miller wrote (1925, id., p. 312 [201]):

> *"The agreement between the results for 1921 and 1925 is striking especially when it is recalled that the interferometer has been rebuilt as to its details, has a different system of illumination and observation, and has been changed to a new site, in a house differently oriented, and that many variations in observational procedure have been tried".*

In this analysis of 1925, the azimuths that Miller considered were *perpendicular* to those considered in his memoir of 1933, but he makes no mention of this in his memoir of 1933.

[6] R. J. Kennedy, 1926, Proc. Nat. Acad. Sci., 12, 621-629 [163], and 1928, Astrophys. J. 68, 367 [198]; A. Piccard and E. Stahel, *L'absence de vent d'éther au Rigi* (Absence of any ether wind on Rigi), CRAS, 28 November 1927, pp. 1198-1200 [216], and 1928, *Réalisation de l'expérience de Michelson en ballon et sur terre ferme* (Implementation of Michelson's experiment in a balloon and on solid ground), Journal of Physics and of Radium, Series VI, Vol. IX, No. 2, pp. 49-60 [217]; see also E. Brylinski, *Sur la vitesse de la Terre* (On the speed of the Earth), CRAS, 19 December 1927, pp. 1458-1459 [89]; A. A. Michelson, F. G. Pease, and F. Pearson, *Repetition of the Michelson-Morley Experiment*, Nature, 19 January 1929, p. 88 [200]; and Georg Joos, 1930, *Die Jenaer Wiederholung des Michelsonversuchs* (The Jena Repetition of the Michelson Experiment), Annalen der Physik, Vol. 7, Issue 4, pp. 385-407 [160].

[7] Miller, 1933, id., pp. 239-240 [203].

However all these experiments, like the previous experiments of Michelson, of Michelson and Morley, and of Morley and Miller, *were confined to limited observations made at a given time for testing specific hypotheses.*

In fact, *none of these experiments* was based upon observations *continued night and day at different periods of the year, as was the case with Miller's experiments in 1925-1926*, which in total comprised 6,400 rotations of the apparatus during which 200,000 readings were taken, over four periods of six to eight days each.[8]

## *E.3 Totally unfounded criticisms*

In summary, all the criticisms disputing *the validity of the observations of Miller were completely unfounded.* Whatever point of view one may take, the speeds and the azimuths measured by Miller's interferometer correspond to *very real* phenomena, independent of any perverse effects, and Miller's observations constitute *inescapable* data.

*Thus it is completely wrong to assert that the Michelson experiment has always yielded negative results, as is everywhere maintained today.*

[8] For example, *no parallels can be drawn* between the experiments of Piccard and Stahel, which were made in a balloon, and the 200,000 observations by Miller (Miller, 1933, id., p. 228 [203]), which were performed in a perfected laboratory with all possible precautions being taken.

# F THE INTERPRETATION OF MILLER'S OBSERVATIONS

What interpretation can be given to Miller's observations as summarized in the eight fundamental Graphs of the speeds and the azimuths, *whose coherence and validity must be considered as indisputable?*

## *F.1 The interpretation given by Miller to his observations*

Miller deduced from his observations that they can be interpreted as resulting from the combined effects of the orbital velocity of the Earth and a cosmic velocity of the solar system, of which he deduced the magnitude, the declination, and the right ascension from the classical formulas of astronomy.[1]

*In his theoretical analysis, Miller only considered the deviations*

$$(1) \qquad A^* = A - \overline{A}$$

*of the azimuths from their mean values, and he supposed that the deviations $A^*$ and the speeds $v$ measured by the interferometer only depended upon the velocity of the Earth,* which is the sum of its orbital velocity and its cosmic velocity.

---

[1] Miller, 1933, id., pp. 222-238 [203].

The characteristics of this cosmic velocity are as follows:

$$v_c = 208\text{km/sec} \qquad \delta_c = -70°33' \qquad \alpha_c = 4\text{h}54\text{mn}$$

where $\delta_c$ and $\alpha_c$ are its declination and right ascension (Miller, 1933, id., p. 234, and Table 4, p. 233 [203]).

The global velocity of the Earth is equal to the sum of its orbital velocity of 30 km/sec and its cosmic velocity of 208 km/sec.

Actually, two *essential* facts remain *utterly unexplained* in Miller's analysis.

First, the averages $\overline{A}$ of the daily azimuths, *which ought to be null according to the astronomical formulas used by Miller, are not; and they vary considerably from one period to another.*

Moreover, Miller was led to suppose that the cosmic and orbital speeds of the Earth were reduced *in the same proportion* by a coefficient $k = 0.0514$.[2]

*Thus, the theory of Miller leaves unexplained, on the one hand the coefficient k of speed reduction, and on the other hand and more importantly, the average daily values* $\overline{A}$ *of the azimuths and their variations from one period to another.*[3]

*Whatever the apparent coherence of Miller's fittings and their approximate validity may be,*[4] *his interpretation and his estimations appear to be fundamentally unacceptable.*

In fact, the astronomical relations used by Miller[5] can only be applied to the *effective* values $A$ of the azimuths, and not to the differences $A^* = A - \overline{A}$. Actually, *the astronomical formulas used by Miller were only applied by him to the differences* $A^* = A - \overline{A}$ *of the azimuths from their average values, whereas they ought only have been applied to the azimuths A themselves.* Accordingly Miller does not take any account of the average deviations $\overline{A}$ of the azimuths.

---

[2] Miller, 1933, id., p. 235, *Table V* [203].

[3] In fact, Miller considered that his theory *could not explain*, on the one hand the coefficient of reduction $k$, and on the other hand the average daily values $\overline{A}$ of the azimuths which were *non-null and variable from one period to another*. In his memoir of 1933, he wrote (Miller, 1933, id., p. 234 [203]):

> "*There are found two facts of observation which are wholly unexplained on this simple theory.*"

[4] Miller, 1933, id., pp. 235-237 [203].

[5] Miller, 1933, id., pp. 225-226 [203].

*From the above, the interpretation given by Miller to his observations cannot be considered as valid. It is in fact totally impossible to represent and explain the observations of Miller, and in particular the average deviations* $\overline{A}$ *of the azimuths, by only considering the total velocity of the Earth, whatever it may be.*[6, 7]

[6] An in-depth critical analysis of Miller's analysis is given in *Chapter IV*, Section C of the *second volume* of this work (p. 29 above).

[7] Nevertheless, however valid may be the criticisms - *which indeed are fundamental and decisive* - that can be made of the theory of Miller and of the estimations he deduced from it, the facts are that *the validity of his observations, in themselves, is completely independent* of such theoretical criticisms, and that they present a *very remarkable* coherence that ought to be interpreted and explained, and that can be.

## *F.2 The interpretation that corresponds to the observational data: an anisotropy of optical space*

### An anisotropy of optical space

1 - In view of the almost perfect perpendicularity, in each epoch, of the hodographs H* to the average deviations ($\bar{\bar{A}}$) of the azimuths $A$ (*Graphs XI and XII*), *the only possible interpretation* of Miller's observations is that they result from an anisotropy of optical space, variable with the epoch, whose average azimuth is, in each epoch, the average deviation ($\bar{\bar{A}}$) of Miller's azimuths.

And, in view of the observational data (*i.e.* the coherence of the observations in sidereal time and the semi-annual and annual periodicities of the characteristic parameters of the observations),[1] this anisotropy must be considered as resulting from astronomical influences, notably the position of the Earth on its orbit.[2]

### An impossibility

2 - What is *certain* is that *it is totally impossible to explain the observational data only by consideration of the velocity of the Earth*, resulting from its orbital velocity and its cosmic velocity.

---

[1] Notably §D.1 and D.5 above.

[2] In fact, as will be described in *Chapter V* (§V.D.3, pp. 455-456 and 458), the average deviations of the azimuths *are correlated with the position of the Earth* on its orbit.

This correlation is very *similar* to the temporal correlation of the average deviation of the average azimuths $\overline{\phi}$ of the plane of oscillation of the paraconical pendulum with the position of the Earth on its orbit.

Indeed, in the most general case, the hodograph of the projection of the velocity of the Earth upon a horizontal plane is *an ellipse that is symmetrical with respect to the meridian* and whose major axis is perpendicular to the meridian, having a minor axis to major axis ratio equal to $\sin\varphi$, where $\varphi$ is the latitude of the location in question.[3]

## Determination of the factors explaining the anisotropy of optical space

3 - As much as the existence of an anisotropy of optical space is a certainty, and as much as its dependence upon astronomical factors and particularly upon the position of the Earth in its orbit is a certainty, a general formulation of this anisotropy currently equally appears to be *utterly premature* in view of the far too limited number of observations available at the moment.

*What is essential is that, whatever this formulation may be, the validity of Miller's observations, their very great coherence, and the existence of an anisotropy of space are completely independent of any hypothesis and any theory of any kind.*[4]

---

[3] *Chapter IV*, Section B of the *second volume* of this work (p. 29 above).

Taking the abscissas $x$ along the meridian positively towards the North, and the ordinates $y$ along the perpendicular to the meridian positively towards the East, the equation of the hodograph is

$$\left(\frac{x - V\sin\delta\cos\varphi}{\sin\varphi}\right)^2 + y^2 - V^2\cos^2\delta$$

where $V$ is the total speed of the Earth and $\delta$ its declination, and where $\varphi$ is the latitude of the observational location.

[4] The three *sections B, C, and D* of *Chapter IV* of the second volume of this work (p. 29 above) are concerned with the theoretical determination of the hodograph of the velocity of the Earth which is the sum of its orbital velocity and its cosmic velocity, with critical analysis of Miller's theory and its estimations, and finally with a brief examination of an interpretation of Miller's observations that is based upon anisotropy of optical space and also upon the hypothesis of a cosmic velocity of the Earth towards the constellation of Hercules.

These analyses are useful for understanding the determination of the velocity of the Earth upon which all the interferometric experiments are founded, Miller's theoretical approach, and the implications of an approach based both on the anisotropy of space and the velocity of the Earth. *But the general conclusions of this Chapter IV are totally independent of the above.*

## The anisotropy of space in the literature concerning Michelson's interferometric experiment.

4 - It is very significant that, in the immense literature about Michelson's interferometric experiment, at least two authors, Henri Poincaré[5] and Richard Birkeland[6] were led to consider *the possibility of an anisotropy of space*.

---

[5] In his Conference on 24 September 1904 during the *Congress of Arts and Science* at St. Louis in the USA, *L'état actuel et l'avenir de la physique mathématique* (The present and the future of mathematical physics) (published in the *Bulletin des Sciences mathématiques* of December 1904, pp. 302-324 [221]), Henri Poincaré declared (pp. 319-320):

> *"Should we not also strive to obtain a more satisfying theory of the electrodynamics of moving bodies?...*
>
> *"Let us therefore take Lorentz's theory...*
>
> *"Instead of supposing that a moving body suffers a contraction in the direction of movement and that this contraction is the same whatever the nature of the body may be and to whatever forces it may be subjected, can we not formulate a simpler and more natural hypothesis? One might imagine, for example, that it is the ether that is modified when it is in movement relative to a material medium that penetrates it, and that, when it is thus modified, it no longer transmits vibrations with the same speed in all directions. It might transmit more rapidly those vibrations that are propagating parallel to the movement of the medium, either in the same direction or in the opposite direction, and less rapidly those that are propagating perpendicularly. The surfaces of the wave would no longer be spheres but ellipsoids, and we could dispense with this extraordinary contraction of the body.*
>
> *"I only give this as an example, because the modifications that might be tried could evidently be varied ad infinitum."*

Poincaré repeated these comments in 1905 and 1908 in *La Valeur de la Science* (The Value of Science) [222] and *Science et Méthode* (Science and Method) [225] (see below, *Chapter VII*, §A.7, pp. 573-574).

[6] It is *very significant* that, in order to explain the "*negative*" results of 1881 and 1887 of Michelson, in 1919 Richard Birkeland proposed an "*anisotropy of the ether*" ('*An Attempt to Explain the Michelson Interference Experiment*', Philosophical Magazine and Journal of Science, XXXVII, January-June 1919, pp. 150-159 [70]).

a - Birkeland supposes that, in the direction making an angle $\varphi$ with the velocity $\vec{v}$ of the Earth, the speed $c_\varphi$ of light is:

$$(1) \qquad c_\varphi = c\left(1 - \frac{1}{2}\frac{v^2}{c^2}\sin^2\varphi\right)$$

and he explains the "*negative*" result of Michelson's experiment in this way.

The light is supposed to have the same frequency in all directions.

b - In fact, it is clear that the hypothesis of a *partial* dragging of the ether by the Earth during its orbital movement is equivalent to an anisotropy such as

$$(2) \qquad c_\varphi = c\left(1 - \frac{\lambda}{2}\frac{v^2}{c^2}\sin^2\varphi\right)$$

where $\lambda$ is a constant.

For Richard Birkeland as for Henri Poincaré, it was a question of explaining the results of Michelson's experiment that were considered as being "*negative*", but the same consideration of an anisotropy of space is quite as valuable for explaining the "*positive*" effects corresponding to Miller's experiments.[7, 8]

---

[7] As Poincaré himself emphasized (note 5 above), there are very many conceivable hypotheses.

Among all these hypotheses, there is one that has never been imagined until now:

- *total dragging of the ether by the Earth during its orbital movement of translation*, whence follows the impossibility of determining this movement of translation by interferometric observations; and - *an anisotropy of space* due to astronomical influences that explains the variations of the speed of light in different directions, brought into evidence by Miller's experiments.

Contrary to an assertion often made, *total ether dragging is not incompatible with a consistent explanation of aberration* (Bouasse, 1925, *Propagation de la lumière* (Propagation of Light), Delagrave, §62, pp. 117-119 [77]).

Rather than speaking of dragging of the ether by the Earth, one might alternatively legitimately conceive of a dragging of the Earth by the ether, or at least of identical movement of the Earth and the ether around the Sun (see note 8 below).

Naturally these are only hypotheses, but, whether or not they are valid, this does not affect the fact that one must consider *the validity and the coherence of Miller's observations as having been perfectly established.*

[8] It is easy to verify that the observational data are *fully compatible* with the hypothesis of a fluid - the ether - which circulates at 30 km/sec around the Sun.

For such a fluid, in fact we have

$$(1) \qquad \frac{d\vec{v}}{dt} = \frac{\partial \vec{v}}{\partial t} + \frac{1}{2}\nabla v^2 + \nabla \times \vec{v} \times \vec{v}$$

(R. Bricard, *Calcul vectoriel* (Vector Calculus), Chapter VII B, Hydrodynamics, Armand Colin, 1929, p. 162 [80]).

Since this movement is constant, we have $\partial \vec{v}/\partial t = 0$. If the ether rotates around the Sun with the Earth, then its acceleration is the same as that of the Earth, and we have, at a point M in the orbit of the Earth and in the first approximation

$$(2) \qquad \frac{d\vec{v}}{dt} = -\frac{GM_S}{d_T^2}\vec{n}$$

counting the acceleration and its components on the unitary vector $\vec{n}$ of the line SM, where $G$ is the constant of gravitation, $M_S$ is the mass of the Sun, and $d_T$ is the distance from the Earth to the Sun.

Since we have $v = \omega d_T$, where $\omega$ is the speed of rotation around the Sun, accordingly

$$(3) \qquad \nabla\frac{v^2}{2} = \nabla\frac{\omega^2 d_T^2}{2} = \omega^2 d_T \vec{n}$$

Finally we have

$$(4) \qquad \nabla \times \vec{v} \times \vec{v} = -2\omega v \vec{n}$$

Now in CGS units (p. 124, note 15, and p. 119, note 5, above)

(5) $G = 6.67 \times 10^{-8}$ $\quad M_S = 1.987 \times 10^{33}$ $\quad d_T = 149.5 \times 10^{11}$

$\omega = \omega_T/365.25 = 0.729 \times 10^{-4}/365.25 = 1.9959 \times 10^{-7}$

$\omega d_T = 29.839\,\mathrm{km/s}$

where $\omega_T$ is the speed of rotation of the Earth. Therefore we have

(6) $-GM_S/d_T^2 = -0.5930\,\mathrm{cm/s^2}$

$\omega^2 d_T = 0.5956\,\mathrm{cm/s^2}$

$-2\omega v = -2\omega^2 d_T = -1.1912\,\mathrm{cm/s^2}$

## An unexpected result

5 - *The most remarkable regularity presented by Miller's observations is certainly the fact that the elliptical fittings of the hodographs are perpendicular to the average directions.*

Frankly speaking, this is a result that is *completely unexpected*, both by the partisans of the relativity theory and by its adversaries. It contradicts *at once both* the theory of relativity, because it demonstrates *the non-invariance of the speed of light,*[9] and the classical theory, because the hodographs *are not symmetric with respect to the meridian* (note 3 above, p. 421).

As far as one can judge, the most plausible interpretation is that the speeds measured by the interferometer have no *direct* relationship with the velocity of the Earth, and that *they only correspond to the anisotropy of space. This is a hypothesis that is consistent with all the observational data.*

---

Equation (1) can thus be written

(7) $-0.5930 = 0.5956 - 1.1912 + \varepsilon$

The difference $\varepsilon = +0.0046$ results from the approximations made.

*It is therefore verified that the ether may be considered as having the same movement around the Sun as the Earth.*

If the absolute values of its three terms are denoted by $a$, $b$, and $c$, then, to the first approximation, Equation (7) may be written as $-a = b - c$ with $a = b$, $c = 2b$. The equation $a = b$ corresponds to the fact that, for the Earth, the force of attraction $a$ is compensated by the centrifugal force $b$.

[9] This invariance implies that the four hodographs (pp. 407 and 408 above) reduce to the point 0 corresponding to the origin.

# G SIGNIFICANCE AND CONSEQUENCES OF MILLER'S OBSERVATIONS

### An affirmative answer to the three fundamental questions

1 - The observations of Miller, which are completely inexplicable in the currently accepted framework of relativity, are characterized by a *very remarkable internal coherence.*

In fact, the preceding analysis[1] allows us to answer *in the affirmative* to the three fundamental questions *with full certainty:*[2, 3]

1. *Miller's observations cannot be attributed to accidental causes or to perverse effects, such as might result from temperature, for example.*
2. *The speed of light varies with its direction.*
3. *From Miller's observations, it is possible to deduce the position of the Earth on its orbit.*

*These conclusions are independent of any hypotheses, and of all theoretical analysis of any sort.*

In fact, most of the results upon which they are based, *and in particular the most significant, were not noticed by Miller.*[4]

---

[1] Section D above, pp. 392-411.

[2] §C.4 above, p. 391.

*The implications* for the *theories of Special and General Relativity* of these totally affirmative answers to the three fundamental questions will be analyzed in Sections A to C of *Chapter VII* below.

[3] We might naturally ask ourselves why, convinced as he was of the validity and the importance of his experiments from the point of view of the theory of relativity, after 1933, Miller did not complete his observations by performing observations during a few supplementary months, as for example June and December. But in 1933 Miller was already 67 years old, and he was to die in 1941. In any case, he would have needed access to further significant financial backing. Moreover, the scientific community was then dominated by the *all-powerful* Theory of Relativity.

[4] Thus Miller cannot be accused of any more or less voluntary bias in his analysis - all the more so, because his own interpretation cannot be considered as valid (§F.1 above, pp. 417-419).

## Miller's observations and the Theory of Relativity

2 - *The very foundations of the theories of special and general relativity rest on a triple postulate: the results of Michelson's experiment, considered as "negative"; the invariance of the speed of light in all directions; and the impossibility of determining the movement of the Earth with respect to the fixed stars by a purely terrestrial experiment.*

However, *in view of the analysis of Miller's observations, it is certain* that it cannot be maintained that interferometric experiments yield a "*negative*" result, that the speed of light *is invariant in all directions*, and that no purely terrestrial experiment can determine the position and the movement of the Earth.

As a consequence, *the theories of special and general relativity, which are based upon postulates invalidated by the observational data, cannot be considered as scientifically valuable.*[5]

## A scandal of contemporary physics

3 - In conclusion, and having analyzed Miller's work *repeatedly in greater and greater depth since 1955*, today I can only *entirely* reaffirm my judgment in my memoir of 1958, *Doit-on Reconsidérer les Lois de la Gravitation?* (Should the Laws of Gravitation be Reconsidered?):[6]

> *"It is at the very least astounding that the results of this article were received with religious silence.*
>
> *"The pure and simple burial of Miller's article seems to me to be one of the scandals of contemporary physics."*

---

[5] The gist of the developments of *Section D* above (pp. 392-411) and of §1 and §2 of this Section was the basis of my article *Les expériences de Dayton C. Miller 1925-1926 et la Théorie de la Relativité* (The Experiments of Dayton C. Miller, 1925-1926, and the Theory of Relativity), published in *La Jaune et la Rouge* (the Review of the *Ecole Polytechnique*), August-September 1996, pp. 29-37 [55].

It is very significant that, after that publication, none of the letters addressed to the editorial department of that magazine or to me personally disputed the *regularities* in Miller's observations of 1925-1926 *that had been brought to light*, or *their very great coherence*.

[6] p. 102, note 38 [23].

# H OVERALL VIEW OF THE OBSERVATIONS OF ESCLANGON AND MILLER

A few conclusions result from the above analyses of the observations by Esclangon and the observations by Miller.

## Connections between different experiments

1 - The observations of Esclangon and Miller, *destined to provide evidence of the movement of the Earth*, present certain connections with my observations of the paraconical pendulum[1] and with my observations of optical sightings at marks.[2]

*All these experiments mutually reinforce one another, as far as their meaning and implications are concerned.*

## Continuous and repeated observations

2 - All these observations were performed *in a continuous manner, at different epochs in the year.*

In fact, the lunisolar components for the paraconical pendulum for which I provided evidence would have remained undetected if I had not proceeded by taking *continuous* observations over several periods of a month, and there would then have been a great temptation to attribute the observed deviations to simple accidental perturbations, as was the case, for example, with the interpretation of Miller's results by certain critics. The same holds for the optical deviations of sightings at marks that I brought into evidence in July 1958.

---

[1] *Chapters I and II* above.

[2] *Chapter III* above.

In reality, the reason that so many mechanical and optical experiments have failed or have been poorly interpreted is that they have been based upon observations over *limited periods*. As Miller so justly underlined:[3]

> *"Probably a considerable reason for the failure is the great difficulty involved in making the observations at all times of day at any one epoch. Very few, if any, scientific experiments require the taking of so many and continuous observations of such extreme difficulty; it requires greater concentration than any other known experiment ....*
>
> *"Professor Morley once said,* Patience is a possession without which no one is likely to begin observation of this kind."

## The existence of an anisotropy of space

3 - Finally, the *existence of an anisotropy of space* that corresponds both to the observations of the paraconical pendulum and to the optical sightings at marks and collimators, is *totally confirmed* by the works of Esclangon and Miller.[4]

---

[3] Miller, 1933, p. 222 [203].

In fact, *it is not sufficient to build suitable apparatus. It is also necessary to know how to use it suitably, and to observe over a long period in a continuous manner.*

[4] Here it is proper to emphasize in particular the *extremely probable, even certain*, connection between the auto-collimation experiments of Esclangon and the optical deviations of sightings at marks that I demonstrated in July 1958 (See above, §B.1.2, note 3, p. 377). The former related to *vertical* deviations and the latter related to *horizontal* deviations.

# Chapter V

# PERIODIC SEMI-ANNUAL AND ANNUAL STRUCTURE OF THE OBSERVATIONS OF THE PARACONICAL PENDULUM, OF THE SIGHTINGS AT MARKS, OF THE OBSERVATIONS OF ESCLANGON AND OF THE INTERFEROMETRIC OBSERVATIONS OF MILLER

*The more a physical theory produces evidence of valid connections, the truer it is.*

Henri Poincaré *

# A THE OBSERVATIONS OF THE PARACONICAL PENDULUM, THE SIGHTINGS AT MARKS, THE OPTICAL OBSERVATIONS OF ESCLANGON, THE INTERFEROMETRIC OBSERVATIONS OF MILLER AND THE MOVEMENT OF THE EARTH ALONG ITS ORBIT

## The results of Michelson's interferometric experiment, claimed to be negative

1 - Foucault's experiment on his pendulum in 1851 demonstrated the rotation of the Earth *using purely terrestrial apparatus.*

The interferometric experiments of Michelson in 1881 and then of Michelson and Morley in 1887 had the objective of demonstrating the position of the Earth on its orbit, as embodied in its velocity. These experiments were interpreted as yielding "*negative*" results, and this conclusion, considered as being "*definitely established*", led Lorentz to his celebrated equations for the transformation of space and time coordinates and to his formula for the composition of velocities. These formulations were improved by Poincaré.[1] Upon the same formulations, Einstein founded his *Special Theory of Relativity.*[2]

---

* Henri Poincaré, 1913, *La Valeur de la Science* (The Value of Science), Flammarion , p. 272 [222].

[1] On this question, see in particular Augustin Sesmat, 1937, *L'Optique des Corps en Mouvement* (The Optics of Moving Bodies), Hermann, Paris, pp. 594-603 [246], and René Dugas, 1950, *Histoire de la Mécanique* (A History of Mechanics), pp. 450-458 [114].

See also below, *Chapter VII*, §A.3, A.4, and A.5.

[2] Einstein, 1905, *Zur Elektrodynamik bewegter Körper* (On the Electrodynamics of Moving Bodies) [123].

See *Chapter VII* below, §A.4, pp. 562-564.

For his part, Dayton C. Miller was convinced that the conclusion that the experiments of Michelson and Morley in 1887 had yielded a "*negative*" result was erroneous, and it was this conviction that led him to set up his experiments of 1925-1926.[3]

## Correlations with the position of the Earth upon its orbit

2 - In fact, for Miller's observations, for my observations with the paraconical pendulum with anisotropic support, for the variations of optical sightings at marks, and for the observations of Esclangon, it is possible to show that all of these are *correlated with the position of the Earth upon its orbit, and moreover this can be done from only the raw experimental data, without appealing to any theoretical interpretation whatever.*[4]

*From this, it follows that it has been possible to demonstrate the displacement of the Earth along its orbit by purely terrestrial experiments.*

To show this is the essential objective of the present Chapter.

---

[3] See above, *Chapter IV*, §C.1, note 3, p. 383.

[4] The previous Chapters I, III, and IV have analyzed the *diurnal and monthly periodicities* of these various observations. But it has seemed to me best to group together the analyses of their semi-annual and annual periodicities in the present Chapter, in view of their obvious importance.

For the paraconical pendulum with isotropic support, see note (8) of §B.1 below, p. 433.

# B SEMI-ANNUAL AND LONG DURATION PERIODICITIES OF THE OBSERVATIONS OF THE PARACONICAL PENDULUM WITH ANISOTROPIC SUPPORT

## *B.1 Semi-annual periodicity of the average monthly azimuths of the paraconical pendulum with anisotropic support*

*Graph I* shows the fitting of the average azimuths of the seven monthly series of observations of the paraconical pendulum with anisotropic support to a semi-annual periodicity.[1] *The average azimuth is at maximum on 2 April at 0h, near the spring equinox.* The correlation has a coefficient of 0.672, and thus is of *very little significance.*[2, 3]

*However*, as long as we are examining the possibility of a semi-annual periodicity, *it is proper to group together the average azimuths that correspond to substantially the same period of the year*, by associating series 1 and 3, series 2 and 6, and series 4 and 5 of *Table I* of *Chapter I*, and by considering the corresponding average values. We thus obtain four values for the four average dates.[4] Correlating them, with the weightings 2, 2, 2, and 1, gives the fitting of *Graph II*, whose correlation coefficient is 0.899. *The maximum still corresponds to 2 April.*

---

[1] *Chapter I*, §A.3, *Table I*, p. 92. The average azimuths $\overline{\phi}$ of the seven monthly series in this Table, in degrees, are 148, 145, 135, 145, 148, 154, and 157.

[2] For an annual periodicity, the correlation is still lower. The correlation coefficient in this case is only 0.581.

[3] For each correlation, I give the correlation coefficient $R$ and the expression $1 - R^2$, the average value $m$ and the amplitude $r$ of the fitting sinusoid, and also the standard deviation $\sigma$ of the residues and the standard deviation $\Sigma$ of the series analyzed. I also give the period considered and the date of the maximum.

For the *statistical significance* of the fittings, see *Chapter VI*, Sections B and C of the *second volume* of this work (p. 30 above).

[4] *I.e.*, with the notations of *Table 1* of *Chapter I*, §A.3, p. 92: 173.5, 336.5, 197.5, and 90.5, taking 0h on 1 January as the origin.

Examination of the residues of the fitting of *Graph I* shows that these residues are characterized by a multi-year periodicity that appears to be close to 5.9 years. *Graph III* shows the fitting of the residues of the fitting of *Graph I* to this period of 5.9 years. The correlation coefficient here is 0.910.[5]

In fact, the period of 5.9 years is very close to very significant periods of the solar system: it corresponds to an eighth of the synodic period of Uranus and Saturn which is 5.67 years; to a third of the synodic period of Jupiter and Saturn which is 6.62 years; to a third of the sidereal rotation of the nodes of the Moon which is 6.22 years; and to half of the sunspot period of the Sun which is 5.56 years - the average of which is 6.01 years $[(5,67+6,62+6,22+5,56)/4]$.[6] *It might therefore be considered that the period of 5.9 years corresponds to the average action of the planets on the average azimuth of the paraconical pendulum.*

*Graph IV* shows a fitting to the average azimuths, taking into account *both the semi-annual period* (whose maximum is on 2 April) *and the period of 5.9 years* (whose minimum was on 10 August 1956). Here, the coefficient of multiple correlation is 0.964.

In summary, *everything happens as though, to the diurnal and monthly actions of the Moon and the Sun upon the movement of the paraconical pendulum with anisotropic support,*[7] *there are added a semi-annual influence corresponding to the movement of the Earth along its orbit, and a planetary influence having a period close to six years.*[8]

---

[5] One might naturally object that the quality of the fitting of the seven values of *Graph III* to four arbitrary parameters (the period of 5.9 years and its phase, its amplitude, and a constant) only has a reduced statistical significance, even though fittings made with seven values chosen at random would in nearly all cases give correlation coefficients much weaker than the coefficient of 0.910.

*But the essential question for appreciating the fittings of Graphs III and IV is the physical significance of the estimates derived for the parameters, in particular as far as the periods and the phases are concerned.*

[6] Research, not yet published, that I have conducted into sunspots has convinced me that they correspond to tides raised on the surface of the Sun by the movements of the planets, and that their *average* period of 11.11 years (very variable over time) can be considered as resulting from the average action of the planets.

[7] Analyzed in *Chapter I* above. Their average values are nil over half a year.

[8] This analysis and the following one are *based upon consideration of the seven monthly periods of observations*, but, for the paraconical pendulum with isotropic support, only two monthly series of observations are available: November-December 1959, and March-April 1960 (*Chapter II* above). The available data (only two months) for the isotropic pendulum is thus *insufficient* to perform a satisfactory analysis of its semi-annual periodicities.

*GRAPH I*

## AVERAGE AZIMUTHS OF THE PARACONICAL PENDULUM WITH ANISOTROPIC SUPPORT

**Semi-annual periodicity**

*Azimuths in degrees, counted positively in the anticlockwise sense from the North*

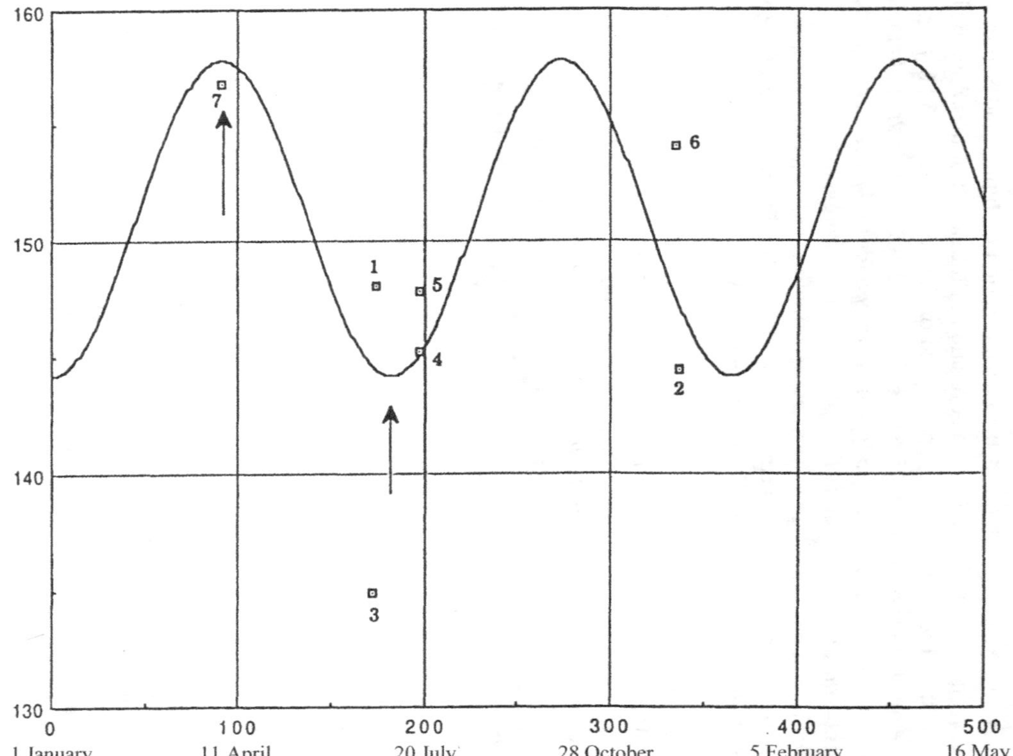


***Legend :*** $\Sigma = 6.55°$; $R = 0.672$; $1\text{-}R^2 = 0.548$; $m = 151.0°$; $r = 6.81°$; $\sigma = 4.86°$; (for notation, see §B.1, note 3)
Period: 182.625 days (6 months); date of the maximum: $s_0 = 91.3$; 0h on 2 April.
The number of each point gives the number of the monthly series in *Table I* of *Chapter I*.

***Sources :*** Calculation $217^{bis}$ (5 January 1995); Graph $13241^{bis}$.

*GRAPH II*

## AVERAGE AZIMUTHS OF THE PARACONICAL PENDULUM WITH ANISOTROPIC SUPPORT

**Semi-annual periodicity**

*Weighted values of the azimuths in degrees, counted positively in the anticlockwise sense from the North*

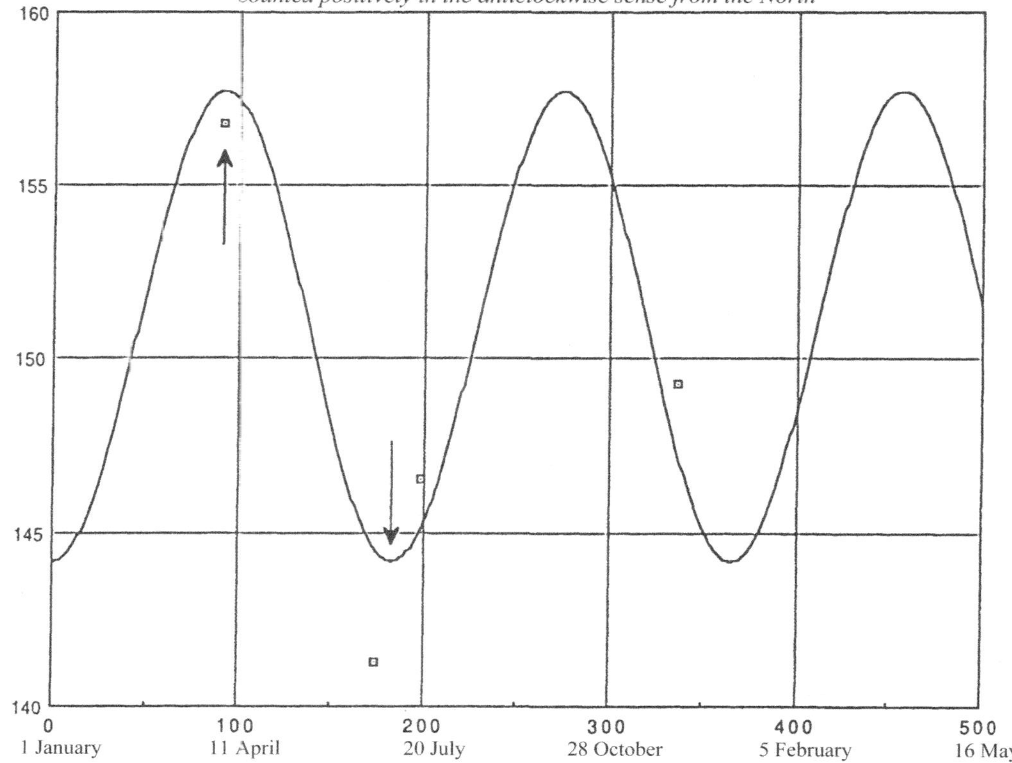


***Legend :*** $\Sigma = 4.87°$; $\mathbb{R} = 0.899$; $1\text{-}R^2 = 0.192$; $m = 150.95°$; $r = 6.77°$; $\sigma = 2.13°$; (for notation, see §B.1, note 3)
Period: 182.625 days (6 months); date of the maximum: $s_0 = 91.3$; 0h on 2 April.
The number of each point gives the number of the monthly series in *Table I* of *Chapter I*.

***Sources :*** Calculation $239^{bis}$ (1 August 1995); Graph $13263^{bis}$.

*GRAPH III*

**AVERAGE AZIMUTHS OF THE PARACONICAL PENDULUM WITH ANISOTROPIC SUPPORT**

**Period of 5.9 years**

**Fitting of the residues of Graph I to a sinusoid of period 5.9 years**

*Azimuths in degrees, counted positively in the anticlockwise sense from the North*

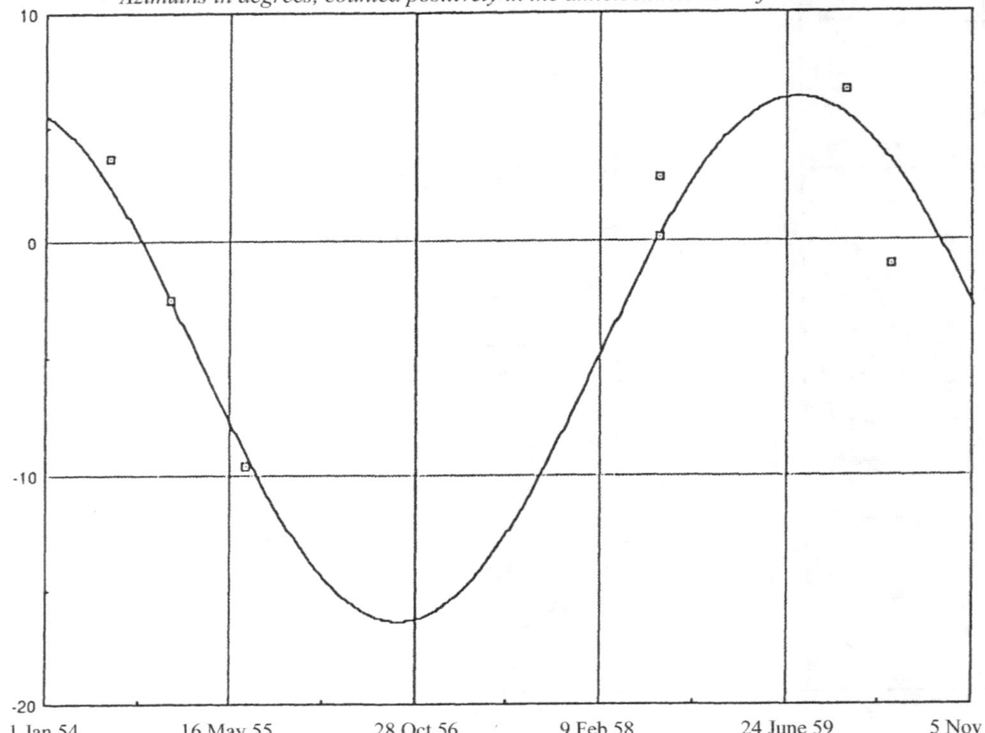


***Legend :*** $\Sigma = 4.85°$; $R = 0.910$; $1\text{-}R^2 = 0.173$; $m = -5.03°$; $r = -11.4°$; $\sigma = 2.02°$; (for notation, see §B.1, note 3)
Period: 2155 days (5.9 years); date of the minimum: $s_0 = 950$, *i.e.* 0h on 10 August 1956 (origin of s: 0h on 1 January 1954)

***Sources :*** Calculation 721 (14 November 1995); Graph 13578.

*GRAPH IV*

## AVERAGE AZIMUTHS OF THE PARACONICAL PENDULUM WITH ANISOTROPIC SUPPORT

**Fitting to a double periodicity of 6 months and 5.9 years**

*Azimuths in degrees, counted positively in the anticlockwise sense from the North*

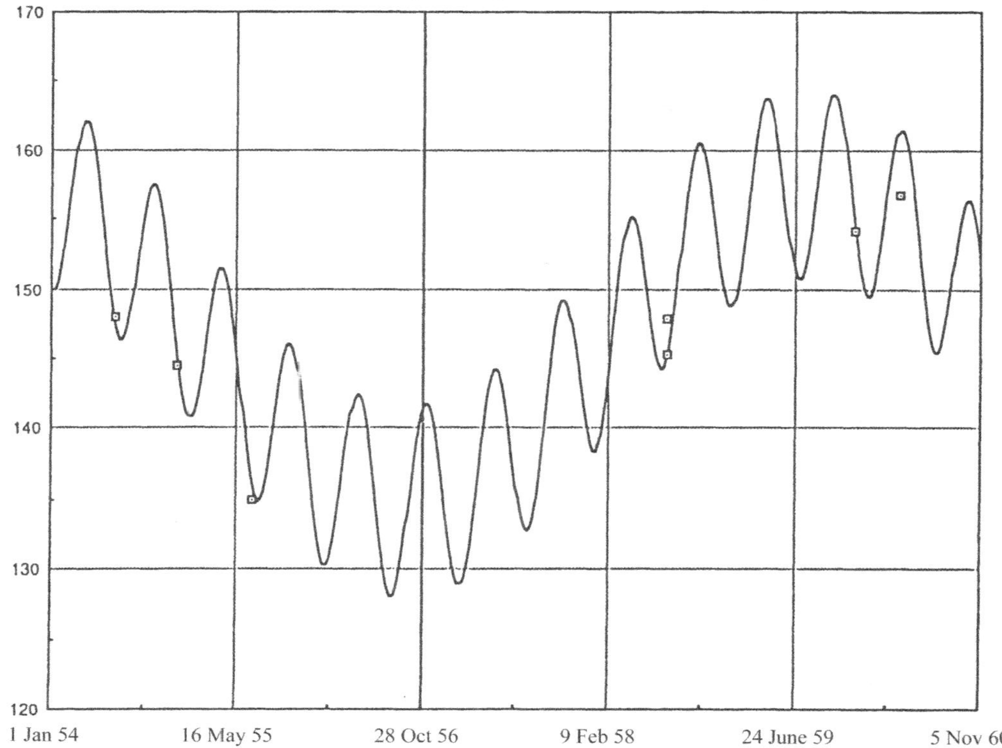


***Legend :*** $\Sigma = 6.55°$; $R = 0.964$; $1\text{-}R^2 = 0.0696$; $m = 146.1°$; $r = 6.81°$; $r' = -11.4°$; $\sigma = 2.02°$; (for notation, see §B.1, note 3)
Periods: 182.625 days (6 months) and 2155 days (5.9 years); date of the minimum of the 5.9 year periodic component: 0h on 10 August 1956.

***Sources :*** Calculation 762 (15 November 1995); Graph 13585.

## *B.2 Semi-annual periodicity of the amplitude of the diurnal component of 24h50m of the paraconical pendulum with anisotropic support*

*Table I* of *Chapter I* shows that, from 1954 to 1960, the amplitudes of the component of 24h and the component of 25h (which is representative of the component of 24h50m) *vary quite considerably*. In fact, there are *very great differences in amplitude* between series 2 and 3, and the five other series.[1]

As for the average azimuths $\overline{\phi}$, it certainly seems that the differences can be explained by *simultaneous semi-annual and planetary influences*, at least in great part.

*Graph V* shows a fitting of the seven amplitude values of the component of 25h to a semi-annual periodicity. The minimum corresponds to 1 March. As for *Graph I*, the correlation is not very significant; its coefficient is 0.534.

However, as for *Graph I* that relates to the average azimuths $\overline{\phi}$, as long as we are examining the possibility of a semi-annual periodicity, it is proper to group together the average azimuths that correspond to substantially the same period of the year, by associating series 1 and 3, series 2 and 6, and series 4 and 5 of *Table I* of *Chapter I* (§A.3), and by considering their average values as corresponding. We thus obtain four values for the four average dates. Correlating these, with the weightings 2, 2, 2, and 1, gives the fitting of *Graph VI*, whose correlation coefficient is 0.907. *The maximum still corresponds to 1 March.*

[1] For the components of 24 and 25 hours (*Table I* of *Chapter I*, §A3, p. 92), the ratios of the *average* amplitudes of series 2 and 3 to the corresponding *average* amplitudes of the series 1, 4, 5, 6, and 7 are respectively:

$$[(10.3+11.7)/2]/[(2+1.4+0.8+2.5+1.8)/5] = 11/1.7 = 6.47$$

for the component of 24h, and

$$[(12.9+14)/2]/[(3.2+2.2+2.1+1.3+1.5)/5] = 13.45/2.06 = 6.53$$

for the component of 25h, which is representative of the component of 24h50m.

These ratios are thus very close to one another, while on the other hand the ratios $R_{25}/R_{24}$ of *Table I* of *Chapter I* (p. 92) are relatively *much more stable*.

*Graph VII* shows the correlation *of the residues of Graph V* (correlation of the amplitude $2R_{25}$[2] with a sinusoid of period six months) with a sinusoid of period 5.9 years *found for the average values of the azimuths*. The correlation coefficient is 0.899. The sinusoid reaches its *maximum* at 31 July 1956, which is a *date very close* to 10 August 1956 corresponding to the minimum of *Graph III* relating to the average azimuths.

*Graph VIII* shows the fitting of the amplitudes $2R_{25}$ while, as for the average azimuths, taking account of the two fittings of *Graphs V and VII* simultaneously, with their respective minimum and maximum of 1 March for the semi-annual period and of 31 July for the period of 5.9 years. The coefficient of multiple correlation is equal to 0.937. The respective minimum and maximum of the two sinusoids are at the dates of 1 March and 31 July 1956, which dates are relatively close to the dates of 2 April and 10 August corresponding to *Graph IV* that shows the fitting of the average azimuths to the two periodicities of 6 months and 5.9 years.

*It is seen that the basic behavior of the fluctuations of the amplitude* $2R_{25}$ *is well represented by two periodicities of 6 months and 5.9 years as for the the average azimuths, and, from that fact alone, the relatively large amplitudes corresponding to the series 2 and 3 of Table I of Chapter I* (§A.3)[3] *are for the greater part explained.*

---

[2] While the correlation of *Graph V* is not very significant, the correlation of the residues of this correlation with a sinusoid of period 5.9 years *is very significant*, because there are seven data items and only three arbitrary parameters - the phase, the amplitude, and a constant.

[3] See Note 1 above.

*GRAPH V*

## PARACONICAL PENDULUM WITH ANISOTROPIC SUPPORT

***Semi-annual periodicity of the amplitude of the 25h component***

*in degrees*

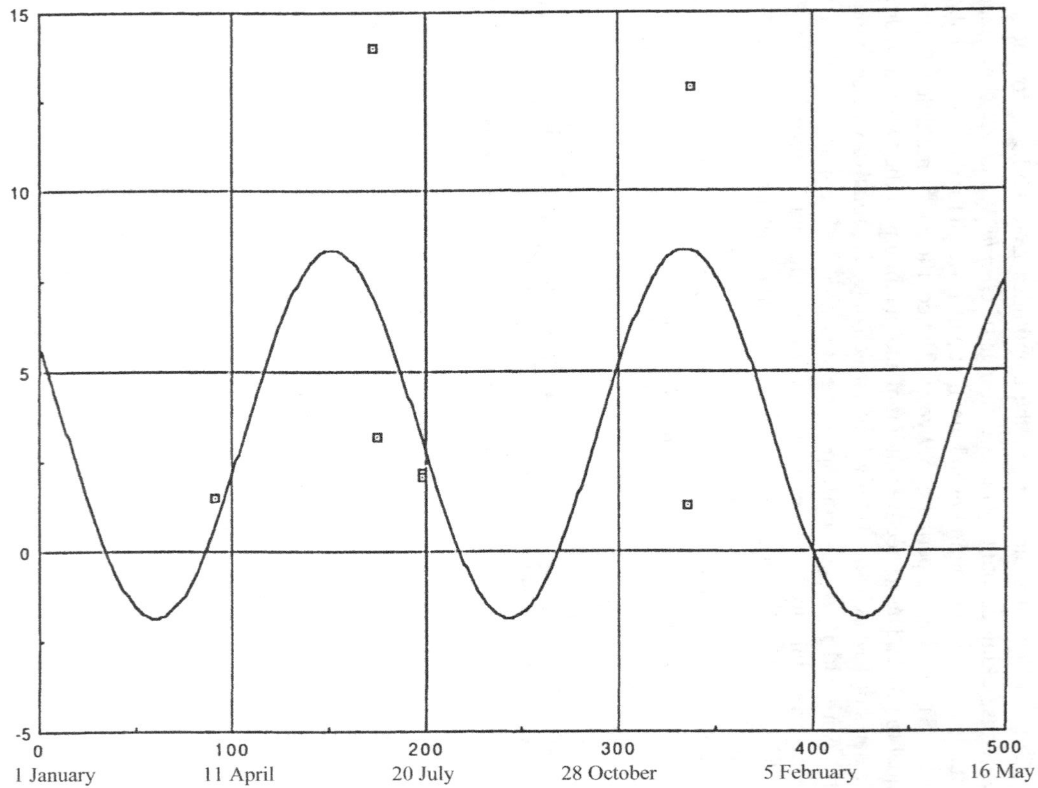


***Legend :*** $\Sigma = 5.18°$; $R = 0.534$; $1\text{-}R^2 = 0.715$; $m = 3.25°$; $r = -5.11°$; $\sigma = 4.38°$; (for notation, see §B.1, note 3)
Period: 182.625 days (6 months); date of the minimum: $s_0 = 59.3$; 0h on 1 March.

***Sources :*** Calculation 372 (7 June 1995); Graph 13411$^{bis}$.

*GRAPH VI*

**PARACONICAL PENDULUM WITH ANISOTROPIC SUPPORT**

***Semi-annual periodicity of the amplitude of the 25h component***

*Weighted values in degrees*

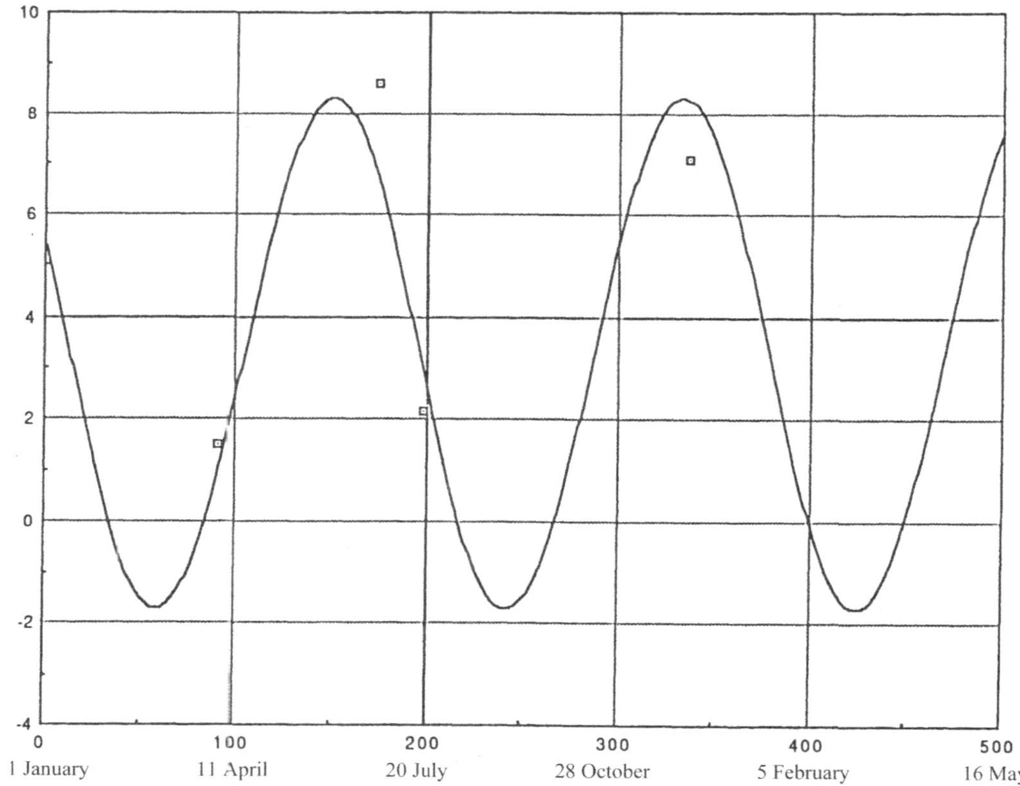


***Legend :*** $\Sigma = 2.99°$: R = 0.907; $1\text{-}R^2 = 0.178$; m = 3.30°; r = -5.02°; $\sigma = 1.26°$; (for notation, see §B.1, note 3)
Period: 182.625 days (6 months); date of the minimum: $s_0 = 59.3$; 0h on 1 March.

***Sources :*** Calculation 440 (1 August 1995); Graph 13518.

*GRAPH VII*

## PARACONICAL PENDULUM WITH ANISOTROPIC SUPPORT

**Amplitude of the 25h component**
**Fitting of the residues of Graph V to a sinusoid of period 5.9 years**
*in degrees*

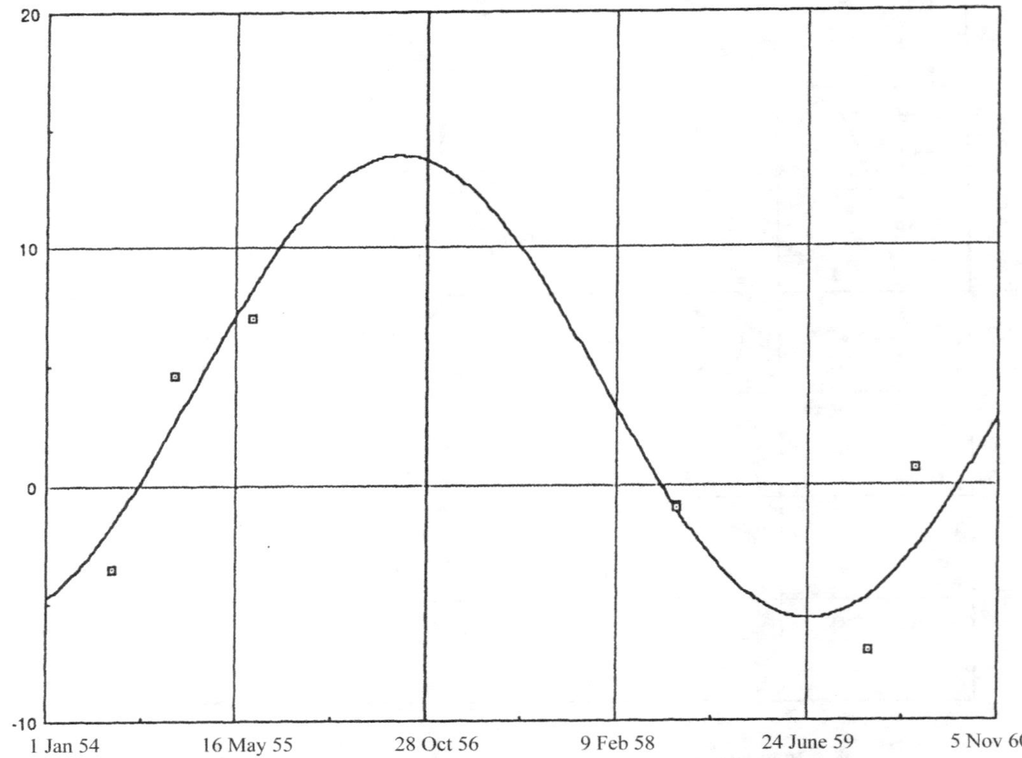


***Legend :*** $\Sigma = 4.38°$; $R = 0.899$; $1\text{-}R^2 = 0.191$; $m = 4.12°$; $r = 9.76°$; $\sigma = 1.92°$; (for notation, see §B.1, note 3)
Period: 2155 days (5.9 years); date of the maximum: $s_0 = 925.5$, *i.e.* 12h on 31 July 1956.

***Sources :*** Calculation 455 (10 August 1995); Graph 13528.

*GRAPH VIII*

**PARACONICAL PENDULUM WITH ANISOTROPIC SUPPORT**

**Amplitude of the 25h component**

**Fitting to a double periodicity of 6 months and 5.9 years**

*in degrees*

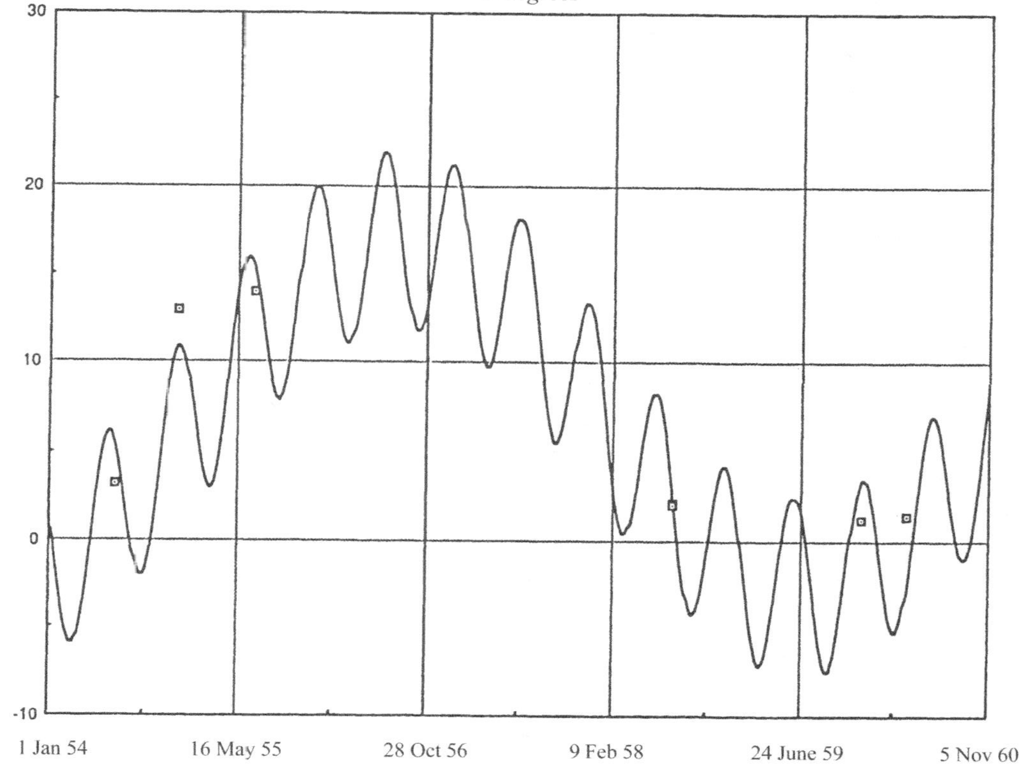


***Legend :*** $\Sigma = 5.18°$; $R = 0.937$; $1\text{-}R^2 = 0.122$; $m = 7.16°$; $r = -5.11°$; $r' = 9.76°$; $\sigma = 1.81°$; (for notation, see §B.1, note 3) Periods: 182.625 days (6 months) and 2155 days (5.9 years); 6 month component: minimum at 0h on 1 March; 5.9 year component: maximum on 31 July 1956.

***Sources :*** Calculation 771 (15 November 1995); Graph 13586.

### B.3 Similarities of the semi-annual and long-term fittings of the average azimuths and the amplitudes of the 25h component

In fact, and although the correlation of the average azimuths $\overline{\phi}$ and the amplitudes $2R_{25}$ of the 25 hour component is not very well defined,[1] it follows from *Table I* that there is a *remarkable structural coherence* between the characteristics of the fittings I through IV and those of the fittings V through VIII.

First, the dates of 2 April which is the maximum of *Graph I*, and of 1 March which is the minimum of *Graph V*, are rather close together.[2] The same holds for the two dates of 10 August 1956 and 31 July 1956 that correspond to *Graphs III and VII*. The maxima and minima are reversed: a maximum for the average azimuths corresponds to a minimum for the amplitudes; and the situation for the period of 5.9 years is reversed: a maximum for the amplitudes corresponds to a minimum for the average azimuths.

In the same way, comparison of the correlation coefficients obtained for the average azimuths and the amplitudes of the 25h component shows that, for each type of dependence, they are very close together. The same holds for the ratios $r/\Sigma$, $r'/\Sigma$, $\sigma/\Sigma$.

This resemblance is still more remarkable because the average azimuths $\overline{\phi}$ and the amplitudes $2R_{25}$ of the 25h component (which is representative of the 24h50m component) *correspond to two phenomena that are totally different in nature.*

---

[1] The correlation coefficient between the series of seven values of $2R_{25}$ and the series of seven values of $\overline{\phi}$ is only -0.794, due to the differences in phase between the two series.

[2] For *Graphs I and V* (pp. 434 and 440), the average of the values of $s_0 (91 + 59.3)/2 = 75.15$ corresponds to the date of 17 March, which is very close to the spring equinox.

*TABLE I*

**STRUCTURAL RESEMBLANCES OF THE FITTINGS OF THE AVERAGE AZIMUTHS AND THE AMPLITUDES OF THE 25H COMPONENTS OF THE PARACONICAL PENDULUM WITH ANISOTROPIC SUPPORT**

| Graph | Period | Series | n | Σ | R | $1-R^2$ | m | r | r' | σ | r / Σ | r'/ Σ | σ / Σ | Phase |
|---|---|---|---|---|---|---|---|---|---|---|---|---|---|---|
| I | 6 months | $\overline{\phi}$ | 7 | 6.55 | 0.672 | 0.548 | 151.0 | 6.81 | | 4.86 | 1.04 | | 0.765 | maximum on 2 April |
| V | 6 months | $2R_{25}$ | 7 | 5.18 | 0.534 | 0.715 | 3.25 | 5.11 | | 4.38 | 0.98 | | 0.846 | minimum on 1 March |
| II | 6 months | $\overline{\phi}$ | 4 | 4.87 | 0.899 | 0.192 | 150.9 | 6.77 | | 2.13 | 1.39 | | 0.437 | maximum on 2 April |
| VI | 6 months | $2R_{25}$ | 4 | 2.99 | 0.907 | 0.178 | 3.30 | 5.02 | | 1.26 | 1.03 | | 0.421 | minimum on 1 March |
| III | 5.9 years | $\overline{\phi}$ Residues of I | 7 | 4.86 | 0.910 | 0.173 | - 5.04 | | 11.4 | 2.02 | | 2.35 | 0.416 | minimum on 10 August 56 |
| VII | 5.9 years | $2R_{25}$ Residues of V | 7 | 4.38 | 0.899 | 0.191 | 4.12 | | 9.76 | 1.92 | | 2.23 | 0.438 | maximum on 31 July 1956 |
| IV | 6 months and 5.9 years | $\overline{\phi}$ | 7 | 6.55 | 0.964 | 0.070 | 146.0 | 6.81 | 11.4 | 2.02 | 1.04 | 1.74 | 0.308 | minimum on 10 August 56 |
| VIII | 6 months and 5.9 years | $2R_{25}$ | 7 | 5.18 | 0.937 | 0.122 | 7.16 | 5.11 | 9.76 | 1.92 | 0.98 | 1.88 | 0.371 | maximum on 31 July 1956 |

***Legend*** : n = number of values in the series; Σ = standard deviation of the series; R = correlation coefficient;
σ = standard deviation of the correlation residues; m = average value in degrees;
r = semi-amplitude of the 6-month component in degrees; r' = semi-amplitude of the 5.9-year component in degrees

### *B.4 Overall view of the semi-annual and long-term periodicities of the paraconical pendulum with anisotropic support*

From the above analysis relating to the average azimuths and the amplitudes of the 25h component of the paraconical pendulum with anisotropic support,[1] certain conclusions can be drawn:

a - First of all, a *very great coherence* is present between the structure of the average azimuths and the structure of the amplitudes of the 25h component, which represents the 24h50m component, both for the phases and for the correlation coefficients of the fittings that were performed, although the average azimuths and the amplitudes of the 25h component correspond to phenomena *whose natures are entirely different.*

b - Differences between the average azimuths and the amplitudes of the sinusoids of period 25h of the different series, *which at first appear inexplicable*, and which for a long time I was not able to elucidate, can be *very simply* explained by the coexistence of two periodicities, one semi-annual and the other of around 5.9 years.

c - Although each of the correlations, in itself, has a relatively low statistical significance, *taken together, they have a very great significance.*[2]

d - Not only does analysis of the diurnal periodicities of 24h and 25h of the paraconical pendulum, which are inexplicable in the framework of isotropy of space, lead to the conclusion of an anisotropy of space, but also the analysis of the periodicities of 6 months and 5.9 years *can only be explained by such an anisotropy.*

e - *From the fact that the movements of the paraconical pendulum with anisotropic support exhibit a semi-annual periodicity, it is possible to deduce that it is possible to determine the position of the Earth upon its orbit by purely terrestrial experiments with the pendulum.*

---

[1] Analogous analyses could be performed for the other elements of *Table I* of *Chapter I* above (p. 92).

[2] The question of the statistical significance of the results obtained is examined deeply in *Chapter VI* (Sections B and C) of the *second volume* of this work (p. 30 above).

# C SEMI-ANNUAL PERIODICITIES OF THE OPTICAL SIGHTINGS AT MARKS AND OF THE OPTICAL SIGHTINGS OF ESCLANGON

## *C.1 Semi-annual periodicity of the optical sightings at marks*

### The observations at IRSID in 1958 and at IGN in 1959

1 - As I have mentioned,[1] two month-long series of observations of optical sightings at marks were performed in July 1958 at IRSID and in February-March 1959 at IGN. Since a very marked correlation was recognized in July 1958 between the observations of the paraconical pendulum and the optical sightings at marks,[2] and the average azimuths of the paraconical pendulum were particularly characterized by a semi-annual periodicity, one is naturally led to ask whether the optical sightings at marks might similarly exhibit a semi-annual periodicity.

### Trends of the averages (N+S)/2 of the sightings at marks at IRSID in 1958 and at IGN in 1959

2 - *Graphs V and XII* of *Chapter III*[3] show, in centesimal seconds, the average deviations $(N+S)/2$ of the North and South sightings at marks at IRSID from 1 to 31 July 1958 and at IGN from 23 February to 25 March 1959, corresponding in both cases to a duration of 31 days. The deviations are reckoned *positively in the clockwise sense.*

---

[1] *Chapter III*, Section A, pp. 332-333 above.

[2] *Chapter III*, §B.3 above, p. 338.

[3] *Chapter III*, §B.5 et C.3 above, pp. 344 and 361.

*To simplify the calculations*, I have considered the two linear trends of the fittings in the two *Graphs V and XII*, and, in each case, have considered the six central values of the first six five-day periods, which are five days apart.

*Graph IX* represents the fitting *to a period of 6 months* of the two series of observations at IRSID and at IGN, which were 6 months and 23 days apart. The correlation coefficient is 0.9991, which is remarkably high, and the minimum of the fitting sinusoid is at *1 April*, near the spring equinox.[4]

## Comparison with the observations of the paraconical pendulum with anisotropic support

3 - Since the azimuths of the paraconical pendulum with anisotropic support were reckoned positively *in the anticlockwise sense* whereas the optical deviations were reckoned positively *in the clockwise sense*, it is seen that the optical deviations (*Graph IX* opposite) are *practically in phase* with the average azimuths of the paraconical pendulum (*Graph II* of §B.1, p. 435), and that the summits of the fittings are at *2 April* in the first case and *1 April* in the second case, when the directions of reckoning of the angles are considered. *The agreement in phase is thus very remarkable.*

*In fact it is just as remarkable as the agreement in phase that has been brought to light between the diurnal components of the movement of the paraconical pendulum with anisotropic support and those of the sightings at marks.*[5]

---

[4] In fact, given that the lenses were differently adjusted in the two cases, the two series of six values are only comparable up to a constant. I have thus searched for the value of the offset $\Delta$ between the two series of azimuths, that maximizes the correlation of the composite series constituted by the observations at IRSID and also by the observations at IGN increased by $\Delta$ centesimal seconds, with a sinusoid of period equal to 6 months. The maximum correlation is obtained for $\Delta = -75$ centesimal seconds.

[5] *Chapter III*, §B.3 above, p. 338.

*GRAPH IX*

**TRENDS OF THE AVERAGES OF THE SIGHTINGS AT MARKS (N+S)/2 at IRSID (1 to 30 July 1958) and at IGN (23 February - 24 March 1959)**

***Semi-annual period***

*Deviations in centesimal seconds reckoned positively in the clockwise sense*

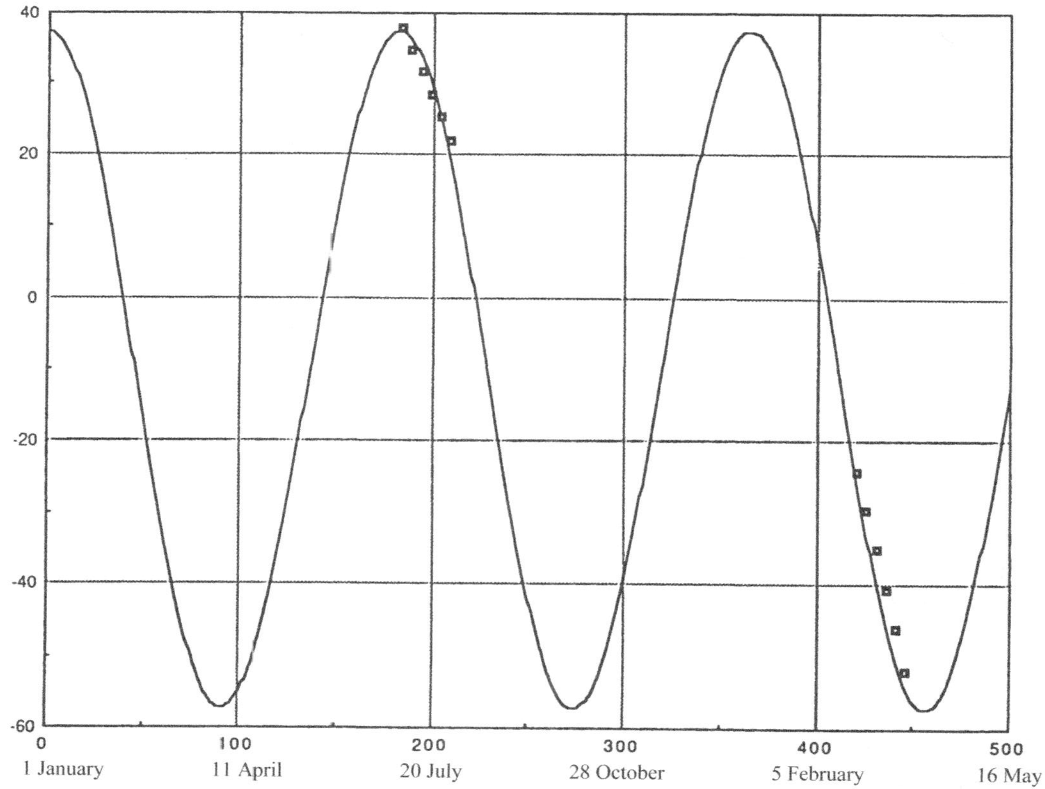


***Legend :*** $\Sigma = 37.25''$; $R = 0.999$ ; $1-R^2 = 0.0018$; $m = -9.84''$; $r = -47.35''$; $\sigma = 1.58''$; (for notation, see §B.1, note 3) Period: 182.625 days (6 months); Date of minimum: $s_0 = 90$; 0h on 1 April.

***Sources :*** Calculation 918 (10 January 1995); Graph 13634.

## C.2 *Semi-annual periodicity of the optical observations of Esclangon*

*Graph X* shows the fitting to a periodicity of six months of the monthly averages of the vertical deviations of Esclangon for the eight months for which sufficient data is available, *i.e.* March to July 1927 and November 1927 to January 1928.[1]

The monthly average is *minimum* on 25 March, *near the spring equinox.* The correlation coefficient is 0.841.[2] This fitting is completely comparable to the fittings of *Graphs II, VI, and IX* above relating to the paraconical pendulum with anisotropic support and to the sightings at marks.

These graphs demonstrate the existence of *a remarkable consistency* between the periodic characteristics of the paraconical pendulum with anisotropic support, those of the sightings at marks, and those of the optical sightings of Esclangon, particularly *as concerns the agreements in phase, with the direction of reckoning of the angles being considered.*[3]

As for the paraconical pendulum, we are thus led to the conclusion that it is possible to determine the position of the Earth upon its orbit *by purely terrestrial experiments.*

---

[1] *Chapter IV*, §B.2 above, pp. 378-381.

Esclangon, 1928, *Sur l'existence d'une dissymétrie optique de l'espace* (On the existence of an optical asymmetry of space), pp. 54-57 [140].

The monthly averages of the optical sightings of Esclangon in thousandths of a minute of arc for March to July 1927 and November 1927 to January 1928, *i.e.* eight monthly values, are as follows: for March to July 1927: -4.03; -4.00; -2.35; 2.11; 2.91; and for November 1927 to January 1928: 0.65; 1.18; and -1.75.

In the months of February, August, September, and October 1927 there were far too few observations to be sufficient, or even no observations.

Since the deviations of Esclangon were measured vertically, their signs are not comparable to the signs of the azimuthal deviations of the paraconical pendulum with anisotropic support and of the sightings at marks.

[2] This correlation is *particularly significant*. In fact, it corresponds to eight data points and three degrees of freedom: the average, the amplitude, and the phase. Each of the data points corresponds, on average, to 17 observations.

[3] If we measure all the azimuths positively in the anticlockwise sense, the maxima of *Graphs II* (§B.l) and IX (§C.l) are respectively at 0h on 2 April and at 0h on 1 April, near the spring equinox on 21 March (see §E.l below). For *Graph X*, the minimum of the vertical deviation corresponds to 0h on 25 March (see §E.l below).

*GRAPH X*

**OPTICAL DEVIATIONS OF ESCLANGON**

***Semi-annual period***

*in thousandths of a minute of arc*
*Deviations reckoned vertically*

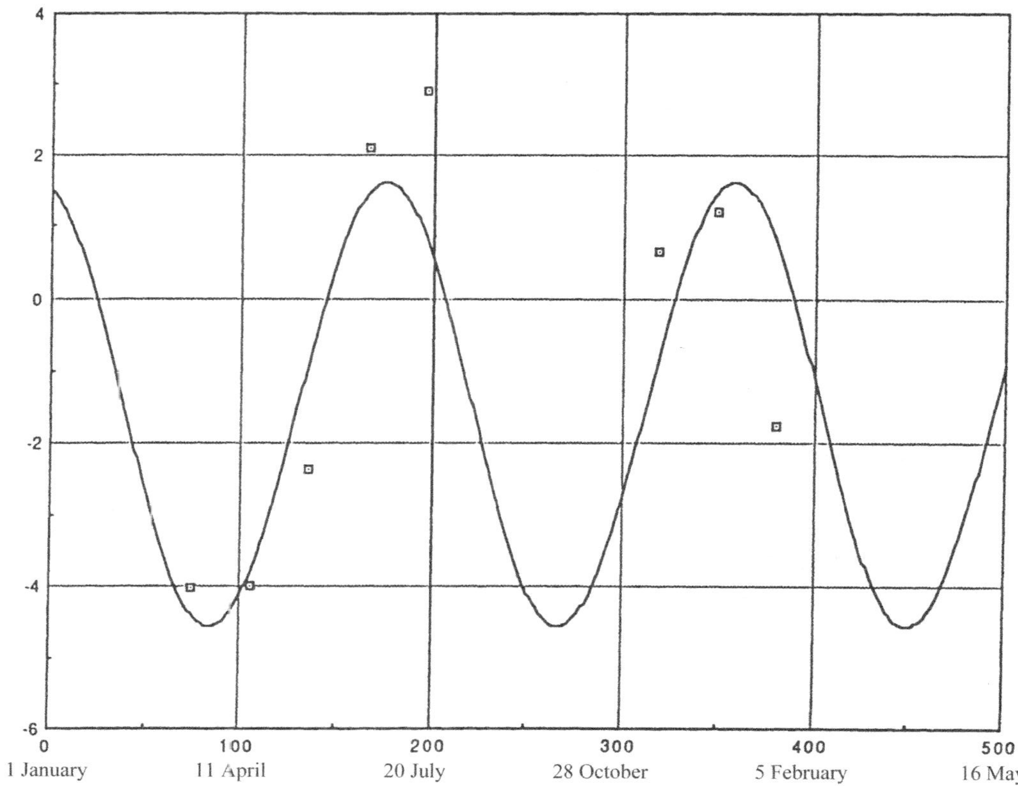


***Legend :*** $\Sigma = 2.55$; $R = 0.841$; $1\text{-}R^2 = 0.293$; $m = -1.49$; $r = -3.09$; $\sigma = 1.38$; (for notation, see §B.1, note 3)
Period: 182.625 days (6 months); Date of minimum: $s_0 = 82.9$; 0h on 25 March.

***Sources :*** Calculation 717 (3 November 1995); Graph 13574.

# D SEMI-ANNUAL AND ANNUAL PERIODICITIES OF THE INTERFEROMETRIC OBSERVATIONS OF MILLER

## *D.1 Determination of the position of the Earth upon its orbit from Miller's observations, independently of any hypothesis*

*In fact, as is the case for the observations of the paraconical pendulum, for the sightings at marks, and for the observations of Esclangon, it is possible to deduce the position of the Earth upon its orbit from the interferometric observations of Miller by direct calculation, without making any theoretical hypothesis about the interpretation to be placed upon those observations.*[1]

This possibility results from harmonic analysis of the characteristic parameters of the *eight fundamental graphs of Miller*.[2] I give some illustrations below.

[1] Introduction to the present Chapter, Section A above, pp. 430-431.

[2] *Graphs III and IV* of *Chapter 4*, §C.3 above, pp. 388-389.

## D.2 Direct estimation of the characteristic parameters of the eight fundamental graphs of Miller.

The *most significant* parameters characterizing the eight fundamental Graphs of Miller are, for each period of observations: the maximum and minimum speeds $v_M$ and $v_m$, the average values $\overline{A}$ of the azimuths $A$, and the maximum amplitudes $A_M^*$ of their variations around their average values.

*Table II* opposite gives *the direct estimates of these parameters that*, in June 1995, *I made graphically* from photographic enlargements of the eight fundamental graphs of Miller that summarize his observations.[1]

These are estimates that have subsequently been used in *all* my calculations, and for drawing *all* the Graphs of this Chapter relating to Miller's observations.[2, 3]

[1] These estimations have been made simply from copied Graphs (*without any calculations*), and their precision is of course only relative, especially for 8 February 1926.

One can only regret that Miller did not publish the values corresponding to the moving averages of his Graphs, as Esclangon did (whom Miller cited); he could have done so in two pages. He would only have needed to reproduce 160 values, 20 for each Graph, along with their sidereal times.

[2] I made these estimations on 29 June 1995 and *never modified them during my further calculations*, and naturally, when I made them, I was not able to foresee what the results of the temporal and structural correlations of *Tables III, IV, V, VII, VIII, and X* would be, see below.

*The results obtained are all the more significant.*

[3] Here are Miller's estimations, *as they result from his overall analysis, explicitly or implicitly* (see *Chapter IV*, Sections A and C of the *second volume* of this work, p. 29 above). They are particularly very different in the case of the series of observations of 8 February 1926. (Particularly see the estimates of $v_M$ and $\overline{A}$, Miller 1933, id., p. 230 et 235 [203]).

*Table II**

*Miller's estimates of the characteristic parameters*

| | $v_M$ | $v_m$ | $A_M$ | $A_m$ | $\overline{A}$ | $A_M^*$ |
|---|---|---|---|---|---|---|
| 1 April 1925 | 10.1 | 7.63 | 58.0 | 22.0 | 40 | 18.0 |
| 1 August 1925 | 11.2 | 6.27 | 44.5 | -24.5 | 10 | 34.5 |
| 15 September 1925 | 9.6 | 4.44 | 89.0 | 21.0 | 55 | 34.0 |
| 8 February 1926 | 9.3 | 7.23 | 7.8 | -27.8 | -10 | 17.8 |

*TABLE II*

## FUNDAMENTAL GRAPHS OF MILLER

### *GRAPHIC ESTIMATIONS OF THE SPEEDS AND AZIMUTHS*

*Speeds (in km/sec)*

| | $v_M$ | $v_m$ |
|---|---|---|
| 1 April 1925 | 10 | 7,8 |
| 1 August 1925 | 11,6 | 6,5 |
| 15 September 1925 | 9,8 | 4,2 |
| 8 February 1926 | 10 | 7,3 |

*Azimuths (in degrees)*

| | $A_M$ | $A_m$ | $\overline{A}$ | $A^*_M$ |
|---|---|---|---|---|
| 1 April 1925 | 60 | 20 | 40 | 20 |
| 1 August 1925 | 45 | - 20 | 12,5 | 32,5 |
| 15 September 1925 | 90 | 20 | 55 | 35 |
| 8 February 1926 | 15 | - 40 | -12,5 | 27,5 |

***Legend :*** $v_M$ and $v_m$ : maximum and minimum values of the speed

$A_M$ and $A_m$ : maximum and minimum values of the azimuths

$\overline{A} = (A_M + A_m)/2$

$A^*_M = (A_M - A_m)/2$

$A^* = A - \overline{A}$

(Latitude of Mt. Wilson: $\varphi = 34°13' = 34.217°$; Miller, 1933, p. 230)

***Sources :*** These estimates were deduced *graphically* from photographic enlargements of *Graphs III and IV* of Miller (*Chapter IV*, §C.3), *independently of any hypothesis*. These graphs correspond to the averages of the observations over 6 or 8 days (Miller, 1933, id., p. 213).

These estimates were made on 29 June 1995, and they have been used in *all* the calculations of this *Chapter V*, unless otherwise noted.

## *D.3 Semi-annual and annual periodic structure of the characteristic parameters of the fundamental Graphs of Miller*

For lack of space, there naturally can be no question of analyzing the periodic structure of the characteristic parameters of the fundamental Graphs of Miller in detail. I shall limit myself to commenting *very briefly* upon the *semi-annual and annual periodic structure* of some of the particularly *significant parameters*: the speeds $v_M$ and $v_m$, the average azimuths $\overline{A}$, the semi-amplitudes $A_M^*$, and some indications derived from these.[1]

### Average azimuths $\overline{A}$. A dominant semi-annual periodicity

1 - *Graph XI* shows the average azimuths fitted to a semi-annual periodicity. The correlation coefficient is 0.840. The peak corresponds to *6 April*. This is *near the spring equinox*.[2]

---

[1] For this analysis, it is very regrettable that Miller's four periods are not better spread over the year. *No observations are included in the period from 15 September to 8 February, i.e. nearly five consecutive months.*

Certainly the choice of the dates of 1 April, 1 August and 15 September 1925, and 8 February 1926, was made due to preconceived ideas or practical requirements. In fact, it would have been much simpler and much more effective to employ four periods centered around 21 March, 21 June, 21 September, and 21 December.

This was indeed suggested by Michelson and Morley (1887, *On the Relative Motion of the Earth and the Luminiferous Ether*, American Journal of Science, p. 341 [199]):

> *"The experiment will therefore be repeated at intervals of three months, and thus all uncertainty will be avoided".*

That was there the very reason, but, *contrary to many assertions*, this repetition was never performed.

[2] The annual periodicity is relatively negligible, the correlation coefficient of 0.447 being rather insignificant.

The date of the maximum of the semi-annual sinusoid fitted to the average azimuths $\overline{A}$, which is 6 April, is relatively close to that corresponding to the average azimuths $\overline{\phi}$ of the paraconical pendulum with anisotropic support, which is *2 April*. The correlation coefficient of 0.840 is also of a comparable order of magnitude to that of the weighted average azimuths of the paraconical pendulum, which is 0.900 (*Graph II* above).[3]

*It is very remarkable that the average azimuths $\overline{A}$, which Miller considered to be inexplicable, are characterized by a semi-annual periodicity which Miller did not recognize and which is completely analogous to that of the average azimuths of the paraconical pendulum with anisotropic support.*[4]

## Semi-amplitude $A_M^*$. A dominant annual periodicity

2 - *Graph XII* shows the fitting of the semi-amplitude $A_M^*$ of the oscillations of the azimuth $A$ *to an annual periodicity*. The correlation coefficient is −0.973, and the minimum of the fitting corresponds to 22 April.[5]

---

[3] In fact, since Miller's angles are reckoned *positively in the clockwise sense* whereas for the pendulum they are reckoned *in the anticlockwise sense*, the phases are opposite.

As we will indicate (*Chapter V* of the *second volume* of this work), it is possible to establish *a theoretical correspondence relationship* between the average azimuths $\overline{\phi}$ of the paraconical pendulum with anisotropic support and the average azimuths $\overline{A}$ of Miller.

[4] This realization is sufficient by itself to eliminate the objections raised against Miller's experiments that assert them to be void of any real significance, and to result from local perturbations like temperature (*Chapter IV*, §E.1 above, pp. 412-413).

[5] The periodicity of six months, which corresponds to a correlation coefficient of 0.706, is noticeably less marked. Its peak corresponds to 26 February.

The standard deviation of the residues of this correlation is equal to 4.26, which is approximately three times the standard deviation of the correlation corresponding to the annual periodicity, which is only 1.32.

### The difference $v_M - v_m$. A dominant annual periodicity

3 - *Graph XIII* shows the fitting *to an annual periodicity* of the difference $v_M - v_m$, between the maximum and minimum speeds. The correlation coefficient is $-0.9995$ and the minimum of the fitting corresponds to 22 March, *which is almost exactly the spring equinox.*

### The ratio $v_m/v_M$. A dominant annual periodicity

4 - *Graph XIV* shows the fitting *to an annual periodicity* of the ratios $v_m/v_M$ of the minimum and maximum values of the speeds. The correlation coefficient is 0.992 and the peak of the fitting corresponds to 7 April, near the spring equinox.

*GRAPH XI*

**OBSERVATIONS OF MILLER**
**AVERAGE AZIMUTHS $\bar{A}$**
**Semi-annual periodicity**
*in degrees, reckoned positively in the clockwise sense*

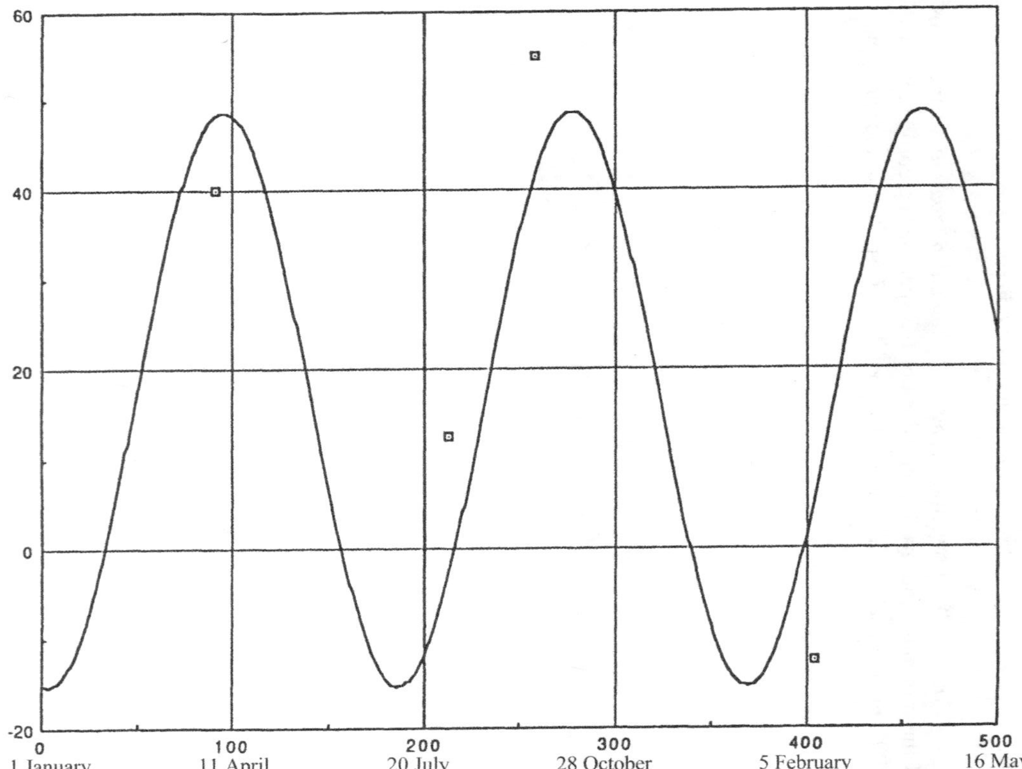


***Legend*** **:** $\Sigma = 25.9°$; $R = 0.840$; $1\text{-}R^2 = 0.294$; $m = 16.63°$; $r = 31.95°$; $\sigma = 14.0°$. Period: 182.625 days (6 months); date of maximum: $s_0 = 95.9$; 12h on 6 April. (See the *Legend* of §B.l, note 3 above).

***Sources*** **:** Table II; Calculation 461 (11 August 1995); Graph 13529.

*GRAPH XII*

**OBSERVATIONS OF MILLER**
**SEMI-AMPLITUDE $A^*_M$ OF THE AZIMUTHS A**
**Annual periodicity**
*in degrees*

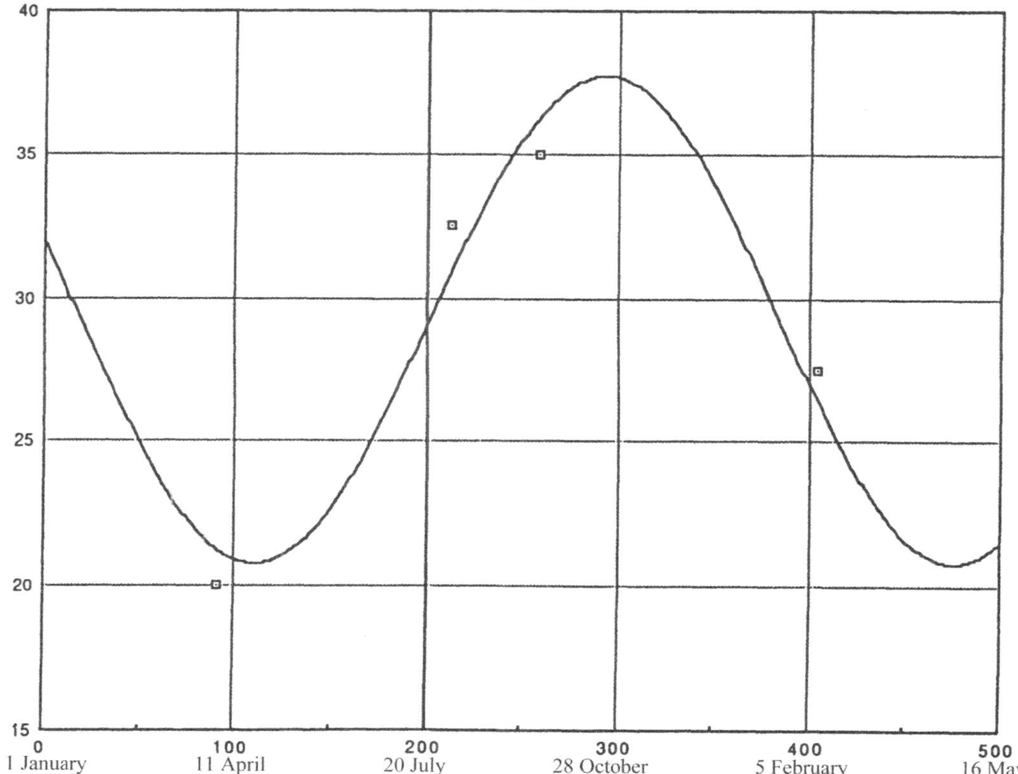


***Legend :*** $\Sigma = 5.73$; $R = -0.973$; $1-R^2 = 0.053$; $m = 29.3°$; $r = -8.47°$; $\sigma = 1.32°$. Period: 365.25 days (1 year); date of minimum: $s_0 = 110.8$; 0h on 22 April. (See the *Legend* of §B.1, note 3 above).

***Sources :*** Table II; Calculation 459 (18 August 1995); Graph 13530.

*GRAPH XIII*

**OBSERVATIONS OF MILLER**

**DIFFERENCE $v_M$-$v_m$ BETWEEN MAXIMUM AND MINIMUM SPEEDS**

**Annual periodicity**

*in km/s*

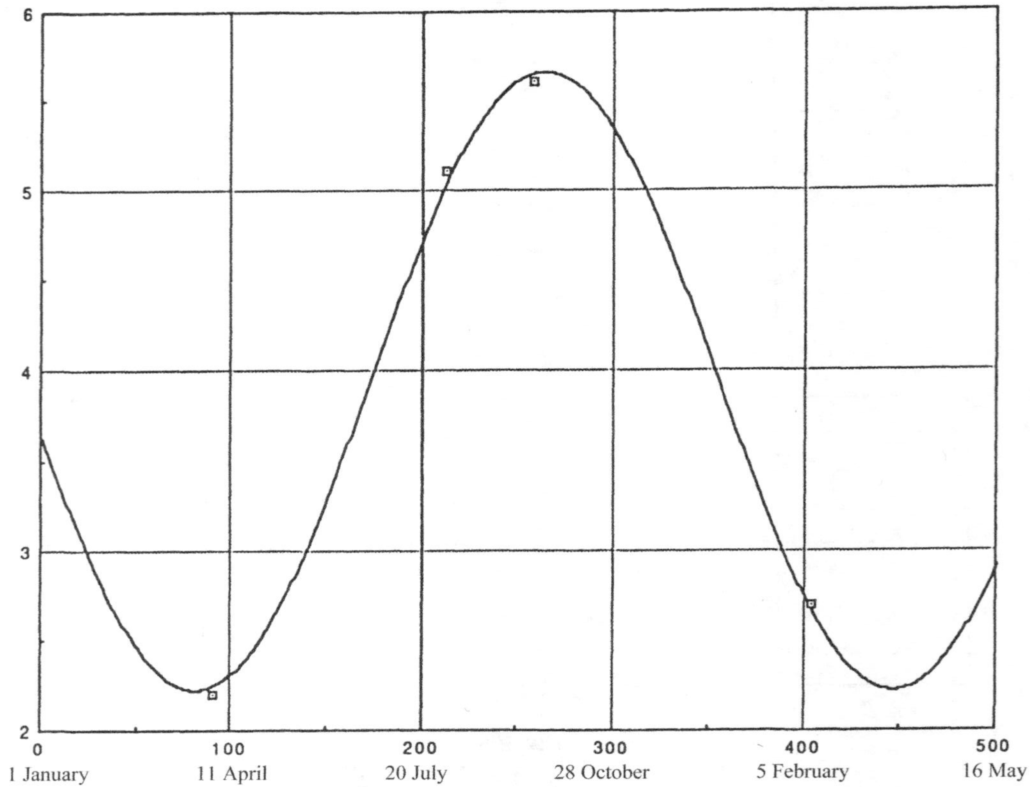


***Legend :*** $\Sigma$ = 1.47; R = 0.9995; 1-$R^2$ = 0.0011; m = 3.94; r = -1.72; $\sigma$ = 0.049. Period: 365.25 days (1 year); date of minimum: $s_0$ = 80.9; 12h on 22 March. (See the *Legend* of §B.1, note 3 above).

***Sources :*** Table II; Calculation 758 (10 July 1995); Graph 13581.

*GRAPH XIV*

**OBSERVATIONS OF MILLER**
**RATIO BETWEEN MINIMUM AND MAXIMUM SPEEDS**
**Annual periodicity**

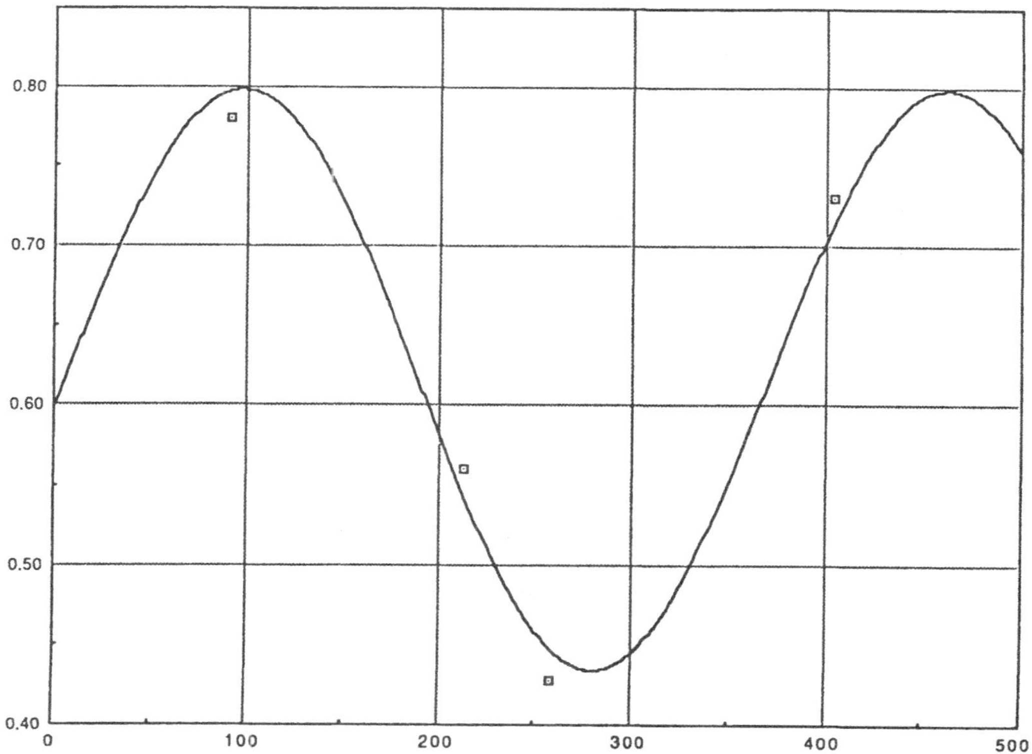


***Legend*** **:** $\Sigma = 0.139$; $R = 0.992$; $1\text{-}R^2 = 0.0159$; $m = 0.616$; $r = 0.182$; $\sigma = 0.0176$. Period: 365.25 days (12 months); date of minimum: $s_0 = 96.3$; 12h on 7 April. (See the *Legend* of §B.l, note 3 above).

***Sources*** **:** Table II; Calculation 759 (10 November 1995); Graph 13582.

## D.4 A double periodicity

We see that Miller's observations are characterized by either a dominant semi-annual periodicity (*Graph XI*) or by a dominant annual periodicity (*Graph XII, XIII, and XIV*), according to the parameter considered.

*This double periodicity appears to correspond to two different and distinct influences: the influence of anisotropy which is characterized by the average azimuths* $\overline{A}$ *and whose dominant periodicity is semi-annual, and the influence of the movement of the Earth along its orbit whose periodicity is annual.*

*Table III* below gives, for various different significant parameters, the dominant periodicities and their characteristics. Four of them correspond to an annual periodicity, and nine to a semi-annual periodicity. This Table suggests certain observations:

a - When there is a dominant annual periodicity, its maximum or minimum is *generally* close to 21 March, and *not to 21 June*. This is a *very remarkable* circumstance.[1]

b - While the phases $s_0$ of the semi-annual or annual periodicities vary, their average 81.8 is remarkably close to the value $s_0 = 79.5$ which corresponds to 12h on 21 March, in other words is within two days of the date of *the spring equinox*.[2]

c - A dominant periodicity of 6 months corresponds to the average value $\overline{A}$ of the azimuths, whereas a period of 12 months corresponds to the semi-amplitude $A_M^*$ of the variations of the azimuths. Each of these two periodicities corresponds to one of the two influences above.

[1] In the case of a semi-annual periodicity, the sinusoid considered in the calculations may equivalently be considered as having its maximum on 21 March or 21 June. The consideration of one or the other of these dates involves a simple change of sign.

The case of an annual periodicity is *entirely different*.

[2] For that matter, the deviations from this value of 79.5 can be explained by the influences of the second annual or semi-annual periodicity, at least in part.

d - The fact that the maximum speed $v_M$ is characterized by a dominant period of 6 months shows that the first influence prevails over the second. Conversely, the minimal speed $v_m$ seems to be dominated by the second influence. Similar observations can be made for the other parameters.

e - Consideration of the different results of *Table III* clearly shows that there is *a relatively complex interaction* between the two influences.

f - Overall, the correlation coefficients are *relatively high*, and in fact they are yet more significant because none of them corresponds to an isolated observation but rather to *an average of 6 or 8 observations*, since the eight fundamental *graphs* of Miller result from the averages of observations over a period of 6 or 8 days.[3]

[3] It seems to be of interest to give the results corresponding to the four indicators $v_M^2 \cos^2 \overline{A}$, $v_M^2 \sin^2 \overline{A}$, $v_m^2 \cos^2 \overline{A}$, $v_m^2 \sin^2 \overline{A}$. These indicators may be considered as mixed, since they involve, at the same time, the speeds $v_M$ and $v_m$ and also the average deviation $\overline{A}$ of the azimuths.

The coefficient of the first of these correlations is 0.99981, corresponding to a periodicity of 6 months, and the date of the minimum is *12 March*, near the spring equinox.

This result is *striking, to say the least.* In fact, if one takes Miller's estimates (Note 3 of §D.2 above) instead of my estimates in *Table II* above, one finds a correlation coefficient of 0.9999994 *with a minimum on 12 March.*

This coefficient, *virtually equal to unity*, led me to conceive that it might reflect an underlying identity *originating in the fitting methods used by Miller* for the observational data of his model. That led me to proceed to the *direct* estimates of *Table II* above, *which are totally independent of any theory, and in any case are totally independent of the estimates corresponding to the model considered by Miller* (*Chapter IV*, §F.l, p. 417 above).

*TABLE III*

**OBSERVATIONS OF MILLER DOMINANT PERIODICITIES**

| Series | P in months | R | $1-R^2$ | $s_o$ | date |
|---|---|---|---|---|---|
| $v_M$ | 6 | - 0.851 | 0.275 | minimum $s_o = 66.8$ | 0h on 8 March |
| $v_m$ | 12 | 0.946 | 0.105 | maximum $s_o = 112.6$ | 12h on 23 April |
| $\overline{A}$ | 6 | 0.840 | 0.294 | maximum $s_o = 95.4$ | 12h on 6 April |
| $A^*_M$ | 12 | - 0.973 | 0.053 | minimum $s_o = 110.8$ | 0h on 22 April |
| $v_M + v_m$ | 6 | - 0.944 | 0.110 | minimum $s_o = 62.3$ | 12h on 4 March |
| $v_M - v_m$ | 12 | - 0.9995 | 0.0011 | minimum $s_o = 80.5$ | 12h on 22 March |
| $v_m / v_M$ | 12 | 0.992 | 0.016 | maximum $s_o = 96.3$ | 12h on 7 April |
| $\overline{A} + A^*_M$ | 6 | 0.749 | 0.438 | maximum $s_o = 76.0$ | 0h on 18 March |
| $\overline{A} - A^*_M$ | 6 | 0.920 | 0.154 | maximum $s_o = 109$ | 0h on 20 April |
| $v^2_M \cos^2 \overline{A}$ | 6 | 0.99981 | 0.00038 | minimum $s_o = 70.7$ | 12h on 12 March |
| $v^2_M \sin^2 \overline{A}$ | 6 | 0.944 | 0.110 | maximum $s_o = 72.7$ | 12h on 14 March |
| $v^2_m \cos^2 \overline{A}$ | 6 | 0.787 | 0.381 | minimum $s_o = 66.4$ | 12h on 8 March |
| $v^2_m \sin^2 \overline{A}$ | 6 | 0.993 | 0.013 | minimum $s_o = 43.5$ | 12h on 13 February |
| Averages | $\lvert\overline{R}\rvert = 0.918$ | | $\overline{(1-R^2)} = 0.150$ | $\overline{s}_o = 81.8$ | 12h on 23 March |

***Legend :*** P = dominant periodicity

R = coefficient of multiple correlation of the parameter considered with the two functions of s: cos(2πs/T) and sin(2πs/T), where s is the date in question in days, from 0h on 1 January.

$s_0$ = date of the maximum or the minimum of the fitting, reckoned in days from 0h on 1 January.

Semi-annual periodicities: average $\overline{s}_0$ of the $s_0$: $\overline{s}_0 = 73.6$; 15 March.

Annual periodicities: average $\overline{s}_0$ of the $s_0$: $\overline{s}_0 = 100$; 11 April.

***Sources :*** Data in *Table II*, p. 454.

Calculations 767, 766 , 461, 459, 765, 758, 759, 757, 763, 450, 472, 473, 474. (9 August - 17 November 1995).

## *D.5 Structural interdependences between the speed and azimuth observations of Miller.*

There are *particularly prominent structural interdependences* between the different characteristics of the eight fundamental Graphs of Miller.

As an illustration, *Table IV* shows several multiple correlations between characteristic parameters of the maximum and minimum speeds $v_M$ and $v_m$ of Miller's Graphs, and the average deviations $\overline{A}$ and the amplitudes $A_M^*$ of the variations of the azimuths $A$.

*Overall, it is the interdependence between the indicators of the speeds and the maximal deviation* $A_M^*$ *that is the most striking.*

The multiple correlation coefficients are particularly elevated, and correspond to structural interdependences *whose existence appears undeniable*. These regularities *alone* are sufficient to sweep away any attribution of Miller's observations to fortuitous causes.

*TABLE IV*

**OBSERVATIONS OF MILLER**

**STRUCTURAL INTERDEPENDENCES**

***Speeds $v_M$ and $v_m$ as functions of the angles $\bar{A}$ and $A^*_M$***

| Series | Σ | R | 1-R² | m | f | g | f/g | D |
|---|---|---|---|---|---|---|---|---|
| $v_M - v_m$ | 1.47 | 0.974 | 0.050 | 3.81 | 0.41 | 1.34 | 0.31 | $A^*_M$ |
| $(v_M-v_m)/(v_M+v_m)$ | 0.11 | 0.999982 | 0.000022 | 0.22 | 0.046 | 0.095 | 0.48 | $A^*_M$ |
| $v_m^2 / v_M^2$ | 0.17 | 0.997 | 0.0051 | 0.43 | - 0.055 | - 0.155 | 0.36 | $A^*_M$ |
| $v_M^2 / (v_M+v_m)^2$ | 0.069 | 0.9997 | 0.00060 | 0.37 | 0.030 | 0.059 | 0.51 | $A^*_M$ |
| $v_m^2 / (v_M+v_m)^2$ | 0.041 | 0.9997 | 0.00066 | 0.15 | - 0.016 | - 0.036 | 0.44 | $A^*_M$ |
| | **Average of R = 0.994** | | | | **Average of 1-R² = 0.0113** | | | |

***Legend :*** Quantities considered: see *Table II* of §D.2.
The angles $\bar{A}$ and $A^*_M$ are considered *in standardized values*:
$(\bar{A} - \bar{\bar{A}})/\Sigma_{\bar{A}}$, and $(A^*_M - \bar{A}^*_M)/\Sigma_{A^*_M}$
$\bar{\bar{A}}$ = average value of $\bar{A}$ (calculation 491 of 28 August 1995) = 2.003
$\bar{A}_M$ = average value of $A_M$ (calculation 494 of 28 August 1995) = 29.957°
R = coefficient of multiple correlation of the series considered, with standardized values of $\bar{A}$ and of $A^*_M$.
m, f, and g respectively represent the average values and the regression coefficients corresponding to the angles $\bar{A} - \bar{\bar{A}}$ and $(A^*_M - \bar{A}^*_M)$.
D indicates the influence that is dominant in the correlation.

***Sources :*** Data of *Table III.*
Calculations 627$^{ter}$, 545$^{ter}$, 551$^{ter}$, 552$^{ter}$, and 553$^{ter}$ (27 October 1995).

## D.6 *Essential aspects of the semi-annual and annual structure of Miller's observations*

The above results enable us to arrive at several *absolutely essential* conclusions:

a - Miller's observations, as presented in the set of eight Graphs in his Memoir of 1933,[1] have a *very marked* semi-annual or annual structure. *This structure was not perceived by Miller.*

This periodic structure is, *by itself*, enough to eliminate all the criticisms that were leveled against Miller's observations and *that denied them any real significance.*[2]

b - The average $\overline{A}$ of the azimuths, *which Miller could not explain*, has a semi-annual periodicity, *very similar* to that of the average azimuths $\overline{\phi}$ of the paraconical pendulum with anisotropic support.

c - The average amplitude $A_M^*$ of the variations of the azimuths $A$ and the difference $v_M - v_m$ of the maximum and minimum speeds exhibit *marked annual periodicity*, and this by itself confirms the influence of the position of the Earth along its orbital trajectory.

d - The fittings corresponding to the semi-annual and annual periodicities in general have maxima and minima at dates *quite close to the spring equinox.*

e - Miller's observations are characterized by *very strong structural interdependences* between the observed speeds and azimuths.

---

[1] §C.3 of *Chapter IV* above, pp. 388-389.

[2] §E.1 of *Chapter IV* above, pp. 412-413.

f - The very strong correlations that are observed[3] seem *even more remarkable* in view of the fact that the graphical estimates of the essential characteristic parameters $\nu_M$, $\nu_m$, $\overline{A}$ and $A_M^*$ from Miller's eight fundamental Graphs can only be considered as approximate.[4]

g - *In total one conclusion appears absolutely certain, independently of any hypothesis and any theoretical interpretation. The observations of Miller incontestably present a very strong internal consistency, and this internal consistency must be considered as being directly related to the position of the Earth upon its orbit.*

---

[3] *Tables III and IV* above. See also *Tables VII, VIII, and X* of §E.2 and §E.3 below, pp. 475, 476, and 482.

[4] Particularly for 8 February 1926. See Note (1) of §D.2 above.

I cannot too strongly underline the fact that all the above calculations were made *subsequently* to the estimates of *Table II*, which were made on 29 June 1995 (see note 2 of §D.2 above, p. 453).

Analogous results can be obtained from the characteristic parameters of the elliptic fittings of *Chapter IV*, §D.4, pp. 404-408 (See the *second volume* of this work, *Chapter IV*, p. 29 above).

# E COMPARED SEMI-ANNUAL AND ANNUAL PERIODIC STRUCTURES FOR THE OBSERVATIONS OF THE PARACONICAL PENDULUM WITH ANISOTROPIC SUPPORT, FOR THE SIGHTINGS AT MARKS, FOR THE OPTICAL OBSERVATIONS OF ESCLANGON, AND FOR THE INTERFEROMETRIC OBSERVATIONS OF MILLER

## *E.1 Phases of the semi-annual periodic components of the four series of observations compared*

With the convention of reckoning the azimuths of all the angles *positively in the anticlockwise sense*, *Table V* brings out the agreements in phase of the different semi-annual fittings of the average azimuths of the paraconical pendulum, of the sightings at marks, of the observations of Esclangon, and of the average azimuths of Miller.[1]

The two first fittings are almost exactly in phase, even though they relate to *very different* phenomena and to *very different* epochs.

The average of the fittings of the average azimuths of Miller and of the vertical deviations of Esclangon is almost exactly in phase with the average of the two other fittings.[2]

Overall, in view of the very different phenomena considered and of their *very different* epochs, the observed agreements of phase are *extremely remarkable.*

---

[1] I remind the reader that, for the paraconical pendulum with isotropic support, the available data is insufficient to permit a meaningful analysis of the semi-annual periodicity (see §V.B.l, note 8, above, p. 433).

[2] I also remind the reader that, in *Graph I* (p. 434), the angles are reckoned *positively in the anticlockwise sense*. However, in *Graphs IX and XI* (pp. 449 and 458) they are reckoned *positively in the clockwise sense*.

*TABLE V*

## PARACONICAL PENDULUM WITH ANISOTROPIC SUPPORT
## SIGHTINGS AT MARKS
## OBSERVATIONS OF ESCLANGON AND MILLER

***Fittings to the semi-annual period of 182.625 days***
***Comparison of the phases of the fitting sinusoids***

*Fittings to azimuths reckoned positively in the anticlockwise sense*

| Fittings | q | Series | R | 1-$R^2$ | Date of maximum | $s_o^*$ |
|---|---|---|---|---|---|---|
| Average azimuths of paraconical pendulum 1954-1959 *Graph II* | 1 | $\bar{\phi}$ | 0.899 | 0.192 | 0h on 2 April | 91.3 |
| Sightings at marks July 1958 February-March 1959 *Graph IX* | 2 | (N+S) / 2 | 0.9991 | 0.0018 | 0h on 1 April | 90 |
| | | | | | Date of minimum | |
| Deviations of Esclangon 1927-1928 *Graph X* | 3 | E | 0.841 | 0.293 | 0h on 25 March | 82.9 |
| Average azimuths of Miller 1925-1926 *Graph XI* | 4 | $\bar{A}$ | 0.840 | 0.294 | 0h on 6 April | 95.9 |
| Date of the spring equinox: 12h on 21 March; $s_o^*$ = 79.5 | | | | | | |

***Legend :*** All the angles of series 1, 2, and 4 are reckoned positively in the anticlockwise sense. Esclangon's deviations are reckoned in the vertical sense.

Averages of the $s_o^*$ : (1+2) / 2 = 90.65 (3+4) / 2 = 89.40
(1+2+3+4) / 4 = 90.03

***Sources :*** *Graphs II, IX, X, and XI* above, pp. 435, 449, 451, and 458.

## *E.2 A double periodicity centered upon 21 March*

Because 21 March appears as a specific date in all the observational series, and the *semi-annual and annual* fittings have maxima or minima near that date,[1] it is interesting to determine the fittings corresponding *to a double periodicity, both semi-annual and annual, and centered upon 21 March.*

These fittings are determined by the correlations of the series in question with two sinusoids whose periods are six and twelve months and the dates of whose peaks are *both* 21 March.

### Paraconical pendulum with anisotropic support, sightings at marks, and optical sightings of Esclangon

1 - *Table VI* shows the results of the fittings for the average azimuths and the amplitudes of the 25h component of the paraconical pendulum with anisotropic support (1954-1960), for the sightings at marks at IRSID in 1958 and at IGN in 1959, and for the optical sightings by Esclangon in 1927-1928.

The relatively high coefficients of multiple correlation confirm the existence of semi-annual and annual periodicities centered on 21 March. The semi-annual periodicity is predominant for the average azimuths of the paraconical pendulum and for the sightings at marks.

[1] See in particular *Table III* of §D.4, p. 464, and *Table V* of §E.1, p. 470.

Overall, better fittings are obtained by associating, with a semi-annual sinusoid whose peak is at 21 March, an annual sinusoid whose peak is also at 21 March, rather than an annual sinusoid whose peak is at 21 June.[2]

[2] The following gives a comparison of the results obtained by considering an annual component having its peak at 21 March (line 1) and by considering an annual component having its peak at 21 June (line 2).

*Table VI**

| | | | $\overline{\phi}$ | $2R_{25}$ | M | E |
|---|---|---|---|---|---|---|
| 1 *Table VI* | 21 March | R | 0.876 | 0.965 | 0.995 | 0.937 |
| | | $1-R^2$ | 0.232 | 0.068 | 0.011 | 0.122 |
| 2 | 21 June | R | 0.983 | 0.735 | 0.992 | 0.850 |
| | | $1-R^2$ | 0.033 | 0.459 | 0.016 | 0.277 |

*Overall, the results are better for an annual component centered upon 21 March.*

The date of 21 June gives a better result for the average azimuths $\overline{\phi}$ (weighted values), but for this fitting we have $r/r' = -2.22$ (as against $-2.18$ for the date of 21 March, semi-annual component), so that overall the influence of the date of 21 March appears to be predominant.

*Sources* : Calculations 698, 699, 1200 and 1247 (31 October 1995 - 20 August 1996, 24 January 1997).

*TABLE VI*

**PARACONICAL PENDULUM WITH ANISOTROPIC SUPPORT SIGHTINGS AT MARKS, AND OPTICAL SIGHTINGS OF ESCLANGON SEMI-ANNUAL AND ANNUAL DOUBLE PERIODICITY**

***centered on 21 March***

| Series | Σ | R | $1 - R^2$ | m | r | r' | r/r' | P |
|---|---|---|---|---|---|---|---|---|
| $\overline{\phi}$ | 4.87 | 0.876 | 0.232 | 152 | 9.21 | -4.23 | -2.18 | 6 |
| $2R_{25}$ | 2.99 | 0.965 | 0.068 | 1.30 | -8.28 | 7.98 | -1.04 | ~ |
| Averages: | | | R = 0.920 | | | $1 - R^2 = 0.150$ | | |
| M | 37.2 | 0.995 | 0.011 | 1.87 | -37.3 | -12.4 | 3.00 | 6 |
| E | 2.55 | 0.937 | 0.122 | -0.388 | -1.48 | -2.86 | +0.52 | ~ |
| Averages: | | | R = 0.966 | | | $1 - R^2 = 0.067$ | | |
| General averages: | | | R = 0.943 | | | $1 - R^2 = 0.108$ | | |

***Legend :***

**Σ =** standard deviation of series; R = coefficient of multiple correlation of the series in question, with a sinusoid of period six months and a sinusoid of period twelve months, both having their peaks at 21 March; m = average of the fitting; r = semi-amplitude of the semi-annual component; r' = semi-amplitude of the annual component.

**P =** dominant period in months (a dash indicates that there is no dominant period, corresponding to the condition $1/2 < |r/r'| < 2$).

$\overline{\phi}$ = average azimuth of the paraconical pendulum (weighted values). (*Graph II* of §B.1 above).

**$2R_{25}$** = amplitude of the 25h periodic component of the paraconical pendulum (weighted values). (*Graph VI* of §B.2 above).

**M =** optical sightings at marks (*Graph IX* of §C.1 above).

**E =** monthly averages of the optical sightings of Esclangon from March 1927 to January 1928, i.e. 8 monthly values (see the note 1 of §C.2 above, p. 450). (*Graph X* of §C.2 above, p. 451).

***Sources :*** Calculations 499, 500, 1198, and 1246 (29 August 1995 -19 August 1996, and 23 January 1997).

## Characteristic parameters of the eight fundamental Graphs of Miller

2 - *Tables VII and VIII* show the results of multiple correlations of different indicators deduced from the characteristic parameters of the eight fundamental Graphs of Miller with a double periodicity centered on 21 March.

While in *Table III* above the fittings correspond *to a single period and to an indeterminate phase*, *Tables VII and VIII* correspond to fittings *with two periods of 6 and 12 months, both the reference sinusoids having maximum on 21 March*.

Some of the correlation coefficients are particularly high, and certainly correspond to underlying regularities.[3, 4]

This is particularly the case for the difference $v_M - v_m$, to which a particularly high correlation coefficient corresponds (as also shown in *Table III*, p. 464), and *which has only annual periodicity*. It is also *very remarkable* that the *semi-annual* components of $v_M$ and $v_m$ are practically the same, whereas the *annual* component of $v_m$ is more than five times greater than the annual component of $v_M$ (*Table VII* opposite).

---

[3] It seems interesting here to mention the multiple correlation coefficients corresponding to $v_M \cos\overline{A}$ ($R = 0.977; 1 - R^2 = 0.0462$), $v_M \sin\overline{A}$ ($R = 0.903; 1 - R^2 = 0.184$), $v_m \cos\overline{A}$ ($R = 0.99999990; 1 - R^2 = 0.000000019$), $v_m \sin\overline{A}$ ($R = 0.765; 1 - R^2 = 0.415$). The average of these correlation coefficients ($\overline{R} = 0.911$) is somewhat lower than the average of the correlation coefficients corresponding to their squares ($\overline{R} = 0.968$).

For the ratios $r/r'$ for these four correlations respectively, we have the values: $-8.2$, $-3.3$, $-2.6$, and $-9.6$, *which correspond to a very marked predominance of the six month period*.

*Sources*: Calculations 835, 833, 834, and 836 (14 December 1995).

[4] The coefficient of multiple correlation is higher for $\tan\overline{A}$ ($R = 0.983; 1 - R^2 = 0.0338$) than for $\overline{A}$ ($R = 0.943; 1 - R^2 = 0.110$).

*TABLE VII*

## OBSERVATIONS OF MILLER

***Characteristic parameters of the eight fundamental Graphs***

**SEMI-ANNUAL AND ANNUAL DOUBLE PERIODICITY**

***centered on 21 March***

| Series | Σ | R | $1\text{-}R^2$ | m | r | r' | r/r' | P |
|---|---|---|---|---|---|---|---|---|
| $v_M$ | 0.73 | 0.833 | 0.307 | 10.9 | -1.13 | -0.263 | 4.3 | 6 |
| $v_m$ | 1.38 | 0.952 | 0.093 | 6.9 | -1.03 | 1.44 | -0.71 | ~ |
| $\overline{A}$ | 25.9 | 0.943 | 0.110 | 2.0 | 45.6 | -13.3 | -3.43 | 6 |
| $A^*_M$ | 5.7 | 0.944 | 0.109 | 30.0 | -2.2 | -6.1 | +0.36 | 12 |
| Averages : $\overline{R} = 0.918$ $\overline{1\text{-}R^2} = 0.155$ | | | | | | | | |

***Legend :***

$v_M$, $v_m$, $\overline{A}$ and $A^*_M$ : *Table II*

The values of $v_M$, $v_m$, $\overline{A}$, $A^*_M$ correspond to the averages of the observations over 6 or 8 days (Miller, 1933, id., p. 213).

**Σ** = standard deviation of series; R = coefficient of multiple correlation of the series in question, with a sinusoid of period six months and a sinusoid of period twelve months, both having peak at 21 March; m = average of the fitting; r = semi-amplitude of the semi-annual component; r' = semi-amplitude of the annual component.

**P** = dominant periodicity in months (a ~ indicates that there is no dominant period, corresponding to the condition $1/2 < |r/r'| < 2$).

***Sources*** : Data of *Table II* (graphic estimations of 29 June 1995 from the eight fundamental Graphs of Miller). Calculations 582, 583, 491, and 494 (28 August - 13 November 1995).

*TABLE VIII*

## OBSERVATIONS OF MILLER

*Characteristic parameters of the eight fundamental Graphs*

**SEMI-ANNUAL AND ANNUAL DOUBLE PERIODICITY**

**centered on 21 March**

*Mixed correlations*

| Series | Σ | R | 1-R² | m | r | r' | r/r' | P |
|---|---|---|---|---|---|---|---|---|
| $v_M + v_m$ | 1.64 | 0.865 | 0.252 | 17.82 | -2.158 | 1.179 | -1.83 | 6 |
| $v_M - v_m$ | 1.47 | 0.99992 | 0.00016 | 3.98 | -0.096 | -1.705 | 0.057 | 12 |
| $v_m / v_M$ | 0.14 | 0.990 | 0.0205 | 0.64 | -0.041 | 0.160 | -0.27 | 12 |
| $tg\,\overline{A}$ | 0.624 | 0.983 | 0.0338 | 0.321 | 1.123 | -0.359 | -3.13 | 6 |
| $\overline{A} + A^*_M$ | 27.0 | 0.969 | 0.061 | 31.96 | 43.4 | -19.4 | -2.24 | 6 |
| $\overline{A} - A^*_M$ | 26.0 | 0.915 | 0.163 | -27.95 | 47.9 | -7.22 | -6.63 | 6 |
| $v^2_M \cos^2 \overline{A}$ | 36.6 | 0.974 | 0.051 | 114 | -73 | 4.7 | -15 | 6 |
| $v^2_M \sin^2 \overline{A}$ | 25.1 | 0.99990 | 0.00020 | 5.7 | 49 | -11 | -4.6 | 6 |
| $v^2_m \cos^2 \overline{A}$ | 16.7 | 0.9990 | 0.0019 | 45.2 | -25 | 14 | -1.8 | 6 |
| $v^2_m \sin^2 \overline{A}$ | 9.4 | 0.897 | 0.195 | 2.7 | 15.6 | 3.8 | +4.1 | 6 |
| Averages : | $\overline{R} = 0.960$ | | | | $\overline{1-R^2} = 0.080$ | | | |

*Legend :*

Σ, R, $1\text{-}R^2$, m, r, r', r/r', P : see the Legend of *Table VII*.

*Sources :*

Data of *Table II* (graphic estimations of 29 June 1995 from the eight fundamental Graphs of Miller). Calculations 768, 612, 607, 555, 769, 770, 497,492, 489, and 498 (28 August - 15 November 1995).

It is certainly *very significant* that, for the annual period, the phase corresponding to 21 March *in general* gives better results than the phase corresponding to 21 June.[5]

*The predominance of the periodicities of 6 months is obviously a result of a strong influence of anisotropy corresponding to the average deviations* $\overline{A}$ *of the azimuths.*[6]

---

[5] The following shows the results of *Tables VII and VIII* (line 1 below) compared with those observed by associating a sinusoid of period equal to one year having its maximum on *21 June* (line 2) to the 6 month period whose maximum is on 21 March:

*Table VII**

| | | | $v_M$ | $v_m$ | $\overline{A}$ | $A^*_M$ |
|---|---|---|---|---|---|---|
| 1 *Table VII* | 21 March | R | 0.833 | 0.952 | 0.943 | 0.944 |
| | | $1-R^2$ | 0.307 | 0.093 | 0.110 | 0.109 |
| 2 | 21 June | R | 0.9995 | 0.383 | 0.980 | 0.314 |
| | | $1-R^2$ | 0.00091 | 0.883 | 0.039 | 0.901 |

*Table VIII**

| | | | $v_M \cdot v_m$ | $v_m / v_M$ | $v_M^2 \sin^2 \overline{A}$ | $v_m^2 \cos^2 \overline{A}$ |
|---|---|---|---|---|---|---|
| 1 *Table VIII* | 21 March | R | 0.99992 | 0.990 | 0.99990 | 0.9990 |
| | | $1-R^2$ | 0.00016 | 0.0205 | 0.00020 | 0.0019 |
| 2 | 21 June | R | 0.525 | 0.412 | 0.954 | 0.782 |
| | | $1-R^2$ | 0.724 | 0.830 | 0.089 | 0.388 |

The date of 21 June for the annual periodicities gives the best results in only two cases, $v_M$ and $\overline{A}$. Yet it must be emphasized that for $v_M$ we have $r/r' = -1.12$, and for $\overline{A}$ we have $r/r' = 1.80$, which implies for $v_M$ an equal influence of the two dates, and for $\overline{A}$ a greater influence of the date of 21 March for the semi-annual component.

In summary, the influence of the spring equinox appears to be dominant.

*Sources*: Calculations 691, 692, 693,694,696, 695, 702, and 703 (31 October 1995).

[6] Since the interferometer measures the squares of the speeds, we may naturally ask if the temporal correlations are more prominent for the squares $v_M^2$ and $v_m^2$ than for $v_M$ and $v_m$. In fact we have, for the multiple correlations

| | $v_M$ | $v_M^2$ | $v_m$ | $v_m^2$ | $v_M - v_m$ | $v_M^2 - v_m^2$ | $v_m/v_M$ | $v_m^2/v_M^2$ |
|---|---|---|---|---|---|---|---|---|
| $R$ | 0.833 | 0.833 | 0.952 | 0.954 | 0.99992 | 0.985 | 0.990 | 0.991 |
| $1-R^2$ | 0.307 | 0.306 | 0.093 | 0.089 | 0.00016 | 0.029 | 0.0205 | 0.017 |

Overall, the results are *practically equivalent*.

Sources for the calculations relating to $v_M^2$, $v_m^2$, $v_M^2 - v_m^2$, $v_m^2/v_M^2$ : Calculations 705, 706, 708, and 707 (2 November 1995).

## Magnetic activity and temperature

3 - It is interesting to compare these various results with those corresponding to other geophysical series, such as those of magnetic activity or those of temperature.[7]

[7] This is the equivalent of *Tables VI and VII* for the index U of magnetic activity and for the temperatures at Paris, Boston, and San Francisco:
*Table VI***

MAGNETIC ACTIVITY AND TEMPERATURES

SEMI-ANNUAL AND ANNUAL DOUBLE PERIODICITY

*Centered on 21 March*

| Series | $\Sigma$ | $R$ | $1-R^2$ | $m$ | $r$ | $r'$ | $r/r'$ | $P$ |
|---|---|---|---|---|---|---|---|---|
| U | 9.57 | 0.832 | 0.308 | -0.095 | 10.5 | -3.97 | -2.64 | 6 |
| Temperatures at Paris | 5.54 | 0.557 | 0.690 | 11.6 | 0.085 | -4.35 | -0.019 | 12 |
| Temperatures at Boston | 8.73 | 0.554 | 0.693 | 9.26 | -0.26 | -6.84 | 0.039 | 12 |
| Temperatures at San Francisco | 1.83 | 0.734 | 0.462 | 13.2 | -0.61 | -1.80 | -0.34 | 12 |
| Temperatures: Averages: $\overline{R}=0.615$ $\overline{1-R^2}=0.615$ $\overline{r/r'}=0.615$ | | | | | | | | |

*Legend*: For $\Sigma$, $R$, $m$, $r$, $r'$, $P$, see the *Legend of Table VI.*

U = magnetic activity. The twelve monthly values from January to December are the following (Bartels and Chapman, 1940, *Geomagnetism*, Vol. I, pp. 365-366 [61]): -15, -2, +10, 1, 0, -12, -11, 0 , +12, +14, +10, -8. These averages correspond to the period of 59 years from 1872 to 1930.

- *Temperature at Paris*: monthly averages for 1958 at the Observatory of Montsouris.
- *Temperatures at Boston and at San Francisco*: *Smithsonian Physical Tables*, 1934, p. 556 [2]. The period for which the averages are calculated is not given.

*Sources* : Calculations 493, 604, 508, and 509 (28 August - 16 October 1995).

The value U of the magnetic activity has an *entirely different* periodic structure from the periodic structure of the temperatures. The semi-annual periodic component of U is very strongly marked, while for the temperatures it is completely negligible.[8]

While the semi-annual periodicity of the magnetic activity U is quite comparable to those of the series of observations relating to the paraconical pendulum, to the optical deviations of the sightings at marks, to the experiments of Esclangon, and to the experiments of Miller,[9] the situation is quite different for the temperatures considered in different locations. [10]

---

[8] This difference would be even more marked if we were to compare the fittings *corresponding to semi-annual and annual periodicities* for the magnetic activity and the temperatures, with the dates of the peaks of the sinusoids remaining entirely free.

In fact, the temperatures are characterized by a *much more marked* annual periodicity for a sinusoid having its maximum or minimum on 21 June, than for a sinusoid having its maximum or minimum on 21 March.

[9] See in particular *Tables VI, VII, and VIII* above, pp 473, 475, and 476.

[10] The periodic structure of the series relative to the temperatures seems to be *incompatible* with an explanation for the series studied in this work based upon the effects of temperature.

## *E.3 A simplified global approach: consideration of only dominant periodicities of the four series of observations, with their maxima or minima on 21 March*

### Consideration of only dominant periodicities with maxima or minima on 21 March

1 - For lack of space, a deep comparative analysis of the periodic structures of the observations of the paraconical pendulum with anisotropic support, of the sightings at marks, of the optical observations of Esclangon, and of the interferometric observations of Miller cannot be presented here.

The simplest approach is undoubtedly to consider only the dominant periodicities associated with maxima or minima on 21 March. The corresponding results are given in the two *tables IX and X.*

Although the fittings only consider *a single sinusoid of reference having semi-annual or annual period and having its peak on 21 March,* all the correlation coefficients remain relatively high.[1, 2] Their significance is all the greater because the parameters considered do not correspond to isolated observations, but rather to averages of *very numerous observations.*[3]

---

[1] This observation particularly holds for the characteristic indicators of Miller's *Graphs* if we compare *Table X* with *Tables III, VII, and VIII.*

The correlation coefficients for $v_M$, $v_M + v_m$, $\overline{A} + A^*_M$ that are less than 0.8 *may be explained by the fact that the influence of the periodicity of 12 months, which has been omitted, cannot be considered as being of little importance.*

[2] As is indicated in *Chapter VI,* Sections B and C of the *second volume* of this work, p. 30 above, the statistical significance of the complete set of correlation coefficients of *Tables IX and X* is extremely high: the probability of these correlation coefficients *being simultaneously obtained* by chance with the same explanatory cause is in fact less than $10^{-10}$.

[3] It should be remarked here that, in the theory of luni-solar forces, the semi-annual periodic component has a much greater coefficient than that of the periodic annual component. (0.0728 versus 0.018). (Paul Schureman, 1941, Manual of Harmonic Analysis and Prediction of Tides, p. 164 [244]).

*TABLE IX*

## PARACONICAL PENDULUM WITH ANISOTROPIC SUPPORT AND OPTICAL SIGHTINGS OF ESCLANGON SEMI-ANNUAL PERIODICITIES

***Fittings to a sinusoid of period 6 months and having peak on 21 March***

| Series | P | Σ | R | $1 - R^2$ | m | r | σ |
|---|---|---|---|---|---|---|---|
| $\overline{\phi}$ | 6 | 4.87 | 0.849 | 0.278 | 151 | 6.60 | 2.57 |
| $2\,R_{25}$ | 6 | 2.99 | - 0.703 | 0.506 | 3.42 | - 3.35 | 2.12 |
| Averages : | | $\lvert\overline{R}\rvert = 0.776$ | | $\overline{1\text{-}R^2} = 0.392$ | | | |
| M | 6 | 37.2 | - 0.992 | 0.016 | - 0.175 | - 47.9 | 4.81 |
| E | 6 | 2.55 | - 0.836 | 0.302 | - 1. 55 | - 3.12 | 1.40 |
| Averages : | | $\lvert\overline{R}\rvert = 0.914$ | | $\overline{1\text{-}R^2} = 0.159$ | | | |
| General averages : | | $\lvert\overline{R}\rvert = 0.845$ | | $\overline{1\text{-}R^2} = 0.275$ | | | |

***Legend :***

**P , Σ , R , $1\text{-}R^2$, m , r , σ** : see the *Legend of Table VI.*

$\overline{\phi}$ = average azimuth of the paraconical pendulum with isotropic support in degree (*Table I* of *Chapter I*). The correlation is *weighted* as mentioned in §B.1 (data of *Graph II* of §B.1 above).

**$2R_{25}$** = amplitude of the 25h periodic component of the paraconical pendulum (*Table I* of *Chapter I*). The correlation is *weighted* as mentioned in §B.2 (data of *Graph VI* of §B.2 above).

**M** = sightings at marks, in centesimal seconds (Data of *Graph IX* of §C.1 above).

**E** = monthly averages of the optical sightings of Esclangon in thousandths of a minute of arc, from March 1927 to July 1927 and from November 1927 to January 1928, i.e. eight monthly values (§C.2 above, note 1, p. 450).

***Sources :*** Calculations 724, 725, 1201, and 722 (6 November 1995 - 21 August 1996).

*TABLE X*

## OBSERVATIONS OF MILLER
## DOMINANT SEMI-ANNUAL OR ANNUAL PERIODICITIES
### Fittings to a sinusoid of period 6 or 12 months and having peak on 21 March

| Series | P | Σ | R | $1-R^2$ | m | r | σ |
|---|---|---|---|---|---|---|---|
| $v_M$ | 6 | 0.73 | - 0.772 | 0.404 | 10.9 | - 1.15 | 0.46 |
| $v_M + v_m$ | 6 | 1.64 | - 0.607 | 0.632 | 17.6 | - 2.05 | 1.30 |
| $\overline{A}$ | 6 | 25.9 | 0.834 | 0.305 | 2.3 | 44.4 | 14.3 |
| $\overline{A} + A^*_M$ | 6 | 27.0 | - 0.744 | 0.447 | 32.4 | 41.5 | 18.1 |
| $\overline{A} - A^*_M$ | 6 | 26.0 | 0.880 | 0.225 | - 27.8 | 47.1 | 12.3 |
| Averages : $\overline{\vert R \vert} = 0.767$ $\overline{1-R^2} = 0.403$ | | | | | | | |
| $v_m$ | 12 | 1.38 | 0.880 | 0.225 | 6.4 | 1.41 | 0.65 |
| $v_M - v_m$ | 12 | 1.47 | - 0.9994 | 0.0012 | 3.94 | - 1.71 | 0.050 |
| $v_m / v_M$ | 12 | 0.14 | 0.980 | 0.041 | 0.62 | 0.159 | 0.028 |
| $A^*_M$ | 12 | 5.73 | - 0.924 | 0.145 | 28.9 | - 6.2 | 2.18 |
| Averages : $\overline{\vert R \vert} = 0.946$ $\overline{1-R^2} = 0.1030$ | | | | | | | |
| General averages : $\overline{\vert R \vert} = 0{,}847$ $\overline{1-R^2} = 0.269$ | | | | | | | |

*Legend :*

P, Σ, R, $1-R^2$, m, r, σ : see the Legend of *Table VI.*

***Sources :*** Table II, and Calculations 726, 727, 778, 729, 728, 720, 719, 779, and 780 (7-17 November 1995).

## *E.4 A triple conclusion*

All these results, together with those of the previous Chapters, lead *with absolute certainty* to a triple conclusion:

• *First*, there exists *a very great coherence* between the observations of the paraconical pendulum, the optical observations of sightings at marks, the optical observations of Esclangon, and the interferometric observations of Miller.

• *Second*, it is *completely impossible* to attribute this great coherence to random causes.

• *Third*, the four series of observations all demonstrate a *very strong correlation* with the position of the Earth upon its orbit.

## F OVERALL VIEW OF THE DIURNAL, MONTHLY, SEMI-ANNUAL, ANNUAL, AND LONG DURATION PERIODIC STRUCTURE OF THE OBSERVATIONS OF THE PARACONICAL PENDULUM WITH ANISOTROPIC SUPPORT, OF THE SIGHTINGS AT MARKS, OF THE OPTICAL SIGHTINGS OF ESCLANGON, AND OF THE INTERFEROMETRIC OBSERVATIONS OF MILLER

What can be deduced from the periodic structures of the observations of the paraconical pendulum with anisotropic support, of the optical observations of sightings at marks that were performed in the experiments at IRSID and at IGN, of the optical observations of Esclangon, and of the interferometric observations of Miller, analyzed in the three preceding Chapters I, III, and IV, and in this Chapter?

## *F.1 Similarities between the four series of experiments*

First of all it is right to underline that, as contrasted with all known previous experiments, *all these experiments*, with the exception of those of Esclangon, were the only ones that were performed *in a continuous manner day and night without interruption.* As for Esclangon's experiments, in spite of the numerous gaps that they include, they correspond to a very great number of observations made on almost every day during almost an entire year.

The experiments on the paraconical pendulum with anisotropic support were much more extended than those of Miller. Miller's experiments were in fact only carried out during four periods of 6 or 8 days during a single year, whereas the experiments on the paraconical pendulum with anisotropic support were carried out during seven one-month series over several years from 1954 to 1960.[1]

The parameters that characterize each of these series of experiments are thus deduced from a *considerable* number of observations. *The correlations that correspond to them thus are based, not upon isolated measurements, but upon average characteristics each of which has very great significance.*

There were *considerable difficulties in performing these observations* day and night. Commenting upon other interferometric experiments, Miller so truly wrote:[2]

> *"In none of these other experiments have the observations been of such extent and of such continuity as to determine the exact nature of the diurnal and seasonal variations."*

Undoubtedly this explains why Miller's experiments were not repeated, just as my own experiments on the paraconical pendulum have not been repeated by other experimenters.

---

[1] Furthermore it must be underlined that, with the paraconical pendulum with anisotropic support, numerous series of experiments were also performed over shorter periods of one or two weeks.

[2] Miller, 1933, id., p. 240 [203].

Analogous remarks can be made for the series of optical observations performed at IRSID in 1958 and at IGN in 1959. But in these two cases, continuous observations were only made over a single month.

The same holds for the observations of Esclangon, which were carried out during an entire year, but with very many gaps.

In fact, all these experiments were characterized, *not only by the diurnal and monthly periodicities* analyzed in the previous Chapters, *but also by semi-annual and annual, and even planetary periodicities.*[3]

It is very remarkable that neither Miller nor Esclangon thought of determining the semi-annual and annual structure of their results. In fact, the results of the harmonic analyses I have performed upon the experiments of Miller and Esclangon *strongly tend to reinforce their validity and their scientific importance*. They certainly cannot be attributed to purely fortuitous circumstances.

[3] See §B. 1 and B.2 above, pp. 432-443.

## *F.2 Coherence of the periodic structures of the four series of observations*

All the analyses performed, both in the previous Chapters and in this Chapter, bring out *a very great coherence*, not only within each series of experiments, but also between the various series.

• *Thus these analyses lead us to the conclusion that the phases of the semi-annual components are relatively close to the spring equinox on 21 March, and that the same is very generally true for the phases of the annual components.*

The agreements in phase that appear for all the experiments considered are very remarkable. *They could not be accidental, and they undeniably correspond to an underlying reality.*[1]

• *In the same way, both for the experiments of Esclangon and for the experiments of Miller, consideration of sidereal time instead of civil time over the same period of a year* also brings out *very remarkable diurnal sidereal periods.*[2]

• And, in the same way, consideration of the periodicities of 6 months and 5.9 years has enabled us to explain *the differences over time* in the average azimuths of the paraconical pendulum with anisotropic support and in the amplitudes of the 25h wave (which represents the wave of 24h50m), which for a long time appeared to me to be inexplicable.[3]

• At first sight, each correlation might appear to have only limited statistical significance[4], but in reality this is not true at all, because each value of the parameter in question is actually *the average of a very great number of observations*. In any event, *the ensemble* of the correlations obtained has a *very great* statistical significance, which in reality amounts to certainty.[5]

---

[1] See in particular §B.4 and E.1 above, pp. 446 and 469-470.

[2] *Chapter IV* above.

[3] §B.1 above, pp. 432-437.

[4] This is particularly the case when four average observations are correlated with a sinusoid whose period is 6 or 12 months. In fact, these correlations include three arbitrary parameters: the phase, the amplitude, and a constant.

[5] See *Chapter VI*, Sections B and C, of the *second volume* of this work (p. 30 above).

• In the same way, analysis of the characteristic parameters of Miller's observations brings out a *very strong structural coherence* between the minimum and maximum speeds and the average azimuths $\overline{A}$ and the amplitudes $A_M^*$ of the diurnal fluctuations of the azimuths $A$ observed at each instant.[6]

• In the same way again, *analysis of the hodographs* corresponding to Miller's observations brings out *a very remarkable underlying coherence.*[7]

• *In summary, a hidden order appears under the great diversity of phenomena in question, and relations are uncovered between phenomena which until now have appeared to be totally independent.*

---

[6] §D.5 above, pp. 465-466.

[7] *Chapter IV*, §D.4, pp. 404-408.

## *F.3 Inescapable prolegomena for any synthetic theory of the four series of analyzed observations*

Whatever may be the final overall interpretation of the four series of observations *of the paraconical pendulum with anisotropic support, the sightings at marks, the optical deviations of Esclangon, and Miller's interferometer*, analyzed in this Chapter and in the three preceding Chapters I, III, and IV, as far as their *diurnal, monthly, semi-annual, annual, and longer duration* periodic structures are concerned, the fact is that the temporal and structural correlations brought to light constitute *inescapable* prolegomena for the elaboration of a synthetic and coherent theory of all the observations performed, which correspond to a reality that is *certainly very complex.*[1]

*One fact is incontestable: any theory must be rejected that is based upon foundations that are incompatible with the found temporal periodicities and structural correlations.*

The significance and the implications of the preceding analysis can be particularly illustrated in the light of two aspects: *the anisotropy of space, and the determination of the position of the Earth upon its orbit by purely terrestrial experiments.*

[1] I remind the reader that, due to the lack of sufficient data, no analysis of the semi-annual and annual periodicities of the observations of the paraconical pendulum with isotropic support (*Chapter II*) could be made (see note 8, p. 433 above).

## *F.4 The anisotropy of space*

In all cases - the observations of the paraconical pendulums with anisotropic support and with isotropic support, the sightings at marks and at collimators, the optical observations of Esclangon, and the interferometric observations of Miller - the analysis of the empirical data leads to the same conclusion: *space is anisotropic, and, as far as can be judged, this anisotropy of space results from the influence of the heavenly bodies.*

In the case of the paraconical pendulum, this anisotropy can be explained by supposing that *inertial mass is not the same in all directions.* In the cases of the optical observations, and in particular with the interferometric observations of Miller,[1] the experiments show that *the speed of light is not the same in all directions.* Naturally these two phenomena are only two different aspects *of the same anisotropy of space.*

With regard to contemporary theories, the anisotropy of space is so important that the following *Chapter VI* is entirely devoted to that subject.

[1] I remind the reader that the interferometer makes it possible to measure the difference between the speeds of light in two perpendicular directions.

## F.5 The determination of the position of the Earth upon its orbit by purely terrestrial experiments

The analysis given in this Chapter shows that the characteristic parameters of Miller's Graphs exhibit *a strong correlation with the position of the Earth upon its orbit.* These correlations result from consideration *only* of the observations of the azimuths and speeds deduced from the experiments, independently of *any hypothesis and of any theoretical interpretation.*[1]

*The same conclusion can be made* for the observations of the paraconical pendulum with anisotropic support, from the optical observations of sightings at marks, and from the observations of Esclangon.[2]

From all the evidence, these are *absolutely essential* results from the point of view of contemporary theories, and particularly that of the *Theory of Relativity, one of whose postulates being that no purely terrestrial experiment can determine the position of the Earth upon its orbit.*

*The very foundations of the Theories of Special and General Relativity* in fact rest, in the last analysis, *upon a single experimental result*: the results of the interferometric experiment of Michelson and of subsequent experiments, which are considered as having been "*negative*". In reality, and as follows from the preceding analysis,[3] *Miller's experiments of 1925-1926 demonstrated that this was not the case at all.*

But this conclusion does not derive only from Miller's observations. *For the same reasons, it also* results from the observations of the paraconical pendulum with anisotropic support, from the optical observations of sightings at marks, and from the observations of Esclangon, *due to their semi-annual and annual periodicities.*[4]

---

[1] §D.6 above, pp. 467-468.

[2] Sections B et C above, pp. 432-451.

[3] See §D.3 above, pp. 455-461, and *Chapter IV*, pp. 373-428.

[4] Sections B et C above, pp. 432-451.
For the paraconical pendulum with isotropic support, see note 8 of §V.B.1 above, p. 433.

Here, it is appropriate to repeat what Henri Poincaré wrote about the rotation of the Earth:[5]

> *"If the sky were forever covered with clouds, and if we had no means of observing the heavenly bodies, we might, nevertheless, conclude that the Earth rotates; we should be apprised of this fact by its flattening at the poles, or by the experiment of Foucault's pendulum...*
>
> *"Let us resume our imaginary story: thick clouds hide the heavenly bodies from human beings who cannot observe them and are even ignorant of their existence; how would those men know that the Earth rotates? No doubt, for longer than did our ancestors, they would regard the soil on which they stood as fixed and immovable; they would wait much longer than we did for the coming of a Copernicus. But finally this Copernicus would come. How would he come?"*
>
> *"(From complications to further complications) They would invent something no more extraordinary than the glass spheres of Ptolemy... until the long-awaited Copernicus would tell them: "It is much more simple to admit that the Earth rotates."*

It is clearly the same for all the semi-annual and annual periodicities that I have demonstrated. If the sky remained hidden by thick clouds, these periodicities would one day cause mankind to think that the Earth is moving.[6]

*With regard to contemporary theories, this possibility of determining the position of the Earth upon its orbit by purely terrestrial experiments has such an exceptional importance that a large part of Chapter VII will be devoted to that fact.*

---

[5] Henri Poincaré, 1905, *La science et l'hypothèse* (Science and Hypothesis), id., pp. 138-141 [220].

[6] But this would undoubtedly be a much easier task than in the case of the Foucault pendulum. The speed of rotation of the plane of oscillation of a pendulum is in fact equal to $\omega \sin \lambda$, where $\omega$ is the speed of rotation of the Earth and $\lambda$ is the latitude. From this it follows that, at the latitude of IRSID in Saint Germain, the period of rotation of the Foucault pendulum is not equal to 24h, but to $24/\sin 48.9° = 31.85$h.

As far as I know, this is *the only periodic physical phenomenon for which the period of the effect is considerably different from the period of the cause*.

This means that it would be rather difficult for Poincaré's Copernicus to deduce the rotation of the Earth in 24h from the local rotation of the plane of oscillation of a Foucault pendulum at a given latitude.

# Chapter VI
# THE ANISOTROPY OF SPACE

*"The conclusion of all this is that one must efface one's own opinion as well as those of others in front of the decisions of experiment...*

*"The art of scientific investigation is the cornerstone of all the experimental sciences. If the facts that serve as the basis of reasoning are badly established or erroneous, everything will collapse or become false; and this is why, most usually, errors in scientific theories owe their origin to errors of fact...*

*"The theories that represent the assemblage of our scientific ideas are undoubtedly indispensable for representing science. They should also serve as a point upon which new investigative ideas can find purchase. But these theories and these ideas are by no means immutable truths; one must always be ready to abandon them, to modify them, or to change them as soon as they no longer represent reality. In a word, theory must be modified to adapt it to nature, not nature to adapt it to theory."*

Claude Bernard *

## A VERY SIGNIFICANT CONNECTIONS BETWEEN APPARENTLY VERY DIFFERENT PHENOMENA

All the experiments examined in this memoir lead irresistibly to the same conclusion: *the space in which matter swims is not isotropic*. This conclusion is *imposed* by the observational data.[1, 2]

For all these observations - *i.e.* the observations of the paraconical pendulum with anisotropic support and with isotropic support, the optical sightings at marks and at collimators, the optical observations of Esclangon, and the interferometric observations of Miller - the following facts hold: *they all correspond to very real phenomena and cannot be attributed to perverse effects; they are all totally inexplicable in the framework of currently accepted theories; they all show agreements in phase in their periodic structure; and they all imply anisotropy of space*. These four characteristics have continued to *appear* in all the foregoing developments.

The agreements in phase that appear must certainly betray *a deep underlying unity*. There cannot be a gravitational space, an optical space, an electromagnetic space, etc. There can only be one and the same space, and *this fact is the explanation of the agreements in phase detected between phenomena that, at first sight, appear very different*.

---

* Claude Bernard, 1865, *Introduction à l'étude de la médecine expérimentale* (Introduction to the study of experimental medicine), Garnier-Flammarion, 1966, pp. 72, 42, and 73 [69].

[1] Certain people may say that this is only a hypothesis. But this hypothesis is absolutely analogous to the hypothesis that the Earth is a spheroid. It is *inescapable*.

[2] As I have already done in the preceding Chapters, in the following I will be lead to refer to numerous citations.

I do not do this in order to place myself under any sort of authority, but simply *because these citations correspond very exactly to my own thoughts*, and because the custodians of official thought will find it difficult to contest their validity.

The *daily* variations of the azimuths and speeds as measured *by Miller* with his interferometer are characterized, for example, by a *coherent underlying* structure when consideration of sidereal time is substituted for that of civil time.[3, 4] The same is true for the experiments *of Esclangon* at the Observatory of Strasbourg in 1927-1928.[5]

---

[3] *Chapter IV*, § D.1 above, pp. 392-393.

[4] It may even be possible to establish a direct relationship between Miller's unexplained average azimuths $\overline{A}$, and the average azimuths $\overline{\phi}$ of the paraconical pendulum with anisotropic support.
See *Chapter V* of *the second volume of this work* (p. 29 above).

[5] *Chapter IV*, § B.2 above, pp. 378-381.

# B THE INTERPRETATION OF THE OBSERVATIONAL DATA

## B.1 Three guidelines

Ever since I have been reflecting upon the implementation of my own experiments on the paraconical pendulum with anisotropic support and with isotropic support, upon the optical deviations of sightings at marks, upon the optical deviations of sightings at marks and at collimators that followed, upon the experiments of Esclangon and Miller and their results, and upon the deep nature of the physical problems to which they correspond, three guiding ideas have been borne in upon me more and more strongly: *the existence of an anisotropy of space; the determination of this anisotropy of space due to astronomical influences; and the existence of an intermediate medium.*

### The anisotropy of space

1 - The first guiding idea was that, *whatever the cause may be*, the fundamental result suggested by my own experiments on the paraconical pendulum with anisotropic support and with isotropic support at Saint-Germain (IRSID) and at Bougival from 1954 to 1960, by the experiments on sightings at marks performed at IRSID in 1958 and by the experiments on sightings at marks and at collimators at IGN in 1959, by the experiments of Esclangon in 1926-1927, and by the interferometric experiments of Miller in 1925-1926, as well as by a great number of other experiments,[1] is the existence of *an anisotropy of space*.

*Thus for the paraconical pendulum with isotropic support, for example, a consequence of this anisotropy is that there exists, at any moment, a privileged direction towards which the plane of oscillation of the pendulum tends to displace itself while being observed over a period of fourteen minutes, and this direction varies over time.*

[1] See Allais, 1958, *Doit-on reconsidérer les lois de la gravitation ?* (Should the laws of gravitation be reconsidered?), § C, pp. 102-103 [23].

## Astronomical influences and the anisotropy of space

2 - The second guiding idea was that this spatial anomaly results from astronomical influences, *notably*:

- the relative movements of the Sun and the Moon due to the diurnal rotation of the Earth;
- the position of the Earth on its orbital trajectory around the Sun;
- the influences of the movements of the planets;
- the overall movement of the solar system with respect to the fixed stars.

## Existence of an intermediate medium

3 - The third guiding idea was the hypothesis of *the existence of an intermediate medium*[2] that serves to support all physical phenomena, and whose structural modifications and movements entail an anisotropy of space and influence various physical phenomena, *resulting in phase agreements for all those phenomena.*[3]

## Hypotheses and Experiment

4 - *A priori*, these three guiding ideas *can merely be viewed as hypotheses*, but *the scientific value* of my experiments on the paraconical pendulum with anisotropic support and with isotropic support, of my experiments on optical deviations of sightings at marks, of the experiments at IGN on optical deviations of sightings at marks and at collimators, of the optical experiments of Esclangon, and of the interferometric experiments of Miller, *is totally independent of those guiding ideas*. In all events, they cannot fail to suggest useful reflections to the reader.

---

[2] On this intermediate medium, see § B.4 below, pp. 506-509.

[3] It was this conviction that led me, in 1958, to perform my optical experiments on sightings at marks in parallel with my experiments with the paraconical pendulum.

*In fact, the results that were obtained verified my conviction in an outstanding fashion, although nobody had previously thought of making such a linkage.*

See above, note 1, pp. 332-333.

These guiding ideas may appear completely natural, or on the contrary very debatable, but the validity and the value of the experiments are *totally independent* of this judgment. As Max Planck emphasized long ago:[4]

> *"The scientific value of precise experiments is independent of their theoretical interpretation."*

Claude Bernard underlined the same idea with very great force:[5]

> *"It is always essential to distinguish two things in experimental criticism: the fact of experiment, and its interpretation. Before everything, science insists that everybody must agree on the facts, because they constitute the basis upon which reasoning is to be performed. As for interpretations and ideas, they may vary, and it is even a good thing that they should be in dispute, because such discussions lead to other researches being undertaken and to new experiments being performed."*

---

[4] Max Planck, 1925, *Initiations à la Physique* (Introduction to Physics), Flammarion, 1941, p. 256 [218].

[5] Claude Bernard, 1865, *Introduction à l'étude de la médecine expérimentale* (Introduction to the Study of Experimental Medicine), Garnier-Flammarion, 1966, p. 263 [69].

## *B.2 The anisotropy of space*

### Observations on the paraconical pendulum with anisotropic support and with isotropic support

1 - The diurnal periodic effects detected in the movements of the paraconical pendulum with anisotropic support and of the paraconical pendulum with isotropic support are from twenty to a hundred million times greater than those deduced from current theories, and are *totally inexplicable* by those theories.[1]

Still, is it necessary to reject those theories entirely? *Certainly not.*

The experiments that I performed with the paraconical pendulum, for example, have made it possible to verify *comprehensively and quantitatively* the value of the rotation of the plane of oscillation of the pendulum under the influence of the rotation of the Earth, *provided that its trajectory remains plane.*[2]

In any case, the detected anomalies of the paraconical pendulum are very small, of the order of $2 \times 10^{-6}$ radians per second, and the fact that they have never previously been observed *simply means that current theories need to be supplemented to take account of them.*

In fact, as early as 1955, I was easily able to show that a very small relative difference of the inertial mass in two orthogonal directions, of the order of $10^{-6}$, variable with time in magnitude and direction, *would be able to explain the effects detected* during the experiments with the paraconical pendulum with anisotropic support.[3] For this it is enough, for example, that the influence of the Sun and of the Moon should bring about an anisotropy of space that is variable with time in direction and intensity, and that there should result *an anisotropy of inertial space*, variable with time in direction and intensity.

---

[1] *Chapter I*, § B.3, above, pp. 130-141, and *Chapter II*, § F.2, above, pp. 284-287.

[2] *Chapter I*, § A 4 and E.2, above, pp. 93-95 and pp. 173-175.

[3] *Chapter I*, § F.3, above, pp. 206-212, and *Chapter II*, Section I, above, pp. 320-325.

This is naturally only a hypothesis, but the hypothesis is *very simple* and very appealing.[4]

It follows that the movements of the paraconical pendulum are determined, *at one and the same time*, by the structure of the pendulum, by the influence of the support, and by the variations of the field of gravitation and inertia *in the space swept by the pendulum*.[5]

We remark that the movement of the paraconical pendulum is not only marked by diurnal and monthly periodicities, but also by semi-annual periodicities whose phases are near to the spring equinox.

## The deviations of optical sightings at marks

2 - For the lunar component of 24h50m, the amplitude observed at IRSID of the sinusoid fitted to the angular deviations is of the order of 1", *i.e.* $1.57 \times 10^{-6}$ radians.

In any event, the experiments performed at IGN in 1959 *definitively* showed that *neither deformation of the ground nor relative movement of the pedestals* could explain the effects detected.[6]

The optical anomalies that were detected *can only be explained by the influence of a medium in the space intermediate to the apparatus, in other words by an anisotropy of space*.

As for the pendulum, the observed optical deviations exhibit diurnal and monthly periodicities, and also semi-annual periodicities whose phases are all near to the spring equinox.

---

[4] In fact, *a priori*, isotropy of inertial space would appear not to be very *unlikely*. It would nevertheless look likely only for theoreticians blinded by "*well-established truths*".

[5] *Chapter I*, Section E, above, pp. 171-196, and *Chapter II*, Section I, above, pp. 320-325.

[6] *Chapter III*, § C.4.2, above, pp. 364-365.

## The optical observations of Esclangon

3 - Esclangon's object in his experiments was to detect an anisotropy of space, and the effects seen were of the order of $0.35 \times 10^{-6}$ radians per second, *i.e.* of an order of magnitude comparable to those of the amplitudes of the periodicities of the paraconical pendulum with anisotropic support and with isotropic support, and to those of the optical deviations of the sightings at marks and at collimators.[7]

The optical observations of Esclangon exhibit a semi-annual periodicity whose phase is also near to that of the spring equinox.[8]

## The interferometric observations of Miller

4 - The amplitude of the diurnal variations of the azimuths $A$ *observed* by Miller was of the order of 50° in twelve hours, which corresponds to a variation of $2.1 \times 10^{-5}$ radians per second, *i.e. to an effect of the same order of magnitude as for the paraconical pendulum.*[9]

The average azimuths $\overline{A}$ of Miller vary from one epoch to another. They exhibit a semi-annual periodicity. Here also, *anisotropy* of space is the only explanation that can be given for the *non-null* values of the average azimuths $\overline{A}$ *and their variations over time.*

The other characteristic parameters of Miller's observations, *i.e.* the observed maximum and minimum speeds $v_M$ and $v_m$, and the amplitudes $A_M^*$ of the deviations of the azimuths around their averages $\overline{A}$, are similarly marked by semi-annual and annual periodicities whose phases are also near to the spring equinox.

---

[7] *Chapter IV*, § B.2, above, pp. 378-381.

[8] *Chapter V*, § C.2, above, pp. 450-451.

[9] *Chapter IV*, § D.3, above, pp. 395-399.

The average value of the amplitudes $|2r|$ of the diurnal oscillations of the azimuth $A$ is $2(24.3 + 17.5 + 31.2 + 30.7)/4 = 51.85°$ (*Table I*, p. 397).

## The speed of light

5 - The observations of Miller show that variations over time of the speed of light depending upon its direction certainly correspond to the anisotropy of space. In view of the indications above, such an anisotropy will be of the order of $10^{-5}$.[10] *This in no way contradicts the known experimental results.*

Having put together all the experimental results, A. Kastler[11] effectively tells us for instance that the speed of light has a value of 299,792.3 km/sec with a relative uncertainty less than 3 km/sec, *i.e.* with an uncertainty of the order of $10^{-5}$.

## A general conclusion

6 - In fact, all the observations analyzed in the five preceding Chapters lead *to the same conclusion: "Space is anisotropic"*.[12]

---

[10] $8/300{,}000 = 2.67 \times 10^{-5}$ (*Table II* of § IV.D.4 above).

[11] A. Kastler, 1959, *Optique* (Optics), Masson, p. 30 [162].

[12] Here it does not appear useless to quote a particularly significant passage of Einstein *on the anisotropy of space*:

> *"The thought of Mach comes to its full flowering in the ether of the theory of general relativity. According to this theory, the metric properties of the spatio-temporal continuum are different in the neighborhood of each spatio-temporal point, and are determined by the matter that is located outside the region considered.*
>
> *"This spatio-temporal variation of the relation between rulers and clocks, or the conviction that empty space is neither physically homogenous nor isotropic - which obliges us to represent its state by ten functions, the gravitational potentials $g_{\mu\nu}$ - these facts, I say, have definitely departed from the conception that space is physically empty."*

Einstein, 1920, *L'Ether et la théorie de la relativité* (The Ether and the Theory of Relativity), 1921, Gauthier-Villars, p. 12 [124].

In any case, the orders of magnitude of the anisotropy of space implied by the general theory of relativity (*Chapter VII* below, § A5.2, note 7, pp. 566-567) are much smaller than those corresponding to the experiments on the paraconical pendulum with isotropic support (*Chapter II*, Section I, above, p. 321).

## Tensor formulation of the anisotropy of space

7 - To the anisotropy of space it should necessarily correspond a tensor formulation of the equations of gravitation and electromagnetism in a three-dimensional space.[13, 14]

Naturally, in the current state of available information, and *in view of the experimental data which as yet is completely insufficient, this tensor formulation can only be outlined.*

---

[13] The anisotropy that we are here brought to visualize differs from that of the theory of general relativity in that it only applies to the spatial coordinates; in other words, in the Einstein notation (*Chapter VII* below, § A.5.1, equation 1, p. 565), we have

$$(1) \qquad g_{14} = g_{24} = g_{34} = 0$$

[14] For illustration, the Hély equation that is a generalization of the Lorenz equation

$$(1) \qquad \Delta\varphi - \frac{1}{c^2}\frac{\partial^2\varphi}{\partial t^2} - 2\frac{k_0}{c}\frac{\partial\varphi}{\partial t} - k_0^2\varphi + 4\pi K\delta = 0$$

becomes in the case of spatial anisotropy

$$(2) \qquad \frac{1}{\sqrt{|g|}}\partial_i\left(\sqrt{|g|}g^{ij}\partial_j\varphi\right) - \frac{1}{c^2}\frac{\partial^2\varphi}{\partial t^2} - 2\frac{k_0}{c}\frac{\partial\varphi}{\partial t} - k_0^2\varphi + 4\pi K\delta = 0$$

(See the *Introduction*, § B.3, note 4, above, and Section C below, § C.1.4, C.1.5, and C.2).

## *B.3 Astronomical influences and the anisotropy of space*

The astronomical influences upon the observations corresponding to the experiments analyzed in *Chapters I through V* above are *undeniable.*

*The observations of the paraconical pendulum* with anisotropic support and with isotropic support exhibit *diurnal* periodicities directly linked with the relative movements of the Sun and the Moon with respect to the diurnal rotation of the Earth. They also present *sidereal* monthly *lunar* periodicities. The average monthly azimuths of the paraconical pendulum with anisotropic support also exhibit semi-annual periodicities *connected to the position of the Earth upon its orbit.* Moreover, they present a periodicity of the order of 5.9 years that is connected with planetary movements.[1]

The optical deviations of the sightings at marks and at collimators detected at IRSID and IGN exhibit diurnal periodicities, notably of 24h and 24h50m, a sidereal monthly lunar periodicity, and a semi-annual periodicity phase is near to the spring equinox, *i.e.* is connected to the position of the Earth upon its orbit.[2]

The observations of Esclangon present a *sidereal diurnal* periodicity and a semi-annual periodicity whose phase is near to the spring equinox.[3]

*The interferometric azimuths and speeds of Miller* exhibit a *sidereal diurnal* periodicity, as in the case of the optical observations of Esclangon. The azimuths and speeds of Miller are similarly characterized by semi-annual and annual periodicities whose phases are near to the spring equinox.[4]

---

[1] *Chapter I*, § A5, C.2, C.3, and E.5, *Chapter II*, Sections E, F, and G, and *Chapter V*, Sections B, E and F above.

[2] *Chapter III*, Sections B, C, and D, and *Chapter V*, Sections C, E, and F above.

[3] *Chapter IV*, § B.2, and *Chapter V*, Sections C, E and F above.

[4] *Chapter IV*, § D.3 and D.5, and *Chapter V*, Sections D, E, and F above.

Thus, the anisotropy of space appears to be *directly connected with astronomical influences* whose periodicities include sidereal semi-diurnal and diurnal periods, sidereal monthly periods, semi-annual and annual periodicities, and planetary periodicities of several years.

## *B.4 The incontestable existence of an intermediate medium*

The greatest physicists, for example Newton, Fresnel, Faraday, and Maxwell, were convinced that *no action at a distance could be conceived without the existence of an intermediate medium.*

The fact that certain physical phenomena propagate through space led very early to the hypothesis that space is not empty, but is constituted of an intermediate medium, the *"ether"*, *which is the natural support of all those phenomena.*[1]

### A pure quarrel about words

1 - However, since the advent of the *Theory of Relativity*, a somewhat fanatical dogmatism on the part of certain of its partisans has excluded the concept of the *"ether"* from the domain of science. Anybody who dares speak of the *"ether"* nowadays is considered as being an ignoramus and a retarded mentality, and he loses all credit in scientific circles - although in reality those who criticize him employ the same concept of an intermediate medium under the cloak of other language, such as for example *"field"*, *"associated fluid"*, *"probability fluid"*, *"pilot fluid"*, *"quantum fluid"*, etc.

Most often people simply speak of a *"field"*. Thus Einstein and Infeld wrote:[2]

> *"The electromagnetic field is, for the modern physicist, as real as the chair on which he sits...*

---

[1] On the concept of the ether and its applications, see the two remarkable works of Edmund Whittaker, 1951 and 1953, *History of the Theories of Aether and Electricity*, Vol. I, *The Classical Theories* [279], and Vol. II, *1900-1926* [280]. Whittaker refers all the way back to Descartes (1596-1650) for the introduction of the ether into science (Vol. I, p. 6).

See also René Dugas, 1951, *L'éther optique et gravifique au sens de Newton* (The optical and gravitational ether in the sense of Newton) [115].

For reference, also see Marie-Antoinette Tonnelat, *Ether*, Encyclopædia Universalis, Vol. 6, 1968, pp. 655-658 [268].

[2] Albert Einstein and Leopold Infeld, 1938, *The Evolution of Physics* [136].

The last paragraph of this text essentially relates to Maxwell, but Maxwell himself only followed the concepts of Fresnel and Faraday, about forty years earlier; and Newton had already expressed his conviction of the existence of an intermediate medium, the *"ether"*, a hundred and fifty years before Faraday (see the two quotations from Newton below, in § D.1 and E.1, pp. 518 and 536).

> *"It needed great scientific imagination to realize that it is not the charges nor the particles but the field in the space between the charges and the particles which is essential for the description of physical phenomena...."*

*But anybody can see that "the field" described in this way is exactly the "ether" in the sense of all the nineteenth century physicists.* In fact, whatever semantic expressions may be used, *in every case the discussion relates to an intermediate medium* that fills all space and supports the actions of gravitation and light and electromagnetic waves, via which *all actions at a distance* are transmitted, and which Newton and Fresnel, and following them all the physicists of the nineteenth century, denoted by the word *"ether"*.

This is therefore a pure quarrel about words, *completely sterile.* I myself think that the word *"ether"* is actually the most appropriate, and I hope that the reader will not consider the use of this word to be a provocation on my part. As Vilfredo Pareto emphasized,[3] *"One should never argue about words"*.

Einstein himself wrote:[4]

> *"Physical space and the ether are only two different expressions for one and the same thing; the fields are the physical states of space"...*
>
> *"The theoretical spirit cannot support the idea that there are two structures of space, each independent from the other, one of them being that of metric gravitation and the other being electromagnetic. The conviction is compelling that these two types of field must correspond to a unitary structure of space."*

## The properties of the ether

2 - Throughout the nineteenth century, efforts were made to give a coherent and satisfying mechanical representation of the ether. *But all these efforts failed.* The physical laws of gravitation, of electromagnetism, and of optics, were relatively simple and clear, but all the properties attributed to the ether for explaining them *simultaneously* appeared inconsistent and contradictory.[5] From this, some people came to the conclusion that an ether whose properties could not be formulated precisely in an indisputable manner could not exist.

---

[3] Vilfredo Pareto, *Traité de sociologie générale* (Treatise on General Sociology), Payot, 1917, p. 10 [212].

[4] Albert Einstein, *Comment je vois le monde* (How I see the world), Flammarion, 1939, pp. 222 and 230 [129].

[5] Some, for example, thought that this medium was unmoving and isotropic, others that it was anisotropic and capable of deformations and relative displacements.

But such a conclusion is *completely unacceptable.* We should not conclude that the ether does not exist, just because *up until now* it has not been possible to formulate a coherent model of its properties. Here, we should meditate upon this reflection of Claude Bernard: "*We must believe that, in Nature, what seems absurd according to our theories is not always impossible.*"[6]

*In reality, the properties of the ether are on the one hand the indefinite equations of gravitation, and on the other hand the indefinite equations of electromagnetism,* respectively representing the gravitational field and the electromagnetic field in vacuum. As yet nobody has succeeded in establishing complete liaison between these fields, but this is an obstacle that sooner or later will be overcome, because *it is inconceivable that one and the same medium could serve to support two different fields that had no fundamental mutual internal coherence.*

*Certainly we do not know the real nature of the ether or of its constitution, but we know some of its manifestations, gravitational, electromagnetic, and optical.* It therefore cannot be said that the ether is an unreal concept and metaphysical in nature, *because it manifests itself everwhere and at all times via very real phenomena.*

[6] Claude Bernard, 1865, *Introduction à l'étude de la médecine expérimentale* (Introduction to the study of experimental medicine), 1966, id., p. 71 [69].

**Related questions**

3 - *As far as the existence and the nature of this intermediate medium are concerned*, some comment here upon is appropriate:

- *the properties of isotropy or of anisotropy* of this intermediate medium; and
- *the claimed oppositions* in contemporary theories: - *the claimed opposition between the two aspects of light, wave and particle; - the claimed opposition between the continuity that seemingly characterizes the ether and the discontinuity that characterizes the quantum theory; - and the claimed opposition between the deterministic concepts and the indeterministic concepts of physics.*

The following *sections C and D* attempt precisely this analysis.

# C APPARENT ISOTROPY OR REAL ANISOTROPY - AN ILLUSTRATION

## *C.1 The evolution of the theory of potentials*

### Laplace's Equation

1 - In 1782, Laplace showed that in vacuum Newtonian potentials satisfy the equation

$$(1) \qquad \Delta\varphi = \frac{\partial^2\varphi}{\partial x^2} + \frac{\partial^2\varphi}{\partial y^2} + \frac{\partial^2\varphi}{\partial z^2} = 0$$

### Poisson's Equation

2 - Thirty years later, in 1813, Poisson showed that in the most general case, when the point $x, y, z$ is located in matter of density $\delta\,(x, y, z)$, Laplace's equation should be replaced by the equation

$$(2) \qquad \Delta\varphi + 4\pi K\delta = 0$$

where $K$ is a constant. The integral of this relation is expressed as

$$(3) \qquad \varphi\,(X, Y, Z) = \int K \frac{\delta(x, y, z)}{r} \mathrm{d}v$$

where $r$ is the distance between the points $m(x, y, z)$ and $M(X, Y, Z)$, and where the integral is taken over all space.

## Lorenz's Equation

3 - In 1867, Ludwig Lorenz proposed the replacement of Poisson's equation by the equation

$$(4) \qquad \Delta\varphi - \frac{1}{c^2}\frac{\partial^2\varphi}{\partial t^2} + 4\pi K\delta = 0$$

where $c$ is a constant, so that the solution is

$$(5) \qquad \varphi(X,Y,Z,t) = \int K\frac{\delta(x,y,z,t-\frac{r}{c})}{r}\mathrm{d}v$$

The potential $\varphi$ is a retarded potential that takes account of a speed of propagation $c$.[1]

## Hély's Equation

4 - In 1948, Jean Hély generalized Lorenz's equation by considering the equation[2]

$$(6) \qquad \Delta\varphi - \frac{1}{c^2}\frac{\partial^2\varphi}{\partial t^2} - 2\frac{k_0}{c}\frac{\partial\varphi}{\partial t} - k_0^2\varphi + 4\pi K\delta = 0$$

of which the solution is

$$(7) \qquad \varphi(X,Y,Z,t) = \int K\delta(x,y,z,t-\frac{r}{c})\frac{e^{-k_0 r}}{r}\mathrm{d}v$$

where the integral is taken over all space, and $r$ represents the distance between the points $M(X,Y,Z)$ and $m(x,y,z)$.

[1] For the equations of Laplace, Poisson, and Lorenz, see particularly Whittaker, *History of the Theories of Aether and Electricity, The Classical Theories* [279], Nelson, 1951, pp. 61 and 268.

[2] Jean Hély, *Les fondements théoriques de l'électrodynamique* (The theoretical foundations of electrodynamics) - 1950, p. 29 [157] - The demonstration of the property pointed out by Jean Hély is immediate, and is performed following the same principles as the demonstration of the potentials of Lorenz. (This demonstration was given by Mr. Hély in September 1948 in the mimeographed text of a conference at the *Ecole Nationale Supérieure du Génie Maritime* (National Higher School of Maritime Engineering) *"Sur l'électrodynamique rationelle"* ("On rational electrodynamics"), p. 23).

For $k_0 = 0$, we naturally recover Lorenz's equation (4).

In the isotropic and Euclidean space $E_i$ of $x,y,z$, the propagation of perturbations takes place in straight lines with the constant speed $c$ and a constant rate of attenuation $k_0$. The potentials are thus both *retarded and damped.*

Equation (6) holds for a space that is isotropic and Euclidean, for which we have the metric element

$$(8) \qquad \mathrm{d}s^2 = \mathrm{d}x^2 + \mathrm{d}y^2 + \mathrm{d}z^2$$

**My generalization of the Hély equation in the case when space is anisotropic**

5 - The equations of Laplace, Poisson, Lorenz, and Hély are valid for a space that is isotropic and Euclidean. In November 1957 I generalized the Hély equation for the case of a space that is anisotropic and non-Euclidean by considering the equation (written in tensor notation)[3]

$$(9) \qquad \frac{1}{\sqrt{|g|}}\partial_i\left(\sqrt{|g|}g^{ij}\partial_j\varphi\right) - \frac{1}{c^2}\frac{\partial^2\varphi}{\partial t^2} - 2\frac{k_0}{c}\frac{\partial\varphi}{\partial t} - k_0^2\varphi + 4\pi K\delta = 0$$

This equation has as a solution the expression

$$(10) \qquad \varphi(X_1,X_2,X_3) = \int \sqrt{|g|}K(x_1,x_2,x_3)\delta(x_1,x_2,x_3,t-\frac{r}{c})\frac{e^{-k_0 r}}{r}\mathrm{d}v$$

with

$$(11) \qquad r = \int_M^m \mathrm{d}s$$

this integral being calculated along the straight line $L$ that connects the points $M$.

[3] It is known that in a three-dimensional space the Laplacian

$$(1) \qquad \Delta\varphi = \frac{1}{\sqrt{|g|}}\partial_i\left(\sqrt{|g|}g^{ij}\partial_j\varphi\right)$$

which is explicitly written as

$$(2) \qquad \Delta\varphi = \frac{1}{\sqrt{|g|}}\sum_{ij}\frac{\partial}{\partial x_i}\left(\sqrt{|g|}g^{ij}\frac{\partial\varphi}{\partial x_j}\right)$$

is a pure scalar, *invariant* under all transformations of coordinates. And $g$ is defined by Equation (14)

and $m$ in the space of the $x_i$, and $\mathrm{d}s$ being given in tensor notation by the equation

$$(12) \qquad \mathrm{d}s^2 = g_{ij}\mathrm{d}y^i\mathrm{d}y^j$$

We have for the volume element

$$(13) \qquad \mathrm{d}v = \sqrt{|g|}\mathrm{d}x_1\mathrm{d}x_2\mathrm{d}x_3$$

with

$$(14) \qquad g = \begin{vmatrix} g_{11} & g_{12} & g_{13} \\ g_{21} & g_{22} & g_{23} \\ g_{31} & g_{32} & g_{33} \end{vmatrix}$$

In Equations (9) and (10), the coefficients $c$ and $k_0$ are considered to be constants. The $g_{ij}$ can be interpreted as the squares of indices of refraction, and, when $k$ and $n$ are variable, we have

$$(15) \qquad \frac{k}{k_0} = \frac{n}{n_0} \qquad \text{with } n_0 = 1$$

$n_0 = 1$ corresponds to the index of refraction in vacuum *at the surface of the Earth.*

In this case Equation (9) is written as

$$(16) \qquad \frac{1}{\sqrt{|g|}}\partial_i\left(\sqrt{|g|}g^{ij}\partial_j\varphi\right) - \frac{n^2}{c^2}\frac{\partial^2\varphi}{\partial t^2} - 2\frac{k_0 n^2}{c}\frac{\partial\varphi}{\partial t} - k_0^2 n^2\varphi + 4\pi K\delta = 0$$

While in the isotropic space $E_i$ of Hély perturbations are propagated along straight lines with the constant speed $c$ and the constant rate of attenuation $k_0$, in the anisotropic space $E_a$ perturbations are propagated along curved trajectories with variable speeds and variable rates of attenuation.[4]

[4] Naturally it is only possible to pass by a linear transformation from the anisotropic space $E_a$ to the isotropic space $E_i$ of § 1.4 under restrictive conditions of integrability.

Transposing the well-known conditions of the theory of relativity in four-dimensional space-time to a three-dimensional space, we see that, in the case considered, the conditions are obtained by canceling the six distinct components of the Ricci tensor obtained by contraction of the third order Riemann-Christoffel curvature tensor, which, in view of the identities of Bianchi, gives three independent conditions that the nine coefficients $g_{ij}$ must satisfy.

## *C.2 The anisotropy of space*

The presentation given above of the five successive forms of the potential equation calls for a few comments.

**The equation of Lorenz is not the only one imaginable.**

1 - It would be a mistake to claim that Ludwig Lorenz's equation (4) represents the definitive and unchangeable form of the indefinite equation of potentials. *The Hély equation (6) shows that a generalization is possible.* This equation is rather natural, and if $k_0$ is sufficiently small it reduces to Lorenz's equation in the first approximation.

**The generalization of the potential equation in the case of anisotropy**

2 - Equations (9) and (16) show how the Hély equation can be generalized in the case of a three-dimensional anisotropic medium.

*If space is anisotropic, it is absolutely unnecessary to consider four-dimensional space-time.*[1]

---

[1] In fact as I have indicated (*Introduction*, § B.3, note 4, p. 55, and *Chapter I*, § G.5.2, pp. 227-228), Equation (9) of § C.l above was the object of a *Note* on 4 November 1957 on my part to the Academy of Sciences [11], *which was refused.*

This *Note* of 4 November 1957 was the basis of an application to the interpretation of the terrestrial magnetic field in my *Note* of 24 October 1957 [14] and of two applications to the interpretation of anomalies of gravity in my *Notes* of 5 November 1957 [13] and March 1960 [39] (see above, *Introduction*, § B.3.3, note 4, p. 55).

## An anamorphosis of space

3 - The passage from Equation (6) of Hély for an isotropic space to Equation (16) that generalizes it to an anisotropic space can be interpreted as an anamorphosis of space.[2]

It may be the case that actual space is anisotropic and non-Euclidean, while its isotropic and Euclidean representation that we construct for ourselves corresponds to Equation (4) of Lorenz or to Equation (6) of Hély.

According to this concept, light propagates in straight lines at constant speed in the *fictitious* space corresponding to the $ds^2$ of Equation (8) which is *isotropic and Euclidean*, whereas, in the case of *actual* space that is *anisotropic and non-Euclidean*, light propagates along curved trajectories with variable speed.[3]

## The definition of time

4 - In a conception such as the above, how can time be defined? If we consider the isotropic and Euclidean space corresponding to Equation (6) of Hély, time can be defined *by the condition that Equation (6)*, which corresponds to propagation of light in straight lines at constant speed, *should be valid.*

---

[2] A very simple image of anamorphosis of a three-dimensional Euclidean space is a ball of rubber that we squeeze. To straight lines in the initial configuration correspond curved lines in the squeezed ball.

[3] *All these developments here only correspond to a simple illustration*. Deeper analysis of the Hély equation (Equation 6 of § C.1.4 above), of its implications, and of its generalization in the case of spatial anisotropy (Equations 9 and 16 of § C.1.5 above) would be outside the scope of this work, and will be the object of a future publication.

Here I only shall say that this analysis leads to a value of $\delta$ that is *non-zero in vacuum*.

When an anisotropic and non-Euclidean space corresponding to Equation (16) and to the $ds^2$ of Equation (12) is considered, time is *implicitly* defined by the validity of Equations (16) and (12); in other words, in the final analysis, by reference to periodic phenomena.[4]

## The anisotropy of space

5 - As it has been shown,[5] the anisotropy of space corresponds to the data from observation. *It is not a hypothesis. It is a brute fact.*

The very brief suggestions above make it possible to define a theoretical approach that can cope with the anisotropic structure of space.

---

[4] Such a concept in fact corresponds to the definition by Eddington of time intervals that are equal with reference to cyclic phenomena (Eddington, 1924, *The Mathematical Theory of Relativity*, § 1.3, p. 12 [121]).

[5] § B.2 above, pp. 499-503

# D THREE OPPOSING CLAIMS IN CONTEMPORARY THEORIES

## *D.1 The two aspects of light, wave and particle*

There *cannot be contradiction of any sort* between the two aspects of light, wave and particle. The particle aspect corresponds with the movement of a particle, while the wave aspect corresponds to the vibrations of the *"ether"* that accompany this particle, whether or not the particle is considered as a local singularity in the *"ether"*.

That was in fact the point of view of Newton, who wrote, three centuries ago in 1675,[1] in a remarkable text that has indeed proved prophetic in the light of further developments of physics in the 20th century:

> *"Let us suppose that light rays consist of corpuscles emitted by luminescent bodies in all directions; they must excite vibrations in the ether... as necessarily as stones do so in the water into which they are thrown...*
>
> *"If I had to form a hypothesis, it would be that light is something that can excite vibrations in the ether...*
>
> *"Light is neither the ether, nor a vibratory movement of the ether, but something different propagated from luminous bodies...*
>
> *"Do not bodies act on light at a certain distance, and do not they bend its rays? and is not this action stronger (ceteris paribus) at a lesser distance?"*

---

[1] René Dugas, 1954, *La Mécanique au 17ème siècle* (Mechanics in the 17th century), Dunod, pp. 392, 395, and 405 [116].

According to such a conception, the double aspect of light, particle and wave, results from the *simultaneous coexistence of photons and of an intermediate medium, the "ether", two factors that are distinct but that act upon one another.*

Photons can also be considered as *singular local regions* in the ether, in symbiosis with the regions of the ether that surround them.

## D.2 Continuity and quanta

*Similarly, the model of a continuous ether and the quantum theory, which is essentially based upon discontinuity, should not be considered as mutually contradictory.*

In fact, current quantum theory is based upon a great number of hypotheses, *certain ones of which do not correspond to any experimental necessity*, and upon certain deductions that seem quite debatable.

### Schrödinger's Equation

1 - *For example, there is no real demonstration of Schrödinger's equation from any specified hypotheses*, although it plays an absolutely essential role in *quantum mechanics.*[1]

---

[1] The partial differential equation of Schrödinger for a particle is written (E. Durand, 1970, *Mécanique Quantique, Equation de Schrödinger* (Quantum Mechanics, Schrödinger's Equation), Masson, p. 16, equation 68 [120]) as:

$$(1)\qquad \frac{h^2}{8\pi^2 m}\Delta\psi(x,y,z,t)+\frac{h}{2\pi i}\frac{\partial\psi(x,y,z,t)}{\partial t}-U(x,y,z,t)\psi(x,y,z,t)=0$$

It defines the function $\psi$, *where $i$ is the imaginary symbol* and $h$, $m$, and $U$ respectively stand for Planck's constant, the mass of the particle under consideration, and the potential energy at the point $x,y,z$ at the moment $t$. The operator $\Delta$ is the Laplacian

$$(2)\qquad \Delta\psi=\frac{\partial^2\psi}{\partial x^2}+\frac{\partial^2\psi}{\partial y^2}+\frac{\partial^2\psi}{\partial z^2}$$

The solution $\psi$ of Equation (1) is a complex function, and we consider the real function

$$(3)\qquad \rho=\psi^*\psi$$

where $\psi^*$ is the function conjugate to $\psi$, under the condition

$$\int_v \rho\,\mathrm{d}v=\int_v \psi^*\psi\,\mathrm{d}v=1$$

the integration being performed over all space.

The function $\rho$ is *interpreted* as the density of the *"probability"* of presence of the particle in question at the point $x,y,z$ at the moment $t$.

*This is only an interpretation in view of the experimental results, supposed to be correctly represented by Schrödinger's equation.* It is only possible to say that, *following this interpretation*, the number of particles observed in a volume $\mathrm{d}v$ at the point $x,y,z$ at the moment $t$ is equal to the quantity $\rho\mathrm{d}v$, $\rho$ being considered as being proportional to the square of the amplitude of Fresnel's waves.

The introduction of the imaginary symbol $i$ into Schrödinger's equation (1) only underlines *the artificial character* of that equation.

In fact, Schrödinger's equation is only the end result of a series of successive inductions that are more or less intuitive and quite artificial,[2] and *whose only justification is the verification of some of their consequences.*

Thus, Schrödinger's equation appears as a pure and simple hypothesis, whose only justification is derived from the agreement of some of its consequences with the observational data.[3] But the fact that certain consequences of this equation have been verified *in particular cases cannot mean in any way* that this equation adequately represents real phenomena *in the most general case.*[4]

## A claimed contradiction

2 - In fact, it is completely impossible to go along with Einstein when he writes:[5]

> *"At first sight it does not seem at all possible to deduce, from a field theory that operates by means of differential equations, that a mechanical system is only capable of permanently having discrete energy values or states, as experiment shows."*

---

[2] See for example E. Schrödinger (1932), *Mémoires sur la Mécanique Ondulatoire* (Memoirs on Wave Mechanics), Alcan, 1933, pp. 161-165 [243]; and de Broglie, 1953, *Eléments de théorie des quanta et de mécanique ondulatoire* (Elements of quantum theory and wave mechanics), Gauthier-Villars, § IX.3-IX.5, pp. 151-160 [85].

[3] See particularly Georges Guinier, 1949, *Eléments de Physique Moderne Théorique* (Elements of Modern Theoretical Physics), Bordas, p. 46 [153], and Durand, 1970, id., pp. 15-18 and 60-61 [120].

[4] In fact there is an immense distance between the equations of quantum mechanics, which represent the results of observations, and the sometimes extravagant interpretations that are given to them.

The celebrated theorem of von Neumann, according to which it is impossible to give an interpretation of the laws of quantum mechanical probability by appealing to hidden variables, is often cited in support of the indeterministic concept. In fact this demonstration is *erroneous* (see Fer, *L'irréversibilité, fondement de la stabilité du monde physique* (Irreversibility, foundation of the stability of the physical world), Gauthier-Villars, 1977, p. 39 [142]; also see de Broglie, *Jalons pour une nouvelle microphysique* (Milestones for a New Micro-Physics), Gauthier-Villars, 1978, pp. 31-32 [87]).

[5] Albert Einstein, *Comment je vois le monde* (How I see the world), Flammarion, 1939, p. 193 [129].

In a two-dimensional space, differential equations can indeed admit *limit cycles* as solutions. The same is obviously true for a three-dimensional space.[6]

In fact, the theoretical perspectives opened up by the theory of *non-linear* differential equations, and particularly by the theory of limit cycle dynamics, have been *quite insufficiently* explored to date.

It will certainly only be possible to explain the existence and the properties of the various particles *as local singularities and limited configurations of the ether via further exploration of non-linear differential equations.*[7]

---

[6] In a completely different domain, that of economy, I have showed that, in the framework of a *non-linear* model that allows for *stable limit cycles*, it is possible to transit from one regime to another as a consequence of perturbations from outside, precisely like those that, in the case of the physical world, could result from almost periodic fluctuations of the ether.

See Allais, 1954, *Explications des cycles économiques par un modèle non linéaire à régulation retardée* (Explanation of economic cycles by a non-linear model with delayed regulation), Metroeconomica, Vol. VIII. April 1956. pp. 4-83 [6]; and 1955, *Explications des cycles économiques par un modèle non linéaire à régulation retardée, Mémoire complémentaire* (Explanation of cycles by a non-linear model with delayed regulation, Complementary Memoir), CNRS, Dynamic Models in Economy, 1956, pp. 259-308, § 29-30 and 32, and pp. 276-277 [7].

An analogous explanation for quantum mechanics was suggested by Francis Fer, 1977, *L'irréversibilité, fondement de la stabilité du monde physique* (Irreversibility, foundation of the stability of the physical world), Gauthier-Villars, Chapter IV, *Limit cycle dynamics as an example of quantum systems*, pp. 42-45 [142].

[7] For the somewhat fascinating literature on non-linear differential equations and limit cycles, see particularly: - Henri Poincaré, *Les méthodes nouvelles de la Mécanique céleste* (New Methods of Celestial Mechanics), 3 vol., Gauthier-Villars, 1892-1899 [219]; - Balth. van der Pol, *On "Relaxation Oscillations"*, Philosophical Magazine, July-December 1926, pp. 1978-1992 [229]; - A.A. Andronov and S.E. Khaikin, 1937, *Theory of Oscillations*, Princeton University Press, 1949, 358 p. [58]; - N. Krylov and N. Bogolyubov, 1943, *Introduction to Non-Linear Mechanics*, Princeton University Press, 1947, Annals of Mathematical Studies, no. 11, 106 p. [166]; - N.F. Minorsky, 1947, *Introduction to Non-Linear Mechanics*, J.W. Edwards, Ann Arbor, 464 p. [204]; - Y. Rocard, *Dynamique générale des vibrations* (General Dynamics of Vibrations), Masson, 1949, Part I, Chapters XIV and XVII [233]; - A.A. Andronov, A.A. Vitt, and S. E. Khaikin, 1966, *Theory of Oscillators*, Pergamon Student Edition, 816 p. [59]

## D.3 Causality and indeterminism

*It is also no longer tenable to oppose, as being contradictory, the model of an ether that behaves in a deterministic manner, and indeterministic theories that reject the principle of causality.*[1]

A simple examination of the experimental data of quantum mechanics and of the models that represent it has led many thinkers to consider the ideas of Laplace and Poincaré on chance and determinism as being outdated; and people who cast doubt on certain interpretations of the theory of quantum mechanics, however aberrant they may be in their extreme formulations, are considered as being retrograde mentalities - even strong personalities have not always had the courage to resist.[2]

---

[1] For lack of space, here I must limit myself to a few *very brief* general observations, and must refer for more developed explanation to my two memoirs of 1983 and 1988: 1983, *Fréquence, Probabilité et Hasard* (Frequency, Probability, and Chance), Journal de la Société Statistique de Paris (Journal of the Statistical Society of Paris), Vol. 124, pp. 70-102 and 144-221 [43], and 1988, *Phénomènes Aléatoires et Modèles Fréquentiels. Réalité et Théorie. Prolégomènes pour une Révision des Théories admises* (Random Phenomena and Frequency Models. Reality and Theory. Prologomena for a Revision of Accepted Theories) [47]. The reader will find a summary of my memoir of 1983 in my memoir of 1982, *Frequency, Probability, and Chance*.

See also Allais, 1982, *The Foundations of the Theory of Utility and Risk*, pp. 8-27, and 73-89 [44].

[2] This judgment on my part may seem somewhat excessive, at least at first glance, so I take the liberty of citing one significant example here. Commenting upon the 5th Solvay Conference on Physics in October 1927, Louis de Broglie wrote (1953, *La Physique quantique restera-t-elle indéterministe ?* (Will Quantum Physics Remain Indeterministic?), pp. 12, 13, 15, 21, and 22 [86]):

> *"Mr. Schrödinger could not follow me because he does not believe in the existence of particles. Messrs. Bohr, Heisenberg, Born, Pauli, Dirac, etc... have been developing the purely probabilistic interpretation that I have already described above as being the currently orthodox interpretation. The president of the Conference, Lorentz, could not accept such an interpretation, and strongly reaffirmed his conviction that theoretical physics should remain deterministic and should continue to employ clear images in the classical framework of space and time. Einstein criticized the probabilistic interpretation and himself raised somewhat disturbing objections...*
>
> *"I returned to Paris very troubled by these discussions... I discouraged myself, and rallied to the purely probabilistic interpretation of Bohr and Heisenberg. After twenty-five years, I have adopted it as the basis of my teaching and have expounded it in my books and in conferences...*
>
> *"After twenty-five years almost all physicists have rallied to the purely probabilistic interpretation of Bohr and Heisenberg. There are however a few notable exceptions: scientists as eminent as Messrs. Einstein and Schrödinger have always refused to accept it...*
>
> *"Undoubtedly some people may accuse me of unfaithfulness, having seen me abandon my first attempts and promote the interpretation of Bohr and Heisenberg in all my writings for twenty-five years... But a reply is possible... a serious one.*
>
> *"The history of the sciences shows us that the progress of science has been constantly fettered by the tyrannical influence of certain ideas that have ended up being considered as dogmas. For this reason, it is appropriate periodically to give very deep examination to principles that have in fact been admitted without further discussion."*

## The deterministic conception

1 - Laplace has explained the respective roles of determinism and chance in perfectly clear language:[3]

> *"All events, even those that seem not to follow the great laws of Nature due to their small scale, are consequences of them as necessary as the movements of the Sun. In ignorance of the bonds that unite them to the entire system of the universe, we made them depend upon final causes or upon chance, according as to whether they happen and succeed one another regularly or without apparent order; but these imaginary causes have been repeatedly pushed back by the boundaries of our knowledge, and disappear entirely in the light of healthy philosophy, which only sees them as being the expression of our proper ignorance of the real causes...*
>
> *"We therefore must visualize the present state of the universe as being the effect of its previous state, and as the cause of the next that will follow. An intelligence that at one given instant could know all the forces that animate nature and the respective locations of the entities of which nature is composed, and that also was vast enough to subject this data to analysis, could include in one formula both the movements of the greatest bodies in the universe and those of the lightest atom: nothing would be uncertain for such an intelligence, and the future would be as available to its view as the past. The human spirit offers a feeble hint of such an intelligence by the perfection to which it has brought astronomy."*

---

In view of such an explanation, and remembering that Louis de Broglie received the Nobel Prize in Physics in 1929 with all the authority that this accolade gives, we may well ask ourselves what are the possibilities for anyone to contend effectively against the kind of tyranny exerted by the conceptions of the dominant *"establishment"*.

[3] Pierre-Simon Laplace (1749-1827), 1814, *Essai philosophique sur les probabilités* (A philosophical essay on probabilities), Gauthier-Villars, 1921, I, pp. 2-3 [169].

## The indeterministic conception

2 - According to the indeterminists *there is no such thing as any causal relation, and "chance", which here means "pure contingency"* according to an excellent expression of Louis de Broglie,[4] *is the ultimate reality as physical phenomena unfold.*[5]

As Max Planck writes:[6]

> *"For the indeterminists... the absence of regularity must be sought behind every apparent regularity, and a law in statistical form is the only kind that can be fully satisfactory by itself."*

In fact, here it is appropriate to distinguish between statistical distributions of two types: *distributions that result from the simultaneous interaction of a great number of complex causes, and those that are considered not to derive from any cause*. This distinction is absolutely *crucial.*

---

[4] Louis de Broglie, 1953, *La Physique quantique restera-t-elle indéterministe ?* (Will Quantum Physics Remain Indeterministic?), Gauthier-Villars, p. 15 [86].

[5] Max Born wrote (*Atomic Physics*, Blackie, 1956, pp. 95-102 [75]):

> *"According to this view, the entire course of events is determined by the laws of probability: to a state in space there corresponds a definite probability, which is given by the de Broglie wave associated with the state. A mechanical process is therefore accompanied by a wave process, the guiding wave, described by Schrödinger's equation, the significance of which is that it gives the probability of a definite course of the mechanical process ....*
>
> *"It is clear that the dualism, wave-corpuscle, and the indeterminateness essentially involved therein, compel us to abandon any attempt to set up a* deterministic theory. *The* law of causation, *according to which the course of events in an isolated system is completely determined by the state of the system at time $t = 0$, loses its validity, at any rate in the sense of classical physics."*

[6] Max Planck, 1925, *Initiations à la Physique* (Introduction to Physics), Flammarion, 1941, Chapter X, p. 235 [218].

As Louis de Broglie also so justly wrote:[7]

*"As Einstein often underlined, the question that arises is to know finally whether the current interpretation that establishes the Schrödinger wave Ψ, which is of statistical character, is a "complete" description of reality, in which case one must admit indeterminism and the impossibility of representing realities on the atomic scale in a precise manner in the framework of space and time; or whether, on the contrary, this interpretation is "incomplete" and hides behind itself, like the old statistical theories of classical physics, a reality that is completely determined and can be described in the framework of space and time by variables that are hidden from us, in other words that escape our experimental determination."*

We should therefore conclude with Max Planck:[8]

*"Some indeterminists have considered themselves authorized to eliminate the principle of determinacy from physics definitively. However, a deeper examination shows that this conclusion, which rests upon the identification of the representative universe of the physicists and the universe of our senses, is at least premature."*

Does not such a viewpoint risk being abandoned in its turn *because there is always the possibility that causual laws may be discovered?* As Bertrand Russell so correctly writes:[9]

*"It is very rash to erect a theological superstructure upon a piece of ignorance which may be only momentary....*

*"To prove that a given set of phenomena is not subject to laws is essentially and theoretically impossible. All that can be affirmed is that the laws, if any, have not yet been discovered.*

*"We may say, if we choose, that the men who have been investigating the atom are so clever that they must have discovered the laws if there were any. I do not think, however that this is a sufficiently solid premise upon which to base a theory of the universe."*

---

[7] Louis de Broglie, 1953, id., p. 21.

[8] Max Planck, 1925, *Initiations à la Physique* (Introduction to Physics), Flammarion, 1941, Chapter X, p. 259 [218].

[9] Bertrand Russell, 1931, *The Scientific Outlook*, Chapter V, p. 105 [239].
As for the Heisenberg uncertainty principle, Bertrand Russell adds quite rightly:

*"The use to which the principle of indeterminacy has been put is largely due to an ambiguity in the word* "determined"*. In one sense a quantity is determined when it is measured, in the other sense an event is determined when it is caused. The Principle of Indeterminacy has to do with measurement, not with causation. The velocity and position of a particle are declared by the principle to be undetermined in the sense that they cannot be accurately measured... There is nothing whatever in the Principle of Indeterminacy to show that any physical event is uncaused."*

The matter could not be better expressed.

*For the significance of the uncertainty principle*, see the *very suggestive* demonstrations of Max Born, *Atomic Physics*, 1956, id. [75], and of Niels Bohr, 1960, *Atomic Physics and Human Knowledge* [72]. See also K. Popper, 1967, *Quantum Mechanics without "The Observer"*, in *Quantum Theory and Reality*, Ed. Bunge, pp. 7-44 [230].

Indeed, to claim that no causal explanations exist because men who have been judged to be eminent have not discovered them, is a completely wanton postulate. As Vilfredo Pareto so excellently puts it, *does not the history of science reduce to the history of the mistakes of competent men?*

*In fact, the impossibility of prediction is relative to man. It does not correspond to an intrinsic property of nature.*[10]

---

[10] For this reason it seems to be to be impossible to adopt the point of view of Planck (Max Planck, 1925, *Initiations à la Physique* (Introduction to Physics), Flammarion, 1941, p. 229 [218]) when he writes:

> *"An event is causually conditioned when it can be predicted with certainty."*

For me, this definition of causality is *too restrictive*. Indeed certain phenomena, as for example the drawing of a determinate lot from an urn, cannot be predicted although nevertheless they are characterized by causality.

In any event, the history of science shows that *the same facts can be explained by entirely different theories.*[11]

In reality, indeterministic theories are only based upon major confusions, in particular: - *semantic confusions; - an erroneous interpretation of statistical distributions; - a complete ignorance of the real determinism of apparently indeterministic phenomena.*

**Semantic confusions**

3 - First of all, the semantics currently used can only lead to errors. *The variables called "random" have nothing random about them.* They are simply variables that have a distribution, *and the theories that consider them are entirely deterministic.*

The same word *"random variables"* covers *entirely different* realities: *frequency variables* of mathematical models, of which *all* the values are in fact considered (explicitly, or generally and unhappily, implicitly) as being realized *simultaneously*; and *empirical random variables* given by observation, *one single value* being realized at each instant for each of them.

Undoubtedly, never has a word been so badly chosen to represent the concept considered, as that of the *"random variable"* introduced by Cantelli and Kolmogorov. Perhaps never has the use of such a word been able to suggest so many false ideas.

---

[11] Notably, this is the case for the experiment of Fizeau in 1851, verified *both* by the theory of Fresnel in 1818, and by the Theory of Relativity (see below, *Chapter VII*, § C.5, pp. 601-602).

In fact, the theory of Kolmogorov of additive ensembles and independent variables *is a deterministic theory*, because *chance plays no part in it*. It is based entirely on the deduction of the mathematical consequences of postulates that are admitted at the start, and the character of this deduction is fundamentally *entirely deterministic*. No role is left for *"chance"* in these deductions.

It is indeed remarkable that such a *deterministic* theory can represent phenomena that are *considered to be random* to a very good approximation, but it cannot follow that the nature of this theory is random and that its phenomena are purely random; and it is a real abuse of the facts to call the variables considered by Kolmogorov "random". The same remark can be made *for all models that are termed "probabilistic"*.[12]

## The representation of deterministic structures by statistical distributions. Theorem T

4 - According to an assertion that unhappily is much too widespread, if observations are distributed according to the normal law, this is because they are a result of chance, and they do not actually correspond to any structural regularity. According to this assertion, *"chance" and indeterminism are present as soon as there is a statistical distribution*. But this distinction collapses as soon as it can be proved that phenomena that are *entirely deterministic* can be characterized by well defined statistical distributions.

[12] On all these points, see my memoir of 1982, *Fréquence, Probabilité et Hasard* (Frequency, Probability, and Chance) (note 1 above, p. 523).

In fact I showed in 1982 that the sum of the ordinates of sinusoids of incommensurable periods converges to the normal law, and that therefore *phenomena that are totally deterministic can show all the appearances of chance.* This is what I have called *Theorem (T).*[13]

*"Chance"* cannot therefore be defined only by reference to the existence of a statistical distribution, because at least in the case of *Theorem (T)*, for example, we *simultaneously* have absence of *"chance"* and a normal distribution.

In fact, apart from the perspectives that it opens upon the interpretation of numerous physical phenomena, *Theorem (T)* also allows us to verify that the *central limit theorem* is not only applicable to quantities of the type suggested by the expression *"random variable"*, but that it is also very applicable to the sums of functions that are *entirely deterministic.*

Indeed, how can it be explained that some phenomenon or another can be characterized by some law of statistical distribution or another, if precisely there does not exist a hidden order relating to statistical distributions of a well-determined type?

The simple fact that a variable is distributed according to a well-determined statistical law and that this law is *stable in space and time* excludes indeterminism, *because it precisely implies an underlying interdependence in space and time, incompatible with the postulate of indeterminism.*

---

[13] See *Appendix II, Theorem (T)*, in my memoir of 1982, *Fréquence, Probabilité et Hasard* (Frequency, Probability, and Chance).

Herman Wold, who presided over the meeting in Oslo at which I presented my Communication on *Theorem (T)* in 1982, emphasized its full importance, and he did not hesitate to declare that, after three or four years, this theorem would appear in all the textbooks of the theory of probability. This prediction however ignored the power of *"well-established truths"*.

*Theorem T* was the object, on 30 May 1983, of a *Note* on my part to the Academy of Sciences, *"Sur la distribution normale des valeurs à des instants régulièrement espacés d'une somme de sinusoïdes"* ("On the normal distribution of values of a sum of sinusoids at regularly spaced instants"), CRAS, Vol. 296, series I, pp. 829-830 [45].

## Apparent indeterminism and real determinism

5 - *Many phenomena only become completely incomprehensible* if one rejects the existence of an intermediate medium, whether one calls it the *"ether"* as Newton, Fresnel, and Maxwell did, or something completely different, and if one does not take into account the fact that this medium is a vibrational medium through which propagate an extremely large number of waves whose frequencies occupy a very wide range. The association of the hypothesis that this medium is real with the idea that it has a very great number of vibrational components, *can in fact make us understand, or can even completely explain, with the application of Theorem (T), the production of all the appearances of chance in a mass of phenomena that are considered to be non-deterministic, but which in fact are not random at all. In reality, the fact that we have a statistical distribution absolutely does not mean that a non-deterministic phenomena is being considered.*

In view of all the debates about determinism and causality, it is naturally essential to be in full agreement on the exact meaning of the words *"determinism"* and *"causality"*. For me, *determinism and causality are present if the state of nature at the instant t is entirely determined from the states of nature at all previous instants. "Pure contingency"*, that is to say *indeterminism*, implies that the state of nature at the instant t is *independent, totally or partially*, of the states of nature at different previous instants.

Thus, in the case of trajectories of particles, the question that must be asked is the determination of the point at which a specified particle will arrive at a given instant. The problem is analogous to that relating to the throw of a die whose construction assures conditions of symmetry. Each of the faces has an *intrinsic frequency* of 1/6, but only some particular face will appear in a specific throw.

What Richard Feynman, for example, emphasizes in his discussion of experiments related to the passage of photons emitted from the same source through two holes, is that there is a *fundamental difference* between the appearance of a specific face in the throw of a die, and the appearance of some circumstance or another in quantum mechanics. In the first case, Feynman declares himself ready to admit that what we lack in order to make a precise prediction is a sufficiently detailed knowledge of the circumstances of the throw, it having been specified that we are dealing with a phenomenon completely governed by causality. But in the second case there is no longer causality of any type; *"chance"* is omnipresent, and the *ultimate reality* is the existence of *"probability waves"*. *Chance cannot be eliminated* and cannot be *"explained"*, in other words *cannot be referred to a hidden deterministic order.*[14] Consideration of Young's two-hole experiment is very illuminating here.

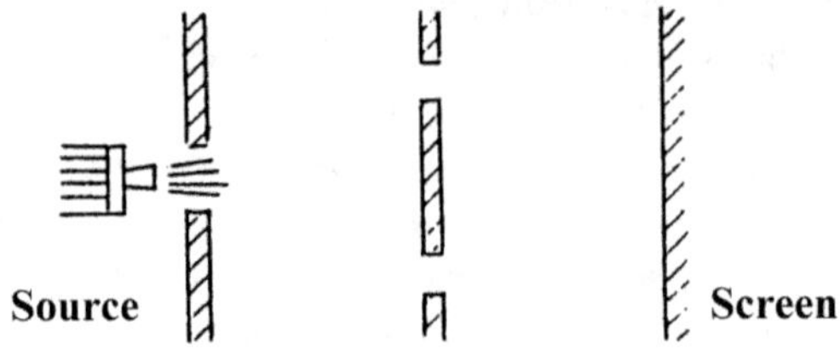


Commenting upon Young's two-hole experiment, Feynman writes:[15]

*"When we throw a die up into the air... we are ready to admit that we do not have sufficiently detailed knowledge to make a precise prediction; we therefore calculate the chances of getting one result or another. But what we argue here is that the probabilities are there from the beginning: chance is present in the fundamental laws of physics... It is not the absence of detailed knowledge that prevents us making predictions...*

*"It is not our ignorance of the internal workings, of the internal complications, that drags the appearance of probabilities into nature. It seems that they are intrinsic, which someone has explained in this way: 'Nature herself doesn't know which hole the electron is going to pass through.'*

*"A philosopher once wrote: 'For science to be able to exist, it is necessary that the same causes always produce the same effects.' Well, that's not what happens."*

[14] R. Feynman, 1950, *The concept of Probability in Quantum Mechanics*. Second Berkeley Symposium [144], and 1965, *The Character of Physical Law*, The M.I.T. Press, Chapter VI [145]; French translation, 1980, *La Nature de la Physique*, Seuil Editions, pp. 151-176.

[15] *id.*, pp. 145-147.

In fact, the analysis of the phenomena to which Feynman's argument applies, with all the authority that obtaining the 1965 Nobel Prize in Physics gives him, can only be acceptable if we reject the hypothesis of the existence of a vibratory *"ether"* in the sense of Newton, Fresnel, and Maxwell, *of which Feynman's analysis takes no account.* If on the contrary we admit the existence of this medium, then it seems quite possible, indeed elementary, to explain what Feynman considers as being inexplicable, and for which he saw no "explanation" other than the absence of any causal link.[16]

The denial of the existence of *"hidden parameters"* in quantum mechanics amounts to denying, in the case of throwing a die, the existence of different parameters that determine the appearance of one face or another.

*The partisans of indeterminism maintain, however, that the two phenomena are different in nature.* In the case of a die, they admit, following the classical point of view, that knowledge of all the conditions of the throw would make it possible to determine the face that appears. In the case of the particle, on the other hand, they maintain that it is impossible to specify the causes for a specific localization of the particle in question, and that therefore there is *total absence of causality, so that indeterminism results. "Chance", in other words "pure contingency", must constitute the ultimate and irreducible explanation of the phenomenon.*

It is indeed certain that this deduction appears inevitable *from the moment that we admit the model that the partisans of indeterminism consider.* But, precisely, the question is to know whether this model is indeed appropriate for *complete* representation of the observational data.[17]

---

[16] The experiments commented upon by Feynman derive from the experiments in 1909 of G.I. Taylor (Proc. Cambridge Phil. Society, 15, 114, 1909 [259]). For the interpretation of these experiments, see in particular Max Planck, 1925, *Initiations à la Physique* (Introduction to Physics), Flammarion, 1941, pp. 237-242 [218]; Max Born, 1956, *Atomic Physics*, pp. 79-80 and 101-102 [75]; Jean-Pierre Vigier, 1956, *Structure des micro-objets dans l'interprétation causale de la théorie des quanta* (Structure of micro-objects in the causal interpretation of the quantum theory), Gauthier Villars, pp. 4-12 and 104-105 [273]; Niels Bohr, 1958, *Atomic physics and human knowledge*, Wiley [72]; and Louis de Broglie, 1978, *Jalons pour une Nouvelle Microphysique* (Milestones for a New Microphysics), Gauthier-Villars, Chap. 2, pp. 36-38 [87].

[17] Suppose that for example photons travel through a medium, the vibratory *"ether"* of Newton, Fresnel, and Maxwell, and that Schrödinger's equation (§ D.2.1 above) effectively represents the principal part of the wave function corresponding to the movement of the *"ether"* associated with the movement of the photons, but that vibrations, other than those due to the particles considered, propagating through the *"ether"* involve a relatively small corrective term $\delta\psi$. For an ensemble of particles the average value of $\delta\psi$ is substantially null, and the distribution of the photons occurs in practice with a *frequency density* $\rho$; but for an isolated particle and at a given instant, this term $\delta\psi$ will decide the effective trajectory that will be observed - just as in the throw of a die, the conditions of the throw will on average have a null effect over a large number of throws, while still having a significant and determinative influence for a particular throw.

Therefore there is no need to invoke "chance" here, even on the sub-quantum level.

*The same fluctuations* $\delta\psi$ can explain why, at some given instant, certain atoms rather than others can disintegrate *due to* the particularly heightened intensity of the global fluctuations of the *"ether"* at a given instant in the regions occupied by the atoms under consideration.

In fact Jean Thibaud produced evidence for *periodic fluctuations over time of alpha decays*, suggesting that these decays are not produced entirely by chance. (1941, *Les caractères systéma-*

In fact, the *fundamental* error committed in all the indeterministic analyses of quantum mechanics *is the neglect of external influences propagated through the medium in which the experiments in question are performed.*

In the current state of our knowledge and in the light of the experimental facts, *it thus appears impossible to assert that there is necessarily a fundamental difference between the a priori indeterminacy in the appearance of some face in the throw of a die, and the indeterminacy resulting from consideration only of Schrödinger's equation for the movement of a particle, and that the latter indeterminacy is an intrinsic property of nature, all causality being excluded no matter what the future evolution of physics may be.*

---

*tiques de la distribution dans le temps des désintégrations alpha* (Systematic characterics of the time distribution of alpha decays), Annales de Physique, 2nd series, Vol. 15, April-June 1941, pp. 225-257 [261]; - 1944, *Sur les caractères systématiques de certaines distributions* (On the systematic characters of certain distributions), CRAS, Vol. 218, 1944, pp. 873-874 [262]; -1952, *La théorie des groupes et les distributions radioactives* (The theory of groups and radioactive distributions), in Louis de Broglie, *Physicien et Penseur* (Physicist and Thinker), Albin Michel, collection "*Les Savants et le Monde*" ("Scientists and the World"), 1952, 497 p. [263]

Thus it is impossible to say with Einstein and Infeld that *"there is not the slightest trace of any law governing the individual behavior of atoms"* in the phenomena of radioactivity (Albert Einstein and Leopold Infeld, 1938, *The Evolution of Physics*, p. 277 [136]).

## The crucial question

6 - In the final analysis, the choice between determinism and non-determinism comes down to the choice between two conceptions of the universe: *the deterministic one* characterized by laws that man could discover and make explicit, or by statistical regularities which are effects of a hidden order that it has not yet been possible to make explicit; *and the indeterministic one*, according to which everything, both the explicit laws that we have discovered and the statistical regularities that we notice, fundamentally results from only *"chance"* in the sense of this conception, that is to say, from *"pure contingency"*.

The *crucial* question is thus to determine whether or not the known facts compel us to the conception of statistical distributions *that do not result from any causal effect.*

In reality only the interpretation of certain models, based upon observations that have been insufficiently analyzed, has been presented as being counter to the deterministic conception of the universe, and, in reality, *none of the experimental phenomena* that are the object of quantum mechanics compel us to such a conclusion.

The choice *is not* between laws of nature that are essentially deterministic and statistical laws, because, to repeat yet again, statistical laws can result from totally deterministic phenomena, and here *Theorem (T)* justifies Laplace's conception of chance.

The *real choice is* between a world governed by fundamental laws based upon causality and deterministic in character, and a world dominated by *"pure contingency"*. Naturally we cannot decide *a priori*, but we can do so *a posteriori*. In fact, our *direct, tangible, and permanent* experience of the physical world leads us to the conclusion of causality, and we can assert that, in the current state of the matter, no decisive proof *based upon experiment* has been advanced that can permit us to reject the Laplacian concept of causality and *"chance"*.

# E THE REALITY OF THE ETHER

## *E.1 Conceptions of the ether*

That certain conceptions of the *"ether"* from the 17th to the 19th century were abusively simplistic, that they were revealed as mutually incompatible, and that as yet nobody has succeeded in specifying precisely and in a coherent manner the *indefinite equations* of the movement of this medium, capable of serving as the basis for building a unitary theory of physics, in agreement simultaneously with the three groups of phenomena that constitute *gravitation, electromagnetism, and quanta, does not and cannot mean that this medium does not exist*; because, in one manner or another, *under one name or another*, the theories of today - as those of yesterday - *inevitably* lead us to that conception, because *without the hypothesis of this medium, the material support for all physical phenomena, everything becomes incomprehensible, in particular everything that concerns action at a distance.*

That was in fact the point of view of Newton - who nevertheless was the creator of the concept of action at a distance - which he expounded in his celebrated letter to Bentley:[1]

> *"That one body may act upon another at a distance through a vacuum, without the mediation of anything else, by and through which their Action and Force may be conveyed from one to another is to me so great an Absurdity, that I believe no man who has in philosophical matters a competent Faculty of thinking, can ever fall into it".*

---

[1] Edmund Whittaker, 1951, *History of the Theories of Aether and Electricity*, I - *The classical theories*; II - 1900-1906, Nelson, 1951-1953, Vol. I, p. 28 [279], [280].

Arthur Koestler cites a more complete text (*Les Somnambules* (The Sleepwalkers), Calmann-Lévy, 1960, pp. 323 and 485 [164]):

> *"It is inconceivable that inanimate brute matter should, without the mediation of something else which is not material, operate upon and affect other matter, without mutual contact...*
>
> *And this is one reason why I desired you would not ascribe 'innate gravity' to me. That gravity should be innate, inherent, and essential to matter, so that one body may act upon another at a distance, through a vacuum, without the mediation of anything else, by and through which their action and force may be conveyed from one to another, is to me so great an absurdity, that I believe no man who has in philosophical matters a competent faculty of thinking can ever fall into it. Gravity must be caused by an agent acting constantly according to certain laws; but whether this agent be material or immaterial, I have left to the consideration of my readers."*

According to Whittaker (p. 28), Newton *"conjectured that the density of the aether might vary from place to place, and that bodies might tend to move from the denser parts of the medium to the rarer."*

This conviction was also shared by Maxwell who wrote at the conclusion of his *Treatise on Electricity and Magnetism:*[2]

> *"All theories lead us to conceive a medium in which propagation takes place; and, if we admit the hypothesis of this medium, I believe that it must hold a predominant place in the framework of our research, and that we must try to combine in our minds and to imagine all the details of its action: that is the object that I have constantly held before myself all through this Treatise."*

In view of all the knowledge we possess today, *it is not possible any longer to consider*, as did Einstein in 1905, *that the ether constitutes a superfluous hypothesis,*[3] because we are certain that gravitational and electromagnetic fields *really* exist, and these fields could not have any effective existence *without some material support.*

---

[2] Maxwell (1831-1879), French translation, Gauthier-Villars, Vol. II, 1889, p. 562 [194].

[3] Einstein, 1905, *Sur l'électrodynamique des corps en mouvement* (On the Electrodynamics of Moving Bodies), Gauthier-Villars, 1925, p. 3 [126].

> *"We can see that the introduction of a "luminiferous ether" becomes superfluous by the fact that our conception makes no use of a "space at absolute rest" endowed with specific properties, and does not make a velocity vector correspond to a point in empty space where electromagnetic processes take place."*

Here it is however proper to underline that the qualification "superfluous" *only* relates to the existence of a *"space at absolute rest"*.

In fact, Einstein himself wrote in 1920:[4]

*"The line to adopt at first glance ... seemed to be the following: - There is no such thing as the ether. The electromagnetic fields are not states of a medium but independent realities, which cannot be reduced to terms of anything else and are bound to no substratum, any more than are the atoms of ponderable matter...*

*"Electromagnetic fields appear as ultimate and irreducible realities, and at first sight it seems superfluous to postulate a homogeneous and isotropic ether, of which fields should be considered as representing the states...*

*"More attentive reflection, however, teaches us that this denial of the ether is not necessarily required by the principle of special relativity...*

*"To deny the existence of the ether means, in the last analysis, denying all physical properties to empty space. But such a view is inconsistent with the fundamental facts of mechanics...*

*"The facts have definitely deviated from the conception that space is physically empty...*

*"The main thing that matters is that, in addition to observable objects, another imperceptible entity has to be regarded as real, in order for us to be able to regard acceleration, or rotation, as something real...*

*"We may sum up as follows: According to the general theory of relativity space is endowed with physical qualities; in this sense, therefore, an ether exists. In accordance with the general theory of relativity space without an ether is inconceivable."*

---

[4] Einstein, 1920, *L'éther et la théorie de la relativité* (The ether and the theory of relativity), Gauthier-Villars, 1921, pp. 9, 11, 12, 11, and 15 [124].

In view of this citation, we see to what extent certain judgments by relativist authors appear unjustified.

Thus J.L. Synge (*Relativity, The Special Theory*, North Holland, 1956, pp. 4 and 161 [256]) wrote:

*"Out of the work of Lorentz and Poincaré the special theory of relativity emerged, Einstein (1905) clearing up philosophical difficulties by destroying the concept of ether...*

*"The (Michelson) result has been of tremendous historical importance because it helped to loosen the hold of the ether-theory, but that does not concern us since the concept of the ether is dead and gone".*

In the same vein, Ugarov (*Special Theory Relativity*, Mir Editions, 1974, p. 294 [269]) wrote:

*"Special relativity has put an end to the hypothesis of the ether... The principle of relativity definitively bars the way to any possibility of detecting the ether experimentally."*

*In fact the ether, which constitutes a physical reality, should be considered neither as being homogeneous and isotropic, nor as being deprived of all movement..*

Its *real* nature still escapes us, in particular due to the fact that we are still incapable of elaborating a unitary theory of gravitation, electromagnetism, and quanta. *But this incapacity only demonstrates the incapacity of our minds, and cannot be considered as being a proof of the non-existence of the ether.*

*It seems completely impossible to imagine that actions at a distance can take place without the existence of an intermediate medium that transmits them. A space without ether is totally inconceivable.*[5]

The ether is the material support of the totality of all gravitational, electrical, magnetic, and optical phenomena. *We know certain laws of its manifestations, but currently we really know nothing of its ultimate constitution.*

---

[5] For once, I am in complete disagreement with Henri Poincaré when he wrote: *"It is not important to us whether the ether really exists: that is an affair for the metaphysicians. This hypothesis is convenient for the explanation of phenomena"* (1902, *Science and Hypothesis*, Flammarion, 1927, pp. 245-246 [220]).

Certainly, the existence of the ether is a convenient hypothesis, but it is an *inevitable* hypothesis, and research into its properties is certainly a condition for realization of very great progress, *and in any case for the construction of a unitary theory* of physics that can assemble gravitation, electromagnetism, and quanta into one and the same homogenous and coherent ensemble.

On the concept of *"convenience"*, see below, p. 554, the very pertinent reflections of Paul Painlevé.

Contrary to a too common opinion, the experiments of Michelson, Morley, and Miller did not demonstrate the non-existence of the *"ether"*, but simply that the hypothesis that the *"ether"* at every point of space is fixed with respect to the fixed stars cannot be retained.

Even though the concept of the ether appeared to have been excluded from the domain of science since the advent of the theory of relativity, it reappeared little by little under other terminology: - that of an *"associated fluid"* (Paulette Fevrier, *L'interprétation physique de la mécanique ondulatoire et des théories quantiques* (The physical interpretation of wave mechanics and quantum theories), Gauthier-Villars, 1956, pp. 114-115 and 151-198 [143]); - that of a *"hydrodynamic representation"* (id. pp. 165-166); - that of a *"pilot fluid"* (J.P. Vigier, *Structure des micro-objets dans l'interprétation causale de la théorie des quanta* (Structure of micro-objects in the causal interpretation of the quantum theory), Gauthier-Villars. 1956, p. 110 [273]); - that of a *"relativistic gas"* (J.L. Synge, *The Relativistic Gas*, North Holland, 1957 [257]); - that of a *"probability fluid"* (E. Durand, *Mécanique quantique: I, Équation de Schrödinger* (Quantum Mechanics, *I,* Schrödinger's Equation), Masson, 1970, pp. 67-68 [120]); that of a *"quantum fluid"* (F. Fer, *L'irréversibilité: fondement de la stabilité du monde physique* (Irreversibility, foundation of the stability of the physical world), Gauthier-Villars, 1977, Chapter V, p. 56-79 [142]); etc.

*Our situation corresponds to the myth of Plato's cave.* The shadows that we perceive on the walls are more and more precise, but *the ultimate nature of the reality* they represent still escapes us, and it will probably continue to escape us for the complete foreseeable future.[6]

[6] *In fact, the more our knowledge advances, the more incomprehensible the world in which we live appears to be.*

No more than twenty-five centuries ago, we could not imagine that the world had a beginning and will have an end. But the two contrary hypotheses are also both unimaginable.

In the same way, a finite universe is just as inconceivable as an unlimited universe.

In the same way again, that there is a limit to the infinitely small, or that there is no such limit, both represent equally inconceivable propositions.

It is a fact that here we are *inevitably in the domain of the unverifiable and in the domain of metaphysics.*

## *E.2 An experimental evidence*

The *most striking* underlying regularity in the observations of Miller is certainly *the perpendicularity of the hodographs of the velocities to the average direction of the azimuths.*[1]

This is an observational datum that is *incontestable and inescapable*, and it has *considerable* implications: indeed, *at one stroke*, it invalidates both *the classical theory and the theory of relativity.*[2]

This way, what this regularity, emphasized by the interferometric observations of Miller, shows *is not what was expected, i.e.* a direct correlation with the orbital and cosmic velocity of the Earth, *but a different phenomenon, an anisotropy of space.*

A great number of hypotheses may be considered for interpretation of this regularity in Miller's observations. According to the one undoubtedly most likely, the movement of translation of the Earth drags the ether along with it completely, and there is also an anisotropy linked to astronomical phenomena. According to this interpretation, the velocity of the Earth does not exert any *direct* influence on the interferometric observations.[3] *The observed velocity differences only correspond to the anisotropy of space.*[4]

---

[1] See above, *Chapter IV*, § D.4.3, pp. 404-408, and *Graphs X and XI*, pp. 407-408.

[2] It invalidates the classical theory because the hodographs are not symmetrical about the meridian (*Chapter IV* above, § 6.2.2., p. 421), and it invalidates the theory of relativity because it demonstrates the non-invariance of the speed of light (*Chapter IV* above, § F.2.5, p. 424).

[3] In support of this interpretation, we may consider that the hodograph $H_T$ corresponding to the velocity of the Earth is symmetrical with respect to the meridian, while the observed hodograph $H$ and its fitting to an ellipse $H^*$ appear to be perpendicular and symmetric with respect to the average azimuth $\overline{A}$ of Miller, which is different from the meridian.

In fact, and *a priori*, any influence of the hodograph $H_T$ upon the hodographs $H$ and $H^*$ should compromise the perpendicularity and the symmetry of the hodograph $H$ with respect to the azimuth $\overline{A}$, but we see that this is not the case.

[4] See above, *Chapter IV*, § F.2.4, and F.2.5, notes 7 and 8, pp. 422-424.

## *E.3 The foundations of a unitary theory of physics*

These considerations clearly show the directions in which we should proceed in the search for a unitary theory.

The fact is that the ether *is constantly manifesting itself by phenomena that can be represented by perfectly definite laws*, not rigorously, but to a certain approximation which, in the case of gravitation, is of the order of $10^{-6}$.

Researching the properties of the ether from *a priori* considerations is an enterprise that has always failed. By contrast, if one is completely convinced that the properties of the ether are the indefinite equations that characterize gravitation, electrical and magnetic phenomena, and optical phenomena, then research into the structure of the ether *amounts to a search for a synthetic formulation that includes all these relations in a homogenous and coherent structure.*[1, 2]

---

[1] I am personally convinced that it is possible to derive the formulation of gravitation from the formulation of electromagnetism.

Certain interesting attempts have already been made, notably those of L. Décombe: *Théorie électronique de la gravitation* (Electronic theory of gravitation), CRAS, 17 March 1913 [103]; - *Les pellicules sphériques électrisées. Calcul direct de la constante de la gravitation en fonction des constantes d'Avogadro, de Faraday, de Rydberg et de Planck* (Small electrified spherical shells. Direct calculation of the constant of gravitation as a function of the constants of Avogadro, Faraday, Rydberg, and Planck), CRAS, 24 November 1924 [104]; *Les pellicules électrisées et les séries spectrales* (Electrified shells and spectral series), CRAS, 5 December 1927 [105].

See also G. Darrieus: *Sur une relation entre la constante de la gravitation et les autres constantes fondamentales* (On a relation between the constant of gravitation and the other fundamental constants), CRAS, 19 July 1926 [100].

See also, for example, F. Prunier, 1932, *Essai d'une physique de l'éther* (Essay on a physics of the ether) [231]; and Henri Varcollier, 1949, *Fondements de l'explication électromagnétique de la gravitation universelle* (Foundations of an electromagnetic explanation of universal gravitation) [271].

[2] As perfect as Newton's laws of universal gravitation may appear to be, *it is a fact that they cannot explain the very numerous regularities that are present in the solar system* - in particular, the regularities that characterize the trajectories of the planets and that seem to correspond to a system of stationary waves.

On these regularities, see particularly: - Gaussin, *Lois concernant la distribution des astres du système solaire* (Laws concerning the distribution of the bodies of the solar system), CRAS, vol. 90, 1880, p. 518 [149]; - Gaussin, *Lois concernant la distribution des astres du système solaire* (Laws concerning the distribution of the bodies of the solar system), CRAS, vol. 90, 1880, p. 593 [150]; - Belot (E.), *Formule applicable aux durées de rotation directe des planètes et du soleil* (Formula applicable to the direct rotation periods of the planets and the Sun), CRAS, vol. 143, 1906, p. 1126 [65]; Belot (E.), *Sur la distance des satellites d'Uranus et de Jupiter* (On the distance of the satellites of Uranus and Jupiter), CRAS, April 1907, p. 885 [66]; - Delauney, *Lois des distances des satellites du soleil* (Laws of the distances of the satellites of the Sun), Gauthier-Villars, Paris, 1909 [106]; - Butavand (F.), *Les lois empiriques du système solaire et les harmoniques tourbillonaires* (The empirical laws of the solar system and vortex harmonics), Gautier-Villars, Paris, 1913 [90]; - Ollive (F.), *Sur le système solaire* (On the solar system), CRAS, vol. 157, 1913, p. 1501 [206]; - Blagg (M.A.), *On a suggested substitute for Bode's Law, Royal Astronomical Society*, Vol. 73, 1913, p. 414 [71]; - Demozay (L.), *Relations remarquables entre les éléments du système solaire* (Remarkable relations between the elements of the solar system), Gauthier-Villars, Paris, 1919 [108]; Delauney, *Problèmes astronomiques* (Astronomical Problems), Gauthier-Villars, 1920 [107]; - Vilar (A), *Notes sur les distances des planètes au soleil* (Notes on the distances of the

*With such a conception, all particles, and generally speaking all matter, only represent local singularities in the ether and particular solutions of this synthetic formulation.*

This was also the profound conviction of Einstein when he wrote:[3]

> *"It would naturally be considerable progress if we could succeed in uniting the gravitational field and the electromagnetic field in a unique representation. Only in this way could the era of mathematical physics, inaugurated by Faraday and Maxwell, culminate in a satisfying result. Then the opposition of ether and matter would disappear, and all physics would represent the same coherent system of ideas."*

and as he further wrote:[4]

> *"A really rational theory ought to deduce the elementary particles (electron, etc.), and not posit them* a priori*."*

One cannot but follow J.P. Vigier when he writes:[5]

> *"I cannot prevent myself believing that, in conformity with the dream of Descartes who idealized theoretical physics as the geometrical study of matter in movement, nature can be reduced to a unique substance, material, geometrically describable, whose successive forms, in perpetual transformation, can account for the prodigious diversity of elementary phenomena."*

---

planets from the Sun), Jouve, Paris, 1923 [275]; - Belot (E.), *La naissance de la Terre et de ses satellites* (The birth of the Earth and its satellites), Gauthier-Villars, 1931 [67]; - Bourgeois (P.) and Cox (J.F.), *Sur la répartition des inclinaisons et des excentricités des orbites des petites planètes* (On the distribution of the inclinations and the eccentricities of the orbits of the minor planets), CRAS, T. 198, 1934, p. 53 [79]; - Prunier (F.), *Quelques observations et expériences nouvelles* (Some observations and new experiments), Chapter I, *Archives des Sciences Physiques et Naturelles* (Archives of the Physical and Natural Sciences), Institute of Physics of the University of Geneva, 1946 [232].

See particularly : - Butavand, 1913; - Demozay, 1919; - Prunier, 1946.

[3] Einstein, 1920, *L'éther et la théorie de la relativité* (The ether and the theory of relativity), 1921, id., p. 15 [124].

[4] Einstein, Letter of 10 September 1952, *Correspondence with Michele Besso, id.*, p. 283 [135].

[5] J.P. Vigier, 1953, *Physique relativiste et Physique quantique* (Relativistic Physics and Quantum Physics), in Louis de Broglie, *La Physique quantique restera-t-elle indéterministe ?* (Will Quantum Physics Remain Indeterministic?), Gauthier-Villars, p. III [86].

*This is a concept with which I entirely agree.* Contemporary physics is currently passing through a period of transition, even a crisis, since the experimental data can currently appear to be contradictory. But sooner or later, in essential matters, physics will surmount this crisis.

*The current crisis of physics, if crisis there be, is for the greatest part nothing other than a crisis of intelligence.*

## *E.4 The ether, the inescapable explanatory factor*

*Whatever point of view one takes, the ether appears to be an inescapable element for explanation.*

To limit ourselves here to the movement of the Earth, it is a fact that the experiments of Foucault and of Michelson and Gale have demonstrated that it rotates, but if it rotates, *it rotates with respect to something local* as soon as we have excluded the incredible hypothesis of actions at a distance, and *this local something can be nothing but the surrounding ether.*

In the same way, all the phenomena analyzed in the first five chapters of this work demonstrate *a very strong correlation of the observations with the position of the Earth on its orbit. None of these phenomena can be conceived as anything other than effects transmitted via the ether as an intermediary.*

It is the same with the association of waves with photons, with the profound nature of quanta, and with the indeterministic appearance of certain phenomena. *They are only inexplicable if one rejects the existence of an intermediate medium, the ether.*[1]

As Eddington wrote:[2]

> *"In any case the physicist does not conceive of space as void. Where it is empty of all else there is still the aether. Those who for some reason dislike the word 'aether', scatter mathematical symbols freely through the vacuum, and I presume that they must conceive some kind of characteristic background for these symbols. I do not think any one proposes to build even so relative and elusive a thing as force out of entire nothingness."*

---

[1] See above, § D.3.5, pp. 531-534.

[2] A.S. Eddington, *The Nature of the Physical World*, Cambridge University Press, 1927, p. 137 [122].

If so many errors have been perpetrated in the past, if science has been led astray into dead ends, it is because she has too often allowed herself to be guided by *a priori* views, and too often she has valued those views over objective examination of the facts. But as Pareto wrote:[3]

> *"It is false to believe that it is possible exactly to discover the properties of concrete facts by reasoning upon ideas that we have conceived for ourselves* a priori *before those facts, without modifying these concepts by comparing, a posteriori, their consequences with the facts."*

Poincaré wrote in 1905, in *"The Value of Science"*:[4]

> *"We may be permitted to hope that the complication of physical phenomena conceals us... some kind of simple, still unknown cause."*

*This simple cause - is it not this intermediate medium, the ether, whose properties determine all physical phenomena?*

---

[3] Vilfredo Pareto, 1909, *Manuel d'économie politique* (Manual of Political Economy), Giard, 1927, p. 13 [213].

[4] Henri Poincaré, 1905, *La Valeur de la Science* (The Value of Science), id., p. 162 [222].

# Chapter VII
# THE OBSERVATIONAL DATA AND THE NECESSITY FOR REVISION OF CERTAIN POSTULATES OF CONTEMPORARY THEORIES

*"The most beautiful mathematical theories are only speculations in the air as long as they have not found a solid point of application in the results of experiment."*

Max Planck *

The objective of the present Chapter is to present, as briefly as possible, the observational data that are incompatible both with the theory of relativity and classical theories.

*Section A* reviews the discussions throughout the 19th century and during the earlier part of the 20th century *concerning establishing connection between optical phenomena and the dual movements of the Earth with respect to the fixed stars - its rotation and its translation along its orbit.*

*Section B* reviews the experimental data of *Chapters IV and V*, from which, *contrary to assertions too often advanced*, it follows that Michelson's interferometric experiment *did not yield negative results*, and that Miller's observations of 1925-1926 showed, on the one hand that the speed of light is not invariant in all directions, and on the other that its variations are strongly correlated with the position of the Earth along its orbit. The very foundations of the theory of relativity are thus invalidated by the experimental data.

*Section C* presents *certain critical observations* on the development of the theory of relativity, which is a grandiose construction, but whose foundations are in reality uncertain and fragile.

*Section D* briefly reviews the anomalies observed *in relation to the classical theories of mechanics and optics*, which correspond to an anisotropy of space demonstrated by the observations of the paraconical pendulums with anisotropic support and with isotropic support, by the deviations of optical sightings at marks and collimators, by the optical observations of Esclangon, and by the interferometric observations of Miller. All these anomalies lead to the same conclusion: *the space in which matter swims is not isotropic*, and the magnitude of this anisotropy is *totally inexplicable* in the framework of currently accepted theories.

---

* Max Planck, 1925, *Initiations à la Physique* (Initiation into Physics), Flammarion, 1941, p. 171 [218].

*Section E* reviews the *fundamental principles of scientific analysis* which should be applied in the re-examination of both the theory of relativity and classical theories.

Finally, *Section F* points out that, independently of any hypothesis and any theory and *only as a result of the experimental data*, currently accepted theories *must be re-examined and revised.*

# A THE BIRTH OF THE THEORY OF RELATIVITY - PRECONCEIVED IDEAS

Nothing is more interesting in connection with analysis of physical reality than the discussions that took place at the end of the 19th century and at the start of the 20th on the subject of the influence of the movement of the Earth upon mechanical, optical, and electromagnetic phenomena,* and that led to the formulas of Lorentz,[1] then to Poincaré's analysis,[2] then to the special theory of relativity of Einstein,[3] and later to his theory of general relativity.[4, 5]

---

* For all these discussions, see in particular the memoirs of Mascart, 1872-1874 [191, 192], of Oliver Lodge, 1893 [182], and of Sesmat, 1937 [246], already cited in *Chapter IV* above, §C.l, note 1, p. 382.

See also: - Chwolson, 1914, *Traité de Physique* (Treatise on Physics), Vol. V, Book III, *Champ magnétique variable* (Varying Magnetic Fields), Chapter IV, *The fundamentals of the electron theory*, and Chapter V, *The principle of relativity*, pp. 175-262 [94]; - René Dugas, 1950, *Histoire de la Mécanique* (History of Mechanics), Book IV, Chapter X, *Discussion of Newtonian principles*, and Book V, Chapters I and II, *Special and general relativity*, pp. 419-519 [114]; - Edmund Whittaker, 1951 and 1953, *History of the Theories of Aether and Electricity*, Vol. I, Chapter XIII, *Classical Theory in the Age of Lorentz*, pp. 386-428, and Vol. II, Chapter II, *The Relativity Theory of Poincaré and Lorentz*, pp. 27-77 [279]; and Henri Bouasse, 1925, *Propagation de la lumière* (Propagation of light), Delagrave, pp. 1-156 [77].

1 H.A. Lorentz (1853-1928), 1895, *Versuch einer Theorie der elektrischen und optischen Erscheinungen in bewegter Körpern* (Attempt at a theory of electrical and optical phenomena in moving bodies) [184]; and 1904, *Phénomènes électromagnétiques dans un système en mouvement avec une vitesse quelconque inférieure à celle de la lumière* (Electromagnetic phenomena in a system moving with any velocity smaller than that of light) [185].

2 See particularly: - Henri Poincaré (1854-1912), 24 september 1904, *L'état actuel et l'avenir de la physique mathématique* (The present and the future of mathematical physics), Bulletin des sciences mathématiques (Bulletin of Mathematical Science), 1905, pp. 302-334 [221]; - 1905, Note to the Academy of Sciences, 5 June 1905, *Sur la dynamique de l'électron. Oeuvres de Henri Poincaré* (On the dynamics of the electron, Works of Henri Poincaré), Vol. IX, Gauthier-Villars, 1954, pp. 489-493 [227]; - 1905, *Sur la dynamique de l'électron* (On the dynamics of the electron), Mathematical Circle of Palermo, 23 July 1905, Works, id., pp. 494-550 [223].

3 See particularly Albert Einstein (1879-1955), 1905, *Zur Elektrodynamik bewegter Körpern* (On the Electrodynamics of Moving Bodies), Annalen der Physik (Annals of Physics), 17, pp. 891-921, (article received on 30 June 1905 and published on 26 September 1905 by the German review *Annalen der Physik*) [123]; French translation, *Sur l'électrodynamique des corps en mouvement*, Gauthier-Villars, 1925 [126].

See also: - Peter Michelmore, *Einstein*, Encyclopaedia Britannica, Vol. VI, 1975, pp. 510-514 [195]; - Peter Gabriel Bergmann, *Relativity*, Encyclopaedia Britannica, Vol. XV, 1975, pp. 581-589 [68].

4 Einstein, 1916, *Die Grundlage der allgemeinen Relativistätstheorie* (The Foundations of the General Theory of Relativity), French translation, *Les fondements de la théorie de la relativité générale*, Hermann, 1933 [127]; - 1916, *La théorie de la relativité restreinte et générale* (The theories of special and general relativity), Gauthiers-Villars, 1976 [134]; - 1920, *L'éther et la théorie de la relativité* (The ether and the theory of relativity), Gauthier-Villars, 1921 [124].

5 With Zeeman, Lorentz received in 1902 the second Nobel Prize in Physics for his work on magnetism; and Einstein received in 1922 the Nobel Prize in Physics for 1921 (reserved in 1921), "*for his services to Theoretical Physics, and especially for his discovery of the law of the photoelectric effect*".

*In fact, all these discussions were dominated, and indeed vitiated, by preconceived ideas about the nature of the ether, its movements, and its deformations.*[6]

[6] The historical development of all these theories is *extremely complex*. Very many authors contributed. The reader will find a detailed exposition in the references given in note (1) above (p. 550). See also the references of note (1) of § A.5 below (p. 565).

*Here I limit myself to what appears essential from the general viewpoint of the present work.*

## *A.1 The experimental results in 1900*

In 1900 the totality of available experimental results, in particular *the Foucault pendulum experiment in 1851 and the Michelson and Morley interferometric experiment in 1887*, had led to a double conclusion:

- *the first, corresponding to the Foucault experiment, was that the rotational motion of the Earth had been demonstrated by a purely terrestrial experiment;*
- *the second, corresponding to the supposed "negative" result of the Michelson and Morley experiment in 1887, was that it was impossible to demonstrate the translational motion of the Earth by a purely terrestrial experiment.*[1]

The experiment of Michelson and Morley in 1887 did actually demonstrate a speed of the order of 8 kilometers per second,[2] but *this result was considered to be of the same order of magnitude as the possible error, and was attributed to observational error.*

*In any case, it was considered to be a* "well-established" *result that all attempts to detect the translational motion of the Earth had failed.*[3]

[1] This double conclusion *rested on a double hypothesis*: that the ether is isotropic and immobile with respect to the fixed stars, and that it participates neither in the rotational motion of the Earth nor in its translational motion with respect to the fixed stars.

[2] Miller. 1933, id., p. 206. See *Chapter IV*, §E.2, above, pp. 414-415.

Michelson (1852-1931) received the *Nobel Prize in Physics* in 1907 for his interferometric work.

[3] Thus in 1905, in "*La Valeur de la Science*" (The Value of Science), Poincaré wrote: "*All attempts to measure the speed of the Earth with respect to the ether have yielded negative results.*" [222]

## *A.2 The rotation of the Earth and its privileged reference frame*

• The fact cannot be too strongly insisted upon that, in all the discussions on the effects of the movements of translation and rotation of the Earth, *nobody has ever argued at any time that the rotation of the Earth should inevitably be attached to a privileged reference frame, i.e. to a Galilean reference frame of classical mechanics.*

From the optical point of view, the rotation of the Earth *was demonstrated* by the interferometric experiment of Michelson and Gale in 1925.[1] The average of the observations *corresponded exactly, within 2.6%, to the theoretical value.*

The passionate discussions and interpretations relating to the optical effects of the movement of translation of the Earth are in *sharp contrast* with the commentaries upon the optical effects of its rotation.

• However in 1902, in *Science and Hypothesis*, Henri Poincaré asked the question. He wrote:[2]

> *"Here we find a very important and even slightly disturbing question. I have said that the principle of relative motion was not for us simply a result of experiment; and that* a priori *any contrary hypothesis would be repugnant to the mind.*
>
> *"But, then, why is the principle only true if the motion of the movable axes is uniform and in a straight line? It seems that it should impose itself upon us with the same force if the motion is varied, or at any rate if it reduces to a uniform rotation...*
>
> *"Is there any meaning in saying that the Earth rotates? If there is no absolute space, can a thing turn without turning with respect to something; and, on the other hand, how can we admit Newton's conclusion and believe in absolute space?...*

---

[1] Michelson and Gale, 1925, *The Effect of the Earth's Rotation on the Velocity of Light*, The Astrophysical Journal, April 1925, pp. 137-145 [197].

[2] Poincaré, 1902, *Science and Hypothesis*, Flammarion, 1927, pp. 137-141 [220].

> *"It is more convenient to suppose that the earth rotates, because thereby the laws of mechanics are expressed in much simpler language*
>
> *"That does not prevent absolute space - that is to say, the reference to which we must refer the earth to know if it really does rotate - from having no objective existence."*

However, in reality that was indeed *a rather excessive position, indeed an untenable one*.

In fact, in 1905, in *The Value of Science*,[3] Poincaré accepted that the proposition that the Earth rotates is truer than the proposition that it is fixed, in the sense that only the former makes possible a coherent explanation of multiple phenomena both cosmic and terrestrial, and he wrote:

> *"A physical theory is the truer, the more true relations it brings to light.*
>
> *"In saying the earth rotates, I affirm that all these phenomena are intimately related, and that statement is true and remains true, even though there is not and cannot be any absolute space."*

As Paul Painlevé underlined in 1904:[4]

> *"The same principle that makes us affirm the existence of the Earth also makes us affirm its absolute rotation: the rotation of the Earth has roughly the same certainty, or more exactly has the same order of certainty, as the Earth's existence itself. Another way of putting this, which only appears paradoxical, is*: if it is a convention to say that the Earth rotates, it is similarly a convention to say that it exists, and both of these conventions are justified for identical reasons."

It could not be expressed better.[5]

---

[3] Poincaré, 1905, *The Value of Science*, Flammarion, 1927, pp. 272-273 [222].

[4] Paul Painlevé, 1904, *The Axioms of Mechanics and the Principle of Causality*, in *Les Axiomes de la Mécanique* (The Axioms of Mechanics), Gauthiers-Villars, 1955, p. 79 [208].

[5] In the texts of both Poincaré and Painlevé, *the meaning of the adjective "absolute" needs to be made more precise*.

If absolute space is the space linked to the fixed stars, then translation and rotation of the Earth correspond to relative movements. In fact, Painlevé's commentaries suggest that a reference frame is absolute if it is stationary *with respect to the totality of bodies in the universe* (Painlevé, 1904, id., p. XIV). Such a definition is evidently not operational. In practice, an absolute reference frame is a reference frame whose axes have directions that are fixed with respect to the fixed stars.

Certainly only relative movements exist and absolute movement does not; but the Earth has to turn with respect to something, and, if action at a distance is excluded, *this something can only be the medium with respect to which the Earth rotates locally,* i.e. *the ether.*

## A.3 *The movement of translation of the Earth and Poincaré's principle of relativity*

### Poincaré's principle of relativity

1 - *According to the classical theory* there exists a privileged system of reference axes, *considered to be absolute: the axes of Copernicus*, whose directions are fixed with respect to the distant stars (so-called "*fixed*" stars) and whose origin is the center of gravity of the solar system. *These axes are considered as being linked to the ether*, which itself is considered *as immobile* with respect to the "*fixed stars*". And *Galilean axes* may be thought of as axes that move in *uniform and rectilinear translation* with respect to the Copernican axes.[1]

The *admitted* impossibility of determining *the movement of translation of the Earth along its orbit* from Michelson's experiment led Henri Poincaré, *in 1899*, to cast doubt on the possibility of demonstrating the absolute movement of the Earth by optical experiments, and to formulate the *Principle of Relativity*, according to which optical phenomena only depend upon the relative movement of the material bodies involved.[2]

---

[1] Paul Painlevé, 1922, *Les axiomes de la mécanique* (The Axioms of Mechanics), Gauthier-Villars, 1955 [208].

[2] In his work of 1899, *Electricité et Optique* (Electricity and Optics) (Gauthier-Villars, 1954) [228], Henri Poincaré wrote:

> *"Thus, a supplementary hypothesis has been imagined. All bodies undergo a shortening of their length by a factor of* $\frac{1}{2\times10^9}$ *in the direction of movement of the Earth.*
>
> *"This strange property would seem like a helping hand given by Nature to make it impossible for the movement of the Earth to be revealed by optical phenomena. I cannot be satisfied by this... I consider it very likely that optical phenomena only depend upon the relative movements of the bodies in question... and that, rigorously." (p. 536)*
>
> *"Experiment has revealed a crowd of facts that can be summarized in the following formulation: It is impossible to demonstrate the absolute movement of matter, or better, the relative movement of ponderable matter with respect to the ether; it is only possible to demonstrate the movement of ponderable matter with respect to other ponderable matter." (p. 613).*

According to an article by Lodge in *Nature* on 16 June 1892, the contraction of bodies was first suggested by George Francis FitzGerald (1851-1891). (See Whittaker, *The History of Aether and Electricity, The Classical Theories*, 1951, id., p. 404 [279]).

In 1902, in *Science and Hypothesis*, Poincaré again advances an analogous point of view.[3, 4]

## Lorentz's memoir of 1904

2 - *On 27 May 1904*, in a memoir published by the Amsterdam Academy of Sciences, Lorentz was led to establish a correspondence between the variables $x, y, z, t$ of classical mechanics and the variables $x', y', z', t'$ corresponding to a trihedral $x', y', z'$ in uniform rectilinear motion at velocity $\vec{v}$ with respect to the trihedral $x, y, z$. The time t' attached to the trihedral $x', y', z'$ is termed "*local time*" by Lorentz. To this formulation there corresponds a contraction of the body along the direction of the velocity $\vec{v}$, as well as a timing $t'$ specific to the frame $x', y', z'$.

*From this formulation, the impossibility of detecting the movement of the Earth by Michelson's experiment follows.*[5]

---

[3] Henri Poincaré wrote:

> *"There is no absolute space, and we only conceive of relative movements... There is no absolute time: to say that two periods are equal is an assertion that has no intrinsic meaning, and that can only acquire a meaning by a convention... We have no direct intuition of the simultaneity of two events occurring in two different places... Mechanical facts might be enunciated with reference to a non-Euclidean space..."*

(Henri Poincaré, 1902, *Science and Hypothesis*, Flammarion, 1927, p. 111; see also pp. 201-202 [220]).

[4] For *Poincaré's Relativity Principle*, see the very remarkable analysis by Jules Leveugle, *Poincaré et la Relativité* (Poincaré and Relativity), La Jaune et la Rouge (the magazine of the Ecole Polytechnique), April 1994, pp. 31-50 [173]; also see his response to the observations presented: *A propos de Poincaré et la Relativité* (Concerning Poincaré and Relativity), La Jaune et la Rouge, November 1994, pp. 7-14 [174].

[5] Lorentz, 1904, *Phénomènes électromagnétiques dans un système en mouvement avec une vitesse quelconque inférieure à celle de la lumière* (Electromagnetic phenomena in a system moving with any velocity smaller than that of light) [185], (note 2 above of the *Introduction* to *Section A*, p. 550).

According to Lorentz, there is an essential difference between the two systems $x, y, z, t$ and $x', y', z', t'$. In his Homage to Poincaré, he wrote (p. 686):

> *"In the first - this is my thinking - coordinate axes having a fixed position in the ether and what may be called true time are used; in the second, by contrast, auxiliary magnitudes are employed."*

(Lorentz's Homage to Poincaré, *Deux Mémoires de Henri Poincaré sur la Physique mathématique* (Two Memoirs of Henri Poincaré on mathematical physics), Acta Mathematica, 1921, Vol. 38 [186]; this memoir was reproduced in *Oeuvres de Poincaré* (Works of Poincaré), Vol. IX, Gauthier-Villars, 1954, pp. 683-695 [227]).

The two memoirs of Poincaré analyzed by Lorentz are his memoir of Palermo on the *Dynamics of the "electron"* (*Works*, IX, id., pp. 494-550), and his memoir of 1911 *"Upon the quantum theory"* (*Works*, id., pp. 626-653). Lorentz's commentaries are *essential* for appreciation of Poincaré's contributions to relativity theory.

## Poincaré's memoir of St. Louis

3 - In his appearance in *September 1904* at an international conference in St. Louis, USA on "*The present and the future of mathematical physics*",[6] Poincaré presented "*The Principle of Relativity as a fundamental principle of physics*", fully comparable with the principles of conservation of energy, of degradation of energy, of the equality of action and reaction, of the conservation of mass, and of least action". According to Poincaré:[7]

> *"From the Principle of Relativity, the laws of physical phenomena should be the same for a fixed observer and for an observer in uniform movement, so that we neither have nor can have any way of determinining whether or not we are involved in such a movement."*

In conclusion of his memoir of St. Louis, Poincaré wrote:[8]

> *"Perhaps we should construct a completely new mechanics which we can only glimpse, in which, with inertia increasing with speed, the speed of light cannot be exceeded."*

---

In a conference in 1910, Lorentz wrote (Whittaker, 1953, id., vol. II, p. 36 [280]):

> *"The concept (which the present author would dislike to abandon) that space and time are something completely distinct and that a "true time" exists (simultaneity would then have a meaning independent of position..."*

During the Mt. Wilson Conference of 1927 (*Chapter IV* above, p. 382, note 2) he also declared (*The Astrophysical Journal*, December 1928, p. 350 [198]):

> *"I introduced the conception of a local time which is different for different systems of reference which are in motion relative to each other. But I never thought that this had anything to do with the real time. This real time for me was still represented by the old classical notion of an absolute time, which is independent of any reference to special frames of co-ordinates. There existed for me only this one true time. I considered my time transformation as being only a heuristic working hypothesis."*

[6] This presentation was published in the *Bulletin des Sciences mathématiques* (Bulletin of Mathematical Sciences) of December 1904 (note 3 above of the *Introduction* to *Section A*, p. 550) [221]. It was not reproduced in *Volume IX* of the *Works* of Poincaré.

[7] Bulletin of Mathematical Sciences, id., p. 306.

[8] id., p. 324.

**Poincaré's Memoir of Palermo, and his Note to the Academy of Sciences**

4 - *In July 1905* Poincaré composed a memoir entitled "*On the Dynamics of the Electron*" as a presentation to the Mathematical Circle of Palermo, and after reflection on Lorentz's memoir of 1904.[9]

A summary of this memoir was the subject of a *Note* to the Academy of Sciences of Paris, *Sur la Dynamique de l'Electron* (On the Dynamics of the Electron), on 5 June 1905.[10]

According to Poincaré:

> *"The essential point, established by Lorentz, is that the equations of the electromagnetic field should not be altered by a certain transformation, which I shall call by the name of Lorentz."*

In fact, Poincaré gave the Lorentz transformation the following form, which was destined to become classic:

$$(1) \qquad x' = k(x - vt) \qquad y' = y \qquad z' = z$$

$$t' = k\left(t - \frac{vx}{c^2}\right) \qquad k = 1/\sqrt{1 - \frac{v^2}{c^2}}$$

where $v$ is the speed of the trihedral $x', y', z'$ with respect to the trihedral $x, y, z$, $c$ is the speed of light, and $t'$ represents the local time according to Lorentz's expression. Poincaré designated the totality of these relations by the expression "*the Lorentz transformation*".[11]

---

[9] Henri Poincaré, July 1905, *Sur la Dynamique de l'Electron* (On the Dynamics of the Electron), Rendiconti del Circolo mathematico di Palermo (Proceedings of the Mathematical Circle of Palermo), Vol. 21, pp. 129-176, 1906 [224].

This memoir is reproduced in the Works of Henri Poincaré, Vol. IX, pp. 494-550 [227].

For the analysis of this memoir by Lorentz, see note (5) above.

[10] Comptes Rendus de l'Académie des Sciences (Proceedings of the Academy of Sciences, 5 June 1905, pp. 1504-1508 [223] (*Works of Henri Poincaré*, Vol. IX, pp. 489-493 [227]).

[11] In Poincaré's note, and for simplification, the speed of light is taken as being unity, so that $c = 1$ in Formula (1). The ratio $v/c$ is termed $-\varepsilon$ by Poincaré (p. 1505).

According to this transformation, a speed $v$ of translation causes a relative *contraction* parallel to the direction of movement. Moreover, the time $t'$ is retarded as compared to the time $t$.

Poincaré shows that it is possible to go from the coordinates $x', y', z', t'$ to the coordinates $x, y, z, t$ using the same formulas (1) by changing $v$ to $-v$, and that *the totality of all these transformations constitutes a group.*[12]

It immediately follows from the transformation (1) that the speed of light cannot be exceeded, as Poincaré had announced at the end of his conference in St. Louis.

• In his memoir of Palermo of July 1905,[13] Poincaré also shows that *the principle of composition of speeds* is an immediate consequence of the Lorentz transformation:

$$(2) \qquad u' = \frac{u+v}{1+\frac{uv}{c^2}}$$

where $u$ and $u'$ are, for example, the speeds of an electron in the reference frames $x, y, z, t$ and $x', y', z', t'$, and where $v$ represents the speed of the reference frame $x', y', z', t'$ with respect to the reference frame $x, y, z, t$.[14] For $u = c$, also $u' = c$.

In the memoir of Palermo of July 1905,[15] Poincaré pointed out that a linear transformation of the *Lorentz group* does not alter the quadratic form

$$(3) \qquad x^2 + y^2 + z^2 - c^2 t^2$$

[12] CRAS, id., 1905, p. 1505.

[13] Section 4, *The Lorentz Group* (*Works*, id., p. 513 [227]).

[14] In fact, Lorentz had adopted a different formula

$$u' = (u+v) / \left[1 + \left(v^2/c^2\right)\right]$$

where the product $uv$ of speeds in Poincaré's formulation is replaced by the square $v^2$ of the relative speed $v$. Only Poincaré's formula is correct (for this point, see Sesmat, 1937, *L'optique des corps en mouvement* (Optics of a Moving Body), id., p. 595 [246]).

In Einstein's memoir of 1905 (§5), the formula for the composition of velocities is the same as that of Poincaré.

[15] id., *Works*, Vol. IX, p. 513.

In his memoir on Poincaré's works, Lorentz wrote:[16]

*"It was the considerations that I published in 1904 that permitted Poincaré to write his article, in which he attached my name to the transformation that I had not explored fully... Later I was able to see in Poincaré's memoir that I could have obtained a further great simplification. Since I did not notice this, I did not establish the principle of relativity as being rigorously and universally true. By contrast Poincaré obtained perfect invariance... and formulated the postulate of relativity, a term that he was the first to employ."*

---

[16] Note 5 above (*Works of Henri Poincaré*, Vol. IX, p. 687 [227]).
The article by Poincaré upon which Lorentz is commenting is the memoir of Palermo of July 1905, *Sur la Dynamique de l'électron* (On the Dynamics of the Electron).

## A.4 *Einstein's Theory of Special Relativity*

### The memoir of 1905

1 - Einstein's memoir of 1905, *On the Electrodynamics of Moving Bodies,*[1] is essentially based upon the *Principle of Relativity*, and it deduces the implications for the electrodynamics of moving bodies, and in particular for the dynamics of the electron.

Although Einstein's article of 1905 *contains no reference* to previous works,[2] in fact it continues in a direct line from the previous work of Lorentz and Poincaré, and arrives at the same conclusions, in particular with regard to the explanation of the "*negative*" results of the experiments undertaken to show the movement of the Earth.

In fact in his first part which relates to kinematics, Einstein arrives *exactly*, except for the notation, at Equations (1) and (2) above for the Lorentz transformation and for the formula for composition of speeds, *upon which the entire theory of special relativity is based.*[3]

### Two postulates

2 - Einstein's demonstration is based upon two postulates: *the principle of relativity, and the principle that the speed of light is constant*:[4]

As in the case of the works of Lorentz and Poincaré, Einstein's article is based upon the proposition that, up to 1905, nobody had succeeded in detecting the uniform rectilinear movement of the Earth upon its trajectory by a purely terrestrial experiment.

---

[1] Note 4 of the *Introduction* to *Section A* above, p. 550.

For the theory of special relativity, particularly see Costa de Beauregard, *La théorie de la relativité restreinte* (The Theory of Special Relativity), Masson, 1949 [95]. Rather strangely, this work says not a word about the theory of general relativity.

[2] Notably, it contains no reference to the works of Michelson, Lorentz, and Poincaré.

[3] See below §6, pp. 562-572.

[4] pp. 8 and 9 of the French translation in 1925 by Gauthier-Villars of Einstein's 1905 memoir [126].

As Chwolson so excellently wrote:[5]

*"Einstein's* theory basically consists of replacing the words *"nobody has succeeded" by the words "it is impossible to succeed". This substitution completely changes the sense and the character of the former verbal expression.*

*"Nobody has succeeded" is rather like a historical* fact, *an unexpected result of numerous experimental researches. One may try to explain this fact, for example by introducing new hypotheses like those of Fitzgerald and Lorentz.*

*"It is impossible to succeed"* is an a priori assertion, an axiom or a postulate, on which we may try to construct a new conception of the universe*; but there can be no question of demonstrating this axiom or trying to explain it. By adopting it, we are committed to taking it as the principal foundation upon which Physics is built; we must force ourselves to derive all its possible consequences, and, if it seems to be feasible, we must experimentally verify the exactness of the deductions obtained.*

*"Einstein founded his entire theory on* two postulates. *The first can be formulated in the following way:* The world in which we live is so constructed that no observation in any system S, for example upon the Earth, can give any evidence of uniform rectilinear movement of that system, or, *a fortiori*, can be used to determine the speed of this movement.

*"In other words,* the laws of the phenomena that take place in any system are independent of that system, provided that it is not accelerated...

*"Einstein's second postulate may be expressed as follows:* in whichever (unaccelerated) system we measure the speed of light, and under whatever conditions we perform that measurement, we will always obtain the same numerical value for the speed sought.*"*[6]

---

[5] Chwolson, 1914, id. (note 1 above of *Section A*, p. 550), pp. 235-236 [94].

[6] Chwolson writes very fairly (p. 262):

*"The introduction of clocks into the expression of the principle of relativity cannot be of any use, explains nothing, and can only embarrass the intellect or lead to misunderstandings; the notion of the time is completely foreign to the question of relativity. A clock is a* physical *instrument... It seems to me to be completely impossible to say how such a physical instrument will behave in the circumstances contemplated by the theory of relativity. It cannot be determined* a priori *what action will be exerted upon this instrument by the relative velocity. A preliminary critical examination of the question of time is needed, but it does not seem that any such examination has been made.* When we try to reconcile the reasoning made on this point of view by different authors, we meet with very evident contradictions.*"*

### The formulation of Minkowski

3 - In 1908 Minkowski (who died in 1909) gave the theory of special relativity *a completely new mathematical form.*[7] This formulation amounts to defining the space-time metric by the expression:[8]

$$(1) \qquad ds^2 = dx^2 + dy^2 + dz^2 - c^2 dt^2$$

*which is an invariant expression for the Lorentz transformation.*

*According to Minkowski the ideas of space and time, as considered to be independent and self-contained, must be abandoned, and only their union has any meaning.*

Minkowski reunites space and time into *an indivisible unity* called the *universe*; this universe is expressed in geometric language as a *four dimensional space*, in which time plays the role of the fourth dimension.[9]

---

[7] On the theory of Minkowski, particularly see: - Chwolson (1914, id., pp. 254-258 [94]); - René Dugas (1950, id., pp. 468-473 [114]); - Whittaker (1953, id., pp. 64-68 [280]).

[8] Whittaker (1953, id., p. 64, note 2) truly remarks that this expression had already been considered by Poincaré in his memoir of Palermo of July 1905 (Works, id., Vol. IX, p. 541 [227]) (See above §A.3.4, p. 560).

[9] Putting $icdt = du$ here, expression (1) becomes:

$$ds^2 = dx^2 + dy^2 + dz^2 + du^2$$

which is a symmetric expression already considered by Poincaré, and which is the analog in a *continuum* of four dimensions of the Euclidean spatial distance. But it appears that this is only a mathematical trick.

## *A.5 Einstein's General Theory of Relativity*

The objective of the theory of general relativity[1] is to answer the question raised by the contrast between the interpretations related to the movement of rotation of the Earth and those related to its movement of translation along its orbit. It considers *all possible reference frames, not only Galilean ones.*[2]

### The tensorial formulation of the theory of general relativity

In the last analysis, the theory of general relativity comes down to substituting, in Minkowski's Equation (1) above, the relation

(1) $ds^2 = \sum_{\sigma\tau} g_{\sigma\tau} dx_\sigma dx_\tau$

*where the $g_{\sigma\tau}$ are functions of the $x_\sigma$ that no longer depend upon the orientation and the state of movement of the local system of coordinates.*[3] The expression for the $ds^2$ is supposed to be invariant for all linear transformations of the coordinates.[4]

---

[1] Albert Einstein : - 1916 a, *Les Fondements de la Théorie de la Relativité* (The Foundations of the Theory of General Relativity), Hermann, 1933 [127]; - 1916 b, *la Théorie de la Relativité Restreinte et Générale, Exposé Elémentaire* (The Theory of Special and General Relativity, Elementary Exposition), Gauthier-Villars, 1954 [133].

See also Einstein: 1920, *L'Ether et la Théorie de la Relativité* (The Ether and the Theory of Relativity), Gauthier-Villars, 1921 [124]; - 1921 a, *La Géométrie et l'Expérience* (Geometry and Experiment), Gauthier-Villars, 1934 [128]; - 1921b, *Quatre Conférences sur la Théorie de la Relativité* (Four Conferences on the Relativity Theory), Gauthier-Villars, 1925 [125]; - 1951. *Sur le Problème Cosmologique - Théorie de la Gravitation Généralisée* (On the Cosmological Problem - The Theory of Generalized Gravitation), Gauthier-Villars [130]; - 1953, *La Relativité et le Problème de l'Espace* (Relativity and the Problem of Space), Gauthier-Villars, 1954 [132].

On the theory of general relativity, see particularly: A.S. Eddington, 1923, *The Mathematical Theory of Relativity*, Cambridge University Press, 1960 [121]; - von Laue, 1920-1922, *La Théorie de la Relativité* (The Theory of Relativity), Vols. I and II, Gauthier-Villars, 1924-1926 [276]; - H. Weyl, 1921, *Temps, Espace, Matière, Leçons sur la Théorie de la Relativité Générale* (Space, Time, Matter, Lessons on the Theory of General Relativity), Blanchard, Paris, 1922 [278]; - Max Born, 1920, *La Théorie de la Relativité d'Einstein et ses Bases Physiques, Exposé élémentaire* (Einstein's Theory of Relativity and its Physical Basis, Elementary Exposition), Gauthier-Villars, 1923 [73]; - E.T. Bell, *La Mathématique Reine et Servante des Sciences* (Mathematics: Queen and Servant of Science), Payot, 1953, Chapter X, *A Metric Universe*, pp. 173-198 [64]; - Lichnerowicz (A.), *Théories relativistes de la gravitation et de l'électromagnétisme* (Relativistic Theories of Gravitation and Electromagnetism), Masson, 1955 [178].

The work of Eddington is *particularly suggestive*. The reader will find a *very condensed* explanation of the Theory of Special and General Relativity in Lichnerowicz (A.), *Eléments de calcul tensoriel* (Elements of the Tensor Calculus), Colin, 1951, Chapters VII and VIII, pp. 159-213 [177].

[2] Einstein, 1916 a, id., pp. 7-16.

[3] Einstein, 1916 a, id., p. 17.

[4] Einstein also postulates that the trajectory of a free particle in the space-time continuum is a geodesic of that space, and he accepts that the trajectory of light in the space-time continuum is such that $ds^2 = 0$ along that trajectory (see particularly Eddington, 1923, §47, p. 104 [121]).

*From expression (1) for $ds^2$, all the theory of general relativity develops as a tensor formulation.*[5]

### The $ds^2$ of Schwarzschild

2 - The problem of integrating Einstein's equations in the general case *has never been resolved.* One solution was found by Schwarzschild in the case of an isolated non-rotating spherically symmetric body. In spatial polar coordinates, the expression for $ds^2$ that generalizes the $ds^2$ of Minkowski is the following:

(2) $$ds^2 = -\frac{dr^2}{\gamma} - r^2 d\theta^2 - r^2 \sin^2\theta d\varphi^2 + \gamma dt^2$$

with

(3) $$\gamma = 1 - \frac{2Gm}{c^2 r}$$

where $m$ is the mass of the body, $r$ the distance from the point in question to the center of the body, $G$ the coefficient of universal gravitation, and $c$ the speed of light.[6]

The coefficient $\gamma$ represents the correction relative to Newtonian mechanics. The correction $2Gm/c^2r$ is extremely small.[7]

---

[5] For the tensor calculus used in the theory of relativity, see particularly Pierre Bricout, *Microénergétique* (Microenergetics), Vol. I, Gauthier-Villars, 1933, pp. 3-47 [82]; - Léon Brillouin, *Les Tenseurs en Mécanique et en Elasticité* (Tensors in Mechanics and Elasticity), Masson, 1938 [83]; - Lichnerowicz (A.), *Algèbre et Analyse linéaires* (Linear Algebra and Analysis), Masson, 1947 [176]; - Thiry (R.) and Appell (P.), *Eléments de Calcul tensoriel* (Elements of Tensor Calculus), Gauthier-Villars, 1955 [264]; - Bauer (E.), *Champs de vecteurs et de tenseurs* (Vector and Tensor Fields), Masson, 1955 [62].

[6] See in particular Eddington, 1924, §38, pp. 81-85, and G. Darmois, 1932, *La Théorie Einsteinienne de la gravitation* (The Einsteinian theory of gravitation), Hermann, p. 13 [99]. For a detailed analysis, see: Jean Chazy, *La théorie de la relativité et la mécanique céleste* (The relativity theory and celestial mechanics), Gauthier-Villars, Vol. 1,1928, et Vol. II, 1930 [92].

[7] - *At the surface of the Earth*, in CGS units (*Chapter I*, §B.2.6, note 15, p. 124):

$m = 5.97 \times 10^{27}$ $r = 6.37 \times 10^{8}$

$G = 6.67 \times 10^{-8}$ $c = 3 \times 10^{10}$

whence

$2Gm/c^2r = 1.39 \times 10^{-9}$

- *At the surface of the Sun*, we have

$m = 1.99 \times 10^{33}$ $r = 0.695 \times 10^{11}$

whence

$2Gm/c^2r = 4.24 \times 10^{-6}$

- *For the action of the Moon at the surface of the Earth*, we have

$m = 7.34 \times 10^{25}$ $r = 3.84 \times 10^{10}$

whence

$2Gm/c^2r = 2.83 \times 10^{-13}$

- *For the action of the Sun at the surface of the Earth*, we have

$m = 1.99 \times 10^{33}$ $r = 1{,}49 \times 10^{13}$

whence

$2Gm/c^2r = 1.98 \times 10^{-8}$

Equations (2) and (3) are used for calculating the precession of the perihelion of Mercury[8, 9] and the deviation of light near the Sun.[10]

---

[8] See in particular Eddington, 1924, §39, pp. 85-90; Darmois, 1932, pp. 16-17; and especially W.M. Smart, *Celestial Mechanics*, Longmans, 1953, §2-04, pp. 14-15, and §15-11, pp. 243-246 [250].

[9] In Newtonian mechanics, the orbit of a body of mass m around the Sun of mass $M$ is defined in polar coordinates $(r, \theta)$ by the two equations (Smart, 1953, id., p. 14)

(1) $\frac{d^2r}{dt^2} - r\left(\frac{d\theta}{dt}\right)^2 = -G\frac{(M+m)}{r^2}$

(2) $r^2\frac{d\theta}{dt} = na^2\sqrt{1-e^2} \qquad n = 2\pi/T$

where $a$ is the semi-major axis of the elliptic trajectory, $T$ is the period of revolution, and $G$ is the constant of gravitation. From (1) and (2), we deduce

(3) $\frac{d^2u}{d\theta^2} + u = G\frac{(M+m)}{h^2}$

with

(4) $u = 1/r \qquad h^2 = G(M+m)\,a\left(1-r^2\right)$

The second term of (3) is deduced from the second term of (1) by multiplication by the factor $-\left(r^2/h^2\right)$

From the theory of general relativity, we deduce (Smart, 1953, id., p. 243)

(5) $\frac{d^2u}{d\theta^2} + u = G\frac{(M+m)}{h^2} + 3G\frac{(M+m)}{c^2}u^2$

where $c$ is the speed of light. From (4) and (5), we have

(6) $\lambda = \frac{3G(M+m)}{c^2}u^2/G\frac{(M+m)}{h^2} = 3\frac{h^2}{c^2}u^2 = \frac{3G(M+m)a\left(1-e^2\right)}{c^2r^2}$

and, from (1), (3), and (5), everything happens as though equation (5) can be written

(7) $\frac{d^2r}{dt^2} - r\left(\frac{d\theta}{dt}\right)^2 = -(1+\lambda)\,G\frac{(M+m)}{r^2}$

In the case of the Sun and Mercury, we have in CGS units

(8) $M = 1.99\times10^{33} \qquad m = 3.24\times10^{26} \qquad a = 5.79\times10^{12} \qquad e = 0.2056$

whence, for the order of magnitude of $\lambda$

(9) $\lambda \sim 3GM/c^2a$

The correction is thus of the same order of magnitude as that for the coefficient $\gamma = 1 - 2GM/c^2a$ of Schwarzschild (equation 2, p. 566 above).

From (9), we therefore have

(10) $\lambda \sim 3\times6.67\times10^{-8}\times1.99\times10^{33}/\left(9\times10^{20}\times5.79\times10^{12}\right) = 7.64\times10^{-8}$

We see that the relative correction $\lambda$ to the attractive force due to the theory of relativity (equation 7) is extremely small.

[10] Eddington, 1924, id., §41, pp. 90-91.

## The particular case when the $g_{\sigma\tau}$ are constant

3 - *Following Einstein, the theory of general relativity includes special relativity as a special case.*

In a situation in which the coefficients $g_{\sigma\tau}$ are constant, the quadratic form (1) can always be written in the form of equation (1) of §A.4 (p. 564), and the theory of special relativity is applicable.[11, 12]

---

[11] See particularly Einstein, 1916 a, id., p. 45.

Eddington (1924, id., §36, pp. 76-81) summarizes the cases in which the $g_{\sigma\tau}$ are constant, or not, as follows:

> "*A region of the world is called flat or homoloïdal if it is possible to construct in it a* Galilean *frame of reference.*
>
> "*When the $g_{\sigma\tau}$ are constant,* $ds^2$ *can be reduced to the sum of four squares and Galilean coordinates can be constructed. Thus an equivalent definition of flat space-time is that it is such that coordinates can be found for which the $g_{\sigma\tau}$ are constants ...*
>
> "*When the space-time is not flat we can introduce coordinates which will be approximately Galilean in a small region round a selected point, the $g_{\sigma\tau}$ being not constant but stationary there; this amounts to identifying the curved space-time with the osculating flat space-time for a small distance round the point...*
>
> "*When the Riemann-Christoffel tensor vanishes, we can adopt Galilean coordinates throughout this region. When it does not vanish we can adopt coordinates which agree with Galilean coordinates at a selected point ... These are called natural coordinates at the point. Either Galilean or natural coordinates can be subjected to Lorentz transformations*".

[12] *Naturally, it goes without saying that my explanation here is limited to matters that appear quite indispensable from the point of view of the general concept of the present work.* It certainly is not my objective to give an exhaustive exposition of the theory of general relativity.

## A.6 Einstein and his predecessors

In fact, Einstein's fundamental article of 1905 *does not include any reference* to any experimental researches, in particular to the experiment of Michelson and Morley in 1887, or to any of the theoretical analyses in the literature, in particular to Lorentz's 1904 memoir or to Poincaré's analyses from 1899 to July 1905, the general lines of which I have described.[1]

In the French translation of 1925 of Einstein's article of 1905, *On the Electrodynamics of Moving Bodies*, a note is added in the introduction in which Einstein declares that, at the time, he had no knowledge of Lorentz's memoir of 1904.[2] In the same way, in §3 of the memoir, an additional note comments that the Lorentz transformation can be deduced from the invariance of the quadratic form (3) of §A.3 above.[3] *Neither of these notes was present in the original German article of 1905.*

The absence of any references in Einstein's original article of 1905, *which has been underlined by all commentators*, is at the least shocking, even if Einstein's age of 26 at the time is taken into account.

[1] Einstein's article of 1905 was reproduced in 1913 in the collective work: H.A. Lorentz, A. Einstein, and H. Minkowski, 1913, *Das Relativitätsprinzip* (The Principle of Relativity), Leipzig and Berlin [187].

[2] See this note (Gauthier-Villars, 1925, p. 2):

"*The memoir of H.A. Lorentz entitled* Electrodynamic phenomena in a system moving with any velocity smaller than that of light *(Proceedings Acad. Sci., Amsterdam, Vol. VI, 1904, p. 809) was unknown to me at the time I wrote this memoir.*"

[3] See this note (Gauthier-Villars, 1925, p. 17):

"*The equations of the Lorentz transformation can be deduced in a more direct manner by supposing that, according to them, the equation*

$$\xi^2 + \eta^2 + \zeta^2 - v^2\tau^2 = 0$$

must have as a consequence this other equation

$$x^2 + y^2 + z^2 - v^2t^2 = 0$$

But no reference is made here to Poincaré, who was the first to note this invariance (§A.3.4 above, p. 560).

Subsequently, at several times, Einstein reiterated his ignorance at the time of Lorentz's 1904 article, and even of the Michelson-Morley experiment of 1887.[4] In any case, up to 1955, he never made any reference to the numerous analyses by Poincaré.

In fact, Einstein's memoir presents disturbing analogies with the writings by Poincaré that preceded it.[5] *The least that can be said is that Poincaré's writings were definitely prior to Einstein's 1905 article* which founded special relativity. All the fundamental equations of the theory of special relativity of Einstein's 1905 memoir can be found in the prior works of Poincaré: the principle of relativity, the Lorentz transformation, and the formula for composition of velocities.[6]

---

[4] These statements, at the least surprising, have been the subject of very numerous commentaries. See particularly T. Kahan, 1959, *Sur les origines de la théorie de la relativité restreinte* (On the Origins of the Theory of Special Relativity), Revue d'Histoire des Sciences (Review of Scientific History), April-June 1959, pp. 159-165 [161]; R.S. Shankland, *Conversations with Albert Einstein*, American Journal of Physics, Vol. 31, p. 47, 1963 [248]; G. Holton, *Einstein and the "Crucial Experiment"*, American Journal of Physics, Vol. 37, p. 968, 1969 [158]; Yoshimasa A. Ono, *Translation of a Lecture given by Einstein in Kyoto on 14 December 1922, How I created the theory of Relativity*, Physics Today, August 1982, pp. 45-46 [207]; John Stachel, *Einstein and Ether Drift Experiments*, Physics Today, May 1987, pp. 45-47 [253].

[5] See the articles by Jules Leveugle in 1994, §A.3 above, note 4, p. 557.

See also C. Marchal, 1995, *The Theory of Relativity, Einstein or Poincaré*, 4th Alexander von Humboldt Colloquium [190].

[6] The Lorentz transformation and all the related developments only assumed their final form in Poincaré's thinking after a slow maturation through a progressive process of successive approximations. They were the conclusion of ten years of reflection and reciprocal exchange of views between Lorentz and Poincaré. It could hardly have been otherwise in view of the complexity of the subject.

In fact, it was not totally without ground that Renaud de la Taille was brought to write (*Science et Vie* (Science and Life), April 1995, pp. 114-119):

> *"Thus an independent researcher, who had never previously published anything on the subject, was able to rediscover, almost between one day and the next, what two scientists of the class of Lorentz and Poincaré had only been able to establish after ten years of effort."*

Einstein's statements about the origins of the theory of relativity varied somewhat over the course of time.[7]

In May 1955, shortly before his death, Einstein specified the conditions in which his memoir of 1905 was composed:[8, 9]

> *"It is beyond doubt that, if one looks retrospectively at its evolution, the theory of special relativity was ripe in 1905. Lorentz had already discovered that the transformation that subsequently received his name plays an essential role in the analysis of Maxwell's equations, and for his part Poincaré had penetrated more deeply into the nature of those equations.*

---

[7] On 14 December 1922, in a conference held at the University of Kyoto, he declared:

> *"It was more than seventeen years ago that I had an idea of developing the theory of relativity for the first time. While I cannot say exactly where that thought came from, I am certain that it was contained in the problem of the optical properties of moving bodies ...*
>
> *"When I first thought about this problem, I did not doubt the existence of the ether or the motion of the Earth through it ...*
>
> *"While I was thinking of this problem in my student years, I came to know the strange result of Michelson's experiment. Soon I came to the conclusion that our idea about the motion of the Earth with respect to the ether is incorrect, if we admit Michelson's null result as a fact. This was the first path which led me to the special theory of relativity. Since then I have come to believe that the motion of the Earth cannot be detected by any optical experiment, though the Earth is revolving around the Sun".*

(Yoshimasa A. Ono, 1982, id., p. 46).

By contrast, in his interview with Shankland on 4 February 1950, he declared that before 1905 he had no knowledge of the experiment of Michelson and Morley in 1887.

Shankland wrote (1962, id., p. 48) :

> *"When I asked him how he had learned of the Michelson-Morley experiment, he told me that he had become aware of it through the writings of H.A. Lorentz, but* only after 1905 *had it come to his attention! "Otherwise", he said, "I would have mentioned it in my paper"*

[8] *Technische Rundschau*, Bern, no. 20, 6 May 1955; text cited by T. Kahan, 1959, *Sur les origines de la relativité restreinte* (On the Origins of Special Relativity) [161], see note 4 above, p. 570.

[9] In fact, examination of the relations that may have existed between Einstein's 1905 memoir and the antecedent writings of Poincaré *is not relevant to the present work.*

In any case, this matter only relates to *the paternity of a formulation that was based upon insufficiently analyzed experimental facts, and that has led physics astray into mistaken paths for many decades.*

In reality, in view of *Chapters I through V* above, the formula of Lorentz and the law of composition of velocities are *totally contradicted* by the experimental data.

See *Section B* below, *The Theory of Relativity and Experiment.*

I must be allowed to add that Henri Poincaré was always somewhat reserved with regard to the Lorentz transformation (see §A.7 below, pp. 573-574). In any case to give, *as an analyst*, a correct expression for Lorentz's transformation *does not imply that one accepts it.*

Here two completely different questions must be clearly distinguished: - the principle of relativity; and the Lorentz transformation.

In fact Poincaré was *always* convinced of the principle of relativity, which he was the first to formulate. *By contrast, he never really accepted the Lorentz transformation.*

*"For my part, at that time I was acquainted only with Lorentz's important work of 1895 on Maxwell's electromagnetic theory and with his Essay on the theory of electrical and optical phenomena in moving bodies, but not with the later work of Lorentz nor with the subsequent research by Poincaré. In that sense my work of 1905 was independent.*

*"What was new in my memoir was the discovery of the fact that the implications of the Lorentz transformation go deeper than its connection with Maxwell's equations and implicate the nature of space and time in general. What also was new was the realization that Lorentz invariance was a general condition for any physical theory."*

## *A.7 The alternative hypothesis of an anisotropy of space*

In reality, Henri Poincaré never ceased to be troubled by the implications of Lorentz's formula. Already in his *Conference in St. Louis* of *December 1904*, he underlined that, for explaining the "*negative result*" of the Michelson and Morley experiment of 1887, a simpler hypothesis could be conceived - the hypothesis of an anisotropy of space.[1] As early as 1905, in *The Value of Science*, he further wrote on "*The Role of the Analyst*":[2]

*"Among so many ruins, what remains standing?...*

*"Confronted by this general collapse of principles, what attitude should mathematical physics adopt? And first, before becoming too agitated, it is necessary to ask oneself if all this is really true...*

*"A prejudicial question therefore arises, which it seems that only experiment can resolve. We have therefore no option but to hand over the matter to the experimenters, and, while waiting for them to settle the debate conclusively, we should not worry ourselves with these disturbing problems, but should continue our work calmly as though the principles were still unchallenged...*

*"And still, is it really true that we cannot do anything to rid science of these doubts? It must be admitted, it is not only experimental physics that has given birth to them; mathematical physics has also made its own contribution...*

*"It was the theoreticians who brought up all the difficulties raised by the propagation of light through a moving medium; without them, it is probable that nobody would have encountered those problems. Well, since they have done their best to get us into trouble, they ought also to help us get out of it.*

*"They must submit to criticism all these new views that I have just outlined to you; and they must not abandon their principles until they have made a loyal effort to save them. What can they do in this sense?*

*"Before everything, it is necessary to evolve a more satisfying theory of the electrodynamics of moving bodies? It is there, above all, that the difficulties accumulate... While we can pile up hypotheses, we cannot satisfy all the principles at once; until now, we have only been able to succeed in saving some on the condition of sacrificing others; but we have not yet lost all hope of obtaining better results. Let us therefore take the theory of Lorentz, let us turn it in every sense; let us modify it little by little, and perhaps everything will come out in good order.*

[1] See above, *Chapter IV*, §F.2.4, note 5, p. 422.

[2] Henri Poincaré, 1905, *The Value of Science*, id., pp. 200-203 [222].

*"Thus, instead of supposing that moving bodies undergo a contraction in the direction of movement and that this contraction is the same whatever be the nature of these bodies and the forces to which they are subjected, could not we form a simpler and more natural hypothesis? For example, we could imagine that it is the ether that is modified when it is in relative movement with respect to a material medium that penetrates it, and that, when it is thus modified, it no longer transmits perturbations with the same speed in all directions. It might transmit more quickly those perturbations that are propagating parallel to the movement of the medium, both in the same direction and in the opposite direction, while transmitting less quickly those that are propagating perpendicularly. The surfaces of waves would no longer be spheres but ellipses, and we could avoid this extraordinary contraction of all bodies.*

*"I only give this as an example, because the modifications that we might try could evidently be varied ad infinitum."*

In *Science and Method* in 1908, Henri Poincaré returned to the hypothesis of anisotropy of space:[3]

*"According to Lorentz and Fitzgerald, all bodies moving along with the Earth experience a deformation...*

*"Can we demonstrate this deformation? Evidently not...*

*"The question arises as to what is the use of the Lorentz-Fitzgerald hypothesis, if it cannot be verified by any experiment? The fact is that my explanation is incomplete: I have only spoken of measurements that we can make with a measuring rod, but it is also possible to measure a length by the time that light takes to traverse it, on condition that we admit that the speed of light is constant and independent of the direction.*

*"Lorentz could account for the facts by supposing that the speed of light is greater in the direction of movement of the Earth than in the perpendicular direction. He preferred to postulate that the speed is the same in these various directions, but that bodies are smaller in some directions than others. If the surfaces of light waves underwent the same deformations as material bodies, we would not have noticed the Lorentz-Fitzgerald deformation."*

This double quotation (added to that of 1904) is *very significant*. It shows at least that Poincaré was not very satisfied with the Lorentz-Fitzgerald contraction and with Lorentz's concept of local time.

[3] Henri Poincaré, 1908, *Science and Method*, pp. 98-100 [225].

In any case, it shows that all the discussions at the start of the century *neglected an essential hypothesis*, that of deformation of the ether due to displacement of bodies, *in other words the hypothesis of an anisotropy of space*.[4, 5]

In fact, in 1913, Richard Birkeland proposed an explanation of the "*negative*" result of the Michelson experiment *based upon the hypothesis of an anisotropy of optical space*.[6] His analysis *entirely confirmed* Poincaré's views regarding the possibility of explaining the "*negative*" result of Michelson's experiment by an anisotropy of space.

---

[4] However Faraday, and subsequently Maxwell with his tensor formulation, appealed to an anisotropy of space.

A particular case of the general theory of relativity could have led to an expression for $ds^2$ *preserving Lorentz's concept of absolute time* (see note 5 of §A.3.2 above, p. 557). Such an expression formulating the anisotropy of space may be written

$$(1) \qquad ds^2 = \sum_{i,j} g_{ij} dx_i dx_j - c^2 dt^2$$

the summation being only over the coordinates $x, y, z$.

In my memoir of 1957 on the interpretation of the speed of light, I considered such an expression, starting from the equation

$$(2) \qquad \frac{1}{\sqrt{|g|}}\partial_i\left(\sqrt{|g|}g^{ij}\partial_j\varphi\right) - \frac{1}{c^2}\frac{\partial^2\varphi}{\partial t^2} - 2\frac{k_0}{c}\frac{\partial\varphi}{\partial t} - k_0^2\varphi + 4\pi K\delta = 0$$

which generalizes Hély's equation to the case of an anisotropy of space.

$$(3) \qquad \Delta\varphi - \frac{1}{c^2}\frac{\partial^2\varphi}{\partial t^2} - \frac{2k}{c}\frac{\partial\varphi}{\partial t} - k^2\varphi + 4\pi K\delta = 0$$

for a Euclidean space (see above, *Chapter VI*, §C.1.5, p. 512-514).

[5] A certain number of works have broken away from Einstein's concepts of time and contraction of bodies with speed, in the direction of the anisotropy outlined by Poincaré.

One of the most notable is the work of Pierre Dive, "*Ondes ellipsoïdales et Relativité*" ("Ellipsoidal Waves and Relativity") (Gauthier-Villars, 1950) [112]. In the introduction to this work, he explicitly refers to Poincaré's analysis of 1905 in "*The Value of Science*", which he cites (as I myself have above, p. 573).

On ellipsoidal waves, also see Henri Varcollier, 1949, *La théorie de la propagation ellipsoïdale et ses possibilités. Relativité, Quanta, Gravitation* (The theory of ellipsoidal propagation and its possibilities. Relativity, Quanta, Gravitation) [272].

[6] Richard Birkeland, *An Attempt to explain the Michelson Interference Experiment*, Philosophical Magazine and Journal of Science, Vol. XXXVII, January-June 1919, pp. 150-156 [70]. Birkeland was professor of mathematics at the *Technical High School* of Trondheim. (See above, *Chapter IV*, §F.2.4, note 6, p. 422).

Naturally Birkeland's article *is only one illustration*, among many, of the theoretical possibilities offered by consideration of an anisotropic space.

In any view of the matter, the Lorentz transformation is based upon a dual proposition:

1 - The experiment of Michelson gave "*a negative result*".

2 - The ether is considered *to be totally immobile with respect to the fixed stars*.

Now, this second proposition is founded upon pure hypothesis. If the ether locally accompanies the displacement of the Earth in its movement of translation, the Michelson experiment could never yield anything but a negative result.[7]

In fact Birkeland, like Poincaré, addressed himself to explaining the "*negative result*" of the Michelson experiment, but it is clear that *the same approach can also explain the anisotropy of space demonstrated by Miller's observations, even with the hypothesis that the ether is completely entrained by the Earth.*[8]

---

[7] Contrary to a very frequent assertion, such a displacement of the ether with the Earth is in no way contradicted by the phenomenon of aberration (see in particular Bouasse, 1925, *Propagation de la Lumière* (Propagation of Light), §62, pp. 117-119 [77]).

The incompatibility of a partial or total entrainment of the ether by the Earth with the phenomenon of aberration was notably supported by Einstein. In his account of his interview of 17 November 1950 with Einstein, Shankland wrote (1963, id., p. 52):

> *"He also reminded me that any "drag" would be inconsistent with aberration".*

[8] *Chapter IV* above, §F.2, p. 423.

# B THE THEORY OF RELATIVITY AND EXPERIMENT

## *B.1 A legend: the "negative" result of Michelson's experiment*

1 - It is *an established truth*, taught *everywhere* nowadays in *all* the universities of the world, that all the interferometric experiments that have ever been performed, starting with Michelson's celebrated experiment, have yielded *totally negative results*.

It was in order to explain this *negative* result that Lorentz presented his hypotheses of contraction of bodies according to their speeds, and of local time, and his law of composition of velocities, and that, as generally believed, Einstein formulated his Theory of Special Relativity and then his Theory of General Relativity.[1]

Upon exploring the literature, it is impossible not to be struck by the *absolutely extraordinary* volume of theoretical analysis founded upon the claimed *"negative"* results of the interferometric experiments of Michelson and his successors, in view of the *almost total* lack of any deep analysis of those interferometric experiments, *and in particular of Miller's experiments*. Many articles and many theoretical works have been written *that, in reality, are founded upon misunderstanding or even denial of the observed facts*.[2]

---

[1] See *Section A* above.

[2] I can only revert here to the judgement of Claude Bernard, quoted at the head of *Chapter VI*, p. 393.

For the scientific method, one cannot read and meditate enough upon his work of 1865, *Introduction à l'étude de la Médecine expérimentale* (Introduction to the Study of Experimental Medicine) [69].

2 - Thus, for example, with rarely seen dogmatism, in 1967 A. Foch did not hesitate to write in the revised text of the work of G. Bruhat, *Mécanique* (Mechanics):[3]

> *"No displacement (of the fringes) has ever been observed...*
>
> *"Contrary to the implications of classical conceptions of time and space, the movement of the Earth in the universe is not apparent in any terrestrial experiment...*
>
> *"Whichever Galilean frame is used, the speed of light in a vacuum always has the same value in all directions...*

Similarly in a recent Russian work,[4] we read:

> *"Michelson's experiment has been repeated many times with always increasing precision... but the result obtained by Michelson, or as is often said the negative result of Michelson's experiment, remains untouched. Therefore its correctness is no longer in doubt."*

But these are not isolated texts. *Such citations can be multiplied. Everywhere* it is admitted *without any reservation* that *neither practically nor theoretically* can any purely terrestrial experiment reveal the speed of the Earth, or even simply its position along its orbit.

Even as penetrating a theoretician as Max Born, who was awarded the 1954 Nobel Prize in Physics, wrote in 1923:[5]

> *"Cannot one imagine arrangements that would make it possible to detect the movement of the Earth and of the resulting "ether wind"?"*
>
> *"A great number of experiments for detecting this movement have been imagined and performed. They all teach us that it has never been possible to detect the slightest influence of any ether wind with experiments using light sources...*

---

[3] Masson, 1967, pp. 695-696 [88].

[4] V. Ugarov, 1974, *Théorie de la relativité restreinte* (The Theory of Special Relativity), Editions Mir, p. 35 [269].

[5] Max Born (1882-1970), 1922, *La théorie de la relativité d'Einstein et ses bases physiques* (Einstein's Relativity Theory and Its Physical Basis), Gauthier-Villars, 1954, pp. 129, 225, 63, and 216 [73].

It is quite true that this text was written *well before Miller's Memoir of 1933*. But, as far as I know, Max Born did not modify this point of view *after* the publication in 1933 of this Memoir of Miller's.

*"All the experimental researches... have proved that movement with respect to the ether cannot be demonstrated by any known physical experiment...*

*"In fact, all mechanical phenomena take place upon the Earth as though its rapid movement of translation did not exist, and this law is general and is applicable to any system of bodies that is in uniform rectilinear movement through Newton's absolute space...*

*"The laws of Mechanics are expressed with respect to a coordinate system moving in a uniform and rectilinear manner through absolute space, exactly as they are expressed with respect to a coordinate system that is stationary in space...*

*"The relativity principle that is valid for Mechanics may be extended... to all electromagnetic phenomena."*

3\. - This is just what Einstein never ceased to underline:

*"In experiments performed upon the Earth, we never perceive the terrestrial translator motion."*[6]

*"If the relativity principle were not valid, we should expect that the direction of movement of the Earth would be involved in the laws of nature at every moment, and as a consequence that the behavior of physical systems should depend upon their orientation in space relative to the Earth...*

*"But, in spite of the most careful observations, it has never been possible to observe such an anisotropy in terrestrial physical space, in other words any physical non-equivalence between different directions. This is a very weighty argument in favor of the principle of relativity."*[7]

*"Is this law of the constancy of the speed of light valid in every inertial system? If it were not, then a specific inertial system, or more exactly a specific state of movement (of a reference body) would be distinguished from all others. But all mechanical facts and optical facts in our experience speak against this idea.*

*"For these reasons, it has become necessary to consider the validity of the law of the constancy of the speed of light for all inertial systems as being a principle."*[8]

---

[6] Einstein, *Comment je vois le monde* (How I see the World), Flammarion, 1939, p. 160.

[7] Einstein. *La théorie de la relativité générale et restreinte* (The General and Special Theories of Relativity), Gauthier-Villars, 1954, p. 17 [133].

[8] Einstein, *Conceptions scientifiques, morales et sociales* (Scientific, Moral, and Social Ideas), Flammarion, 1952, pp. 87-88 [131].

*It results from the above that any experiment performed on the Earth's surface that yields evidence (without external reference) either that the speed of light varies according to its direction, or of correlation of observations with the position of the Earth along its orbit, invalidates the very foundations of the theories of special and general relativity.*

## *B.2 The claimed "negative" results of the experiment of Michelson and the experiments of Miller*

The analysis I have presented of Miller's observations[1] leads to a quadruple conclusion:

- *First*, there is *a very great coherence* between Miller's interferometric observations *that is absolutely undeniable*, and it *corresponds to a very real phenomenon.*
- *Second, it is absolutely impossible* to attribute this very great coherence to fortuitous causes or to perverse effects (for example effects of temperature).
- *Third, the speed of light in all directions is not invariant.*
- *Fourth*, Miller's interferometric observations are *strongly correlated* with the position of the Earth along its orbit.

These conclusions are *independent of any hypothesis and any theoretical analysis whatever*. In fact, the greater part of the results upon which these conclusions - and in particular the most important thereof - are based, *were not noticed by Miller*. This only makes them all the more significant.

From the above it follows that it is *completely inaccurate* to consider that the experiment of Michelson, as re-enacted by Miller, yielded negative results.[2]

---

[1] *Chapters IV and V* above.

[2] In any case, it is *completely false* to repeat incessantly that the experiments of Michelson and Morley in 1887 gave a null result, because they demonstrated a displacement of the fringes corresponding to a speed of 8 km/sec (see *Chapter IV*, §E.2.1 above, p. 414).

## *B.3 The fundamental postulates of the theories of special and general relativity invalidated by experiment*

### Three postulates

1 - *At the start*, the very foundations of the theories of special and general relativity *are based upon a triple postulate: the supposed "negative" result of the experiment of Michelson; the invariance of the speed of light in different directions; and the impossibility of detecting the movement of the Earth with respect to the fixed stars by any purely terrestrial experiment.*

However, *in view of the analysis that I have presented of Miller's observations*, and in particular in view of the fittings of the hodographs to ellipses, *it is certain* that it cannot be maintained that the interferometric experiments yielded a "*negative*" result, that the speed of light *is invariant in different directions*, and that no purely terrestrial experiment *can determine* the position of the Earth along its orbit.

From the above, it follows that it becomes impossible to maintain the entire structure of the theory of special relativity. In particular, the postulate of relativity, the law of composition of velocities, the principle of the constancy of the speed of light, the Lorentz contraction, the formulation of local time, the concept of a space-time in which space and time are inseparably linked, and all the mathematical deductions that have been derived from these principles, cannot be considered as being based upon foundations verified by experiment. This is an incontrovertible conclusion. *The disproof given by experiment is categorical and cannot be appealed.*

### An inseparable entity

2 - Since the theory of special relativity is only a particular case of the theory of general relativity and is inseparable from it, the theory of general relativity *must equally be considered as completely invalidated by the experimental data, for the same reasons as in the case of the theory of special relativity.*

The fact was also very clearly asserted by Einstein that the two theories, general relativity and special relativity, *cannot contradict* one another. Thus, in their work *The Evolution of Physical Ideas*, Albert Einstein and Leopold Infeld wrote:[1]

> *"Truly relativistic physics must be applicable to all SCs (systems of coordinates), and, consequently, also to the special case of an inertial SC. The new general laws, valid for all SCs, must, in the special case of an inertial system, reduce to the old known laws.*
>
> *"The problem of formulating the laws of physics for any SC has been resolved by the theory of general relativity: the theory that preceded it and that applies only to inertial systems is called the theory of special relativity. Naturally the two theories cannot contradict one another, because the old laws of the theory of special relativity must always be included in the general laws for an inertial system. But the inertial SC, for which, exclusively, the physical laws were initially formulated, now constitutes only a special limiting case, because all SCs moving with respect to one another in any arbitrary manner are admissible."*

As Einstein underlined:[2]

> *"The principal attraction of the theory (of relativity) is that it constitutes a logical whole.*
>
> *"If a single one of its consequences proved to be inaccurate, it would be necessary to abandon the whole; it appears that no modification would be possible without shaking the entire structure."*

## A universally admitted proposition

3 - *That the very foundation of the Theory of Relativity is based upon the claimed negative result of Michelson's interferometric experiment, is what is universally admitted.* It will be sufficient here to give a few citations.

---

[1] Albert Einstein and Leopold Infeld, *L'évolution des idées en physique* (The Evolution of Physics), Flammarion, 1938, pp. 209-210 [136].

[2] Albert Einstein, *Comment je vois le monde* (How I see the World), 1939, id., p. 213 [129].

## Paul Painlevé (1922)

In his work of 1922 on "*Les Axiomes de la Mécanique*" ("The Axioms of Mechanics"),[3] Painlevé wrote:

> *"That the formulas (of the theory of relativity as deduced from the Lorentz-Einstein transformation) and the hypotheses that they interpret take account of the experiment of Michelson, is certain in advance, because it was that experiment that inspired them."*

## Piccard and Stahel (1928)

In their article "*Réalisation de l'expérience de Michelson en ballon et sur terre ferme*" ("Performing the Michelson experiment in a balloon and on the ground") in the *Journal de Physique*,[4] Piccard and Stahel wrote:

> *"It is known that in 1887 Michelson and Morley undertook their famous experiment that nowadays bears the name "Michelson's experiment" in order to determine the speed of the Earth with respect to the "ether". The result was negative to an accuracy of around 6 km/sec.*
>
> *"In 1905, Morley and Miller repeated the experiment, with the same negative result. Applying himself to this fact, Einstein established the theory of relativity, according to which no determination of relative speed between the observer and the "ether" is possible.*
>
> *"In 1921, Miller repeated his experiments at a greater altitude, and found on Mt. Wilson (altitude 1750 m) an ether wind of 10 km/sec...*
>
> *"The importance of this result is obvious, because it gives a mortal blow to the theory of relativity. It started a furious discussion."*

## Augustin Sesmat (1937)

Commenting upon Miller's experiments, Augustin Sesmat wrote:[5]

> *"We know that Michelson's experiment has been repeated many times; in general the experimenters have confirmed the negative result... Undoubtedly the conclusions of a single experimenter carry little weight when they are opposed to those of all the others; however, the question needs to be examined again, because it can be easily understood that, if any displacement, even one much smaller than that predicted by the classical experimenters, were to be produced by the fact that the Earth has a velocity, then the special theory would collapse at its foundations, and with it the general theory."*

---

[3] Gauthier-Villars, 1955, p. 102 [208].

[4] February 1928, p. 49 [217].

During the Conference of Mt. Wilson in 1927 (p. 382, note 2 above), Miller declared (p. 364): *"It is impossible to state that there is any effect due to altitude."*

[5] Sesmat, 1937, VII, *Essai critique sur la doctrine relativiste* (Critical Essay on the Relativistic Doctrine), Hermann, pp. 431-432 [245].

## Albert Einstein

As Einstein himself wrote in 1925 in the journal "*Science*":[6, 7]

> *"If Dr. Miller's observations were confirmed, the theory of relativity would be wrong. Experiment is the supreme judge."*

*Incontestably the theories of special and general relativity, which are based upon postulates invalidated by the observational data, cannot be considered as scientifically valid.*

## Three confirmations

4 - In fact, the implications of the analysis of Miller's interferometric observations *are confirmed* by the analysis of three other series of experiments: my experiments with the paraconical pendulum, 1954-1960; my experiments in 1958 on optical deviations of sightings at marks, and the subsequent experiments on sightings at marks and at collimators in 1959 by the *Institut Géographique National* (National

---

[6] Cited by E. Carvallo, *La théorie d'Einstein démentie par l'expérience* (Einstein's theory disproved by experiment), Chiron, 1934, p. 5 [91].

[7] But in his *Correspondence with Michele Besso* (Hermann, 1979, p. 127 [135]), Einstein wrote on 25 December 1925:

> *"I also myself believe that Miller's experiments are based upon errors of temperature. I have never taken them seriously."*

The translation is obviously poor. It should read "*are based upon effects of temperature*".

I have searched in vain through the subsequent portions of the *Correspondence with Michele Besso* (which continued until 1955) for even the slightest comment on Miller's memoir of 1933.

However, after his discussions with Einstein on 17 November 1950 and 2 February and 24 October 1952, Shankland (1963, id., pp. 51, 52 and 55) wrote:

> *"He repeated several times... that since the phases found by Miller (which fix the direction in space) were not consistent, this was the strongest argument against the drift reported by Miller...*
>
> *"He said several times however, that he (and also H.A. Lorentz) considered Miller an excellent experimenter and thought his data must be good ...*
>
> *"Einstein also told me that H.A. Lorentz had studied Miller's work for many years and could not find the trouble ...*
>
> *"He emphasized that if there is a systematic effect, however small, it must be explained...*
>
> *"Once again he told me that Lorentz could never explain Miller's result and felt that it could not be ignored, although Einstein was not sure whether Lorentz really believed Miller's result...*
>
> *"Many negative results are not highly important, but the Michelson experiment gave a truly great result which everyone should understand."*

Here I must remind the reader that Miller, Lorentz, and Einstein had no knowledge of the regularities present in Miller's observations, that I have brought to light (*Chapter IV*, Section D, above) *and which definitively validate those observations.*

Geographic Institute); and finally the optical experiments of Ernest Esclangon in 1927-1928. As it results from *Chapter V*, all these lead to the same conclusion as Miller's experiments: *it is possible to demonstrate the position of the Earth along its orbit by purely terrestrial experiments.*

This conclusion impresses itself upon us all the more strongly because *it is not based upon any hypothesis or upon any theory* as to the exact nature of the observed dependences. It results simply from the observed correlations between the series of observations analyzed and the position of the Earth along its orbit, and these correlations constitute *undeniable objective facts that speak for themselves.*

*Accordingly an essential postulate of the Theory of Relativity - that it is impossible to demonstrate the displacement of the Earth along its orbit by purely terrestrial experiments - is proved to be completely unfounded.*

## *B.4 A few commentaries*

The invalidation of the theories of special and general relativity by the experimental data calls for some *absolutely essential* comments.

### Rotation and translation of the Earth

1 - From the "*positive*" result of Miller's experiments and from the regularities that I have demonstrated in his observations, it follows that *there is no distinction to be drawn between the rotation of the Earth and its translation, as the Theory of Relativity does. Both the one and the other* can be demonstrated by purely terrestrial experiments.

Not only was the rotation of the Earth demonstrated by the pendulum experiment of Foucault, but it was also confirmed by the interferometric experiment of Michelson and Gale in 1925.

If we assume that the Earth moves along its orbit through an ether *that is supposed to be isotropic and immobile with respect to the fixed stars*, then *a priori* there is certainly an internal inconsistency between the assertion that it is impossible to demonstrate the movement *of translation* of the Earth by purely terrestrial experiments, and the fact that this possibility has never been doubted as far *as its rotation is concerned. In both cases in fact, these are relative motions.*[1, 2]

---

[1] However, naturally this inconsistency would not exist if the Earth and the local ether were to share the same movement of translation (*Chapter IV* above, §F.2.4 and §F.2.5, notes 7 and 8, pp. 422-423). In that case, it would obviously be impossible to demonstrate the velocity of the Earth with respect to the ether, because that velocity would be nil.

On the other hand, if an anisotropy of space existed that was related to the position of the Earth along its orbit, then it would become possible to demonstrate this position by purely terrestrial experiments.

[2] In fact, the movement of the center of gravity of the Earth in its trajectory along its orbit around the Sun is not really a uniform rectilinear motion.

It includes a rotatory movement around the Sun, *of which no account is taken* by Einstein's formula for composition of velocities.

## The strategy of silence

2 - In order to preserve the theory of relativity, the scientific world has preserved *almost total silence* on the subject of Miller's 1933 Memoir, and when the subject has been mentioned his observations have been attributed to effects of temperature.

As for my experiments on the anomalies of the paraconical pendulum and the optical anomalies that I demonstrated, these never became the target of any critical publication.[3] The world of science was generally content with burying them under a heavy cenotaph of lead, *while at the same time spreading a tide of rumors with the object of discrediting them.*

As a consequence of being very rarely cited, Esclangon's experiments have similarly fallen into oblivion.

*It is so true that Official Science systematically ignores everything that could disturb its certainties.*

## An obvious disproportion

3 - One cannot fail to be somewhat astonished by the *obvious disproportion* between the *enormous* literature on the theory of relativity (thousands of books and articles) and *the almost complete non-existence of any serious or deep discussion* of the purely terrestrial experiments for demonstration of the movement of translation of the Earth, *most particularly in the case of Miller's experiments of 1925-1926 and his memoir of 1933.*[4]

---

[3] A single exception must be made in the case of the paraconical pendulum. In the *Comptes-Rendus de l'Académie des Sciences* (Proceedings of the Academy of Sciences), Jean Goguel thought he could submissively explain the anomalies of the paraconical pendulum by the actions of wind and temperature upon the building (CRAS, 1958, Vol. 246, p. 2340 [152]; see *Chapter I* above, §C.2.3, note 7, p. 148).

In fact, his claimed explanations were *completely swept away* by my crucial experiments in July 1958 at Bougival and Saint-Germain (*Chapter I*, §G.5.3 above, note 9, p. 229) (see the *Complementary Note* annexed to my Memoir of 1958, note 2).

The subsequent refrainment by Goguel (a specialist in geophysics!) to present to the Academy a new *Note* from me is *particularly significant.*

[4] An excellent example is that of the very interesting work of Costa de Beauregard, *La Théorie de la Relativité Restreinte* (The Theory of Special Relativity) (Masson, 1949 [95]). This work does indeed cite Miller's experiments of 1925-1926 (pp. 13 and 165), his article in the Astrophysical Journal in 1928, and his article of 1933, but the author's commentary upon Miller's experiments is limited to rejecting them in three lines (as compared with the 7,000 lines of the work) while drawing parallels between them and the experiments of Kennedy and Illingsworth in 1927 and of Joos in 1930, which were experiments that were only performed at certain specified times (see *Chapter IV* above, §E.2.3, p. 415). *Clearly Costa de Beauregard had not read Miller's 1933 article, or at most had only read it very superficially.*

Nevertheless it must be recognized that Costa de Beauregard is one of the very few authors even to mention Miller's experiments, while almost all other authors pass over them in complete silence.

If needed, one could understand the approach of Lorentz, Poincaré, and Einstein in view of the conviction that was in practice imposed upon the scientific world, *at the beginning of the 20th century*, that it was impossible to demonstrate the movement of translation of the Earth by purely terrestrial experiments. But, after the publication of Miller's review memoir of 1933, it is *really impossible* to understand the refusal to take into account his experiments of 1925-1926, or at least to discuss them deeply.

Certainly Poincaré was long dead, but Einstein - the founder of the theory of relativity - and all those who made powerful contributions to the analysis of the theory of relativity, to its development, and to its diffusion, such as for example Hermann Weyl, Max von Laue, Arthur Eddington, and Max Born, ought not to have ignored Miller's results, which were much more founded than the experiments presented as justification of the relativity theory, such as for example those relating to deviation of light near the Sun.[5]

---

[5] On these experiments, see in particular: - Emile Picard, *La Théorie de la Relativité et ses Applications à l'Astronomie* (The Theory of Relativity and its Applications to Astronomy), Gauthier-Villars, 1922 [214]; - J. Chazy, *La Théorie de la Relativité et la Mécanique Céleste* (The Theory of Relativity and Celestial Mechanics), 2 vol., Gauthier-Villars, 1928 [92]; - G. Darmois, *La Théorie Einsteinienne de la Gravitation. Les Vérifications Expérimentales* (The Einsteinian Theory of Gravitation. Experimental Tests), Hermann, 1932 [99]; - M.A. Tonnelat, *Les Vérifications Expérimentales de la Relativité Générale* (The Experimental Tests of General Relativity), Masson, 1964 (See §C.6 below, pp. 603-607) [267].

In any case, it is at the least curious that none of these works mentions the "*negative*" result of Michelson's experiment as a justification of the Theory of Relativity. The only exception is the short work of P. Bricout, *Ondes et Electrons* (Waves and Electrons) (Armand Colin, 1929, p. 127) [81].

Whatever the validity of the experiments presented in support of the theory of general relativity may be, they cannot be considered as really justifying it, in view of the experimental data corresponding to *Chapters I through V* of the present volume, which *completely invalidate* the postulates of the theory of relativity.

Insofar as the facts presented in support of the theory of relativity can be considered as effectively established, *their explanation must be sought in other directions*. In fact, for example, before the theory of general relativity, numerous explanations had been suggested for the advance of the perihelion of Mercury of 43" per century (See for example §E.1.1 below, pp. 630-631).

## *B.5 Rejection of the theories of special and general relativity as being incompatible with the observational data*

In view of the experimental data, one conclusion is inescapable: *the theories of special and general relativity, which are based upon postulates that are invalidated by the observational data, must be rejected.*[1] Einstein himself explicitly recognized this *in advance*, as is shown by his very explicit declarations that I have recalled.[2]

Whatever dominance the theory of relativity has wielded in the past, today the time has arrived to question it root and branch. In fact, *in the light of experiment, no theory can be considered as being definitive*. According to the scientific philosophy of Einstein himself:[3]

> *"Our conceptions of physical reality can never be definitive. If we wish to act in a logical manner, in as perfect a manner as possible in view of the observable facts, then we must always be ready to modify these conceptions, otherwise termed the axiomatic foundations of physics. In fact, a glance at the evolution of physics allows us to realize that, over time, these foundations have been subject to profound changes."*

---

[1] The rejection of the *Theories of Special and General Relativity* as being incompatible with the observational data *in no way means that all Einstein's contributions should be rejected.*

This rejection simply means that *all* the theoretical developments founded upon hypotheses invalidated by the experimental data must be rejected *as such.*

Those of Einstein's contributions *that appear to be verified by experiment* must naturally be preserved, *but evidently must receive theoretical justification other than from the theory of relativity.*

In physics, many results can be obtained from several different theories. An excellent example is the formula established by Fresnel in 1818 relating to the dragging of light waves, which was verified by Fizeau in 1851. This formula can be deduced *both* from Fresnel's theory of 1818 and also from the theory of special relativity of 1905 (see below, §C.5, pp. 601-602).

[2] §B.3.2 and §B.3.3 above, pp. 583 and 585.

[3] Albert Einstein, 1939, *Comment je vois le monde* (How I see the World), 1939, id., p. 194 [129].

*A theory is only worth what its premises are worth. If the premises are wrong, the theory has no real scientific value. The only scientific criterion for judging the scientific validity of a theory is, in fact, its confrontation with the experimental data.*

The judgment on the history of science expressed by Max Born in 1943 during the apogee of the theory of relativity here finds its application to the theory of relativity itself:[4]

> *"When we examine the history of the sciences, we remark a sort of cycle, periods of experimental progress alternating with periods of theoretical development. Theories have a tendency to become more and more abstract and general. They culminate in principles that at first are repulsed by philosophers, and later are assimilated. As soon as they become elements of the philosophical system, they start to be transformed into dogma and to undergo sclerosis. These tendencies can be observed in the oldest of the exact sciences, Mathematics and Astronomy."*

It could not be expressed better.

[4] Max Born, 1943, *L'expérience et la théorie en physique* (Experiment and theory in physics), Gauthier-Villars, 1955, p. 3 [74].

# C THE DEVELOPMENT OF THE THEORY OF RELATIVITY: AN EXAGGERATED CONSTRUCTION UPON UNCERTAIN AND FRAGILE FOUNDATIONS

In view of the experimental data relating to the paraconical pendulums with anisotropic support and with isotropic support, to the sightings at marks and at collimators, to the observations of Esclangon, and to the observations of Miller, what judgment can be reached concerning the immense and extraordinary development of the theory of relativity?

## *C.1 The foundations of the theory of relativity*

According to almost all the literature, the very foundations of special relativity *rest upon the claimed "negative" result of Michelson's experiment, and upon a postulate that sanctifies it, namely that no purely terrestrial experiment can determine the position of the Earth along its orbit.*[1]

The theory of general relativity appears as an extension of the theory of special relativity, *and the theory of special relativity is only a particular case of the theory of general relativity, from which it is inseparable.*[2]

[1] See §B.3.1 above, p. 582, and §B.3.3, pp. 583-585.

[2] See §B.3.2 and §B.3.3 above, pp. 582-585.

In fact, in 1905 the foundations of the theory of special relativity were both *uncertain and fragile*. Indeed, incidentally, the Michelson-Morley experiment in 1887 and the Morley and Miller experiments of 1902 and 1904 had all detected speeds of the order of 8 km/sec. But they had therefore been considered as inconclusive, and the observed speeds had been attributed to observational error.[3] *That was however only a hypothesis.*

Still further, even supposing that the "*negative*" result of the Michelson-Morley experiment had been definitively established, *alternative hypotheses, such as that of the anisotropy of space, could have been envisaged, and should have been envisaged.*[4]

Today, in view of the analysis of Miller's observations, the very foundations of the theories of special and general relativity *appear to have been totally invalidated by the experimental data.*[5]

---

[3] See Miller, 1933, id., p. 207, and *Chapter IV* above, §E.2.1, p. 414.

[4] See §A.7 above, pp. 573-576, and *Chapter IV*, §F.2, pp. 420-424.

In any case, the hypothesis of a partial or total dragging of the ether by the Earth during its movement of translation cannot be considered to have been definitively eliminated (see notably *Chapter IV*, §F.2.4, notes 7 and 8, p. 423).

[5] Section B above, pp. 577-591.

## C.2 *The Lorentz transformation*

Einstein's memoir of 1905 deduced the Lorentz transformation from the postulate of the constancy of the speed of light in all inertial reference frames K' in uniform movement with respect to an inertial reference frame K.[1]

This transformation implies *a contraction* of all rigid bodies in the direction of movement, whatever form they may have.[2] It also implies *a retardation* of the clocks in the reference frame K' with respect to those in the reference frame K.[3]

If we consider the movement of K with respect to K', the same transformation is applicable, with $x, y, z, t$ and $x', y', z', t'$ respectively replaced by $x', y', z', t'$ and $x, y, z, t$.[4]

*In fact, no matter what anyone might say, the contraction of bodies with their speed and the local time of Lorentz have a common property: that of being totally unverifiable.*

---

[1] Einstein, 1905, *Sur l'électrodynamique des corps en mouvement* (On the Electrodynamics of Moving Bodies), Gauthier-Villars, 1925, §3, pp. 12-20 [126].

[2] Einstein, id., p. 21, and §A.3 and §A.4 above, pp. 556-564.

[3] Einstein, id., p. 22, and §A.3.4 above, pp. 559-561.

[4] This raises some difficulties. In fact, if the rigid bodies of K' contract during the movement of K' with respect to K, then the rigid bodies of K contract equally and in the same proportion in the movement of K with respect to K'. On this difficulty, see Eddington, 1923, *The Mathematical Theory of Relativity*, 1960, id., §10.11, p. 27.

The same difficulty is *also* encountered in the relative retardations of clocks.

In connection with those findings, there is some interest in citing here one of the commentators on Einstein:[5, 6]

*"In drawing all the possible deductions from the message contained in the equation of the Lorentz transformation, Einstein discovered a large number of new and extraordinary truths about the physical universe...*

*"These truths can be described in very concrete terms...*

*"Einstein was led to show unsuspected features of clocks and measuring rods. For example, a clock attached to a moving system runs at a different rate from that of a stationary clock. A standard measuring rod attached to a moving system changes its length along with the speed of the system. The clock shows that the speed increases and the measuring rod contracts in its direction of movement...*

*"An observer moving with the clock and the measuring rod will not notice any change in them, but a fixed observer, that is to say fixed with respect to the moving system, will see that the moving clock runs slower with respect to his stationary clock, and that the moving measuring rod has contracted with respect to his stationary standard of measurement...*

*"The laws governing these contractions are defined by the Lorentz transformation, and are extremely simple. The greater is the speed, the greater is the contraction. A measuring rod moving at a speed of 90% of the speed of light will contract by approximately half; past this limit, the contraction becomes more rapid, and, if the measuring rod could attain the speed of light, it would contract so much as to disappear completely. In the same way, a clock traveling at the speed of light would completely stop...*

---

[5] Lincoln Barnett, 1948, *Einstein et l'univers* (Einstein and the Universe), Gallimard, 1952, pp. 93-101 [60]. Einstein himself prefaced this work with the following comments:

*"It is of the first importance that the wider public should have the possibility of learning - clearly and intelligently - of the efforts and the results of scientific research...*

*"This book by Lincoln Barnett makes a contribution of great value to popular scientific literature. It presents the great ideas of the theory of relativity very well.*

[6] As far as the first line of this citation is concerned, it should always be remembered that Einstein declared that he had no knowledge of the Lorentz transformation when he wrote his memoir of 1905 (see §A.6 above, pp. 569-572).

*"At first sight these facts are difficult to accept, but that is simply because classical physics affirms, without any justification, that an object preserves the same dimensions whether it is moving or stationary, and that a clock runs at the same rate whether it is moving or stationary. Common sense tells us that it must be so. But as Einstein has showed, common sense is really only a residue of the prejudices accumulated in the mind before the age of eighteen. Every new idea that one meets with later on must struggle against this accumulation of "obvious" ideas. It was Einstein's unwillingness to accept as obvious any principle devoid of proof that allowed him to penetrate into the hidden realities of nature more deeply than any thinker before him."*[7]

[7] Barnett adds:

*"According to the Theory of Relativity, the heartbeat of an individual who is moving at a speed close to the speed of light will be relatively slowed down, and so will his breathing and all his other physiological processes. He will not notice his slowing down, because his watch will slow down in the same way. But from the point of view of a stationary observer, he* will age *more slowly."*

At the least, such assertions are rather surprising.

## *C.3 The constancy of the speed of light*

The theory of special relativity is based upon the principle of the constancy of the speed of light. As Einstein wrote:[1, 2]

> *"The second principle upon which the theory of special relativity is based is "the principle of the constancy of the speed of light in a vacuum". This principle asserts that: in a vacuum, light always has a determinate speed of propagation".*

This principle only holds rigorously in the context of the theory of special relativity,[3] but the corrections for the theory of general relativity remain very small.

---

[1] Einstein, 1939, *Comment je vois le monde* (How I See the World), Flammarion, p. 209 [129]. Einstein's first principle is the principle of relativity.

[2] In connection with Michelson's experiment, Lichnerowicz comments as follows upon the principle of constancy of the speed of light as deduced from the "negative" result of Michelson's experiment (1951, *Eléments de calcul tensoriel* (Elements of Tensor Calculus), pp. 161-162 [177]):

> *"There might be some drawbacks in basing a principle of such generality upon the results of only one type of experiment, which an experiment of another type might invalidate.*
>
> *"But in fact Michelson's experiment merely attracted the attention of physicists, in an imperious manner, to a mathematical fact that until then had remained somewhat in shadow, although Poincaré had pointed it out: the equations of Newtonian dynamics and Maxwell's equations of the electromagnetic theory are not invariant under the same group of transformations. There is a conflict between pure dynamics and electromagnetism, and the principle of the constancy of the speed of electromagnetic waves is, in fact, included in Maxwell's electromagnetic equations.*
>
> *"To bridge the conflict between pure classical mechanics and electromagnetism, Einstein proposed to admit the principle of the constancy of the speed of electromagnetic waves, thus preserving the electromagnetic theory of Maxwell while modifying classical dynamics so as to bring it into agreement with electromagnetism."*

[3] Einstein wrote (*La relativité et le problème de l'espace* (Relativity and the Problem of Space), Gauthier-Villars, 1954, p. 84 [132]):

> *"According to the Theory of General Relativity, the often-mentioned law of the constancy of the speed of light in a vacuum, which is one of the fundamental suppositions of the Theory of Special Relativity, cannot claim unlimited validity. Indeed, curvature of a light ray could only be produced if the speed of propagation of light was different in different locations. It might be thought that this consequence overthrows the Theory of Special Relativity, and with it that of the Theory of General Relativity. But this is not the case. We can only conclude that the Theory of Special Relativity cannot claim an unlimited degree of validity; its results are only valid to the extent that one can neglect the influences that gravitational fields exert upon phenomena (upon optical phenomena, for example)".*

In fact, the principle of the constancy of the velocity of light is contradicted by the analysis of Miller's observations.[4]

[4] Miller's observations show the existence of variations of the order of 8 km/sec in the speed of light, according to its direction (*Chapter IV*, §D.4.3, p. 406). The relative difference is of the order of $8/300,000$, *i.e.* of the order of $2.7 \times 10^{-5}$, while in the neighborhood of the Earth the relativistic correction is of the order of $10^{-9}$ (see §A.5.2 above, note 7, pp. 566-567).

## *C.4 The principle of the invariance of natural laws*

Einstein postulated that it should be possible to express the laws of physics *in an invariant manner*, whatever be the system of reference coordinates $x, y, z, t$; *and, on all the evidence, this certainly seems to be necessary*. He wrote:[1]

> *"The laws of physics must be expressed so that they remain valid with respect to reference systems that are moving in any manner whatever...*
>
> *"The general laws of nature must be expressed by equations that are valid for all coordinate systems, in other words that are covariant (in the general sense) with respect to any substitutions whatever."*[2]

But maintaining that Maxwell's equations, *as Maxwell presented them*, are invariant in any coordinate system is quite a different thing. In fact, *nothing* permits us to assert that these laws are effectively invariant in their *current* form.

Actually, if we suppose that the equation of propagation

$$(1) \qquad \Delta\varphi - \frac{1}{c^2}\frac{\partial^2\varphi}{\partial t^2} = 0$$

is invariant for any linear transformation of the coordinates, then we recover the Lorentz transformation.[3] But this is by no means the case if, for example, the equation of propagation (1) is written in the form indicated in my memoir of 1957 on the speed of light.[4]

---

[1] Einstein, 1916, *Les fondements de la théorie de la relativité générale* (The Foundations of the Theory of General Relativity), Hermann, *Théorie de la relativité* (Theory of Relativity), 1933, pp. 10 and 15 [127].

[2] On the meaning of covariance, V. Ugarov (1974, *Théorie de la relativité restreinte* (Theory of Special Relativity), id., p. 293) writes:

> *"Thus, covariance is a property of a system of several equations which, if valid in one reference frame, keep their form in all other reference frames. Invariance is the same property in relation to only one equation. The difference between these two ideas is not often mentioned, and covariance and invariance are used as synonyms."*

[3] Von Laue, 1924, *La théorie de la relativité* (The theory of relativity), Gauthier-Villars, Vol. I, §6, p. 57 [276].

[4] In an isotropic medium this equation, proposed by Jean Hély, in fact is written:

$$(1) \qquad \Delta\varphi - \frac{1}{c^2}\frac{\partial^2\varphi}{\partial t^2} - 2\frac{k}{c}\frac{\partial\varphi}{\partial t} - k^2\varphi + 4\pi K\delta = 0$$

See *Chapter VI* above, §C.1.4, pp. 511-512.

*The postulate of the invariance of the laws of nature is obviously unarguable, but the deductions that are drawn from this postulate in the case of Maxwell's equations are based upon a hypothesis that has not been demonstrated.* In fact, nothing guarantees that they ought not to be *first completed and revised* in order to be considered as corresponding *effectively* to a form that is invariant under all linear transformations of the coordinates.[5, 6]

---

[5] It is at least curious that this remark is not presented anywhere in the literature.

[6] For Maxwell's equations, see Henri Poincaré's introduction to his work of 1899, *Electricité et Optique* (Electricity and Optics) (re-edited in 1954, Gauthier-Villars), pp. III-X) [228].

Notably, Henri Poincaré wrote:

*"It is in electrostatics that my task has been hardest; in fact, that is above all where precision is lacking. One of the French scientists who has the most deeply studied Maxwell's works once said to me: "I understand everything in his book, except what an electrified ball is."*

## C.5 *The formulation by Fresnel in 1818*

*In support* of the formula of the theory of special relativity[1] for addition of speeds, Einstein mentions its verification by Fizeau's experiment of 1851 on the dragging of light waves.[2]

Fizeau's formula, *demonstrated by Fresnel in 1818*, is written:

$$(2) \qquad u' = \frac{c}{n} + v\left(1 - \frac{1}{n^2}\right)$$

where $c$ is the speed of light in a vacuum, $v$ is the speed of moving water, $n$ is its index of refraction, and $u'$ is the observed speed of light in the moving water.

Effectively, Einstein's law of composition of velocities[3] is written here as:

$$(3) \qquad u' = \frac{\frac{c}{n} + v}{1 + \frac{c}{n}\frac{v}{c^2}}$$

which, when developed as a series, gives:

$$(4) \qquad u' = \frac{c}{n} + v\left(1 - \frac{1}{n^2}\right) - \frac{v^2}{nc}\left(1 - \frac{1}{n^2}\right) + \ldots$$

*$v$ is of the order of one meter per second*, so that the first neglected term is of the order of $v^2/c$ (*i.e.* of the order of $10^{-6}$)

---

[1] §C.2 above.

[2] Einstein, 1916, *La Théorie de la relativité restreinte et générale* (The Theories of Special and General Relativity), 1976, id., p. 45. See also Eddington, 1923, The Mathematical Theory of Relativity, id., §6, p. 21.

Neither Einstein nor Eddington mentions that the formula verified by Fizeau in 1851 had been demonstrated by Fresnel in 1818 (*Additional note to the letter to M. Arago*, Annales de Chimie (*Annals of Chemistry*), 1818, volume 9, p. 286 [147]).

For Fizeau's experiment, see in particular Bouasse, 1925, *Propagation de la lumière* (The Propagation of Light), Delagrave, pp. 85-92 [77], and Sesmat, 1937, *L'optique des corps en mouvement* (The Optics of Moving Bodies), Hermann, pp. 529-534 [246].

[3] §A.3.4, p. 560, and §A.4.1, p. 562, above.

Since the law of composition of velocities is certain to be erroneous,[4] here we have *an excellent example of deduction of an exact law from a false formula.*

This shows that *the verification of a theory by experiment does not prove that the theory is correct.* It simply shows that the theory is compatible with experiment, and alternative explanations always remain possible.[5]

As Paul Painlevé so justly wrote:[6, 7]

> *"Fresnel arrived at the formula which Fizeau later so exactly verified via a theory in which matter and the ether that penetrates it are involved simultaneously. This theory was not invented to fit the experiment, since it predicted the experimental facts."*

---

[4] §B.2 and §B.3 above, pp. 581-586.

[5] See below *Section E*, pp. 629-633.

[6] Paul Painlevé, 1922, *Les axiomes de la mécanique* (The axioms of mechanics), 1955, Gauthier-Villars, p. 95 [208].

[7] The supporters of the theory of relativity generally omit to point out that Fresnel determined this formula *thirty-three years before Fizeau's confirmatory experiment*, and that this formula is only approximated in Einstein's theory (equation 4 above).

In fact certain relativistic authors omit to mention the *much earlier* theory of Fresnel, *usually deliberately*.

From this double deduction of Fresnel's equation, it follows that we cannot assert that Fizeau's experiment constitutes a proof of one theory or another, but only that both of those theories are *compatible* with experiment.

## *C.6 The experimental data*

*According to almost all the commentaries in the literature*, the *major* experimental support of the theory of special relativity has *undoubtedly* been the claimed "*negative*" result of Michelson's experiment, *a "negative result" that is completely invalidated by analysis of Miller's observations.*[1]

In parallel with the "*negative*" result of Michelson's experiment, certain experimental results have been presented, both in support of the theory of special relativity, and in support of the theory of general relativity.

### The theory of special relativity and the experimental data

1 - A good overall account of the experimental confirmations of the theory of special relativity was presented in 1955 in an article by Robert Lennuier.[2]

In parallel with Michelson's experiment and the experiment of Fizeau, this article essentially presents three experimental results: *the slowing-down of clocks (the experiments of Ives and Stilwell in 1938 and 1941), the variation of the masses of particles with speed, and the law of equivalence between mass and energy.*

These different results appear to have been experimentally verified, but, *to the extent that they have been, they cannot be explained by the theory of special relativity*, which is based upon the Lorentz transformation and the law of composition of velocities, both of which have been *invalidated by experiment.*[3]

Here, Fizeau's experiment of 1851 is an excellent illustration. It cannot in fact be presented as a justification of the "*law of composition of velocities*", because this law must be considered as having been invalidated by experiment.[4]

---

[1] See Section B above, pp. 577-591.

[2] Robert Lennuier, 1955, *La théorie de la relativité restreinte et l'expérience* (The theory of special relativity and experiment), Encyclopédie Française, Vol. II, *Physics*, Part Three, pp. 2.46.15 - 2.48.6 [172].

[3] §B.3 above, pp. 582-586.

[4] On the other hand, it is perfectly explained by Fresnel's theory of 1818.

## The theory of general relativity and the experimental data

2 - Three experimental results have been presented in justification of the theory of general relativity: *the disparity of 43" per century in the advance of the perihelion of Mercury, unexplained by the Newtonian theory of gravitation; the deviation of light rays near the Sun; and the displacement of the lines in the solar spectrum towards the red.*[5]

In fact, in all these three cases, the effects are *extremely small*, near the limits of errors of observation, and difficult to bring out clearly.[6]

• *For the perihelion of Mercury*, the theory of general relativity leads to an estimate of 42.9" per century, *which is a most remarkable verification* in view of the estimate of 43" given by Newcomb and the estimate of 42" given by Chazy.[7]

In fact, since there is no solution of the two-body problem in general relativity, the perturbations arising from the $ds^2$ of Schwarzschild are added to those due to the trajectories of the planets.

---

[5] On the experimental verifications of the theory of general relativity, see particularly the references given above, §B.4.3, note 5, p. 589.

[6] Thus, for the perihelion of Mercury, the angular deviation of 43" per century that we find is, in CGS units, of the order of $6.6 \times 10^{-16}$ radians per second.

These are quantities of a *much smaller* order of magnitude than those in the anomalies of the paraconical pendulum, in the sightings at marks and at collimators, in the observations of Esclangon, and in the observations of Miller.

[7] As G. Darmois underlines (1932, *La Théorie Einsteinienne de la Gravitation. Les vérifications expérimentales* (The Einsteinian Theory of Gravitation. The Experimental Verifications), id., p. 20):

> *"The uncertainty to which this number 42" is known ought to eliminate the a priori striking character of the coincidence of the* 42"9 *of Einstein and the figure of 43" given by Newcomb."*

The estimates of Newcomb and Chazy correspond to the difference between the observed value of the rotation of the perihelion of Mercury of 572"70 per century and the value calculated from the influences of the other planets (Darmois, 1932, id., p. 19).

However, the theory of relativity gives no explanation for the observed deviations of the nodes of Venus and of the perihelion of Mars.[8]

After a remarkable analysis,[9] Jean Chazy concluded:[10]

> *"In all impartiality, in the current state of Science, the argument cited in favor of the Theory of Relativity relating to the value of the advance of the perihelion of Mercury does not have, and cannot have, the absolute character in which some people believe."*

In any case, other explanations have been proposed, or could be imagined, for explaining the advance of the perihelion of Mercury.[11]

---

[8] Citing Darmois:

> *"This passage from Newton's laws to the theory of relativity introduces no secular inequality: neither into the major axes; nor into the eccentricities; nor into the longitudes of the nodes.*
>
> *"Of the ... secular inequalities considered, the only one that can be checked by observation is that concerning the secular advance of the perihelia."*

(Darmois, 1932, id., p.21).

We inquire, what is shown by observation for the planets other than Mercury:

> *"The line of the nodes of Venus (the line along which the orbit of Venus intersects the plane of the ecliptic) retreats through an angle of the order of 1000" over a century, which is not entirely accounted for by calculation. There remains an unexplained residue of approximately -10" per century, i.e. an advance of the nodes by 10".*
>
> *"The perihelion of the orbit of Mars advances through about 1600" per century, which cannot be exactly accounted for by calculation. There remains an unexplained advance of 8"."*

In connection with these observations:

> *"We see: -1°: that the theory of relativity cannot explain the advance of the node of Venus; -2°: that the value that it gives for the advance of the perihelion of Mars has the desired sign, but remains much too small."*

(Darmois, 1932, id., p. 19).

[9] Jean Chazy, 1928, *La Théorie de la relativité et la Mécanique céleste* (The Theory of Relativity and Celestial Mechanics), Vol. I, pp. 139-236, and in particular pp. 228-236 [92]; see also Eddington, 1923, *The Mathematical Theory of Relativity*, §40, pp. 89-90 [121]; Dugas, 1950, *La théorie de la relativité* (The theory of relativity), id., p. 513; and M.A. Tonnelat, 1964, *Les vérifications expérimentales de la relativité générale* (The experimental tests of general relativity), id., pp. 54-58.

[10] Jean Chazy, id., p. 180.

[11] See particularly Maurice Lévy, *Note to the Academy of Sciences*, CRAS, 17 March 1890 (See below, §E.1, note 4, p. 631) [175].

• For the deviation of light rays near the Sun, the theory of Einstein predicts a deviation of the order of 1.6 to 2". However, according to Esclangon's analysis to which Jean Chazy subscribed:[12]

> *"The sincere conclusion to draw is that these observations are as yet unable to elucidate the question asked. They neither confirm nor invalidate Einstein's law of deviation."*

• For the displacement of the spectral lines toward the red, the discussion is *very complex*. In fact, Sesmat's conclusion appears justified:[13]

> *"While the Einstein effect can be invoked concurrently with other causes in order to explain the displacements of the solar lines, these displacements cannot currently be considered as being a positive proof of the theory."*

• In fact, *none of the three empirical tests of the theory of general relativity can be considered as being really conclusive*. We can only say that these three deductions from the general theory of relativity are not contradicted by experiment.

**Overall view**

3 - *One certainty dominates this entire discussion*. It is *completely inaccurate* to consider that Michelson's experiment, as repeated by Miller, gave a "*negative result*". From this it follows that, to the extent that the theory of special relativity and the theory of general relativity are based upon the "*negative*" result of Michelson's experiment, they must be completely rejected.

It is certain that the speed of light cannot be considered as being invariant in all directions.

---

[12] Jean Chazy, id., p. 255. A detailed analysis is given on pp. 233-256.

[13] Sesmat, 1937, *Théorie relativiste de la gravitation* (Relativistic theory of gravitation), p. 353 [247]. An overall analysis is given on pp. 347-354.

See also Jean Chazy, 1930, id., Vol. II, pp. 30-33; Tonnelat, 1964, id., pp. 103-104; Darmois, 1932, id., pp. 25-29.

It is equally certain that the position cannot be sustained that no purely terrestrial experiment can determine the position of the Earth along its orbit. The observations of the paraconical pendulum with anisotropic support, the sightings at marks and at collimators, and the observations of Esclangon all tend to confirm the observations of Miller.[14]

From the above, and to the extent that the experimental facts presented in support of the theory of relativity can be considered as being verified, *the theories of special and general relativity cannot provide their explanation*, because the foundations of those theories must be considered *as having been invalidated* by experiment.[15] *The explanations of these phenomena must be sought elsewhere.*[16]

---

[14] §B.3.4 above.

[15] §B.3 above.

[16] The claimed verification of the theory of special relativity by Fizeau's experiment in 1851 constitutes an excellent illustration (§C.5 above, pp. 601-602). See also §E.1.1 below, p. 630.

## C.7 *The mask of mathematics*

Overall, the theory of relativity appears to be a grandiose construction encompassing all the phenomena in the universe, but we should not fall victim to the illusion of this construction.

*It cannot be too much repeated: a theory is only worth what its premises are worth. If the premises are wrong, the theory has no real scientific value.*

Certainly, today it is not necessary to justify the necessity and the utility of building rigorous models on the basis of perfectly specified axioms. The thing against which, on the other hand, it is proper to be sedulously on our guard consists of considering that it is sufficient for a theory to be based upon rigorous axiomatization, for it to be scientifically valid. However necessary such an axiomatization may be, *it is really only secondary to critical analysis of the postulates upon which the theory is based, and to confrontation of their implications with the experimental data.*

*In fact, it is really the scientific validity of the postulates that is at the center of any discussion. The rest is only mathematical deductions which in themselves are without interest, other than mathematical interest.*

The very elaborate mathematical machinery of the theory of general relativity has strongly contributed to mystification of the scientific community for several generations. There has been only too strong a tendency to conclude from the rigor of the mathematical reasoning that the conclusions are valid.[1]

---

[1] The invasion of the physical sciences by the pure mathematicians has been as damaging in physics as it has been in economics.

In one sense, the grandiose construction of the theory of relativity is completely comparable to an immense statue of granite supported upon a fragile pedestal of clay.

One of Einstein's commentators did not hesitate to write:[2]

> *"It is the mathematical orthodoxy of the universe that allows theoreticians like Einstein to predict and to discover natural laws, simply by solving equations."*

The somewhat fascinating manipulation of very abstract mathematical symbols that are far removed from reality[3] has only too great a tendency to be substituted for experimentation. For too many of the relativists, mathematics has become a goal in itself. *But, indispensable as it may be, mathematics is only an instrument for analysis of reality, and cannot and ought not to be substituted for reality.*

As Claude Bernard so well underlined:[4]

> *"The art of scientific investigation is the corner-stone of all the experimental sciences. If the facts that serve as the basis of reasoning are badly established or erroneous, then everything will collapse or become false; and it is thus that, most often, the origins of errors in scientific theories are errors of fact...*
>
> *"Men whom I will call systemizers start from an idea based more or less upon observation, which they consider to be an absolute truth. Then they reason logically and without experimentation, and, stepping from consequence to consequence, finally construct a system that is logical but that has no scientific reality whatever...*
>
> *"This is why we sometimes see pure mathematicians falling into errors of this type, although they are very exalted spirits; they simplify too much and reason upon phenomena as they conceive them in their minds, but not as they are in nature...*

---

[2] Lincoln Barnett, 1948, *Einstein and the universe*, id., p. 30.

[3] Here we must remember Bertrand Russell's quip:

> *"Mathematics is the subject in which we never know what we are talking about or if what we are saying is true."*

[4] Claude Bernard, 1865, *Introduction à l'étude de la médecine expérimentale* (Introduction to the study of experimental medicine), id., pp. 42 and 69-71.

> *"When we construct a general theory in our sciences, the only thing of which we can be certain is that, speaking absolutely, all these theories are false. They are only partial and provisional truths that are necessary to us... for advancing in our investigations; they represent no more than the current state of our knowledge, and as a consequence, it will be necessary to modify them."*

What better description could be found for certain mistaken paths, much too frequent, in contemporary theories!

## *C.8 An acute but one-way critical sense*

To Einstein and his successors, we owe critical analyses of the classical theories that are *often very pertinent and very profound*. But the direction of criticism of which they give testimony is unhappily one-way.

It is often too easy to denounce implicit hypotheses in the theory of universal gravitation and its extension, *those hypotheses being often very arguable and insufficiently analyzed*. But the fact is that this theory has made it possible for us to represent the movements of the planets with surprising perfection: the theory of perturbations has, in fact, enabled us to represent these movements with angular inaccuracies not greater than 100 sexagesimal seconds per century.[1]

When Einstein explained the residual advance of the perihelion of Mercury with his own formulas, he accepted without discussion all the conclusions of the classical theory. As Augustin Sesmat wrote:[2]

> *"Does not logic insist that, before claiming to have explained the anomalies by the relativistic theory, we should have calculated them all according to that same theory? Not absolutely, provided that we can establish, proceeding from justified approximations, that the relativistic anomalies, if we could calculate them exactly, would remain almost exactly equal to the classical anomalies. In fact, since the strict application of the relativistic formulas to complicated problems is at present impossible, this is the reasoning process when Einstein's success in the case of Mercury had to be more rigorously justified. No matter that this provides a demanding critic with a new reason for doubt."*

---

[1] See *Chapter I* above, §F.1.3 and §F.2.1, pp. 200-203.

[2] Augustin Sesmat, 1937, *Essai critique sur la doctrine relativiste* (Critical Essay upon the Relativistic Doctrine), Hermann, p. 433 [245].

During the General Conclusions of the Solvay Congress in 1911, Henri Poincaré declared:

> *"(What strikes me) is to see the same theory taking ground sometimes into the principles of ancient mechanics, and sometimes into the new hypotheses that constitute their negation; we must not forget that any proposition can be easily proved, provided only that two contradictory premises are admitted within the proof."*

(*La Théorie du rayonnement et les quanta* (The theory of radiation and quanta), Gauthier-Villars, 1912, p. 451 [251]).

*In fact, up until now, the theory of general relativity has proved incapable of resolving the two-body problem.*[3]

Einstein's reasoning often is based *upon pure sophism*. For example, *on the hypothesis that "clocks of the same construction run at the same speed"*, Einstein wrote:[4]

> *"This convention also contains a physical hypothesis whose validity cannot be doubted, because no empirical proof has ever invalidated it."*

Obviously this cannot be a satisfactory justification; and this is because a proposition, even if it does not appear at a given epoch to be contradicted by experiment, cannot considered for that reason as being true.

Again, in the same way, one cannot deduce from the principle of invariance of natural laws that Maxwell's equations should be considered as being invariant.[5]

Certain deductions from the theory of relativity skirt upon nonsense. Evaluations of the radius and the mass of the universe are examples.[6] *Such speculations are obviously lacking any real objective basis.*

---

[3] As Eddington wrote (1923, id., p. 94):

> *"The problem of two bodies in Einstein's theory remains an outstanding challenge to mathematicians".*

As for the theory of special relativity, it is at the least risky to write, as Georges Darmois did (1932, id., p. 29-30):

> *"It is the extension by Einstein of laws of invariant form to regions in which a gravitational field is present, which constitutes the theory of general relativity...*
>
> *"We start... from an extremely solid theory, special relativity, which must be extended by integration of gravitational phenomena...*
>
> *"The theory of special relativity must be considered as being experimentally based in the most solid manner."*

[4] Einstein, 1916, *La théorie de la relativité restreinte et générale* (The theories of special and general relativity), id., p. 27.

[5] §4 above, pp. 599-600.

[6] Einstein, 1916, *La théorie de la relativité restreinte et générale* (The theories of special and general relativity), id., p. 127; see also Eddington, 1923, *The Mathematical Theory of Relativity*, id., §102, pp. 235-237.

Einstein presented his theory as being based upon *"the perfection and certainty of its fundamentals"*, and he wrote:[7]

> *"The theory of relativity resembles a monument on two levels, which are the theory of special relativity and that of generalized relativity. The first, upon which the second rests, concerns all physical phenomena except gravitation; the theory of generalized relativity gives the law of gravitation and its relations to the other natural forces."*

Notably, Einstein considered the fundamentals of special relativity as being *"powerfully supported by experiment"*.[8] He also wrote:[9]

> *"The development of the situation has shown that for the moment, of all imaginable constructions, a single one has shown itself absolutely superior to all the others."*

But what remains today of all these assertions in view of the observational data?[10] Particularly, how can it be explained that Einstein (1879-1955) never commented in any of his publications upon Miller's observations and his memoir of 1933?[11]

---

[7] Einstein, 1939, *Comment je vois le monde* (How I See the World), Flammarion, p. 208 [129].

[8] Einstein, 1939, id., pp. 208-210.

[9] Einstein, 1939, id., p. 155.

[10] See §3 above, pp. 597-598.

[11] See §B.3 above, note 7, p. 585.

## *C.9 Obvious exaggerations*

Whatever Einstein's real contributions might have been, today it is a bit difficult to understand the acclamations that the theory of relativity stimulated. The foundations and the deductions of the theory of relativity have been considered as being well established and invulnerable truths, and very celebrated personalities have not hesitated to express rather untrammeled praises:

H. Weyl, 1921:[1]

> *"After a great labor of logical criticism during the whole of the last century, a tempest arrived to destroy the (ancient) conceptions of space, time, and matter, thus making room for a freer and more penetrating vision of the world. This revolution was essentially accomplished by the whole-hearted work of a single man: Albert Einstein."*

Max Born, 1923:[2]

> *"The result of Einstein's theory is thus the relativization and objectification of the notions of Space and Time. Today it crowns the edifice of the scientific conception of the Universe..."*
>
> *"The power of the new theory is due to the fact that it comes directly from experiment."*

Paul Langevin, 1931:[3]

> *"In the history of the sciences and in the history of physics in our epoch, Einstein will stand in the first rank. He is and will remain one of the greatest magnitude stars studding the sky of humanity."*

Louis de Broglie, 1949:[4]

> *"(The work of Costa de Beauregard) has the undeniable advantage of flowing directly from the original considerations of Einstein, those profound and fundamental considerations that are so intimately bound to the experimental data."*

---

[1] H. Weyl, 1921, *Temps, Espace, Matière* (Space, Time, Matter), Blanchard, 1922, pp. 1-2 [278].

[2] Max Born, 1921, *La théorie de la relativité d'Einstein et ses bases physiques* (Einstein's Relativity Theory and its Physical Basis), Gauthier-Villars, 1923, pp. XI and 336 [73].

[3] Paul Langevin, 1931, *L'oeuvre d'Einstein et l'astronomie* (Einstein's work and astronomy) [168].

[4] Louis de Broglie, 1949, Preface to the work of Costa de Beauregard, *La théorie de la relativité restreinte* (The theory of special relativity), Masson, 1949, p. 1 [95].

Amid this ocean of dithyrambs,[5] only a few critical voices were raised vigorously. Thus, E. Carvallo wrote in 1934:[6]

> "According to Einstein's theory, no purely terrestrial experiment without exterior references can demonstrate the translation of the Earth. *This is the postulate, which is a natural, but hazardous generalization of the doubtful result of Michelson. Einstein made calculations based upon his postulate, starting from the celebrated transformation of Lorentz.*
>
> *"The audacity of his genius gave his equations physical interpretation, without fearing collision with accepted science. He overthrew Galileo's mechanics, even to the concepts of time and space. The more his conclusions conflicted with common sense, the more they were acclaimed with enthusiasm. With rare exceptions the most eminent geometers, followed by the masters of physics, sucked the great public into the new doctrine. The eclipse of the school of Fresnel, the audacity and the newness of the theory, its mathematical rigor, the great intelligence of the author, his personal charm, were the causes of an unheard-of success...*
>
> *"Will not posterity judge that Relativity for a moment set Science back by three centuries?"*

But the few oppositions that surfaced remained without practical effect, *and an atmosphere of dogmatism and intolerance* developed which unduly retarded the progress of Science.

---

[5] Ben Gurion went so far as to declare:

> *"... even his face resembled that of God, as though the divine spirit was in him, irradiating all those who approached him with this extraordinary manifestation that was divine, human, and cosmic..."*

(Le Monde, 10-11 March 1996, p. 19).

[6] E. Carvallo, 1934, *La théorie d'Einstein démentie par l'expérience* (Einstein's theory disproved by experiment), Chiron, pp. 28-29 [91].

Carvallo was one of the rare physicists who had read Miller's memoir of 1933, *but he did not know how to analyze Miller's observations correctly, any more than did Miller.*

In fact a great number of relativists, including perhaps Einstein himself, were stimulated to adopt the reality of the contractions of bodies with their speeds and the concept of local time by the previous work of Lorentz and Poincaré (see §A.6 above, p. 569-572).

*The paradox here is that all authors, following Einstein, have considered the theory of relativity as being compelled by the experimental data.* But according to Bertrand Russell,[7] *"The fact has been known for a long time that subjective certainty is in inverse ratio to objective certainty."*

How can the development of such a situation, the grip of such a rush, be explained? As Henri Poincaré said,[8] *"Novelties are so attractive, and it is so difficult to seem not to be sufficiently advanced!"* In connection with Heisenberg's theory, Einstein himself underlined:[9] *"For almost everybody, momentary success is more persuasive than considerations of principle, and fashion blinds them, even if only for a certain time."*

*Today, a total re-examination of all the postulates of the theory of relativity is essential. The totality of experimental data makes this necessity inevitable.*

---

[7] Bertrand Russell, *L'esprit scientifique* (The scientific spirit), Janin, 1947, pp. 64-65 [240].

[8] Henri Poincaré, 1908, *Science et Méthode* (Science and Method), id., p. 272.

[9] Albert Einstein, 1949, *Correspondence with Michele Besso*, Hermann, 1979, Letter to Besso of 24 July 1949, p. 238 [135].

On 25 December 1925, he wrote to Besso (id., p. 128) :

> *"The most interesting thing delivered recently by theory, is the theory of Heisenberg-Born-Jordan of quantum states. This is a real calculus of witches, in which infinite determinants (matrices) appear in place of Cartesian coordinates. It is eminently ingenious, and is sufficiently protected against any proof of falsehood by its great complexity."*

It could not have been expressed better, but this is a criticism that Einstein, from many points of view, could have addressed to himself in connection with the theory of general relativity.

# D THE CLASSICAL THEORIES AND THE ANISOTROPY OF SPACE

## *D.1 From the classical theories to the theory of relativity*

It seems to be almost universally admitted today that the classical theories of gravitation, optics, and electromagnetism must be reformulated in the framework of the theories of special relativity and general relativity.

However, it inevitably results from analysis of the observations of Miller[1] and preceding developments[2] that the *very foundations* of the theories of special relativity and general relativity are invalidated by the experimental data.

Must it be concluded that the classical theories must be maintained without any change? That would be a position *incompatible* with the observational data.

---

[1] *Chapter IV*, Sections C through H.

[2] Sections A, B, and C above.

While on the surface of the Earth the corrections due to the theory of general relativity are around $10^{-9}$,[3] the anomalies detected for the paraconical pendulum with anisotropic support, in the sightings at marks and at collimators, in the optical sightings of Esclangon, and in the interferometric observations of Miller are all of the order of $10^{-6}$ to $10^{-5}$, in other words of an order of magnitude at least a thousand times greater than the corrections of the theory of general relativity.[4]

Thus, the order of magnitude of these anomalies is *considerable*. However, from the above discussion, it follows that their existence is unarguable, and that they cannot be attributed to perverse effects.

They all imply *revision* of currently accepted theories, and here it is proper to remind the reader *briefly* of their essential characteristics.

---

[3] *At the surface of the Earth*, the general relativity corrections due to the action of the Earth, the action of the Sun, and the action of the Moon are of the orders of $10^{-9}$, $10^{-8}$, and $10^{-13}$ respectively (§A.5.2, notes 7 and 9 above, pp. 566-568).

[4] See §D.2, §D.3, and §D.4 below, pp. 619-624.

## *D.2 Mechanical anomalies of the paraconical pendulum*

### Paraconical pendulum with anisotropic support

1 - If, for illustration, we consider the diurnal lunar wave of 24h50m, the observed variation of the azimuth of the paraconical pendulum is of the order of $2 \times 10^{-6}$ radians per second, i.e. around twenty million times greater than the variation calculated from the theory of universal gravitation.[1]

The observed variation can be explained by an anisotropy of inertial space of the order of $0.4 \times 10^{-6}$.[2]

• The anomalies of the paraconical pendulum with anisotropic support pose three fundamental questions: are the amplitudes statistically significant? can they be explained by currently accepted theories? can they be attributed to the influence of any known phenomenon? *Absolutely certainly*, the answers to these three questions are yes, no, and no.[3]

• The two crucial experiments of July 1958 in the laboratory of IRSID at Saint-Germain and in the laboratory installed at Bougival in an underground quarry 6.5 km away and with 67 m of chalk and clay covering gave *exactly the same results* for the diurnal lunar component of 24h50m, *as far as amplitudes and phases were concerned.*[4]

---

[1] *Chapter I*, §B.2.1, p. 118, above.

[2] *Chapter I*, §F.3.2, p. 207, above.

[3] *Chapter I*, Section B, pp. 102-141, above.

[4] *Chapter I*, Section C, pp. 141-161, above.

In view of this fact alone, *all* the objections based upon perverse effects such as the temperature or oscillations of the IRSID building are swept away.[5]

• Moreover, during the two total eclipses of the Sun on 30 June 1954 and 2 October 1959, deviations of the plane of oscillation of the paraconical pendulum were observed that are *totally inexplicable* in the framework of currently accepted theories.[6]

**Paraconical pendulum with isotropic support**

2 - Analogous anomalies were observed for the paraconical pendulum with isotropic support, and had greater amplitudes due to the suppression of the anisotropy of the support.

For the diurnal lunar periodic component of 24h50m, the observed amplitude was of the order of $1.4 \times 10^{-5}$ radians per second, *i.e.* about a hundred million times greater than the amplitude calculated from current theory.[7]

Such an anomaly can be explained by an anisotropy of inertial space of the order of $3 \times 10^{-6}$ for the lunar influence alone.[8]

---

[5] *Chapter I*, §E.6, pp. 188-192 above.

[6] *Chapter I*, Section D, p. 162-170 above. The *total* deviations of the plane of oscillation of the paraconical pendulum with anisotropic support were of the same order as the Foucault effect, *i.e.* about $0.55 \times 10^{-4}$ radians per second (p. 164 above).

[7] *Chapter II*, §F.2.2., pp. 284-285 above.

[8] *Chapter II*, §1.1, p. 321 above.

## Agreements of phase

3 - Remarkable agreements in phase for the lunar sidereal monthly periodic component of 27.322 days were observed for the paraconical pendulum with anisotropic support and for the paraconical pendulum with isotropic support, both for the directions of anisotropy and for the deviations of the azimuths of their planes of oscillation from the directions North-South and East-West.[9]

[9] *Chapter II*, Section G, pp. 306-314 above; see particularly *Tables VI, VII, et VIII*, pp. 308, 311, and 313.

## *D.3 Optical anomalies of sightings at marks and at collimators*

**Sightings at marks. IRSID - July 1958. Diurnal cycles**

1 - The experiments on sightings at marks performed in July 1958 at IRSID *in parallel* with the crucial experiments on the paraconical pendulum with anisotropic support performed *simultaneously* at Bougival and at Saint-Germain demonstrated significant diurnal deviations. For the period of 25h, which is representative of the periodicity of 24h50m, the deviations were in the same sense and the amplitudes were of the order of one centesimal second, *i.e.* of the order of $2 \times 10^{-6}$ radians;[1] in other words they were *of the same order of magnitude* as the coefficients ε of anisotropy of the paraconical pendulum with isotropic support.

**Sightings at marks and at collimators. IGN - February-March 1959. Diurnal cycles**

2 - In particular for the periodicity of 25h, diurnal cycles were found having amplitude of order equal to one centesimal second.[2]

**Sidereal monthly lunar periodicities**

3 - At IRSID as at IGN, monthly lunar periodicities were demonstrated, having an amplitude of the order of 5 centesimal seconds, *i.e.* of the order of $8 \times 10^{-6}$ radians.[3]

---

[1] *Chapter III*, Section B, §B.1, §B.2, and §B.3, pp. 334-339, and §B.5, p. 341.

[2] *Chapter III*, Section C, §C.2, pp. 348-353; see particularly *Graph VI*, p. 351.

[3] *Chapter III*, §B.5.3, pp. 342-344; and §C.3, pp. 359-363; see particularly *Graphs V, XII, XIII and XIV*, p. 344, and pp. 361-363.

### Agreements of phase

4 - Remarkable agreements in phase were observed at IRSID for the diurnal periodicities of the North and South sightings, and for the pendulums with anisotropic support at Bougival and at Saint-Germain.[4]

In the same way, remarkable agreements in phase were observed for the lunar sidereal monthly periodicities of 27.322 days of the sightings at marks and at collimators and for the paraconical pendulums with anisotropic support and with isotropic support.[5]

### Deviations that are inexplicable in the framework of current theories

5 - The observed deviations are *totally inexplicable* in the framework of currently accepted theories, and they cannot be attributed to perverse effects.[6]

---

[4] *Chapter III*, §B.2 and B.3, pp. 335-338 above.

[5] *Chapter III*, Section D, pp. 366-370, above; see particularly *Table II*, p. 370.

[6] *Chapter III*, §C.4.2, pp. 364-365, and §E.2, p. 372, above.

## *D.4 Optical anomalies of Esclangon and of Miller*

### Optical anomalies of Esclangon

1 - Analysis of Esclangon's observations has demonstrated a diurnal *sidereal periodicity having an amplitude of the order of* $10^{-6}$ *radians.*[1]

This is *completely inexplicable* in the framework of currently accepted theories.

### Interferometric anomalies of Miller

2 - Analysis of Miller's observations has demonstrated *a diurnal sidereal periodicity in the speeds and in the azimuths.*[2]

A *completely unexpected* result has been demonstrated: the perpendicularity of the hodographs to the average directions of the azimuths, which is an anisotropy that corresponds to an anisotropy of optical space.[3]

This result is *both incompatible with the classical theory*, because the hodographs are not symmetrical with respect to the meridian, *and with the theory of relativity*, because the hodographs correspond to an average difference of speeds between two perpendicular directions of the order of 8 km/sec.[4]

This difference of speeds corresponds to a relative difference of $8/3000000 = 2.67 \times 10^{-5}$, *i.e.* of the order of $3 \times 10^{-5}$ of the speed of light.

The anisotropy of optical space demonstrated by Miller's observations appears to be *completely inexplicable* in the framework of currently accepted theories.

---

[1] *Chapter IV*, §B.2, p. 374, above.

[2] *Chapter IV*, §D.3, pp. 395-399, above.

[3] *Chapter IV*, §D.4, pp. 400-408.

[4] *Chapter IV*, §F.2, note 3, p. 421, and §D.4, Table II, p. 406.

## *D.5 Very remarkable similarities*

Very remarkable similarities are present between the five series of observations analyzed in the five first Chapters: the observations of the paraconical pendulums with anisotropic support and with isotropic support; the optical deviations of sightings at marks and at collimators; the optical observations of Esclangon; and the interferometric observations of Miller.

### Sidereal diurnal and monthly periods

1 - In the first place, all these observations present sidereal diurnal and monthly periodicities having *comparable* phases and amplitudes.

The observed anomalies are of the same order of magnitude: of the order of $10^{-6}$ or $10^{-5}$ .[1]

### Semi-annual or annual periodicities

2 - All the observations present *semi-annual or annual periodicities whose phases are very close to the date of 21 March, which corresponds to the spring equinox.*

Thus they all have *marked correlation* with the position of the Earth along its orbit.[2]

### Continuous observations

3 - All these observations have at least six characters in common: - they have similar periodic structures; - their existence is undeniable; - they cannot be explained by perverse effects; - they are totally inexplicable in the framework of currently accepted theories; - they had remained undiscovered until they were noticed; - they were performed in a continuous manner for long periods, day and night.

---

[1] §D.1, p. 618, above.

[2] *Chapter V* above.

That the regularities demonstrated by these observations have not been detected by previous observations, it is undoubtedly due to their small size, due to their periodic character as a result of which the averages of their effects cancel out, and due to the fact that no other series of experiments had been carried out in a continuous manner, except for astronomical observations.

These regularities correspond to *entirely different phenomena* which *a priori are not connected with one another*, but all of them correspond to an anisotropy of space.

Since all these regularities are *totally inexplicable* in the framework of currently accepted theories, it must be concluded that certain postulates of current theories must be revised, and the theory of universal gravitation is certainly an excellent illustration of this.

## *D.6 The theory of universal gravitation*

### The tests of the theory of universal gravitation and the anomalies of the paraconical pendulum

1 - Of all physical theories, the theory of universal gravitation is certainly *the most impressive*, due to the confirmation of its deductions over several centuries.[1]

However, it seems utterly unable to explain the anomalies that I have demonstrated in the movement of the paraconical pendulum, whose amplitudes are from twenty to a hundred million times greater than those that can be calculated from the currently accepted theory of gravitation.[2]

These deviations are even more astonishing due to the fact that the observed and calculated values of the luni-solar deviations of the vertical are in remarkable agreement.[3]

### The unexplained regularities of the solar system

2 - Moreover, the solar system exhibits very many regularities that still remain unexplained today.[4] They undeniably show that the current theory of gravitation needs to be revised and completed. In any case, they show that the anomalies of the paraconical pendulum are not an isolated case.

---

[1] *Chapter I*, §F.1 and F.2, pp. 197-205, above.

[2] *Chapter I*, §B.2.1, p. 118, and *Chapter II*, §F.22, p. 285 above.

[3] See above, *Chapter I*, §F. 1.3, note 6, p. 200.

[4] See above, *Chapter VI*, §E.3, note 2, p. 542.

These regularities suggest that the solar system contains a system of stationary waves.

In this connection, I should point out that, seen from the Earth, the apparent diameters of the Sun and the Moon are practically identical. The ratios $R/D$ of their radiuses $R$ to their average distances to the Earth are respectively $6.935 \times 10^{10}/1.495 \times 10^{13} = 4.639 \times 10^{-3}$ for the Sun and $1.74 \times 10^{8}/3.844 \times 10^{10} = 4.527 \times 10^{-3}$ for the Moon. We thus have $4.638/4.527 = 1.025$.

## Very simplifying hypotheses

3 - Unarguably, the current theory of gravitation is based upon *very simplifying* hypotheses, such as for example:[5] - gravitational actions propagate *instantaneously over distance*; - empty space is isotropic; - light propagates in a straight line, and its speed is constant at any point in empty space; - the coefficients of attraction of the various heavenly bodies are supposed to be the same for all of them, and equal to the value at the surface of the Earth:[6] $G = 6.66 \times 10^{-8}$.

However, it is evident that these hypotheses cannot be considered as being valid if the propagation of action at a distance takes place via the involvement of a medium whose structure is variable from one point of interplanetary space to another, if the speed of this propagation depends upon the structure of this medium, and if the lines of force are not straight.[7, 8, 9]

---

[5] For the postulates of the current theory of gravitation, see *Chapter I* above, §F.l, pp. 197-200.

[6] The coefficient *G* of universal gravitation has been determined from the experiment of Cavendish (A. Foch, *Mécanique* (Mechanics), Masson, 1967, §268, pp. 357-358 [88]).

[7] See *Chapter VI* above, Section C, pp. 510-517.

*In fact, space could appear to be isotropic while really being anisotropic.*

[8] Nobody seems to be astonished by the values of the densities of the planets and the Sun to which the assumed hypotheses lead. In CGS units these densities are, for example, the following: Earth, 5.52; Jupiter, 1.33; Saturn, 0.71; Sun, 1.41 (*Smithsonian Physical Tables, 1950, id., p. 734*). Thus the density of the Earth is *four times greater* than that of the Sun.

[9] Analogous observations could be presented for the theories of electrostatics and electromagnetism.

Thus, in a discussion on the number and the nature of the fundamental quantities, E. Durand underlines "*our current complete ignorance of the real nature of electric charge*". (*Magnétostatique* (Magnetostatics), Masson, 1968, p. 293 [119])

# E THE FUNDAMENTAL PRINCIPLES OF SCIENTIFIC ANALYSIS

What should we deduce from the fact that the five series of observations analyzed in the present work are *totally inexplicable* in the framework of currently accepted theories, and that thus they contradict those theories?

To answer this question, it seems to me to be necessary to underline three essential propositions that are at the base of all scientific analysis.

## *E.1 The foundations of the scientific method*

### The real meaning of verification of a theory by experiment

1 - First of all, if a theory has consequences that are verified by experiment, *this in no way implies that the theory is entirely correct*. The agreement of the implications of the theory with certain experimental data only means that it is *compatible* with that data - *nothing more*.

This proposition can be illustrated by many examples. Here I shall limit myself to three particularly striking examples.

• In his memoir of 1905 on special relativity, Einstein underlined that the theory he presented immediately made it possible to explain the results of Fizeau's experiment of 1851. However he omitted to remark that Fizeau's experiment made it possible to check the validity of the formulation elaborated by Fresnel in 1818, thirty-three years before.[1] Thus two *incompatible* theories, the theories of Fresnel and of Einstein, make it possible to explain the same phenomenon.[2]

From this double theoretical demonstration of Fresnel's equation, it follows that we cannot deduce that Fizeau's experiment is a proof of the validity of one theory or of the other, but only that each of those theories is compatible with that experiment.[3]

This demonstrates that *the verificiation of a theory by an experiment does not prove that the theory is correct.* It simply shows the compatibility of that theory with that experiment, and alternative explanations always remain possible.

• In the same way, the explanation of the deviation of the perihelion of Mercury by 43" per century is generally presented as being one of the greatest successes of the theory of general relativity. However, the analysis presented by Maurice Lévy on 17 March 1890 to the Academy of Sciences also gives an explanation of this deviation of 43" as a linear combination of the equations of Weber and Riemann.[4] Here as well, an explanation of the phenomenon in question is given by two *entirely different* theories.

*From this also, it follows that the agreement of the consequences of a theory with the experimental data cannot be interpreted as a proof of the validity of the theory.*

• The theory of epicycles gives us a new example. For many centuries it dominated astronomical thought by applying a postulate that was admitted without discussion: nature can only allow circular symmetry. During all those centuries this theory made it possible to forecast the apparent movement of the Sun, of the Moon, and of the planets. It made it possible to predict eclipses with surprising accuracy. However, the discoveries of Kepler and the theory of universal gravitation put an end to the domination exerted by this theory.

---

[1] Fresnel: *Note additionnelle à la lettre à M. Arago* (Additional note to the letter to M. Arago), Annales de Chimie et de Physique (*Annals of Chemistry and Physics*), 1818, vol. 9, p. 128 (or 286) [147].

[2] Moreover, it must be underlined that Fresnel did not know Fizeau's experimental result, whereas Einstein did.

[3] As already mentioned, the partisans of the relativity theory generally omit to mention that Fresnel determined this formula *thirty-three years before Fizeau's experiment that confirmed it*, and that in Einstein's theory this formula is only approached.

In fact certain relativist authors omit to mention the *much earlier* theory of Fresnel, *most often on purpose*.

[4] It is only necessary to take a value of $\alpha = 2$ for Maurice Lévy's coefficient $\alpha$ (p. 549) in order to obtain a value equal to

$$(1+\alpha) \times 14''.4 = 43''.2$$

In his Note of 17 March 1890, Maurice Lévy starts from an estimate of 38" for the deviation of the perihelion of Mercury, so he is led to take $\alpha = -1 + (38/14.4) = 1.63$. With the current estimate of 43", we reach the value of $\alpha = 2$ for $\alpha$, which is an *integer* and seems much more plausible in this case.

Here also, we see that the verification of a theory by the experimental data cannot constitute a proof of the validity of the theory.[5, 6]

## The crucial experiments

2 - *On the other hand if, in one of its hypotheses or one of its consequences, a theory is contradicted by a new item of experimental data, then it cannot be considered as being valid and must be rejected.*

---

[5] For the theory of epicycles, see in particular Pierre Duhem, *Le système du monde. Histoire des doctrines cosmologiques de Platon à Copernic* (The System of the World, History of Cosmological Doctrines from Plato to Copernicus), Hermann, 1959, Vol. I, Chapters III and VIII [118]. See also *La Science antique et médiévale* (Antique and Medieval Science), P.U.F., 1957, Book II, Chapter III, and *La Science Moderne* (Modern Science), P.U.F., 1958, Book I, Chapter II.

[6] Experimental verifications of the expressions in $\sqrt{1-v^2/c^2}$ are often presented as being major verifications of the theory of relativity.

Thus, after my article of August-September 1996 in *"La Jaune et la Rouge"*, *Les expériences de Dayton C. Miller 1925-1926 et la Théorie de la Relativité* (The Experiments of Dayton C. Miller 1925-1926 and the Theory of Relativity) [55], one of my correspondents wrote to me:

> *"In any case, Michelson's experiment is not the foundation of the Theory of Relativity... What happens in a cyclotron is much more significant: in order to make a particle of charge q and "rest mass" $m_0$ circulate at a speed v around a circle of radius R, a magnetic field B normal to the plane of the circle is required whose magnitude is given by a typically relativistic expression: $B = m_0 v/Rq\sqrt{1-v^2/c^2}$*

In fact, *there is nothing "typically relativistic"* about the expression $\sqrt{1-v^2/c^2}$ . This is an expression *that naturally enters into the discussion of retarded potentials and of classical relative (not relativistic) movements.* See for example Oliver Heaviside, *On the Electromagnetic Effects due to the Motion of Electrification through a Dielectric*, Philosophical Magazine, XXVII, 1889, pp. 324-339 [155].

In this article, Heaviside considers the magnetic field $\vec{H}$ created by an electric charge $q$ moving at a velocity $\vec{u}$. He considers the equation

(1) $\Delta\vec{A} - \frac{1}{c^2}\frac{\partial^2\vec{A}}{\partial t^2} + 4\pi\rho\vec{u} = 0$

defining the potential vector $\vec{A}$ at a point $M$ (equation 16, p. 340), where $c$ is the speed of the waves, and he shows that we have (equation 15)

(2) $\vec{H} = \nabla\times\vec{A}$

with (equation 27, p. 331)

(3) $\vec{A} = \frac{qu}{r\sqrt{1-u^2\sin^2\theta/c^2}}$

where $\theta$ denotes the angle of the direction $\overrightarrow{QM}$ with the velocity $\vec{u}$. On the equatorial plane, we have $\sin\theta = 1$.

See also Edmund Whittaker, 1951, *A History of the Theories of Aether and Electricity. The Classical Theories*, Nelson, pp. 307-309 [279].

Heaviside's formulation is *all the more significant* because it appeared more than fifteen years earlier than Einstein's memoir of 1905 on special relativity.

But this is not the only example. Similarly, Lagrange's function $L$ considered by Bernhard Riemann in 1875 is written, in electrostatic units:

(4) $L = -\frac{ee'}{r\sqrt{1-v^2/c^2}}$

(Whittaker, 1951, id., p. 206).

It must be underlined that this conclusion is valid, *no matter how numerous and how precise previous validations of this theory may be. A single experiment is sufficient to contradict the theory. Such an experiment may thus be considered as being crucial.*

Einstein himself and his successors, for example, also relied upon this proposition in order to reject the hypothesis of an *immobile* ether through which the Earth moves, by basing their arguments on the claimed "*negative*" result of Michelson's experiment.

Today the same proposition leads us to reject the theory of special relativity, as soon as it is established that the speed of light is not invariant in all directions.[7]

It goes without saying that the rejection of a theory that has been invalidated by a crucial experiment should not imply that the experimental verifications of the implications of the rejected theory, *if they are valid*, ought to be ignored. Such a rejection simply means that the experimental observations in question *must be explained in some other way.*[8]

## Conditions for the validity of a crucial experiment

3 - Naturally, the rejection of a theory on the grounds of contradictory observations *can only be admitted if the validity of those observations is perfectly established.*

*It is thus this validity that is essential, and not that of previous experiments.*

---

[7] Here we are compelled to recall what Einstein wrote:

> *"The principal attraction of the theory (of relativity) is that it constitutes a logical whole.*
>
> *"If any one of its consequences should be found to be inexact, it would be necessary to abandon the theory; any modification appears to be impossible without destroying the entire edifice."*

Albert Einstein, *Comment je vois le monde* (How I see the World), 1939, Flammarion, p. 213 [129].

[8] Since perfectly established facts are indisputable, *it is certain that some theory exists that explains them and that is compatible with all the observational data.*

## *E.2 Principles of all scientific methodology*

• In fact, three principles *dominate* all scientific methodology

1 - *The verification of a theory by observation is not a proof of its validity. It only shows that the theory is compatible with the observational data.*

2 - *Whatever the number, the quality, and the accuracy of previous verifications of a theory may be, that theory is invalidated if any one of its essential hypotheses or any one of its implications is invalidated by a crucial experiment.*

3 - *For an experiment that is incompatible with a previous theory to be considered as being crucial, it is necessary and sufficient that the observations corresponding to that experiment can be considered as being scientifically incontestable.*

• In view of these three principles, *for example, it would only be possible to refute my conclusions as to the invalidity of the theory of relativity in a valid manner, if it were possible to establish the invalidity of the underlying regularities that I have highlighted in Miller's observations.*[1] *This is the heart of the matter*. In order to refute my conclusions, *it is necessary to show the invalidity* of the regularities that I have demonstrated in Miller's observations.

---

[1] See above, *Chapter IV*, pp. 392-411.

# F A NECESSARY REVISION OF CONTEMPORARY THEORIES

## *F.1 The facts*

In the first five Chapters of this work, significant periodicities have been demonstrated in five series of experiments: - *the periodicities of the observations of the paraconical pendulums with anisotropic support and with isotropic support; - the periodicities of the optical observations of sightings at marks and at collimators; - the periodicities of the optical observations of Esclangon; - the periodicities of the interferometric observations of Miller.*

*All of these periodicities are completely inexplicable in the framework of currently accepted theories. They all present mutual interconnections, and all reinforce one another. They all have significant correlations with the position of the Earth along its orbit. And they all imply an anisotropy of space.*

Among the five series of *absolutely fundamental* observations analyzed in the first five chapters of this work, probably Miller's interferometric experiments are the ones that would be the hardest to perform in a continuous manner. However, Esclangon's experiments would not be easy to set up either.

The anomalies of the paraconical pendulum have the immense advantage of corresponding *to a direct experiment* that is relatively easy to observe and is not very questionable.

But undoubtedly the sightings at marks and at collimators are *the most striking*, since they correspond to a *direct and immediate* observation. Perhaps these are also the experiments *that contradict most directly our experience*, because who could imagine *a priori* that the azimuth corresponding to a sighting at a mark could change with time, *independently of any deformation of the soil and any displacement of the pedestals supporting the apparatus*?

In the most complex case, that of the interferometer, one fact is certain: that *Miller's observations correspond to an underlying and indisputable reality, independently* of any interpretation that may be put upon them.[1]

[1] Undoubtedly, the best illustration that can be given of the reality underlying Miller's observations is *the perpendicularity of the hodographs to the average directions of the azimuths* (*Chapter IV*, §D.4.3, pp. 404-408, above).

## *F.2 The interpretation of the facts*

*The same facts can support different interpretations. These interpretations are always arguable, but, no matter how erroneous one or another interpretation may appear to be, this can never be used as a pretext in order to discard the facts.*

### Miller's observations - Unexpected results

1 - If we consider Miller's observations, *for example*, the interpretation that Miller put upon them was *certainly wrong globally*, and his estimations of the cosmic speed of the Earth cannot be correct, because they took no account whatsoever *of the non-zero values of the average azimuths $\overline{A}$ and of their variations over time.*[1]

Nevertheless it remains true that, *considered in themselves*, his observations have *a very great mutual coherence*, a coherence that actually is much greater than Miller realized, if *one considers, for example, the results relating to the hodographs of the speeds and the results of harmonic analysis of the characteristic parameters of his observations in terms of the semi-annual and annual periodicities that they present.*

Today I consider that a possible interpretation, and perhaps the most probable, is a total dragging of the ether by the movement of translation of the Earth (or a common movement of the Earth and the local ether), together with an anisotropy linked to astronomical phenomena. According to this interpretation, there would not be any *direct* influence of the velocity of the Earth upon the observations of the interferometer.[2, 3] *The observations of Miller would correspond only to an anisotropy of space.*

---

[1] See above, *Chapter IV*, §C.3.2, pp. 388-390.

[2] In support of this interpretation, we may consider that the hodograph $H_T$ corresponding to the velocity of the Earth is symmetrical with respect to the meridian, while the observed hodograph $H$ and its estimation $H^*$ appear to be perpendicular and symmetrical with respect to the average azimuth $\overline{A}$ of Miller (*Chapter IV*, §D.4, pp. 400-408 above).

*A priori*, it seems likely that any influence of the hodograph $H_T$ upon the hodographs $H$ and $H^*$ should compromise the perpendicularity and the symmetry of the hodograph $H$ with respect to the azimuth $\overline{A}$, whereas we observe that this is not the case.

But this is naturally only a hypothesis that remains to be confirmed.

[3] Earlier, I considered successively two other interpretations.

*The first interpretation* was to associate an anisotropy of space corresponding to the average azimuth $\overline{A}$ with the interpretation given by Michelson, according to which the displacement of the fringes corresponds to the ratio $v^2/c^2$ of the square of a speed (in fact, the value of the sum of the cosmic speed and the orbital speed of the Earth) to the square of the speed of light.

*A second interpretation* subsequently seemed very plausible to me, even more plausible: to associate an anisotropy of space *with an anisotropy resulting from the cosmic and orbital velocities of the Earth. In summary, following this interpretation, the interferometer was not measuring speeds. It was only measuring the anisotropies.*

In fact, an in-depth quantitative analysis of Miller's observations led me finally to consider that these two interpretations were rather improbable.

## The danger of preconceived ideas

2 - *In the interpretation of the facts, one should avoid any preconceived ideas.* In fact, *for example*, all the interferometric experiments since 1881 *have been victims of preconceived ideas* on the nature of the ether, on its movements, and on its deformations. Apparently, even though he never ceased to underline the danger of such preconceived ideas, Miller *himself was probably the victim of preconceived ideas* as far as are concerned the interpretation of his own observations and the elaboration of his model.[4]

As Pareto underlined, "*It is false to believe that we can discover exactly the properties of facts by reasoning from ideas that we form* a priori *for ourselves about those facts.*"[5]

[4] Miller himself was blinded by his certitude that the speeds measured by the interferometer could only correspond to the speed of the Earth.

[5] Vilfredo Pareto, 1909, *Manuel d'économie politique* (Manual of Political Economy), Giard, 1927, p. 13 [213].

## *F.3 Dogmatism*

The worst enemy of science *is dogmatism*, the imperturbable assurance of those who are convinced of being possessors of absolute and definitive truth.[1] *In reality, such people are the gravediggers of science.*

In the scientific field, nothing is definitive. "*The ideas and the theories of our predecessors should not be preserved, except insofar as they represent the state of science, but they are evidently destined to change, unless we accept that science should make no further progress, which is impossible.*"[2]

We must remember that, in the long history of science, the greatest spirits have committed errors, or have articulated partial truths that have not been able to stand up to the test of time. *They remain nevertheless great spirits.* Some of them did not hesitate to recognize their errors. That only made them greater.

*The whole history of science always counters dogmatism*, and there are innumerable examples of theories which, at the epoch that they were presented, were considered as being contrary to observational data that were considered as being indisputable, but which later acted as a germ for the discovery of new phenomena.

---

[1] I shall limit myself to a single example. In his work of 1939, *Physique moderne et Philosophie* (Modern Physics and Philosophy), published by Hermann, Jean-Louis Destouches did not hesitate to assert (p. 67) [109]:

> *"No underlying determinism is compatible with wave mechanics: the current theories are essentially indeterministic. We must even abandon all hope of re-establishing determinism...*
>
> *"Rigorous determinism has ceased to reign in physics...*
>
> *"We are left free to believe in an eternally extra-scientific determinism, which cannot be grasped except by transcendental intuition.*
>
> *"Henri Poincaré's opinion upon the necessity for science of absolute and a priori determinism is thus disproved by the facts."*

To these peremptory affirmations, to this insupportable certainty, to this intolerant dogmatism, we are impelled to oppose the following reflection, so full of wisdom, expressed in Paris by Einstein to Paul Montel (Albert Einstein, *Correspondence with Michele Besso*, id., p. 313) in 1922: "*I have helped this science to make a small step forwards. Someone else will come after me and will make another step which will put me in the background.*"

For the deterministic doctrine, see above, *Chapter VI*, §D.3.1, pp. 523-525.

[2] Claude Bernard, 1865, *Introduction à l'étude de la médecine expérimentale* (Introduction to the Study of Experimental Medicine), id., p. 75.

Miller's experiments of 1925-1926 have been *systematically discarded and forgotten*, merely because they contradict "*established truths*".

We cannot but meditate upon this judgment of Claude Bernard:[3]

> *"In science, when we have proposed an idea or a theory, we should not have, as our goal, to preserve it by searching for anything that can support it and discarding everything that could invalidate it. On the contrary, we should examine any facts that appear to overthrow our theory with the greatest attention, because real progress always consists of replacing an ancient theory that covers fewer facts with a newer theory that covers more."*

Evidently, at first sight, this principle may appear to be a mere banality. *But unfortunately it continues to be systematically ignored and violated by the very persons who incessantly proclaim their total respect for the observational data.*

The profundity and the mathematical elegance of a theory, important as they may be, obviously cannot constitute a guarantee of its scientific validity. And the fact that, at some given moment, a majority accept a theory certainly does not constitute a meaningful criterion. "*The validity of a physical law or a theorem of geometry is not to be put to the vote.*"[4] In fact, *the only scientific criterion for judging the respective scientific validity of different theories is to confront them with the experimental data.* It is the application of this *golden rule* that gives an explanation of the innumerable successes of science in the Western world during the last five centuries.

[3] Claude Bernard, 1865, *Introduction à l'étude de la médecine expérimentale* (Introduction to the Study of Experimental Medicine), Flammarion Garnier, 1966, p. 75 [69].

[4] Louis Rougier, 1920, *Les paralogismes du rationalisme* (The Paralogisms of Rationalism), Alcan, p. 49 [235].

## *F.4 An inevitable revision of the fundamentals of current theories*

Whatever interpretations we may give, *the facts impose themselves upon us*, and no theory whatsoever can be preserved if it is in contradiction with the experimental data. As Vilfredo Pareto underlined:[1]

> *"It is always the concrete phenomenon that decides whether a theory must be accepted or rejected. There is no other criterion of the truth of a theory, and there can be no other, than its agreement more or less perfectly with concrete phenomena."*

*Too many theoreticians have all too great a tendency to take no account of facts that contradict their convictions.*

It is indubitable that the *unarguable* existence of the periodic structure of the observations of the paraconical pendulums with anisotropic support and with isotropic support, of the sightings at marks and at collimators, of the observations of Esclangon, and of the observations of Miller, make it necessary to perform *profound revision of certain fundamentals of current theories.*

It cannot be considered that space is isotropic, nor that light is propagated in straight lines in space. In connection with the semi-annual and annual periodic regularities that have been examined, *it is not possible* to sustain the view that the movement of translation of the Earth around its orbit cannot be detected by purely terrestrial experiments, *and this is the case whatever may be the interpretation given to the five series of experiments analyzed in this work, and particularly to the experiments of Miller.*[2]

---

[1] Vilfredo Pareto, 1909, *Manuel d'Economie Politique* (Manual of Political Economy), Giard, 1927, p. 16 [213].

[2] It is no longer possible today to assert, as Max Born did in 1923 (*La théorie de la relativité d'Einstein et ses bases physiques* (Einstein's Relativity Theory and its Physical Basis), Gauthier-Villars, p. 216 [73]):

*"These experiments produced without exception a negative result. There could no longer be any doubt that a motion of translation through the ether cannot be detected by an observer sharing in the motion. Thus the principle of relativity which holds for mechanics is also valid for all electromagnetic phenomena."*

*The fundamental error committed in the interpretation of the results of Michelson's experiment is only too evident. From a result that was smaller than expected, it was concluded that there was no effect.* As Carvallo wrote:[3]

> *"The result found was too small. Naturally but wrongly, the conclusion was that it was null, and that the small displacement found was an experimental error. It was stated that the experiment yielded a negative result.*
>
> *"Moreover, that was an extension, a generalization of the false result. Every other experiment, directed towards the same utopia, should also give a negative result. That was Einstein's postulate, a postulate that conformed to the current pre-judgment."*

When in 1933 Miller's very circumstantial analysis of his observations in 1925-1926 was published, it was immediately declared everywhere to be of no value, and was attributed to the effects of temperature.[4] Such blindness is at the least surprising when it concerns *the very foundations* of the theory of relativity.[5]

---

[3] Carvallo, 1934, *La théorie d'Einstein démentie par l'expérience* (Einstein's Theory Disproved by Experiment), id., pp. 6-7.

[4] See particularly *Chapter IV*, §E.1 above, pp. 412-413.

[5] As I have already pointed out, Augustin Sesmat so justly wrote:

> *"Michelson's experiment, it is known, was repeated many times; in general the experimenters confirmed the negative result... Undoubtedly the conclusions of a single experimenter carry little weight when they are opposed to those of all the others; however, still the question should be examined again, because, as is easily understood, if a displacement, even much smaller than the displacement anticipated by classical physics, occurs due to the fact that the Earth has a speed, then the special theory would collapse at its base, and with it the general theory."*

(Sesmat, 1937, VII, *Essai critique sur la doctrine relativiste* (Critical Essay on the Relativistic Doctrine), Hermann, pp. 431-432 [245])

This is so right. It is necessary again to underline that Miller's experiments were *the only ones* that were carried out *in a continuous manner at several times of the year* (*Chapter IV*, §E.2.3, pp. 415-416 above). Not only should Miller's experiment be repeated, but this ought to be done *in a continuous manner at several times of the year*, as was the initial project of Michelson and Morley, according to which the experiments should be repeated at intervals of three months (Michelson and Morley, 1887, *On the Relative Motion of the Earth and the Luminiferous Ether*, id., p. 341). As Sesmat also wrote (id., p. 441):

> *"(Physicists) know perfectly well that, since the negative experiment of Michelson provoked the relativist movement, another experiment of the same type but this time positive would be enough to provoke a movement in the opposite direction."*

But in view of the above analyses (*Chapters IV and V*), we may consider it *as certain* that Michelson's experiment, suitably implemented, would not yield negative results, and that any repetition of Miller's experiments would also lead to positive results.

In fact, all the periodicities detected in the five series of observations that have been analyzed in the first five Chapters are *totally inexplicable* in the framework of current theories. These theories must accordingly be revised and completed.[6]

It is certain that the contraction and the local time of Einstein's memoir of 1905, i.e. the foundations of the theory of special relativity, cannot be maintained. In any case, space and time appear *to have natures that are irreducibly different.*[7]

As far as the deductions from the theory of relativity that have been the object of experimental tests are concerned, it is equally *certain* that other explanations will eventually be found. Whatever may transpire with all these phenomena, the experimental data constitute assets *that can be neither eliminated nor bypassed.*

---

[6] These conclusions are *completely independent* of the existence of an intermediate medium, *the ether*, even if the existence of such a medium is extremely likely.

Naturally if, *as all the known facts irresistibly suggest*, an intermediate medium exists that serves for the propagation of all actions at a distance, then it is certainly *implausible* to suppose that the medium has no relative movement and that it is never the object of any deformation.

*From this follows the very great likelihood of an anisotropy of space, and in particular of an anisotropy of the inertial space corresponding thereto.*

Naturally, the fact that it is possible to explain the anomalies of the paraconical pendulum by an anisotropy of inertial space *does not prove* the actual existence of that anisotropy, but it proves *that everything takes place as though inertial space was anisotropic.*

Yet again, it is *very remarkable* that the optical anomalies which I demonstrated at IRSID in 1958 and which were confirmed in 1959 at IGN *were of the order of a centesimal second*, in other words of the order of $10^{-6}$ radians, *i.e.* of precisely of the same order of magnitude as that of the coefficient of anisotropy indicated by the analysis of the luni-solar anomalies of the paraconical pendulum.

[7] The displacements of a moving body in space can take place in one direction or the other, but time flows *in an irreversible manner.*

## *F.5 One criterion only. The absolute supremacy of experimental data over theoretical concepts*

All the testimony of the most penetrating spirits leads to the same conclusion *as to the absolute supremacy of experimental facts over theoretical conceptions.* I must cite here the testimony of some of the most eminent among these:

Max Planck:[1] *"The most beautiful mathematical theories are only speculations in the air, if they have no anchor of solid support in the determinations of experiment."*

Max Born:[2] *"Only observable facts are physical realities."*

Vilfredo Pareto:[3] *"All sciences have progressed when men, instead of arguing about the principles, have argued about the results."*

Albert Einstein:[4] *"No theory has ever been found to be useful and fertile via a purely speculative route."*

Louis de Broglie:[5] *"True physical reality is only found in the totality of experimental results."*

Henri Poincaré:[6] *"Experiment is the only source of truth: only experiment can teach us something new; only experiment can give us certainty."*

It cannot be repeated too often: *However imposing the edifice of its mathematical deductions may be, no theory can be imposed when it is based upon postulates contradicted by experiment.*

---

1 Max Planck, *Initiations à la Physique* (Initiation into Physics), Flammarion, 1941, p. 171 [218].

2 Max Born, *La théorie de la relativité d'Einstein et ses bases physiques* (Einstein's Relativity Theory and its Physical Basis), Gauthier-Villars, 1920, p. 291 [73].

3 Vilfredo Pareto, *Traité de sociologie générale* (Treatise on General Sociology), Payot, 1917, p. 3 [212].

4 Albert Einstein, Letter of 28 August 1918, *Correspondence with Michele Besso*, Hermann, 1979, p. 82 [135].

5 Louis de Broglie, 1953, *La physique quantique restera-t-elle indéterministe ?* (Will Quantum Mechanics Remain Non-Deterministic?), id., p. 96.

6 Henri Poincaré, 1902, *La Science et l'Hypothèse* (Science and Hypothesis), id., p. 167.

# Chapter VIII
# A PLAN FOR SIMULTANEOUS EXPERIMENTS

*I advise those who wish to learn the art of scientific prediction not to apply abstract reason to the task, but to decipher the secret language of Nature from the documents of Nature: experimental facts.*

Max Born *

# A ON REPETITION OF THE FIVE SERIES OF EXPERIMENTS ON THE PARACONICAL PENDULUMS WITH ANISOTROPIC AND ISOTROPIC SUPPORT, ON OPTICAL SIGHTINGS AT MARKS AND COLLIMATORS, ON THE OPTICAL SIGHTINGS OF ESCLANGON, AND ON THE INTERFEROMETER OF MILLER

## *A.1 The observed periodicities*

From the preceding discussion it follows that the existence of periodicities that are *totally inexplicable* in the framework of currently accepted theories has been established in five series of experiments: the periodicities of the movements of the paraconical pendulums with anisotropic and with isotropic support; the periodicities of sightings at marks and at collimators; the periodicities of Esclangon's observations; and the periodicities of Miller's observations.

### The periodicities of the paraconical pendulum

1. - The periodicities of the paraconical pendulum with anisotropic support and of the paraconical pendulum with isotropic support have been demonstrated by the experiments that I conducted from October 1953 to June 1960 at IRSID. *Their existence is certain.*

* Max Born, 1943, *L'Expérience et la Théorie Physique* (Experiment and Physical Theory), Gauthier-Villars, 1955, p. 51. [74]

This certainty is founded upon the analysis of seven month-long series of observations of the asymmetric paraconical pendulum with anisotropic support performed at IRSID and at Bougival from 1954 to 1960, upon the analysis of two month-long series of observations of the asymmetric paraconical pendulum with isotropic support performed at IRSID in November-December 1959 and in March-April 1960, and *in particular upon the analysis of the extremely remarkable - one may say decisive - agreements* observed in the periodic structures of the observations performed in identical conditions *during the crucial experiments of July 1958 at Saint-Germain and at Bougival* upon the movement of two asymmetric paraconical pendulums with anisotropic support.[1]

## The periodicities of optical sightings at marks and at collimators

2. - Optical sightings at marks were the object of two successive sets of experiments: at IRSID under my direction in June-July 1958, and at IGN in February-March 1959 under the direction of Claude Palvadeau, a young geographical engineer. At IGN the sightings at marks were accompanied by sightings at collimators.[2]

---

[1] See *Chapters I, II, and V* above.

[2] *Chapter III* above.

Moreover, *trials of continuous registration* of sightings at marks were performed at Palaiseau and at IGN from April 1992 to January 1993 by Michel Kasser, chief geographer engineer (see above, *Chapter III*, note 4, p. 333, and § E.l, note 1, p. 371).

These trials by Michel Kasser in 1992-1993 were performed in *very unfavorable* conditions and with *quite insufficient* means. Nevertheless, they enabled it to be established that the "Wild NA 3000" telescopes (made by Leica) chosen by Michel Kasser were adequate for continuous registration of the observations, and that they can be used for North-South and South-North sightings *at very short distances (about three meters) when mounted on a single very rigid support made from granite*. Although these NA 3000 telescopes were revealed by experience to be *very sensitive to variations in temperature*, appropriate analysis of the *simultaneous* observations at Palaiseau and at IGN in January 1993 made it possible to demonstrate their agreement with the observations during the previous experiments of 1958 and 1959 at IRSID and at IGN.

See the analysis of these trials in *Volume 2* of this work, *Chapter III*, Section B.

The value of these two series of experiments at IRSID and at IGN was *very unequal.* The existence of the diurnal optical periodicities seen at IRSID in July 1958 is *a certainty* due to their similarity, as far as the results of harmonic analysis are concerned, to those relating to the crucial experiments upon the paraconical pendulum with anisotropic support performed simultaneously at Bougival and at Saint-Germain.[3]

However, the results obtained in 1959 by Claude Palvadeau at IGN were much tainted by the very high number of observers and by the defective method that was utilized for eliminating the personal equations of the observers. *Still, these experiments were able to establish with certainty the existence of very significant monthly drifts, in the same sense as one another for the North-South and the South-North sightings at marks, but in the opposite sense to one another for the sightings at collimators, and with twice the absolute amplitudes of the sightings at marks.* These characteristics are sufficient in *themselves* to eliminate any possible influence related to deformation of the ground and movement of the pedestals.[4] Moreover, the observed drifts included significant monthly and semi-annual periodicities.[5]

*Overall, the existence of the diurnal, monthly, and semi-annual periodicities demonstrated in the sightings at marks and at collimators is a certainty.*

## The periodicities of the optical sightings of Esclangon

3. - The observations of Esclangon, which he performed with numerous interruptions during 1926, certainly do not have the same value as the four other series: the observations of the paraconical pendulum with anisotropic support and with isotropic support, the optical sightings at marks and at collimators, and the interferometric experiments of Miller.

---

[3] *Chapter III*, § B.3 above.

[4] See above, *Chapter III*, § C.4.2.

[5] *Chapter III*, § III.C.3; and *Chapter V*, § C.l.

However, the existence of diurnal and semi-annual sidereal periodicities demonstrated by the observations of Esclangon must be considered *as being certain*, considering their coherence *in sidereal time* on the one hand, and the results of their harmonic analysis on the other.[6]

## The periodicities of the interferometric observations of Miller

4. - The interferometric experiments pursued by Miller in 1925 and 1926 over four periods of six to ten days provided a huge number of observations whose details are unfortunately no longer available, but the *essential* results are presented in eight fundamental graphs.[7]

The analysis of these *graphs* brings out a *very remarkable* coherence in Miller's observations *as considered in sidereal time*, and *the validity of Miller's observations, their internal coherence, and their independence from any local perturbation may be considered as an absolute certainty*.[8]

---

[6] *Chapters IV and V* above.

[7] *Chapter IV*, § C.3 above.

The goal of the experiments was to demonstrate the absolute movement of the Earth with respect to the fixed stars. From the observations performed, and by employing a model for analysis, Miller thought he was able to give an estimate of the cosmic speed of the Earth in relation to its orbital speed.

However Miller's analysis *did not take into account the average deviations* $\overline{A}$ *of the azimuths measured with the interferometer, and left them unexplained*. It also supposed that the effective cosmic and orbital speeds of the Earth were reduced by a coefficient of reduction $k = 1/20$, *which was also unexplained*. Although at first sight the results given by Miller's model may appear to be *coherent*, his general analysis *is not acceptable* (*Chapter IV*, § F.l above, pp. 417-419).

[8] *Chapter IV*, Section D, above.

## Connections between the five series of observations

5. - The observations corresponding to the five series of observations considered present *very significant connections* in relation to their orders of magnitude and their phases.[9]

[9] *Chapter VII*, § D.5 above, pp. 625-626.

## A.2 An overall project

From all these indications, and in view of their implications for the foundations of contemporary theories, it follows that there would be the very greatest interest in re-doing with the greatest possible care the experiments corresponding to the five series of observations analyzed in this work, employing all the precision available with currently available technology, and while carrying out *continuous observations during at least four periods of the order of a month within the same year, around the times of the equinoxes and the solstices, and even over several years.*

Once set up, these experiments should be performed *simultaneously in different places*, some near ground level and some deep underground, as was the case in July 1958 at Bougival and Saint-Germain.

From all the available evidence, these five series of experiments would possess *exceptional interest, and would have considerable scientific impact.*

Indeed, they would confirm:

- *the existence of periodic effects that are totally inexplicable in the framework of current theories;*
- *the existence of an anisotropy of space, and of directions of anisotropy that are variable over time;*
- *the existence of agreements in phase between the observed periodicities of the plane of oscillation of the paraconical pendulum with anisotropic support and with isotropic support, the periodicities of the optical sightings at marks and at collimators, the optical periodicities of Esclangon, and the periodicities of the interferometric observations of Miller;*
- *the agreement of the orders of magnitude of the amplitudes of all these periodicities.*

The periodic effects that would be demonstrated notably include: diurnal and semi-diurnal luni-solar periodicities of 24h, 24h50m, 12, and 12h25m; sidereal monthly periodicities of 27.322 days; semi-annual and annual periodicities; and planetary periodicities of several years.

*At present* I can *with complete certainty* make a prediction: if observations of the movement of the paraconical pendulum with anisotropic support and with isotropic support, optical sightings at marks and at collimators like those performed in 1958 and 1959, optical sightings corresponding to the experiments of Esclangon, and finally interferometric observations like those of Michelson and Morley (1887) and of Miller (1925-1926) are performed *in a continuous manner for at least a month, at the same time and in the same place*, then it will be found that the effects seen by Esclangon in 1926 and by Miller in 1925-26 correspond to the anomalies of the movement of the paraconical pendulum with anisotropic support (1954-1960) and with isotropic support (1959-1960) and to the anomalies of optical sightings at marks and at collimators demonstrated in July 1958 and in March-April 1959.

If the experimental arrangements that are employed are *comparable* to those used for the paraconical pendulum with anisotropic support and with isotropic support and the optical sightings at marks and at collimators, to those of the optical sightings of Esclangon, and to those of the interferometer of Miller, then it is *absolutely certain* that the five proposed series of experiments *will confirm the existence of all the previously demonstrated anomalies*, as well as the connections between them.

Here we are compelled to recall what Ampère wrote in his *Mathematical Theory of Electro-dynamic Phenomena*:*

> *"Epochs when phenomena that previously were considered to be due to completely different causes are brought under a single principle have almost always been accompanied by the discovery of a great number of new facts, because the new way of conceiving their causes suggests a multitude of experiments to undertake and explanations to verify."*

---

* André Marie Ampère, 1820-1825, *Mémoire sur la théorie mathématique des phénomènes électrodynamiques, uniquement déduite de l'expérience* (Memoir on the mathematical theory of electrodynamic phenomena deduced only from experiment), Collection of Memoirs related to Physics published by the French Physical Society, Volume III, 1887, p. 118 [57].

# B CONDITIONS FOR IMPLEMENTATION OF THESE EXPERIMENTS

## *B.1 Five separate projects*

When it is considered that I pursued my experiments on the paraconical pendulum with anisotropic support and with isotropic support *during nearly eight years* from October 1953 to June 1960, that the experiments on optical sightings at marks and at collimators and their analysis took *many months*, that the experiments of Esclangon were carried out almost continuously for an entire year, and that the experiments of Miller and their analysis occupied nearly nine years, then it might appear at first sight that my proposition for repeating these five series of experiments is very difficult to implement. Indeed, it does imply *significant* financial resources, personnel resources, and site resources.

However, *at least at the start*, the five proposed series of experiments could be implemented to a great extent *independently of one another*, and they could be performed *by five separate teams. Moreover, the implementation of these projects would be greatly facilitated by the lessons provided by the experiments that have already been performed.*[1]

[1] My work on the paraconical pendulum was the object of a large number of memoirs and working notes on theoretical and experimental analysis, consideration of which would be extremely helpful. Furthermore, there is an immense body of literature concerning work on the Foucault pendulum, and knowledge of this mass of experience will make it possible to avoid many errors.

Experiments on sightings at marks and at collimators have furnished irreplaceable teachings.

The interferometric experiments of Michelson, Morley, and Miller have been the object of many analyses in the literature.

## *B.2 The conditions of success*

For each of the projects, the first condition of success would be to set up a team of three or four *full-time* experimenters *who are highly motivated and have considerable professional experience in research.*

It is not enough to establish suitable experimental setups; it is also necessary to operate them *in a continuous manner*, and to utilize the observed results appropriately. Each series of experiments should be preceded by a preliminary series performed over at least a month's duration, and, subsequently, continuous observations must be considered over long periods - in any case, at least four month-long periods during the same year *at the times of the equinoxes and the solstices.*

*Just as in the case of the anomalies of the paraconical pendulum with anisotropic and isotropic support,* the real difficulty *with the optical anomalies of the sightings at marks and at collimators is not* to demonstrate that they exist, *because in view of the experiments already performed their existence is a certainty. The real difficulty is to invent appropriate apparatus, to implement systems for continuously registering the observations, and to define operational methods of handling and analyzing the data.*[1]

*In fact, first of all, it is imperative to re-demonstrate the regularities that have already been observed using the arrangements previously employed, even if they do not at first seem to be the best possible ones. When one realizes that it is required, above all, to reveal again the anomalies that have been previously recognized, it is necessary, first of all, to keep the experimental arrangements the same, and not to modify them subsequently unless and until such modification appears to be really necessary.*[2]

---

[1] In all cases, the arrangements for continuous registration *should be sedulously checked with classical apparatus*, for example with theodolites in the case of optical sightings.

[2] Here, the example of Esclangon's experiments is full of lessons. Modifications that *a priori* seemed to have no real relevance were enough to make the observed effects disappear (see above, *Chapter IV*, § B.2, note 6, p. 379).

In any case, *it is necessary to avoid any preconceived ideas and any premature theoretical explanation.*[3] *Only the facts to be observed matter.*

We should be particularly aware that each of these five proposed series of experiments offers *great difficulties in implementation. They cannot be improvised or set up in haste.* As Edward W. Morley emphasized[4] and as I have already repeated, "*Patience is a possession without which no one is likely to begin observations of this kind.*"

---

[3] As Miller himself rightly underlines (1933, p. 222 [203]), all interferometric observations before his work had been made *on the basis of preconceived ideas, and this had vitiated all those experiments.*

Apparently Miller himself committed the same error, since he was absolutely insistent on using a model based only upon consideration of the speed of the Earth, *and leaving without explanation* the non-zero values of the average azimuths $\overline{A}$ and their variations over time, as well as the factor $k$ for reduction of the speeds - although these were *major* characteristics of the results obtained (see above, *Chapter IV*, § F.1, pp. 417-419).

[4] Miller, 1933, p. 222.

## B.3 An indispensable progression

Whatever the force of conviction conveyed by the present work as a whole to the reader may be, it is very evident that it will not be sufficient completely to convince the decision makers, and to permit the deployment of the necessary means in terms of personnel, material, and facilities, with corresponding financial resources. *Before everything, it is necessary to have motivated and qualified researchers.*

*Only the evidence of experience* will convince those who have power over the decision to finance the proposed experiments.

Undoubtedly, the best way to proceed is to start with one of the four series of proposed experiments. In fact, of these four series of experiments, the sightings at marks and at collimators appear to be the easiest to perform *in a continuous manner.*[1, 2]

---

[1] In *Chapter VII* (Section B) of the *second volume* of this work (p. 30 above), I present suggestions for implementation of the experiments on optical sightings.

The experiments performed at IRSID in July 1958, at IGN in February-March 1959, and the trials of continuous registration at Palaiseau and IGN in 1992-1993 can provide invaluable information.

The question of continuous registration of sightings at marks *was only partially resolved by Michel Kasser, and the question of continuous registration of sightings at collimators, remain totally unresolved.* They present obvious difficulties.

In any case, verification of the results obtained at IGN in February-March 1959 (*Chapter II*, § 7 above) relating to simultaneous sightings *at marks and at collimators would clearly have considerable importance.*

It would be possible to set up the telescopes and the marks upon a single table made from granite, three to four meters long and around one meter wide. It would also be very desirable to regulate the ambient temperature to within 1/25 of a degree (see my *Note* of 5 December 1994, *Note sur les moyens nécessaires pour la reprise des essais optiques* "Note on the necessary requirements for repetition of the optical trials" [52]).

[2] The experiments relating to sightings at marks and at collimators would necessitate *relatively modest* financing, and their success might make it possible to find support for performing the other experiments.

The experiments upon the paraconical pendulum might present difficulties in the choice of the most appropriate apparatus, in the continuous maintenance of the movement of the pendulum, and in the continuous registration of the observations, but these problems could certainly be overcome.[3]

Interferometric experiments *with continuous registration* of the observations seem to me, a priori, to be much more difficult to implement. The same is probably true of the experiments of Esclangon.

In any case, and whichever type of experiments is concerned, it is absolutely necessary to perform continuous observations during long periods of time of at least a month, and it would be very desirable for it to be possible for the experiments to be continued in a continuous manner for at least a year.[4]

---

[3] For the observations of the paraconical pendulum with anisotropic support and with isotropic support, I think that the considerable experience I have acquired in this field since 1954, both theoretical and experimental, would make it possible to conduct them effectively. *But in both cases - the paraconical pendulum with anisotropic support and that with isotropic support - the choice of the type of pendulum and of the most appropriate suspension could only proceed from preliminary experiments that would undoubtedly take many months.* In both cases, the use of an *asymmetrical* pendulum is undoubtedly the most appropriate (see *Chapter VII*, Section A of the *second volume* of this work, p. 30 above).

There is also room for the use of *short* pendulums with *very great amplitudes*, and for taking account of *perverse* oscillations of the plane of oscillation (*Chapter I*, § F.3, note 15, pp. 209-210, and *Chapter II*, § 1.2, pp. 323-325).

[4] In *Chapter VII* of the *second volume* of this work, I give detailed suggestions for the implementation of these different experiments (p. 30 above).

## *B.4 A set of experiments of exceptional scientific interest*

Whatever the difficulties of implementation might be, from all the available evidence, the repetition of these five series of experiments - on the paraconical pendulum with anisotropic support and with isotropic support, on optical deviations of sightings at marks and at collimators, on the optical sightings of Esclangon, and on the interferometric observations of Miller - *would present exceptional scientific interest.*

These experiments would make it possible to verify *the existence of one and the same anisotropy of space*, that, via different phenomena, makes itself apparent in the form of privileged directions that vary with time as a function of astronomical influences, and particularly as a function of the position of the Earth in its orbit.

*The magnitude of the task to be performed, and the difficulties of all sorts to be overcome, are of the same major level of importance as the scientific issues at stake.*

# Chapter IX
# FIGHTING AGAINST DOGMA

*"People persecute each other because they believe that they know the 'Truth'. ... Energetic people can manufacture 'truth' by persecuting opinions other their own."*

Bertrand Russell *

# A THE TYRANNY OF ESTABLISHED TRUTHS

## *A.1 The makers of truth*

At every epoch, new conceptions are always rejected by the tyrannical power of "*established truth*". From time immemorial, dogmatic and intolerant fanaticism has opposed the progress of science and the revision of the axioms upon which accepted theories depend, when new facts come along to invalidate those axioms.

How can such situations occur? The reason is very simple. So-called "*scientific*" opinion is always blinded by the incessant repetition everywhere of pseudo-truths and by mistaken pre-judgments. In fact, the more dominant ideas are spread and repeated, the more they become rooted in human psychology, so to speak. Erroneous as they may be, due to simple and incessant repetition, in the end they acquire the character of established truths which nobody can call into doubt without putting himself into opposition against the active ostracism of the "*establishment*".

The greatest innovators have been the victims of this process, and Max Planck himself had to face strong obstruction. Here we must recall his own witness:[1]

> *"In the years '89 and '90 of the last century, personal experience taught me what the personal cost is to a searcher who wishes to propagate an idea which he possesses and on which he has reflected deeply. He will find out that the best arguments he can produce towards this goal are of little importance, because his voice does not have enough authority to impose itself upon the scientific world."*

---

* Bertrand Russell, 1928, *Sceptical Essays*, W. W. Norton, p. 63-64. [238]

[1] Max Planck (1858-1947), 1941, *Initiation into Physics*, id., p. 259 [218].

As Bertrand Russell wrote:[2]

> *"The mind of the most rational among us may be compared to a stormy ocean of passionate convictions based upon desire, upon which float perilously a few tiny boats carrying a cargo of scientifically tested beliefs...*
>
> *"As soon as any strong passion intervenes to warp the expert's judgment, he becomes unreliable, whatever scientific equipment he may possess."*

It could not be expressed better. From now on, it cannot surprise us that men who have dedicated themselves for years to deducing with great effort the consequences of poorly interpreted facts and false premises should rise in violence against those who menace what they consider as being their painfully acquired conquests.

[2] Bertrand Russell, 1931, *The Scientific Outlook*, W. W. Norton, p. 4 [239].

## A.2 A significant example

One of the most significant examples of errors committed by the defenders of "*established truths*" who are as blind as they are fanatical, may be that of Saigey who wrote in 1832:*

> *"We will end this chapter by saying a word about the experiments attempted in 1827 by Messrs. Whewell and Airy in the Dolcoath mine in Cornwall... They observed identical and invariable pendulums at ground level and at a depth of 1,220 feet (372 meters)...*
>
> *"The result of a first comparison gave 8.23 seconds of variation over a day between the operation of the two pendulums, the one that was at the bottom of the mine going faster than the other. By a raw calculation, it was concluded that the density of the Earth must be 7.73 times that of the superficial layer, i.e. 20 times that of water; a result that, according to the authors themselves, required verification.*
>
> *"On this subject we will only make one remark: the acceleration of 8.23 seconds is impossible, because, even supposing the density of the superficial layer to be zero, the acceleration would be 86,400 seconds multiplied by the depth of 372 meters and divided by 6,767,600 meters, which gives about 5 seconds for the limit of the acceleration...*
>
> *"From this, we see how erroneous is the observation of these two English scientists...*
>
> *"The observation is accordingly defective, and it is useless to concern ourselves further with it."*

In fact Saigey's calculation is based upon the dogmatic application of the Newtonian potential equation.[1] However, more than a century of observations has confirmed the accuracy of the observations of Airy and Whewell, *in other words the existence of anomalies of gravity, and this validates the experiment and condemns the* a priori *assertion!*

---

* Saigey, 1832, *Petite physique du globe* (Small physics of the Earth), Hachette, pp. 164 and 187 [241].

[1] Saigey's calculation is based upon the dogmatic application of the Newtonian potential equation

$$\Delta V + 4\pi G d = 0 \qquad \vec{g} = \nabla V$$

*i.e.*

$$\frac{d^2V}{dr^2} + \frac{2}{r}\frac{dV}{dr} + 4\pi G d = 0 \qquad g = -\frac{dV}{dr}$$

which gives

$$-\frac{r}{g}\frac{dg}{dr} = 2 - \frac{4\pi G r}{g} d.$$

Saigey would have been better inspired by studying more deeply the experimental results, rather than by coming to a decision in this slashing, wounding fashion by taking account only of the single currently accepted theory. *It is never a good mental attitude to doubt the work of others a priori.* With the passage of time and in the full knowledge of the facts that we today possess, what remains of Saigey's ironic and distrustful assertions? Unfortunately, it would be only too easy to provide many other such examples.

## *A.3 Authority against truth*

How can one struggle against such obstruction by the obscurantist defenders of "*established truths*"? We must repeat incessantly, with Claude Bernard, that science is a perpetual process of becoming, and that it must be modified every time that its propositions are contradicted by experiment:*

> *"The conclusion of all this is that, in front of the decisions of experiment, one must sacrifice one's own opinions as well as those of others...*
>
> *"In one word, theories must not be taught as dogmas or as articles of faith. Such exaggerated belief in theories would give a false idea of science; we would over-burden and enslave the spirit by depriving it of its liberty and suffocating its originality, and by subjecting it to the taste of systems."*

As Stanley Jevons also wrote:[1]

> *"There is ever a tendency of the most hurtful kind to allow opinions to crystallise into creeds. Especially does this tendency manifest itself when some eminent author, with the power of clear and comprehensive exposition, becomes recognised as an authority...*
>
> *"But any man must err, and the best works should ever be open to criticism. If, instead of welcoming inquiry and criticism, the admirers of a great author accept his writings as authoritative, ... the most serious injury is done to truth.*
>
> *"In matters of philosophy and science authority has ever been the great opponent of truth. A despotic calm is usually the triumph of error...*
>
> *"In science and in philosophy nothing must be held sacred."*

---

* Claude Bernard, 1865, *Introduction à l'Etude de la Médecine Expérimentale* (Introduction to the Study of Experimental Medicine), id., p. 72 [69].

[1] Stanley Jevons, 1871, *The Theory of Political Economy*, pp. 265-266 [159].

# B THE SECULAR OBSCURANTISM OF THE MANDARINS OF PSEUDOSCIENCE

## *B.1 Historical permanence*

To say the truth, the hostility that I have encountered is not specific to me. All innovators collide with the incomprehension or bad faith of their contemporaries.

"*When a new idea is introduced into science,*" wrote Richet, "*... objections arise, multiple, harsh and often absurd.*"[1]

In the 17th century the dean of the Faculty of Medicine described the circulation of the blood that had been demonstrated by Harvey as "*paradoxical, useless to medicine, false, impossible, unintelligible, absurd, harmful for human life.*"[2]

"*Was not the great Pasteur himself,*" Auguste Lumiere reminds us, "*according to the expression of Brouardel, 'charged like an accused person' before the Tribunal of the Academy of Medicine, when he was forced to hear his discoveries declared as anti-scientific, ineffective, and dangerous, and that they posed double danger: the social danger of homicide, and the intellectual danger of unreason!*"[3]

---

[1] Auguste Lumière, 1942, *Les fossoyeurs du progrès* (The gravediggers of progress), p. X [188].

[2] Pierre Rousseau, *Histoire de la Science* (History of Science), Fayard, 1945, p. 183 [237].

[3] Auguste Lumière, id., p. X.

Pascal had already written by the 17th century: "*Those who are capable of invention are rare, while those who invent nothing are much greater in number, and accordingly are the stronger.*"[4]

"*Those who are incapable of innovation find it difficult to admit that others can accomplish what they cannot do themselves.*"[5]

In a letter addressed to Pouchet, Broca wrote: "*A new truth that challenges our Masters has no way of defeating their hostility; neither reasoning nor facts can avail.*"[6]

"*The scientific authority of a given time is nothing more than the tyranny of influential positions.*"[7]

"*Official scientists, Professors in Universities, and often the most eminent ones, have too often followed the same errors, fighting against innovations that tend to invalidate classical teachings, or have refused to believe in the reality of discoveries without taking the trouble to check them.*"[8]

"*As Nicolle so justly observed, it is never the Masters who innovate or make discoveries in the branches of science in which they specialized; it is almost always independent researchers who advance our knowledge, and for this very reason they may expect to be treated disdainfully, to be repulsed, and to see themselves discredited.*

"*When new ideas are proposed by personalities who occupy the highest scientific situations, they come to be accepted without the least checking, even if they consist of mistaken solutions.*"[9]

---

[4] Auguste Lumière, id., p. X.

[5] Auguste Lumière, id., p. X.

[6] Auguste Lumière, id., p. XII.

[7] Auguste Lumière, id., p. 68.

[8] Auguste Lumière, id., p. 38.

[9] Auguste Lumière, id., pp. 282-283.

*"Sometimes one finds it hard to believe how far the tactic of suffocating innovators can be taken by conformists who, without much aptitude for making discoveries themselves, do not admit that other people can advance Science, above all when they do not belong to their chapels and are only independent researchers."*[10]

*"In the scientific milieu... the discoveries of innovators are too often judged, not on the facts and the accompanying arguments, but according to the prestige, the personal relations, and the eminent positions of those who present them... The innovator is too often the victim of the unworthy sentiments of the authorities of his time, of their routine habits, of their prejudices or their jealousy."*[11]

*"Elie de Beaumont, a well-informed geologist, made this judgment of Darwin: "He is an intelligent amateur, but that does not mean he is a scientist. It would harm science to open the doors of the Academy to Mr. Darwin."*[12]

It took nearly fifty years for the theory of continental drift, presented in 1912 by the German meteorologist Alfred Wegener, to be recognized by official science.[13]

---

[10] Auguste Lumière, id., p. 298.

[11] Auguste Lumière, id., p. 285.

[12] Louis Rougier, *Scandale à Polytechnique* (Scandal at Polytechnique), Imprimerie des Tuileries (Tuileries Press), July 1959, p. 12 [236].

And Rougier adds (p. 12):

> *"Amateurs! How many great discoveries, how many great inventions are owed to them! Galileo was a mathematician after having pursued studies in medicine; as an amateur, he revolutionized those "two new sciences", astronomy and mechanics. Fermat was the King's adviser in the Parliament of Toulouse; "prince of amateurs", as described by the historian of mathematics E.T. Bell, he was one of the founders of the infinitesimal calculus and of analytical geometry, and he consolidated the theory of numbers upon an immortal platform. Faraday was a bookbinder, and Davy advised him not to abandon an obscure but secure profession; as an amateur, he discovered electrical induction. The monk Mendel was neither a botanist nor a biologist; we owe him the laws of hybridization and of heredity to him as an amateur. Boucher de Perthes was a customs officer; as an amateur, he was the founder of human paleontology. Pasteur was a chemist; as an amateur, he revolutionized biology and medicine. Norbert Wiener was a mathematician; as an amateur, he is the father of cybernetics."*

[13] Alfred Wegener (1880-1930), *La Genèse des Continents et des Océans* ("The Genesis of the Continents and the Oceans") (translation of the third German edition), Albert Blanchard, 1924 [277] - a fascinating work that I discovered by accident in 1935 at a second-hand bookshop in the Latin quarter, and which I devoured with passion while I was still a student at the School of Mines in Paris.

The fact is that, in spite of the very eminent qualities of some of its members, the Academy of Sciences has often committed very regrettable errors.[14]

[14] Auguste Lumière (id., pp. 34-37) wrote:

*"The Academy of Sciences was founded in 1666 by Colbert... It may be considered as being a veritable scientific tribunal to which everyone who occupies himself with science comes to plead for approval for his work.*

*"It is enough to remind ourselves of some of the serious errors committed by this illustrious Company, to show that Colbert completely failed in his ambition, when he tried to form the Academy into an institution destined to encourage Science. On the contrary, it has shown itself hostile to almost all innovators whose discoveries have not been in accord with classical dogmas; it has repulsed precisely those persons who were capable of advancing science, and very often has hampered the march of progress by doing so. Was not its hostility manifested against the ideas of the animal nature of corals, the geological antiquity of man, the existence of aeroliths, the vulcanism of the Auvergne, the possibility of life in the oceanic abysses, the transatlantic telegraph, electrical transport of force, electric illumination by incandescence, steam navigation, the telephone, the phonograph, alternating current generation, and the electrodynamic theory of Ampère? against Darwin, Lamarck, Pasteur, Boucher de Perthes, Geoffroy Saint-Hilaire, etc., etc...?*

*"One may ask oneself why a Company that indeed is composed of very eminent scientists has been able to bar the road to so many men of genius and thus delay scientific progress - facilitation and stimulation of which should be its principal function.*

*"We think that this regrettable lack stems, in the first place, from its constitutional organization that makes it incapable in all fields...*

*"When an author presents to the Academy a work on a botanical subject, for example, the astronomers, the doctors and surgeons, and indeed all the members of sections other than the botanical section are almost completely incompetent...*

*"In principle, only a few of the members are qualified to judge this work...*

*"Very often, in each of the sections, there is one member who drags along the others; in each type of problem there is thus, so to speak, an oracle whose line the entire Company toes.*

*"The innovator who does not have the oracle on his side is lost; and in the immense majority of cases the oracle will in fact be against him, if his conclusions are not in accordance with classical notions or those which the oracle himself holds."*

## *B.2 Two illustrations: Galileo and Kepler*

### The abjuration of Galileo

1 - Faced by the refusal of evidence by his adversaries, Galileo wrote to Kepler on 19 August 1610:[1]

> *"What to do? Should we play the part of Democritus (and laugh), or that of Heraclitus (and cry)? I think, my Kepler, that we should laugh at the distinguished stupidity... What can you say of the highest philosophers of the faculty here (that of Pisa), to whom I have spontaneously offered my work a thousand times, and who... refuse to look at the planets, the Moon, and the telescope? They close their eyes to the light of truth."*

In his letter to the Grand Duchess Christine of Lorraine in 1613, Galileo also wrote:[2]

> *"Some years ago... I discovered in the skies many things that nobody had ever seen before our time... The novelty of these things, and also certain consequences that followed which were in contradiction to the notions of physics commonly accepted by academic philosophers, raised more than one doctor against me...*
>
> *"More fascinated by their doctrines than by the truth, they tried to deny and refute new things that their senses would have shown them, if they had really wanted to look. For this purpose, they hurled various accusations and published numerous writings full of vain arguments...*
>
> *"They should not arrogate to themselves the right to decide upon controversies in disciplines that they have neither studied nor practiced."*

---

[1] Galileo Galilei (1564-1642), *Opere* VI, 118. Florence Edition, 1842-1856 [148].

[2] Arthur Koestler, *Les Somnambules* (The Sleepwalkers), Calmann-Lévy, 1960, pp. 406 and 418 [164].

Galileo's abjuration on 20 June 1633 is justly celebrated:[3]

*"I, Galileo, in the seventieth year of my age, on my knees before Your Eminences, having before my eyes the Holy Scriptures which I touch with my own hands, I abjure, I curse, and I detest the error and the heresy of the movement of the Earth."*

## Kepler, an unrecognized genius

2 - Considering the means available in his time, the three laws of Kepler concerning the trajectories of the planets[4] constituted *one of the greatest discoveries of all history*, and were an exceptional exploit which was little known at the time, even by Galileo.[5]

But the life of Johann Kepler (1571-1630) only consisted of a long series of setbacks.[6] His ideas were too new, too unheard-of, too contrary to established truth for him to be able to expect immediate acceptance.

---

[3] Extracted from the Abjuration of Galileo. The complete text may be found in Bertrand Russell, 1947, *L'esprit scientifique et la science dans le monde moderne* (The Scientific Spirit and Science in the Modern World), editions Janin, pp. 34-35 [240].

For the text of Galileo's condemnation, also see Koestler, id., notes 5 and 6, pp. 559-564.

[4] The first two laws of Kepler were published in 1605, and the last in 1618; they go as follows:

1° Each planet travels in a plane that includes the Sun, and describes an ellipse of which the Sun is a focus;

2° Areas swept by the radius vector are proportional to the time taken to sweep them;

3° The squares of the periods for one revolution are proportional to the cubes of the major axes: if $n$ is the average movement, then this law can be expressed as $n^2 \bullet a^3 = k^2$, where $k^2$ remains constant from one planet to another.

[5] For the three fundamental laws of Kepler relating to the trajectories of the planets, see in particular: W.M. Smart, *Celestial Mechanics*, Longmans, 1952, pp. 1-3 [250]; and Luc Picart, *Astronomie générale* (General Astronomy), Armand Colin, 1931, pp. 101-104 [215].

[6] Koestler, id., fourth part, pp. 213-406.

In his lifetime Kepler experienced nothing but incomprehension and bitterness, but *his glory defies time and will remain immortal.*[7, 8]

[7] On the works of Kepler, see in particular Alexandre Koyré, *La révolution astronomique, Copernic-Képler-Borelli* (The Astronomical Revolution, Copernicus-Kepler-Borelli), Hermann, 1961, Part 2, pp. 107-458 [165].

[8] In fact, science is a human activity consisting of advancing *an ensemble of knowledge that, we have learned from history, is cumulative*. However, as S.F. Mason so justly underlines (*Histoire des Sciences* (History of the Sciences), Armand Colin, 1956, p. 445 [193]):

> *"In reality, only one section of science has been truly cumulative up to the present era - namely its practical techniques, and its facts and experimental laws. When viewed on a long chronological scale, scientific theories have, up till now, been very ephemeral."*

Even though in reality the discovery of experimental facts is *much more important* than formulation of theories, it is such formulation that is generally most productive for their authors. As William Broad and Nicolas Wade wrote (*La souris truquée, enquête sur la fraude scientifique* (Betrayers of the Truth: Fraud and Deceit in the Halls of Science, Inquiry into scientific fraud), Editions du Seuil, 1987, p. 26 [84]):

> *"In reality, the discovery of facts is less well rewarded than the elaboration of theories or of laws for explaining those facts, and this explains the attraction such theories exert."*

This is true even though the same facts can be explained by *quite different* theories, while the facts themselves are *definitive* once they are established.

(See in particular *Chapter VII*, Section E, above).

## *B.3 Present-day science*

Luckily nowadays innovators are not forced to abjure their work, and they are no longer condemned to be burnt alive at the stake, as the monk Giordano Bruno was on 9 February 1600, for having advocated the heliocentric system;[1] *but, while the means have changed, the profound hostility to innovators in the name of dogmas remains the same, still as strong as ever.*

Today one can discern a type of dogmatic and intolerable domination by certain supporters of relativistic theories. For them the theory of relativity has become a type of religion, and it is forbidden to contradict it or even to discuss it.[2]

Just as Einstein and his successors were originally justified in complaining of the opposition to their criticisms of classical theory, so today the intolerance and dogmatism of some holders of relativistic theories are completely insupportable.

As I have previously written:[3]

> *"Innovators themselves also have too much of a tendency to become dogmatic. There are many illustrious scientists who, in the morning of their scientific life, fought against dogmatism and demonstrated by some great discovery the insufficiency of theories supposed to be solidly established, but who, in their twilight years, themselves have assumed, with their disciplines, even more dogmatic and intolerant positions in support of the ideas with which they triumphed at the zeniths of their careers.*
>
> *"In any era and in any science, too many propositions often tend to be presented as solidly established, and the postulates upon which they rest end by acquiring a sort of metaphysical sanctity. Certain theories are presented as acquired truths, even though their fundamentals are at least arguable. Their pretended perfection never seems to be seriously questioned; facts that appear not to be compatible with these imposing constructions are never examined. These theories only become true by the virtue of simple repetition.*
>
> *"Too often the supporters of this or that theory have proved their unlimited power of criticism as regards opposing opinions, while at the same time their confidence in their own theories reveals unlimited naivety."*

---

[1] Pierre Rousseau, 1945, id., p. 174.

[2] See *Chapter VII* above, § C.9, pp. 614-616.

[3] *L'Economique en tant que Science* (Economics as a Science), Revue d'Economie Politique (Review of Political Economy), January 1968, pp. 22-23 [42].

On the metamorphosis of innovators, see note 5 of § G.5 of *Chapter I* above, pp. 226-227.

The mandarins of pseudoscience force one to remember the aphorism of Auguste Detoeuf:[4]

> *"However and by whatever means we may divide an assemblage of people into groups (choice, seniority, examinations, competitions, selection by lots), the proportion of imbeciles in the various groups remains the same."*

We may even add with Bouasse:[5, 6]

> *"Good men are tempted to ascribe all intellectual virtues in the world to the scientist. That he is not worthy of this veneration is a truth so evident that it is useless to emphasize it."*

---

[4] Auguste Detoeuf, 1938, *Propos de Barenton, Confiseur* (The Case of Barenton the Confectioner), Editions du Tambourinaire, p. 137 [110].

I am tempted to add here that in all milieus the proportion of bandits is the same.

[5] Bouasse, 1920, *Des Savants* (Scientists), Preface to *Pendule, Spiral, Diapason* (Pendulum, Spiral, Tuning Fork), II, Delagrave, p. V [76].

[6] The disillusioned propositions of Bouasse are illustrated by the judgment of Pasteur.

> *"Ah! how much trouble and pains, said Pasteur, are required to secure the triumph of truth. It is not a negative, it is a stimulant; the only thing that is painful is bad faith."*

(René Vallery-Radot, *La vie de Pasteur* (The life of Pasteur), Flammarion, 1941, p. 585 [270]).

And he wrote to his wife:

> *"I shall work with rage in my heart. I shall be happy when I can come and read a good memoir, with this cry in my heart: So, you cretins, try to do as well! I speak here of Cretin X..., Cretin Y..., and that many other nullities, parvenus while otherwise deprived (of merits) and by the accidents of fortune."*

(Auguste Lumière, 1942, id., p. 275).

Michael Faraday - an authentic genius - gave an unforgiving portrait of pseudo-scientists, as pretentious as they are incapable:

> *"As when on some secluded branch in forest far and wide sits perched an owl, who, full of self-conceit and self-created wisdom, explains, comments, condemns, ordains and order things not understood, yet full of importance still holds forth to stocks and stones around — so sits and scribbles Mike."*

(J.G. Crowther, *Michael Faraday*, Hermann, 1945, p. 20 [97]).

One can well understand the exasperation of Evariste Galois, when, as a candidate for the Ecole Polytechnique, he finally hurled the eraser cloth of the blackboard at the head of an examiner of dull intelligence.[7]

In the face of so much blind incomprehension, so much obtuse hostility to innovators, new ideas, and discoveries, one can only conclude with Louis Rougier:[8]

> *"Must we therefore conclude that the experience of centuries must be re-learned in each generation? Does not this go to prove that there is no unique and definitive theory, and that the human spirit proceeds by trials and errors, by successive refinements?"*

---

[7] E.T. Bell, *Les grands mathématiciens* (The great mathematicians), Payot, 1939, p. 398 [63]. We owe this pithy judgment to Evariste Galois (id., p 400):

> *"Genius is condemned by bad social organization to be eternally denied justice in favor of servile mediocrity."*

[8] Louis Rougier, July 1959, *Scandale à Polytechnique* (Scandal at Polytechnique), Tuileries Press, p. 12 [236].

# Chapter X
# NEW PERSPECTIVES

*The character of the experimental method is to refer only to itself, because it includes its own criterion, which is Experience. It recognizes no authority other than that of facts.*

Claude Bernard *

*When a revision or a transformation of a physical theory is initiated, almost always it is found at the start that one or more facts cannot be brought into the framework of the theory in its initial form. Indeed, facts are always the key to the vault, and on them depends the stability of every theory, however important it may be.*

*For the theoretician worthy of the name, there is therefore nothing so interesting as a fact that contradicts a theory previously held to be true, and that is when his real work begins.*

Max Planck **

*"All that we know of reality comes from experiment and derives from experiment. Purely logical propositions are completely empty as regards reality...*

*"The result is that our conceptions of physical reality can never be definitive. If we want to agree with the perceptible facts in the most perfect manner logically possible, we must always be prepared to modify those conceptions, which are also known as the axiomatic foundations of physics. In fact, a glance at the evolution of physics tells us that these foundations have undergone profound changes over time."*

Albert Einstein ***

# A TWO COMPLETELY NEW PHENOMENA

## *A.1 Anomalies of the movement of the paraconical pendulum and anomalies of sightings at marks*

My labors in experimental physics from 1953 to 1960 were crowned by *the demonstration of two entirely new phenomena that were totally inexplicable in the framework of current theories of gravitation and geophysics.*

In summary these were: - *the existence of a periodic lunar component of 24h50m* in the movement of the paraconical pendulum with anisotropic support and with isotropic support, *having an observed amplitude of twenty to a hundred million times greater than the amplitude calculated according to the theory of gravitation, whether or not completed by the theory of general relativity;*[1] and - *optical deviations of sightings at marks*, also inexplicable in the framework of current theories, and *distinguished by very significant agreements in phase* with the periodic components of the motion of the paraconical pendulum.[2]

Both of these were *major and authentic discoveries of exceptional interest.*[3]

---

* Claude Bernard, 1865, *Introduction à l'Etude de la Médecine Expérimentale* (Introduction to the study of experimental medicine), id., p. 74 [69].

** Max Planck, 1941, *Initiation into Physics*, Flammarion, p. 40 [218].

*** Albert Einstein, 1930, *Comment je vois le monde* (How I see the World), p. 165 [129].

1 *Chapter I*, § B.2.1, p. 118, and *Chapter II*, § F.2.2, pp. 284-285, above.

2 *Chapter III*, § B.3, p. 338 above.

3 *Introduction*, § C.2, above, p. 60-63.

I fully accept that I am *an auto-didact and an amateur* in physics, if somebody who is not a member of the "*establishment*" can be so described.[4]

But perhaps after all that is an advantage - the advantage of not being a prisoner of "*established truths*" and of seeing everything with new eyes.

Here, Claude Bernard also knew very well how to highlight the advantages and inconveniences for anyone of being an auto-didact and an amateur:[5]

> *"It has often been said that, for making discoveries, it is necessary to be ignorant. This opinion, false in itself, hides nonetheless a nugget of truth. It means that it is better to know nothing than to have in mind fixed ideas based on theories that one is always seeking to verify while neglecting anything that does not agree with them..."*

---

[4] This was incidentally my case in economy in the 1940s and 1950s, and in my Nobel Conference of 9 December 1988 I was able to say

> *"At the time, I was no more than self-taught...*
>
> *"Thus, my 1943 book was the work of a mere amateur, but a passionnate one, and were anyone to be somewhat surprised to learn that the Royal Academy of Sciences of Sweden had expressly referred to the work of an author who called himself an amateur, he would only betray an ignorance of all that amateurs have been able to contribute to science over the centuries. Indeed, within their ranks are to be found such great figures as Fermat, Leibniz, Lavoisier, Mendel, Pasteur, Louis de Broglie, and so many others, who, in their early career, or even, in some cases, throughout their whole life, were only amateurs. Walras and Pareto were, themselves only self-taught, only amateurs.*
>
> *"Amateurs are usually detested by professionals and members of any kind of "establishment"; but they do possess one very exceptional advantage, that of never having been conditioned by university training and the constant repetition of "established truths", and, therefore, of being able to examine every question with a fresh eye, without any preconception and prejudice."*

[5] Claude Bernard, 1865, *Introduction à l'Etude de la Médecine Expérimentale* (Introduction to the Study of Experimental Medicine), id., p. 71.

*"This attitude of spirit is one of the worst, and is highly opposed to invention. Indeed, a discovery is in general an unexpected relation that is not included in theory - otherwise it would have been expected. An ignorant person who does not know that theory is actually, under that circumstance, in the better mental conditions: theory does not hamper him and does not hinder him from seeing new facts which are not perceived by someone who is preoccupied with an exclusive theory.*

*"We hasten to say that this does not mean elevating ignorance to a principle. The more one knows and the more previous knowledge one possesses, the better one's attitude will be for making great and fruitful discoveries. But it is essential to keep one's liberty of spirit."*

I willingly admit that some of my results may be somewhat astonishing and even shocking, and that many people may remain skeptical. Considering the observations of the paraconical pendulum for example, the claim that effects twenty to a hundred million times greater than those predicted by the theory of gravitation have been detected can, at first sight, only provoke incredulity, in view of the constant stream of successes of Newton's theory repeated *over three hundred years*.[6]

Max Born wrote so truly:[7]

*"Newtonian attraction has had to take account of a formidable quantity of facts that, thanks to the perfection of observational methods, have been accumulated over centuries. The fact of its success therein constitutes one of the greatest triumphs of the human spirit...*

*"In an incalculable number of cases, the theory of Newton has thus been tested by new experiments, and has never failed."*

I recognize very willingly that I may be mistaken. *The real kernel of error is to believe that one has the truth, and a person who deceives himself is twice wrong. He is wrong because he is deceived, and he is wrong because he does not know that he is deceived.*

---

[6] See above, *Chapter I*, § F. 1.3, p. 200.

[7] Max Born, 1923, *La théorie de la relativité d'Einstein et ses bases physiques* (Einstein's relativity theory and its physical basis), id., p. 60.

But if this proposition holds for me, it holds equally for those who dispute my results. As I wrote in my memoir of 1958:[8]

> *"Nobody in this world can flatter himself on having a monopoly of truth. Innumerable historical examples should give pause to those who condemn me* a priori *and without any appeal because my conclusions are opposed to their convictions of today.*
>
> *"Yet again, it is possible that I am wrong, but in the actual state of affairs nobody has yet been able to prove it in scientific terms or, other than by purely* a priori *assertions, to show any error in my deductions - which are founded upon more than 220,000 observations and upon consideration of more than 1,500 works and memoirs, and upon five years of work and reflection. At the current stage of the discussion, everything is proceeding as though we were faced with a new phenomenon that is absolutely inexplicable in the framework of classical theory.*
>
> *"If this conclusion were to be contested, it could only be scientifically disputed by reliance upon new experiments or by showing errors in reasoning, but no* a priori *contestation can be considered as being scientific.*
>
> *"Dogmatism and sectarianism are not scientific positions. It is the facts and the facts alone that should decide theories, not the reverse."*

[8] *Faut-il reconsidérer les lois de la gravitation ?* (Should the laws of gravitation be reconsidered?) Perspectives X, 1958, p. 104 [23].

## *A.2 New data on two previous experimental series: The optical sightings of Ernest Esclangon and the interferometric observations of Dayton C. Miller*

The two new phenomena that I demonstrated relating to the paraconical pendulum and to sightings at marks present *striking connections* with the observations resulting from the earlier experiments on optical sightings performed in 1927-1928 by Ernest Esclangon, and with the 1925-1926 interferometric observations of Dayton C. Miller.

In connection with these two series of experiments, and particularly with the experiments of Miller, I have revealed *new regularities* that exhibit *very great internal coherence and that exclude any perverse effect*, for example temperature effects.

These regularities that I made evident are *all the more significant because they escaped the analyses by Esclangon and by Miller, and were missed by everyone who has previously commented on those experiments.*

The harmonic analysis that I undertook demonstrates the existence of *very significant* connections with the anomalies in the movement of the paraconical pendulum with anisotropic support and with isotropic support, and with the anomalies in the optical deviations of sightings at marks and at collimators.

## *A.3 A set of very significant facts*

In the first five chapters of this work I have presented an overall analysis of the observations of the five *very significant* series of experiments corresponding to my own works and to those of Esclangon and Miller: the observations of the paraconical pendulum with anisotropic support; the observations of the paraconical pendulum with isotropic support; the optical sightings at marks and at collimators; the optical observations of Esclangon; and the interferometric observations of Miller.

In view of this analysis, the characteristic periodicities must be considered as being *indeed real and independent of any local perturbing influence*. All these observations are characterized by numerous connections.

*These are the data resulting from observation, completely independent of any hypothesis and of any theoretical interpretation whatsoever; and naturally any theory, whatever it may be, that turns out to be incompatible with these experimental data must be revised and rejected, in whole or in part.*

I hope that this work of mine will be able to stimulate repetition of these five series of experiments by showing their *absolutely fundamental* scientific interest.

In any case, I think that this work will put an end to the rumors that have complacently spread a little everywhere of the invalidity of my experiments on the paraconical pendulum and the optical experiments on sightings at marks, associated by me with them.[1] *These rumors have never ceased to spread but have never been crystallized into publications, and accordingly it has been impossible for me to make any response.* In fact they have considerably contributed to the discrediting of my experiments.

[1] In particular, the results obtained in the *crucial* experiments of July 1958 at Bougival and Saint-Germain *cannot leave any doubt* (see above, *Chapter I*, § C.2.3, p. 148).

Also see *Chapter I*, § G.6, p. 232-236.

# B SUPPORTS AND OPPOSITIONS

## *B.1 Invaluable supports*

I fully appreciate today everything that I owe to all those who, down through the years, have encouraged, helped, and supported me with their friendship.

Now, at the end of this work, I must again express the immense gratitude that I owe to Emmanuel André-Martin, Pierre Ricard, Albert Caquot, and René Dugas, now alas deceased, thanks to whose support from 1953 to 1960 I was able to conduct my experiments on the paraconical pendulum and the associated optical experiments.

It is to Pierre Ricard that I owe the provision of a laboratory at the Institut de Recherche de la Sidérurgie (Institute of Iron and Steel Research) at Saint-Germain with two collaborators, Jacques Bourgeot and Annie Rolland. It is thanks to them that my experiments could be performed with full success.

I must equally thank the members of the Academy of Sciences who, at the time, with Albert Caquot, rendered me very effective support: Pierre Chévenard, Donatien Cot, Jean Coulomb, Joseph Kampé de Fériet, André Léauté, Albert Pérard, René Perrin, Maurice Roy, Pierre Tardi, and René Thiry.

I also particularly thank Guy Berthault, Henry Aujard, and Gilles de La Rochefoucauld my old student and editor, who nowadays give me very effective support.[1]

[1] I take the opportunity of similarly thanking all those who, upon the publication of my recent article, *Les expériences de Dayton C. Miller, 1925-1926, et la théorie de la relativité* ('The experiments of Dayton C. Miller, 1925-1926, and the theory of relativity') in the August-September 1996 issue of the polytechnic magazine *La Jaune et la Rouge* (The Yellow and the Red) [55], rendered me support for which I have been very grateful: Olivier Costa de Beauregard, Claude Friang, Marcel Macaire, Charles Maupas, and Paul-Ernest de Montaigne.

## *B.2 Unrelenting oppositions*

• Starting from 1956, I had to face oppositions that continued to grow without limit, more or less concealed, more or less explicit, and which, in spite of all the efforts of my supporters, finally gained the field in 1959.

I have never been able to identify the reasons clearly. Some people undoubtedly thought that my experiments were liable to bring into question the foundations of the Theory of Relativity, hitherto completely dominant. However I sedulously avoided taking any position whatever on the Theory of Relativity, and I always limited myself to underlining that my results were totally inexplicable in the framework of currently admitted theories. But, in any case, it was indeed true that my experiments were clearly of a nature to call "*established truths*" into question.

Without doubt I was also considered as being an amateur by the professionals, and all amateurs are seen negatively by members of the "*establishment*".[1]

• After more than thirty-five years it is a little difficult today for me to give an effective account of these oppositions and their motives without referring to texts of the time.[2] In fact, only those texts could provide a real view of the reactions excited by the publication of my works on the paraconical pendulum with anisotropic support.[3] Moreover, they are rich in lessons on the scientific method, on the philosophy of the sciences, and on human psychology.

---

[1] After the war, for similar reasons, and with certain exceptions, I met with *very strong opposition* to my work in economy in the milieu of the Faculties.

[2] I give three illustrations of this in *Chapter VIII* of the *second volume* of this work. I have also quoted extensive extracts from my *1958-1960 correspondence* in *Annex I A* thereof (pp. 30-31 above).

[3] It was not possible for me to make any publication of the results of my experiments on the paraconical pendulum *with isotropic support* (see above, *Chapter I*, § G.2.2, p. 216).

*It is true to say that the more ignorant my opponents were, the more fanatical they were.* I could not cope with their relentless campaign, which was the more effective for being subterranean,[4] and which, when expressed overtly, was limited to mere assertions that were never justified or motivated.[5] My most fanatical gainsayers were notable for their *total* ignorance of statistical analysis, and had never performed any experiments of any kind.

I must add that I was completely disheartened at the time by the final refusal of the Academy of Sciences to publish my *Note* of 23 February 1960 presented by Pierre Tardi, "*Existence of periodic components in the variations of readings corresponding to sightings made with a telescope at a fixed mark, correlated with the movements of the paraconical pendulum*", *even though it had already been printed at the Academy press.*[6]

Finally, *due to lack of means*, I was obliged to give up my experiments and to close my two laboratories in Saint-Germain and Bougival on 30 June 1960.

---

[4] I should add that from 1956 to 1960 no Commission of the Academy of Sciences gave me a hearing (see above, *Chapter I*, § G.5.3, pp. 228-231).

[5] Following a visit to my laboratory at IRSID on 28 May 1958, Georges Janson, a civil engineer of Ponts et Chaussées (Bridges and Roads), was able to write to me, even *before* my crucial experiments of July 1958:

> *"Some of those who contradict you leave me with an impression of bitterness.*
>
> *"For years you have been studying a difficult problem in mechanics. You know your subject profoundly, and people who are certainly very qualified try to keep you in check by doing their own mental calculations instead. Is this serious?*
>
> *"If you make a discovery in spite of them they will try, for a change, to say that everything was obvious.*
>
> *"If on the other hand you fail, they will cry to the skies that they were right from the beginning.*
>
> *"This attitude is tainted with an obscure jealousy, and must sometimes provoke in you bitter reflections.*
>
> *"At least, you should know that, around you, there are many unknown figures who admire without reserve your science, your courage, and your disinterestedness.*
>
> *"Galileo and many other illustrious scientists also had to pay a heavy tribute of humiliation to official knowledge."*

[6] As far as I know, in the case of this Note, this obstruction came from a single member of the Academy of Sciences - always the same one, Jean Leray, who was *completely incompetent in the matter* (see *Chapter III*, § B.4, pp. 339-340, above).

## *B.3 The crucial experiments of July 1958*

However the *incontestable* proof of the reality of the anomalies of the paraconical pendulum was furnished by the results of the *crucial experiments* of July 1958 which, with the aid of the *Comité d'Action Scientifique de la Défense Nationale* (National Defense Committee for Scientific Action) and of the *Centre National de la Recherche Scientifique* (National Center for Scientific Research), were performed *simultaneously* in two laboratories at Saint-Germain and at Bougival, *six kilometers apart*, the second being located in a subterranean quarry about sixty meters underground. *The agreement of the observed effects was completely extraordinary.*[1]

*One would have thought that these results would have been sufficient to compel the definitive conviction of the entire scientific world.*

This is all the more so, because the experiments on optical deviations of sightings at marks performed *simultaneously* in the laboratory at Saint-Germain demonstrated the existence of periodic components in the readings corresponding to the optical sightings at marks that were performed, and had *impressive* correspondences in phase with the periodic components of the movements of the paraconical pendulum.[2] *Not only were the directions and the magnitudes of the observed effects completely incompatible with the calculated effects, but also a connection had been established between mechanical and optical phenomena whose basic nature was completely different.*

---

[1] See my two *Notes* to the Academy of Sciences on 3 November and 22 December 1958, *Nouvelles Expériences sur le Pendule Paraconique à Support Anisotrope* ('New Experiments with the Paraconical Pendulum with Anisotropic Support') [25], and *Structure des Mouvements du Pendule Paraconique à Support Anisotrope à Bougival et à Saint-Germain* ('Structure of the Movements of the Paraconical Pendulums with Anisotropic Support at Bougival and at Saint-Germain') [27], and above, *Chapter I*, Section C, pp. 142-161.

In fact, *two months before these crucial experiments*, when I applied myself to making deductions "founded upon more than 220,000 observations and upon consideration of more than 1,500 works and memoirs, and on five years of work and reflection", I was able to assert that "*At the current state of the discussion, everything is proceeding as though we were faced with a new phenomenon that is absolutely inexplicable in the framework of classical theory.*"

(*Anomalies du mouvement du pendule paraconique à support anisotrope* (Anomalies of the movement of the paraconical pendulum with anisotropic support), IRSID, 13 May 1958).

[2] See my *Note* above, received on 23 February 1960 by the Academy of Sciences [38], and *Chapter III*, Section D, pp. 334-338, above.

## *B.4 Cessation of the experiments*

• *Even after the passage of many years*, I cannot understand how institutions whose essential remit was research could not have decided to continue with financing these experiments of really exceptional interest, *since the existence of new phenomena of major importance had definitively been established.*[1, 2]

In hindsight, all the above was really very regrettable; *more than thirty-five years have been wasted* without further analysis of two new phenomena which upon reflection appear *completely astounding*: the observations of the paraconical pendulum, which were totally inexplicable in the context of the theory of gravitation whether or not completed by the theory of general relativity; and the observed deviations of sightings at marks, whose characteristics and connections appear at first sight *absolutely extraordinary*, both in terms of their amplitudes and in terms of the agreements in phase that they exhibit with the periodicities of the paraconical pendulum.

In fact, these phenomena appeared to be capable of leading to *completely new and extremely significant* results, and could have cast doubt upon many received ideas and many "*established truths*".[3]

As Henri Poincaré wrote:[4]

> *"If a new result is to have any value, it must unite elements long since known, but till then scattered and seemingly foreign to each other, and suddenly introduce order where the appearance of disorder reigned. Then it enables us to see at a glance each of these elements in the place it occupies in the whole...*

---

[1] However at least nine members of the Academy of Sciences - Caquot, Cot, Darrieus, Kampé de Fériet, Léauté, Perard, Roy, Tardi, and Thiry - thought that my research ought to be pursued further, and they much favored its continuation (see above, *Chapter I*, § G.6, p. 233).

[2] See above, *Chapter I*, § G.6, pp. 232-236.

[3] After the closure of my laboratory at Saint-Germain in June 1960, I decided to concentrate myself exclusively upon my researches and publications in economy.

I should emphasize that, if the validity of my physics experiments had been fully recognized at the time, I should certainly have been very tempted to concentrate all my activity upon physics. My subsequent career would undoubtedly have been completely altered, and it is very possible that I might not have obtained the *Nobel Prize in Economic Science* in 1988 for my two works, *A la Recherche d'une Discipline Economique, I - L'économie pure* ('In Search of an Economic Discipline: I - Pure Economy'), 1943 [3], and *Economie et Intérêt* ('Economy and Interest'), 1947 [4].

[4] Henri Poincaré, 1908, *Science and Method*, Flammarion, 1927, pp. 24-25 [225]; and 1913, *Dernières Pensées* (Last Thoughts), Flammarion, 1926, p. 336 [226].

> *"If the results are not in agreement with expectation, true scientists... feel... their curiosity highly excited; they know that their efforts, their momentary disappointment, will be repaid a hundredfold, because truth is there, just nearby, still hidden and adorned with the attraction of mystery, but on the point of unveiling itself."*

• To tell the truth, in every domain in which I have worked during the last half-century, both in physics and in economy, I have never ceased to collide with "*established truths*" and against the dogmatism of "*establishments*" of all types which maintain their domination.

These "*established truths*", these dogmas that unceasingly renew themselves, can be compared to the Hydra of Greek mythology, a fabulous serpent with seven heads, each of which, when cut off, was immediately replaced by several others.

# C MY FAITH IN THE FUTURE

## *C.1 The decree of experiment*

1. - Like all new phenomena that appear inexplicable at a given moment in the framework of accepted theories, the anomalies of the paraconical pendulum with anisotropic support and with isotropic support and the optical anomalies of the deviations of sightings at marks that I demonstrated from 1954 to 1960 and in July 1958 respectively, together with the anomalies of optical sightings at marks and at collimators observed at IGN in March-April 1959, with the optical observations of Esclangon in 1927-1928, and with the interferometric observations of Miller in 1925-1926, open *new perspectives* from numerous points of view, notably in terms of the existence of an anisotropy of space that varies over time, the exploration of which, I insist, presents *exceptional scientific interest. Sooner or later experiment will yield the proof.*

From my point of view, after years and years of reflection upon the anomalies of the paraconical pendulum with anisotropic support and with isotropic support, upon the optical anomalies of the deviations of sightings at marks and at collimators, on the optical observations of Esclangon, and upon the interferometric observations of Miller, the repetition of these experiments cannot fail to provide *absolutely essential and coherent* information and answers, and to eliminate many uncertainties about the practical nature of the detected anisotropy of space, *which according to all evidence is of fundamental interest.*

But I also think that this repetition will show how *anti-scientific and obscurantist* was the opposition with which I was confronted more than thirty-five years ago. *This is a certainty.*

However, as previously in 1960, this repetition will come into collision with the blind and fierce opposition of all those who only want to base their thinking upon the credos of "*established truths*".[1]

[1] In fact, after the publication of my article *Les expériences de Dayton C. Miller 1925-1926 et la Théorie de la Relativité* ('The experiments of Dayton C. Miller, 1925-1926, and the theory of relativity') in the August-September 1996 issue of the polytechnic magazine *La Jaune et la Rouge* [55], this bitter opposition reappeared yet again.

See my next work: *L'effondrement de la théorie de la relativité. Implication irréfragable des données de l'expérience* ('The collapse of the theory of relativity. Irrefutable implication of the experimental data'), Editions Clément Juglar, 2004 [56].

## *C.2 Only the future counts*

As I have never ceased to teach in economy over many years, "*only the future counts*".

My ambition today is to publish all my works on physics, theoretical and experimental, in an easily accessible form.

It is also to make it possible for others than myself to repeat the totality of my experiments in physics in a useful manner, *with sufficient facilities and with all the means offered by today's technology*. Perhaps men as clear-sighted as Pierre Ricard was in my time will be able to make such experiments possible.

All these projects are certainly very ambitious - one may say, immense. But perhaps it will be possible to bring them to fruition.

It may seem rather astonishing, and certainly somewhat paradoxical, for me to make projects for the future at eighty-five years of age. But, in large part, one's life is what one makes of it; and perhaps the only way to remain young is to continue devising projects.

Indeed, *whatever the opposition* I encountered may have been, and however hurtful it may have been to me, *the only thing that has really counted for me is the intense satisfaction that I have derived from research, and the results that I have been able to obtain.*

## *C.3 The search for Truth*

These researches have never ceased bringing me great joy, each time new results have arrived to confirm my expectations. As Max Planck wrote once:[1]

> *"All the work and the zeal of the searcher can only appear to him to be a vain and hopeless attempt, if he does not perceive, from time to time, certain natural facts that prove irrefutably to him that the toing-and-froing of his fumbling gropings has finally made it possible for him to make one step further towards truth."*

*For me, the passion for research has been and remains a devouring passion. It has given me the greatest satisfactions of my life.*[2]

Creative imagination and constant preoccupation with synthesis have marked all my researches. Creative imagination is seeing things other than received ideas. Preoccupation with synthesis is uniting in a single coherent structure multiple elements that seem, at first sight, to be disparate or contradictory; it is the demonstration of new agreements between facts that seemed to be unconnected, regularities that were not previously noticed, and invariant relations in space and time.

Here, allow me to recall the words of Pierre Termier, a leading figure of science:[3]

> *"The joy of knowing! Many scientists have tasted it. Some of them, several times in their lives; a few even in a durable and persistent manner..."*

---

[1] Max Planck, 1941, *Initiation into Physics*, id., p. 68.

[2] As Albert Einstein wrote long ago:

> *"It is a happy destiny to be spellbound by work until one's last breath. Otherwise one would suffer too much from human stupidity and folly."*

Letter of 24 July 1949, *Correspondence with Michele Besso*, id., p. 238.

[3] Pierre Termier, *La joie de connaître* (The joy of knowledge), Desclée de Brouwer, 1925, pp. 15-16 [260].

*"The joy of knowledge has consoled them so marvelously for mediocrity, incomprehension, contradiction, hostile idiocy...*

*"To know is one of the reasons for living, and there is no satisfaction that can be compared to that given by scientific research...*

*"The seeker finds immense joys of which other men know nothing... The joy of the scientist or the philosopher, the joy of the artist or the poet. Trying to express them in words is somewhat vain. They are unutterable..."*

As I wrote once:[4]

*"Certainly, nothing is comparable to the inextinguishable passion for research ; to the unquenchable desire to know, understand, clarify, explain ; to the constant will to persist in overcoming every difficulty wherever it is encountered, never to be content with approximation ; to the permanent concern to never lose sight of the whole ; to constantly think about the synthesis.*

*"In reality, nothing comes close to the satisfaction of this construction, the ineffable euphoria of innovation and discovery."*

My faith in the future has only one foundation, which nevertheless is *unshakeable and unassailable*: that *sooner or later the facts will finish by winning out over the theories that deny them. Science is a perpetual becoming. In the end, science will always sweep away "established truths"*. And it is the future that finally will judge works and men.

---

[4] *La philosophie de ma vie*, Review of Political Economy, January-February 1989, p. 51 [49] (French version of my article *My Life Philosophy*, The American Economist, 1989 [50], a text written *before* I received the Nobel Prize in Economic Science in 1988).

Like Kepler long ago,[5] I can say at the end of this work:

> *"I have written my book: it matters little whether it is read by the present age or by posterity. It can wait for its reader."*

*But in the end it will be read, and my contributions will then be fully recognized.*

Saint-Cloud

27 February 1997

[5] Pierre Rousseau, 1945, id., p. 193.

# REFERENCES OF WORKS CITED

# References


1. *Collection de mémoires relatifs à la physique*, Gauthier-Villars, 1889.
2. *Smithsonian Physical Tables*, The Smithsonian Institution, ninth revised edition ed., 1956.
3. M. ALLAIS, *A la recherche d'une discipline économique – Première partie : L'économie pure*, 1943.
4. ———, *Economie et Intérêt*, 1947.
5. ———, *Théorie du pendule conique*. 1955.
6. ———, *Explication des cycles économiques par un modèle non linéaire à régulation retardée*, Metroeconomica, 1956.
7. ———, *Explication des cycles économiques par un modèle non linéaire à régulation retardée, Mémoire complémentaire*, Les Modèles dynamiques en Economie, CNRS, 1956.
8. ———, *Théorie du pendule paraconique*. 1956.
9. ———, *Analyse harmonique des mouvements du pendule paraconique*, Comptes Rendus de l'Académie des Sciences de Paris, 245 (1957), pp. 1875–1879.
10. ———, *Application du test de Schuster généralisé à l'analyse harmonique des azimuts du pendule paraconique*, Comptes Rendus de l'Académie des Sciences de Paris, 245 (1957), pp. 2467–2470.
11. ———, *Mouvement du pendule paraconique et éclipse totale de Soleil du 30 juin 1954*, Comptes Rendus de l'Académie des Sciences de Paris, 245 (1957), pp. 2001–2003.
12. ———, *Observation des mouvements du pendule paraconique*, Comptes Rendus de l'Académie des Sciences de Paris, 245 (1957), pp. 1697–1700.
13. ———, *Sur une interprétation possible des anomalies de la gravité et ses applications*. Nov. 1957.
14. ———, *Sur une interprétation possible du champ magnétique terrestre*. Oct. 1957.
15. ———, *Sur une solution de l'équation aux dérivées partielles ... et sur une interprétation de la constance de la vitesse de la lumière*. Nov. 1957.
16. ———, *Test de Périodicité - Généralisation du test de Schuster au cas de séries temporelles autocorrelées*, Comptes Rendus de l'Académie des Sciences, 244 (1957), pp. 2469–2471.
17. ———, *Théorie du pendule paraconique et influence lunisolaire*, Comptes Rendus de l'Académie des Sciences de Paris, 245 (1957), pp. 2170–2173.
18. ———, *Anomalies des opérations de triangulation et de nivellement. Explication possible et confrontation avec l'expérience*. May 1958.
19. ———, *Anomalies du mouvement du pendule paraconique à support anisotrope*. May 1958.
20. ———, *Application du Théorème de Bour au cas des mouvements terrestres dans le cas le plus général*. Mar. 1958.
21. ———, *Corrections de deuxième approximation du pendule paraconique à support anisotrope et à petites oscillations*. Aug. 1958.
22. ———, *Corrections de première approximation du pendule paraconique à support anisotrope et à petites oscillations*. Aug. 1958.
23. ———, *Doit-on reconsidérer les lois de la gravitation ? (Should the Laws of Gravitation Be Reconsidered?)*, Perspectives X, (1958), pp. 90–106.
24. ———, *Equations du mouvement du pendule paraconique à support anisotrope à petites oscillations. Première et deuxième approximation*. Aug. 1958.
25. ———, *Nouvelles expériences sur le pendule paraconique à support anisotrope*, Comptes Rendus de l'Académie des Sciences de Paris, 247 (1958), pp. 1428–1431.
26. ———, *Recherches théoriques et expérimentales nouvelles sur la Gravitation, Mémoire présenté à la Société Française d'Aéronautique, Prix Galabert 1959*. Dec. 1958.
27. ———, *Structure périodique des mouvements du pendule paraconique à support anisotrope à Bougival et Saint-Germain, en juillet 1958.*, Comptes Rendus de l'Académie des Sciences de Paris, 247 (1958), pp. 2284–2287.
28. ———, *Structure périodique des mouvements du pendule paraconique à suspension anisotrope et influence lunisolaire. Résultats expérimentaux et anomalies. Extraits des Comptes rendus des séances de l'Académie des sciences*, Gauthier-Villars, 1958.

29. ———, *Variations des paramètres osculateurs de l'ellipse décrite dans le mouvement du pendule paraconique à support anisotrope et à petites oscillations. Première et deuxième approximation.* Aug. 1958.
30. ———, *Détermination expérimentale de l'influence de l'anisotropie du support sur le mouvement du pendule paraconique à support anisotrope*, Comptes Rendus de l'Académie des Sciences de Paris, 248 (1959), pp. 764–767.
31. ———, *Détermination expérimentale de l'influence de l'inclinaison de la surface portante sur le mouvement du pendule paraconique à support anisotrope*, Comptes Rendus de l'Académie des Sciences de Paris, 248 (1959), pp. 359–362.
32. ———, *New Theoretical and Experimental Research Work on Gravity, memoir for the annual essay competition of the Gravity Research Foundation.* 1959.
33. ———, *Pendule Paraconique à Suspension Isotrope. Détermination des Variations au Cours du Temps des Caractéristiques de la Corrélation du Mouvement avec l'Azimut.* Nov. 1959.
34. ———, *Pendule Paraconique. Réalisation d'un Support aussi Isotrope que Possible.* Nov. 1959.
35. ———, *Should the Laws of Gravitation be Reconsidered? Complementary Note*, Aero/Space Engineering, 18 (1959), p. 55.
36. ———, *Should the Laws of Gravitation be Reconsidered? Part I-Abnormalities in the Motion of a Paraconical Pendulum on an Anisotropic Support*, Aero/Space Engineering, 18 (1959), pp. 46–52.
37. ———, *Should the Laws of Gravitation be Reconsidered? Part II-Experiments in Connection with the Abnormalities Noted in the Motion of the Paraconical Pendulum with an Anisotropic Support*, Aero/Space Engineering, 18 (1959), pp. 51–55.
38. ———, *Existence de composantes périodiques dans les variations des lectures correspondant aux visées effectuées avec une lunette fixe sur une mire fixe en liaison avec les mouvements du pendule paraconique.* Feb. 1960.
39. ———, *Interprétation des anomalies de la pesanteur comme un effet d'écran des actions gravifiques.* Mar. 1960.
40. ———, *Test de Périodicité: généralisation du test de Schuster au cas de séries temporelles autocorrelées dans l'hypothèse d'un processus de perturbations aléatoires d'un système stable*, in Communication au 33e Congrès de l'Institut International de Statistique, Paris, 1961, Institut International de Statistique. Bulletin de l'ISI, 1962, vol. 39, pp. 143–144.
41. ———, *Les périodicités constatées dans le mouvement du pendule paraconique sont-elles réelles ou non? Généralisation du test de Schuster au cas de séries temporelles autocorrélées*, in Bulletin du Cercle Alexandre Dufour, vol. 120–122, 1967, pp. 80–97, 107–124, and 130–132.
42. ———, *L'économique en tant que science*, Revue d'Economie Politique, Droz, Jan. 1968. pp. 5-30.
43. ———, *Fréquence, probabilité et hasard*, Journal de la Société Statistique de Paris, 124 (1983), pp. 70–102, and 144–221.
44. ———, *Frequency, Probability and Chance*, in The Foundations of the Theory of Utility and Risk with Applications, Bernd F. Stigum and Fred Wenstop ed., D. Reidel Publishing Company, Dordrecht, 1983, pp. 35–84.
45. ———, *Sur la distribution normale des valeurs à des instants régulièrement espacés d'une somme de sinusoïdes*, Comptes Rendus de l'Académie des Sciences de Paris, 296 (1983), pp. 829–830.
46. ———, *Analyse des mouvements du pendule paraconique à suspension isotrope. 20 novembre 1959, 7h. 20 mn. - 15 décembre 1959, 15 h. 20 mn.* June 1985.
47. ———, *Phénomènes Aléatoires et Modèles Fréquentiels. Réalité et Théorie. Prolégomènes pour une Révision des Théories admises*, 1988.
48. ———, *Autoportraits : une vie, une oeuvre*, Montchrestien, 1989.
49. ———, *La Philosophie de ma vie*, Revue d'Economie politique, 99 (1989), pp. 28–54.
50. ———, *My life philosophy*, The American Economist, (1989), pp. 3–17.
51. ———, *Analyse des Observations de l'Institut Géographique National. Février-Mars 1959.* Apr. 1991.

52. ———, *Note sur les moyens nécessaires pour la reprise des expériences optiques avec les lunettes NA*. 1994.
53. ———, *Traité d'Economie Pure*, Clément Juglar, Paris, 1994.
54. ———, *Analyse Empirique et Théorique des Anomalies du Pendule Paraconique*. June 1995.
55. ———, *Les expériences de Dayton C. Miller, 1925-1926, et la théorie de la relativité*, La Jaune et la Rouge, (1996), pp. 29–37.
56. ———, *L'effondrement de la théorie de la relativité. Implication irréfragable des données de l'expérience*, Clément Juglar, 2004.
57. A. M. AMPÈRE, *Mémoire sur la Théorie des phénomènes électro-dynamiques, uniquement déduite de l'expérience, 1820-1825*, vol. III of Collection de Mémoires relatifs à la Physique publiés par la Société Française de Physique, Gauthier-Villars, 1887.
58. A. A. ANDRONOV AND S. E. KHAIKIN, *Theory of Oscillations*, Princeton University Press, 1949.
59. A. A. ANDRONOV, A. A. VITT, AND S. E. KHAIKIN, *Theory of Oscillators*, International series of monographs in physics, Pergamon Student Edition, 1966.
60. L. BARNETT, *Einstein et l'univers*, Gallimard, 1952.
61. J. BARTELS AND S. CHAPMAN, *Geomagnetism*, The International series of monographs on physics, Clarendon Press, 1940.
62. E. BAUER, *Champs de vecteurs et de tenseurs*, Masson, 1955.
63. E. BELL, *Les grands mathématiciens*, Bibliothèque scientifique Payot, Payot, 1939.
64. ———, *La Mathématique Reine et Servante des Sciences*, Payot, 1953.
65. E. BELOT, *Formule applicable aux durées de rotation directe des planètes et du soleil*, Comptes Rendus de l'Académie des Sciences de Paris, 143 (1906), p. 1126.
66. ———, *Sur la distance des satellites d'Uranus et de Jupiter*, Comptes Rendus de l'Académie des Sciences de Paris, (1907), p. 885.
67. ———, *La Naissance de la terre et de ses satellites. Leur évolution cosmique*, Gauthier-Villars, 1931.
68. P. G. BERGMANN, *Relativity*, Encyclopaedia Britannica, XV (1975), pp. 581–589.
69. C. BERNARD, *Introduction à l'étude de la médecine expérimentale*, Baillière, 1865.
70. R. BIRKELAND, *XII. An attempt to explain the Michelson interference-experiment*, The London, Edinburgh, and Dublin Philosophical Magazine and Journal of Science, 37 (1919), pp. 150–156.
71. M. A. BLAGG, *On a suggested substitute for Bode's Law*, Monthly Notices of the Royal Astronomical Society, 73 (1913), p. 414.
72. N. BOHR, *Atomic physics and human knowledge*, Science editions, Wiley, 1958.
73. M. BORN, *La théorie de la relativité d'Einstein et ses bases physiques: exposé élémentaire*, Gauthier-Villars, 1923.
74. ———, *L'expérience et la théorie en physique*, Gauthier-Villars, 1955.
75. ———, *Atomic Physics*, Blackie, 1956.
76. H. BOUASSE, *Pendule, spiral, diapason*, no. v. 2 in Bibliothèque scientifique de l'ingénieur et du physicien, Delagrave, 1920.
77. ———, *Propagation de la lumière: théorie de la réflexion vitreuse et métallique (Propagation of light: theory of the vitreous and metallic reflection)*, Delagrave, Paris, 1925.
78. E. BOUR, *Mémoire sur les mouvements relatifs*, Journal de mathématiques pures et appliquées, (1863), pp. 1–51.
79. P. BOURGEOIS AND J. F. COX, *Sur la répartition des inclinaisons et des excentricités des orbites des petites planètes*, Comptes Rendus de l'Académie des Sciences de Paris, 198 (1934), p. 53.
80. R. BRICARD, *Calcul vectoriel (Vector calculus)*, vol. 112, Armand Colin, Paris, 1929.
81. P. BRICOUT, *Ondes et électrons*, tech. report, Armand Colin, 1929.
82. ———, *Microénergétique*, vol. I, Gauthier-Villars, 1933.
83. L. BRILLOUIN, *Les Tenseurs en mécanique et en élasticité*, Masson, 1938.
84. W. BROAD AND N. WADE, *La souris truquée, enquête sur la fraude scientifique*, Editions du Seuil, 1987.

85. L. D. BROGLIE, *Eléments de théorie des quanta et de mécanique ondulatoire*, no. v. 3 in Traité de physique théorique et de physique mathématique, Gauthier-Villars, 1953.
86. ———, *La Physique quantique restera-t-elle indéterministe? (Will quantum physics remain indeterministic?)*, Gauthier-Villars, 1954.
87. ———, *Jalons pour une nouvelle microphysique*, Collection Discours de la méthode, Gauthier-Villars, 1977.
88. G. BRUHAT AND A. FOCH, *Mecanique: 6e édition revue et completée*, Cours de physique generale, 3, Masson, 1967.
89. E. BRYLINSKI, *Sur la vitesse de la Terre (On the speed of the Earth)*, Comptes Rendus de l'Académie des Sciences de Paris, 185 (1927), pp. 1458–1459.
90. F. BUTAVAND, *Les lois empiriques du système solaire et les harmoniques tourbillonaires*, Gauthier-Villars, 1913.
91. E. CARVALLO, *La théorie d'Einstein démentie par l'expérience*, Etienne Chiron, 1934.
92. J. CHAZY, *La théorie de la relativité et la mécanique céleste*, Gauthier-Villars, 1928.
93. J. CHEVRIER, *Observations du champ électrique de l'air à l'Observatoire de Ksara (Liban) pendant l'éclipse de Soleil du 21 août 1933*, Comptes Rendus de l'Académie des Sciences de Paris, 197 (1933), pp. 1143–1145.
94. O. CHWOLSON, *Traité de physique: Champ magnétique variable*, no. v. 5, pt. 1 in Traité de physique, A. Hermann, 1914.
95. O. COSTA DE BEAUREGARD, *La Théorie de la relativité restreinte*, Collection d'Ouvrages de Mathématiques à l'Usage des Physiciens, Masson, 1949.
96. H. CRAMER, *Mathematical methods of statistics*, vol. 9, Princeton University Press, 1946.
97. J. G. CROWTHER, *Michael Faraday*, Hermann, 1945.
98. A. DANJON, *Astronomie générale: astronomie sphérique et éléments de mécanique céleste*, J. & R. Sennac, 1952.
99. G. DARMOIS, *La théorie einsteinienne de la gravitation: les vérifications expérimentales*, Actualités scientifiques et industrielles, Hermann, 1932.
100. G. DARRIEUS, *Sur une relation entre la constante de la gravitation et les autres constantes fondamentales*, Comptes Rendus de l'Académie des Sciences de Paris, (1926).
101. N. D. DE FONROQUE, *Du Pendule, théorie de ses variations*, Chamerot, 1879.
102. ———, *Sur diverses expériences faites avec un pendule oscillant avec de grandes amplitudes*, Comptes Rendus de l'Académie des Sciences de Paris, 88 (1879), p. 771.
103. L. DECOMBE, *Théorie électronique de la gravitation*, Comptes Rendus de l'Académie des Sciences de Paris, (1913).
104. ———, *Les pellicules sphériques électrisées. Calcul direct de la constante de la gravitation en fonction des constantes d'Avogrado, de Faraday, de Rydberg et de Planck*, Comptes Rendus de l'Académie des Sciences de Paris, (1924).
105. ———, *Les pellicules électrisées et les séries spectrales*, Comptes Rendus de l'Académie des Sciences de Paris, (1927).
106. J. F. DELAUNEY, *Lois des distances des satellites du soleil*, Gauthier-Villars, Paris, 1909.
107. ———, *Problèmes astronomiques*, Gauthier-Villars, Paris, 1920.
108. L. DEMOZAY, *Relations remarquables entre les éléments au système solaire*, Gauthier-Villars, Paris, 1919.
109. J. DESTOUCHES, *Physique moderne et philosophie*, Actualités scientifiques et industrielles. 847, Hermann, 1939.
110. A. DETOEUF, *Propos de O. L. Barenton, Confiseur*, Ed. du Tambourinaire, 1938.
111. P. DEVAUX, *Manifeste scientifique à Polytechnique*, Le Figaro, (1958).
112. P. DIVE, *Ondes ellipsoïdales et relativité*, Gauthier-Villars, 1950.
113. H. C. DUDLEY, *New Principles in Quantum Mechanics*, Exposition Press, N. Y., (1959).
114. R. DUGAS, *Histoire de la Mécanique (A History of Mechanics)*, Bibliothèque scientifique (Neuchâtel, Suisse). Philosophie et histoire, Dunod, Paris, 1950.
115. ———, *L'éther optique et gravifique au sens de Newton*, Revue scientifique, (1952), pp. 85–90.
116. ———, *La mécanique au XVII siècle: des antécédents scolastiques à la pensée classique*, Bibliothèque scientifique, Dunod, 1954.

117. P. DUHEM, *Les théories électriques de J. Clerk Maxwell: Etude historique et critique*, A. Hermann, 1902.
118. ———, *Le système du monde: Histoire des doctrines cosmologiques de Platon à Copernic*, A. Hermann, 1959.
119. E. DURAND, *Magnétostatique*, Masson et Cie, 1968.
120. ———, *Mécanique quantique: Equation de Schrödinger*, Masson, 1970.
121. A. S. EDDINGTON, *The Mathematical Theory of Relativity*, Cambridge University Press, 1923.
122. ———, *The Nature of the Physical World: Gifford Lectures (1927)*, Cambridge University Press, 1927.
123. A. EINSTEIN, *Zur Elektrodynamik bewegter Körper (On the Electrodynamics of Moving Bodies)*, Annalen der Physik, 322 (1905), pp. 891–921.
124. ———, *L'Ether et la théorie de la relativité*, Gauthier-Villars, 1921.
125. ———, *Quatre conférences sur la théorie de la relativité*, Discours de la Méthode, Gauthier-Villars, 1925.
126. ———, *Sur l'électrodynamique des corps en mouvement*, Gauthier-Villars, 1925.
127. ———, *Les fondements de la théorie de la relativité générale ; Théorie unitaire de la gravitation et de l'électricité ; Sur la structure cosmologique de l'espace*, Hermann, 1933.
128. ———, *La Géométrie et l'Expérience*, Gauthier-Villars, 1934.
129. ———, *Comment je vois le monde (How I see the world)*, Flammarion, 1939.
130. ———, *Sur le problème cosmologique: théorie de la gravitation généralisée*, Gauthier-Villars, 1951.
131. ———, *Conceptions scientifiques, morales et sociales*, Flammarion, 1952.
132. ———, *La Relativité et le Problème de l'Espace*, Gauthier-Villars, 1954.
133. ———, *La Théorie de la Relativité Restreinte et Générale, Exposé Elémentaire*, Gauthier-Villars, 1954.
134. ———, *La théorie de la relativité restreinte et générale*, Gauthier-Villars, 1976.
135. ———, *Correspondance avec Michele Besso*, Hermann, Paris, 1979.
136. A. EINSTEIN AND L. INFELD, *The Evolution of Physics*, Simon and Schuster, 1938.
137. E. ESCLANGON, *La dissymétrie de l'espace sidéral et le phénomène des marées*, Comptes Rendus de l'Académie des Sciences de Paris, 183 (1926), pp. 116–118.
138. ———, *Sur la dissymétrie mécanique et optique de l'espace en rapport avec le mouvement absolu de la Terre.*, Comptes Rendus de l'Académie des Sciences de Paris, 182 (1926), pp. 921–923.
139. ———, *Sur la dissymétrie optique de l'espace et les lois de la réflexion*, Comptes Rendus de l'Académie des Sciences de Paris, 185 (1927), pp. 1593–1595.
140. ———, *Sur l'existence d'une dissymétrie optique de l'espace*, Journal des Observateurs, 11 (1928), p. 49.
141. ———, *Recherches expérimentales sur la dissymétrie optique de l'espace*, Comptes Rendus de l'Académie des Sciences de Paris, 183 (1935), pp. 1165–1168.
142. F. FER, *L'irréversibilité: fondement de la stabilité du monde physique*, Gauthier-Villars, 1977.
143. P. FEVRIER, *L'interprétation physique de la mécanique ondulatoire et des théories quantiques*, Gauthier-Villars, 1956.
144. R. P. FEYNMAN, *The concept of probability in quantum mechanics*, in Proceedings of the Second Berkeley Symposium on Mathematical Statistics and Probability, The Regents of the University of California, 1951.
145. ———, *The character of physical law*, MIT Press, 1965.
146. H. FRANÇOIS, *Faut-il reconsidérer les lois de la gravitation ?*, Le Monde, (1958).
147. A. FRESNEL, *Note additionnelle à la lettre à M. Arago*, Annales de Chimie et de Physique, 9 (1818), pp. 286–287.
148. G. GALILEI, *Opere VI*, Florence, 1842.
149. L. GAUSSIN, *Lois concernant la distribution des astres du système solaire*, Comptes Rendus de l'Académie des Sciences de Paris, 90 (1880), p. 518.

150. ———, *Lois concernant la distribution des astres du système solaire*, Comptes Rendus de l'Académie des Sciences de Paris, 90 (1880), p. 593.
151. P. GILBERT, *Mémoire sur l'application de la méthode de Lagrange à divers problèmes de mouvement relatif*, Gauthier-Villars et fils, 1889.
152. J. GOGUEL, *Observations à propos du pendule dit paraconique*, Comptes Rendus de l'Académie des Sciences, 246 (1958), p. 2340.
153. G. GUINIER, *Eléments de physique moderne théorique, préface de Louis de Broglie*, Bordas, 1949.
154. J. HAAG, *Les mouvements vibratoires*, vol. 1, Presses universitaires de France, 1952.
155. O. HEAVISIDE, *On the electromagnetic effects due to the motion of electrification through a dielectric*, The London, Edinburgh, and Dublin Philosophical Magazine and Journal of Science, 27 (1889), pp. 324–339.
156. G. W. F. HEGEL AND A. VERA, *Philosophie de la nature de Hegel*, vol. 1, E. Martinet, 1863.
157. J. HÉLY, *Les fondements théoriques de l'électrodynamique: conférences professées à l'Ecole nationale supérieure du génie maritime*, Les Collections de Montligeon, Editions de Montligeon, 1950.
158. G. HOLTON, *Einstein and the "Crucial" Experiment*, American Journal of Physics, 37 (1969), pp. 968–982.
159. W. S. JEVONS, *The theory of political economy*, Macmillan and co., 1871.
160. G. JOOS, *Die Jenaer Wiederholung des Michelsonversuchs (The Jena Repetition of the Michelson-Morley Experiment)*, Annalen Der Physik, 7 (1930), pp. 385–407.
161. T. KAHAN, *Sur les origines de la théorie de la relativité restreinte*, Revue d'histoire des sciences et de leurs applications, 12 (1959), pp. 159–165.
162. A. KASTLER, *Optique*, Masson, 1959.
163. R. J. KENNEDY, *A refinement of the Michelson-Morley experiment*, Proceedings of the National Academy of Sciences of the United States of America, 12 (1926), p. 621.
164. A. KOESTLER, *Les somnambules : essai sur l'histoire des conceptions de l'univers*, Calmann-Lévy, 1960.
165. A. KOYRÉ, *La Révolution astronomique: Copernic, Kepler, Borelli*, Hermann, 1961.
166. N. KRYLOV AND N. N. BOGOLYUBOV, *Introduction to Non-linear Mechanics*, Annals of mathematics studies, Princeton University Press, 1943.
167. G. LACOSTE, *Les progrès dans nos connaissances physiques des champs ouvrent la voie à de passionnantes recherches*, Sciences et Avenir, (1958), p. 272.
168. P. LANGEVIN, *L'oeuvre d'Einstein et l'astronomie*, Bulletin de la Société Astronomique de France, (1931).
169. P. S. LAPLACE, *Essai philosophique sur les probabilités*, Les maîtres de la pensée scientifique, Gauthier-Villars, 1921.
170. R. LATHAM, *An Interim report on a repeat of the Allais Experiment - the measurement of the rate of increase of the minor axis of a Foucault pendulum using automatic apparatus*, Imperial College Report ICR G 28, Imperial College, London, Jan. 1980.
171. P. D. LATIL, *Le pendule fatal aux lois de la mécanique*, Le Figaro Littéraire, (1968).
172. R. LENNUIER, *La théorie de la relativité restreinte et l'expérience*, 1955.
173. J. LEVEUGLE, *Poincaré et la relativité*, La jaune et la rouge, 494 (1994), pp. 31–51.
174. ———, *A propos de Poincaré et la Relativité*, La jaune et la rouge, 494 (1994), pp. 7–14.
175. M. LÉVY, *Sur l'application des lois électrodynamiques au mouvement des planètes*, Comptes Rendus de l'Académie des Sciences, 110 (1890), pp. 545–551.
176. A. LICHNEROWICZ, *Algèbre et analyse linéaires*, Coll. d'ouvrages de mathématiques, à l'usage des physiciens, publ. sous la dir. de G. Darmois et de A. Lichnerowicz, Masson, 1947.
177. ———, *Eléments de calcul tensoriel*, Collection Armand Colin, Colin, 1951.
178. ———, *Théories relativistes de la gravitation et de l'électromagnétisme*, Masson, 1955.
179. M. LION, *Observations relatives à l'intensité du magnétisme terrestre, faites à Beaune, pendant l'éclipse du 28 juillet*, Comptes Rendus de l'Académie des Sciences de Paris, 33 (1851), pp. 202–204.

180. ———, *Sur des changements de l'intensité magnétique coïncidant avec la durée d'une éclipse*, Comptes Rendus de l'Académie des Sciences de Paris, 34 (1852), pp. 202–204.
181. M. Lion and D. Diamilla Muller, *Observations relatives à l'action des conjonctions écliptiques sur les éléments du magnétisme terrestre.*, Comptes Rendus de l'Académie des Sciences de Paris, 74 (1872), pp. 199–200.
182. O. J. Lodge, *Aberration Problems. A Discussion concerning the Motion of the Ether near the Earth, and concerning the Connexion between Ether and Gross Matter; with Some New Experiments*, Philosophical Transactions of the Royal Society of London. A, 184 (1893), pp. 727–804.
183. A. C. Longden, *On the irregularities of motion of the Foucault pendulum*, Physical Review, 13 (1919), p. 241.
184. H. A. Lorentz, *Versuch einer Theorie der elektrischen und optischen Erscheinungen in bewegter Körpern*, tech. report, Leiden, 1895.
185. ———, *Electromagnetic phenomena in a system moving with any velocity smaller than that of light*, Proceedings of the Academy of Sciences of Amsterdam, (1904), pp. 809–831.
186. ———, *Deux mémoires de Henri Poincaré sur la physique mathématique*, Acta mathematica, 38 (1921), pp. 293–308.
187. H. A. Lorentz, A. Einstein, and H. Minkowski, *Das Relativitätsprinzip*, Teubner, Leipzig, 1913.
188. A. Lumière, *Les Fossoyeurs du progrès: les mandarins contre les pionniers de la science*, L. Sézanne, 1942.
189. M. Macaire, *L'univers de la relativité générale*, Voies nouvelles, May 1996.
190. C. Marchal, *The Theory of Relativity, Einstein or Poincaré*, 1995.
191. M. Mascart, *Sur les modifications qu'éprouve la lumière par suite du mouvement de la source lumineuse et du mouvement de l'observateur - Première partie (On the modifications experienced by light due to movement of the luminous source and movement of the observer - First part)*, Annales scientifiques de l'Ecole Normale Supérieure, 1 (1872), pp. 157–214.
192. ———, *Sur les modifications qu'éprouve la lumière par suite du mouvement de la source lumineuse et du mouvement de l'observateur - Deuxième partie (On the modifications experienced by light due to movement of the luminous source and movement of the observer - Second part)*, Annales scientifiques de l'Ecole Normale Supérieure, 3 (1874), pp. 363–420.
193. S. F. Mason, *Histoire des sciences*, Armand Colin, 1956.
194. J. Maxwell, *Traité d'électricité et de magnetisme (A Treatise on Electricity and Magnetism, French translation)*, vol. 2, Gauthier-Villars, 1889.
195. P. Michelmore, *Einstein*, Encyclopaedia Britannica, VI (1975), pp. 510–514.
196. A. A. Michelson, *The relative motion of the Earth and of the luminiferous ether*, American Journal of Science, XXII (1881), pp. 120–129.
197. A. A. Michelson and H. G. Gale, *The Effect of the Earth's Rotation on the Velocity of Light, II*, The Astrophysical Journal, 61 (1925), p. 140.
198. A. A. Michelson, D. C. Miller, and R. J. Kennedy, *Conference on the Michelson-Morley experiment held at the Mount Wilson observatory Pasadena, California February 4 and 5, 1927*, The Astrophysical Journal, LXVIII (1928), pp. 341–402.
199. A. A. Michelson and E. W. Morley, *On the relative motion of the Earth and the luminiferous ether*, American Journal of Science, Series 3 Vol. 34 (1887), pp. 333–345.
200. A. A. Michelson, F. G. Pease, and F. Pearson, *Repetition of the Michelson-Morley Experiment*, Nature, 123 (1929), p. 88.
201. D. C. Miller, *Ether-drift experiments at Mount Wilson*, Proceedings of the National Academy of Sciences of the United States of America, 11 (1925), pp. 306–314.
202. D. C. Miller, *Significance of the ether-drift experiments of 1925 at Mount Wilson.*, Science (New York, NY), 63 (1926), pp. 433–443.
203. ———, *The ether-drift experiment and the determination of the absolute motion of the earth*, Reviews of Modern Physics, 5 (1933), pp. 203–242.
204. N. F. Minorsky, *Introduction to Non-Linear Mechanics*, J.W. Edwards, Ann Arbor, 1947.

205. C. NORDMANN, *Recherches sur le champ électrique terrestre, exécutées à l'occasion de l'éclipse totale du 30 août 1905*, Comptes Rendus de l'Académie des Sciences de Paris, 142 (1906), pp. 40–43.
206. F. OLLIVE, *Sur le système solaire*, Comptes Rendus de l'Académie des Sciences de Paris, 157 (1913), p. 1501.
207. Y. A. ONO, *How I Created the Theory of Relativity.*, Physics Today, 35 (1982), pp. 45–47.
208. P. PAINLEVÉ, *Les Axiomes de la Mécanique (The Axioms of Mechanics)*, Gauthier-Villars, 1922.
209. ———, *Cours de mécanique*, vol. 1 of Cours de mécanique professé à l'Ecole polytechnique, Gauthier-Villars, 1930.
210. C. PALVADEAU, *Rapport sur des Expériences de Contrôle de la Stabilité des Visées Optiques*, tech. report, IGN, Aug. 1959.
211. PANISETTI, *Expériences sur les oscillations du pendule immobile*, Revue Cosmos, 8 (1856), pp. 503–504.
212. V. PARETO, *Traité de sociologie générale*, Payot, 1917.
213. ———, *Manuel d'économie politique*, 1909, Giard, 1927.
214. E. PICARD, *La Théorie de la relativité et ses applications à l'astronomie*, Gauthier-Villars, 1922.
215. L. PICART, *Astronomie générale*, Armand Colin, 1931.
216. A. PICCARD AND E. STAHEL, *L'absence de vent d'éther au Rigi (Absence of any ether wind on Rigi)*, Comptes Rendus de l'Académie des Sciences de Paris, 185 (1927), pp. 1198–1200.
217. ———, *Réalisation de l'expérience de Michelson en ballon et sur terre ferme (Implementation of Michelson's experiment in a balloon and on solid ground)*, Journal de Physique et le Radium, 9 (1928), pp. 49–60.
218. M. PLANCK, *Initiations à la physique (Introduction to Physics)*, Bibliothèque de philosophie scientifique, Flammarion, 1941.
219. H. POINCARÉ, *Les méthodes nouvelles de la mécanique céleste*, Gauthier-Villars, 1892.
220. ———, *La Science et l'Hypothèse (Science and Hypothesis)*, Flammarion, 1902.
221. ———, *L'état actuel et l'avenir de la physique mathématique (The present and the future of mathematical physics), 1904 St. Louis Conference*, Bulletin des sciences mathématiques, 28 (1904), pp. 302–324.
222. ———, *La Valeur de la Science (The Value of Science)*, Flammarion, 1905.
223. ———, *Sur la dynamique de l'electron*, Comptes Rendus de l'Académie des Sciences, 140 (1905), p. 1504.
224. ———, *Sur la dynamique de l'électron (On the dynamics of the electron), 23 July 1905*, rendiconti del Circolo mathematico di Palermo, Mathematical Circle of Palermo, 1906.
225. ———, *Science et Méthode (Science and Method)*, Flammarion, Paris, 1908.
226. ———, *Dernières pensées*, Bibliothèque de philosophie scientifique, E. Flammarion, 1913.
227. ———, *Oeuvres de Henri Poincaré*, Gauthier-Villars, 1951.
228. ———, *Electricité et optique: leçons professées à la Sorbonne en 1888, 1890 et 1899. La lumière et les théories électrodynamiques*, Cours de la Faculté des sciences de Paris, Cours de physique mathématique, Gauthier-Villars, 1954.
229. B. V. D. POL, *On "Relaxation Oscillations"*, The London, Edinburgh, and Dublin Philosophical Magazine and Journal of Science, 2 (1926), pp. 978–992.
230. K. R. POPPER, *Quantum mechanics without "the observer"*, in Quantum Theory and Reality, Springer, Bunge ed., 1967, pp. 7–44.
231. F. PRUNIER, *Essai d'une physique de l'éther*, Blanchard, 1932.
232. ———, *Quelques observations et expériences nouvelles et leurs conséquences pour les théories de la physique*, Archives des Sciences Physiques et Naturelles, Institut de Physique de l'Université de Genève, 1946.
233. Y. ROCARD, *Dynamique générale des vibrations*, Collection d'ouvrages de mathématiques à l'usage des physiciens, Masson, 1949.
234. J. ROUCH, *Observations du champ électrique de l'atmosphère, faites à Monaco pendant l'éclipse partielle du Soleil du 20 juin 1954*, Comptes Rendus de l'Académie des Sciences de Paris, 239 (1954), pp. 465–466.

235. L. ROUGIER, *Les Paralogismes du rationalisme, essai sur la théorie de la connaissance, Thèse pour le doctorat ès lettres présentée à la Faculté des lettres de l'Université de Paris*, F. Alcan, 1920.
236. ———, *Scandale à Polytechnique (Scandal at Polytechnique)*, Imprimerie des Tuileries, Paris, 1959. Reproduced in: Un Savant Méconnu - Portraits d'un Autodidacte, Editions Clément Juglar, 2002.
237. P. ROUSSEAU, *Histoire de la science*, Les Grandes études historiques, Fayard, 1945.
238. B. RUSSELL, *Sceptical Essays*, W. W. Norton & co, 1928.
239. ———, *The Scientific Outlook*, W. W. Norton & co, New York, 1931.
240. ———, *L'esprit scientifique et la science dans le monde moderne, translated from The Scientific Outlook*, Janin, 1947.
241. J. F. SAIGEY, *Petite physique du globe*, Hachette, 1832.
242. E. J. SAXL AND M. ALLEN, *1970 Solar Eclipse as "Seen" by a Torsion Pendulum*, Phys. Rev. D, 3 (1971), pp. 823–825.
243. E. SCHRÖDINGER, *Mémoires sur la mécanique ondulatoire*, Alcan, 1933.
244. P. SCHUREMAN, *Manual of harmonic analysis and prediction of tides*, Special publication, U.S. Department of Commerce, Coast and Geodetic Survey, Washington, 1941.
245. A. SESMAT, *Essai critique sur la doctrine relativiste*, vol. VII of Systèmes de référence et mouvements, Hermann, Paris, 1937.
246. ———, *Optique des Corps en Mouvement (The Optics of Moving Bodies)*, Systèmes de référence et mouvements, Hermann, Paris, 1937.
247. ———, *Théorie relativiste de la gravitation*, vol. V of Systèmes de référence et mouvements, Hermann, 1937.
248. R. S. SHANKLAND, *Conversations with Albert Einstein*, American Journal of Physics, 31 (1963), pp. 47–57.
249. R. S. SHANKLAND, S. W. MCCUSKEY, F. C. LEONE, AND G. KUERTI, *New analysis of the interferometer observations of Dayton C. Miller*, Reviews of Modern Physics, 27 (1955), p. 167.
250. W. SMART, *Celestial Mechanics*, Longmans, 1953.
251. E. SOLVAY, P. LANGEVIN, AND M. DE BROGLIE, *La théorie du rayonnement et les quanta. Rapports et discussions de la réunion tenue à Bruxelles, du 30 octobre au 3 novembre 1911*, Gauthier-Villars, 1912.
252. H. SPENCER JONES, *The rotation of the earth, and the secular accelerations of the sun, moon and planets*, Monthly Notices of the Royal Astronomical Society, 99 (1939), pp. 541–558.
253. J. STACHEL, *Einstein and ether drift experiments*, Physics Today, 40 (1987), pp. 45–49.
254. N. STOYKO, *L'attraction luni-solaire et les pendules*, Bulletin de la Société Astronomique de France, 14 (1949), p. 36.
255. R. SUDRE, *L'énigme de la gravitation*, Revue des Deux Mondes, (1958).
256. J. L. SYNGE, *Relativity: The Special Theory*, Series in physics, North-Holland Publishing, 1956.
257. ———, *The Relativistic Gas*, Series in physics, North-Holland Publishing Company, 1957.
258. P. TARDI, *Traité de Géodésie*, vol. I, Gauthier-Villars, Paris, 1951.
259. G. I. TAYLOR, *Interference fringes with feeble light*, in Proceedings of the Cambridge Philosophical Society, vol. 15, 1909, pp. 114–115.
260. P. TERMIER, *La joie de connaitre: suite de a la gloire de la terre : souvenirs d'un géologue*, Bibliothèque française de philosophie, Desclée de Brouwer, 1925.
261. J. THIBAUD, *Les caractères systématiques de la distribution dans le temps des désintégrations alpha*, Annales de Physique, 15 (1941).
262. ———, *Sur les caractères systématiques de certaines distributions*, Comptes Rendus de l'Académie des Sciences de Paris, 218 (1944), pp. 873–874.
263. ———, *La théorie des groupes et les distributions radioactives*, in Louis de Broglie: physicien et penseur, Les savants du monde / dirigée par André George, Ed. Albin Michel, 1952.
264. R. THIRY AND P. APPELL, *Eléments de calcul tensoriel: applications géométriques et mécaniques*, Traité de mécanique rationnelle, Gauthier-Villars, 1955.

265. F. TISSERAND, *Traité de mécanique céleste. 1. Perturbations des planètes d'après la méthode de la variation des constantes arbitraires*, vol. 1, Gauthier-Villars, 1889.
266. ———, *Traité de mécanique céleste. 3. Exposé de l'ensemble des théories relatives au mouvement de la lune*, vol. 3, Gauthier-Villars, 1894.
267. M.-A. TONNELAT, *Les vérifications expérimentales de la relativité générale*, Masson et Cie, 1964.
268. ———, *Ether*, Encyclopaedia Universalis, 6 (1968), pp. 655–658.
269. V. A. UGAROV, *Special Theory of Relativity*, Mir, 1974.
270. R. VALLERY-RADOT, *La Vie de Pasteur*, Flammarion, 1941.
271. H. VARCOLLIER, *Fondements de l'explication électromagnétique de la gravitation universelle*, Librairie de l'Université, 1949.
272. ———, *La Théorie de la propagation ellipsoïdale et ses possibilités. Relativité, quanta, gravitation*, Librairie de l'Université, 1949.
273. J.-P. VIGIER, *Structure des micro-objets dans l?interprétation causale de la théorie des quanta*, Grands problèmes des sciences, Gauthier-Villars, 1956.
274. J. VIGNAL, *Nivellement de précision*, Publications Techniques de l'institut Géographique National, Paris, 1948.
275. A. VILAR, *Notes sur les distances des planetes au soleil.*, Paris, Jouve & cie, 1923, 1 (1923).
276. M. VON LAUE, *La théorie de la relativité*, vol. 1 and 2 of La théorie de la relativité, Gauthier-Villars, 1924.
277. A. WEGENER, *La genèse des continents et des océans*, Collection de monographies scientifiques étrangères, Librairie Scientifique Albert Blanchard, 1924.
278. H. WEYL, *Temps, espace, matière, leçons sur la théorie de la relativité générale*, Blanchard, 1922.
279. E. WHITTAKER, *History of the Theories of Aether and Electricity, The Classical Theories*, vol. 1 of A history of the theories of aether and electricity, Thomas Nelson, 1951.
280. ———, *History of the Theories of Aether and Electricity, 1900-1926*, vol. 2 of A history of the theories of aether and electricity, Thomas Nelson, 1953.
281. V. S. M. V. D. WILLIGEN, *Le Pendule Foucault au Musée Teyler*, in Archives du Musée Teyler, vol. 1, Nijhoff, Haarlem, 1868, p. 342.

# INDEX OF NAMES

# Index


**A**
AIRY, George Biddell, 120, 126, 173, 207, 662
ALLEN, Mildred, 169
AMPERE, André-Marie, 60, 61, 652, 668
ANDRE-MARTIN, Emmanuel, 45, 64, 683
ANDRONOV, Aleksandr Aleksandrovich, 522
APPELL, Paul, 126, 200, 566
ARAGO, François, 601, 630
ARMAND, Louis, 71
AUJARD, Henry, 684
AVOGADRO, Amedeo, 542

**B**
BARNETT, Lincoln, 595, 596, 609
BARTELS, Julius, 130, 478
BAUER, Edmond, 566
BEAUMONT, Elie de, 667
BELL, Eric Temple, 565, 667, 674
BELOT, Emile, 542, 543
BEN GURION, David, 615
BENTLEY, Richard, 536
BERGERON, Paul, 66, 68, 222, 232
BERGMANN, Peter Gabriel, 550
BERNARD, Claude, 493, 494, 498, 508, 577, 609, 638, 639, 664, 675, 677, 678
BERTHAULT, Guy, 42, 66, 684
BESSO, Michele, 543, 585, 616, 638, 643, 693
BIANCHI, Luigi, 514
BIRKELAND, Richard, 422, 423, 575, 576
BLAGG, Mary Adela, 542
BOGOLYUBOV, Nikolay Nikolayevich, 522
BOHR, Niels, 523, 526, 533
BORELLI, Giovanni Alfonso, 671
BORN, Max, 35, 37, 523, 525, 526, 533, 565, 578, 589, 591, 614, 616, 640, 643, 645, 646, 679
BOUASSE, Henri, 77, 423, 550, 576, 601, 673
BOUCHER DE CREVECOEUR DE PERTHES, Jacques, 667, 668
BOUR, Edmond, 54, 119, 121, 215
BOURGEOIS, Paul, 543
BOURGEOT, Jacques, 45, 53, 64, 83, 162, 217, 221, 314, 348, 349, 683
BRAUN, Wernher von, 42, 68, 232, 233
BRICARD, Raoul, 423
BRICOUT, Pierre, 566, 589
BRILLOUIN, Léon, 566
BROAD, William, 671
BROCA, Paul, 666
BROGLIE, Louis de, 7, 8, 65, 69, 227, 228, 373, 374, 521, 523–526, 533, 534, 543, 614, 643, 678
BROUARDEL, Paul, 665
BRUHAT, Georges, 126, 578
BRUNO, Giordano, 672
BRYLINSKI, Emile, 415
BUTAVAND, Férréol, 542, 543
BUYS-BALLOT, Christophorus, 90, 96, 98, 100, 105–114, 131–133, 138, 140, 144–147, 150–153, 155, 184, 335, 348, 351–353

**C**
CANTELLI, Francesco, 528
CAQUOT, Albert, 51, 57, 64–67, 69, 76, 142, 215, 220, 232, 233, 683, 688
CARVALLO, Emmanuel, 585, 615, 641
CAVENDISH, Henry, 628
CENTRE NATIONAL DE LA RECHERCHE SCIENTIFIQUE (C.N.R.S.), 9, 66, 142, 216, 218, 220–225, 228, 230, 232, 233, 687
CERCLE ALEXANDRE DUFOUR, 52, 80, 160, 201, 215, 218, 219

CHAPMAN, Sidney, 478
CHAZY, Jean, 566, 589, 604–606
CHEVENARD, Pierre, 67, 683
CHEVRIER, Jean, 164
CHRISTINE OF LORRAINE, 669
CHRISTOFFEL, Elwin, 514, 568
CHWOLSON, Orest D., 550, 563, 564
COLBERT, Jean-Baptiste, 668
COMITE D'ACTION SCIENTIFIQUE DE LA DEFENSE NATIONALE (C.A.S.D.N.), 66, 68, 142, 218, 220–223, 232, 687
COPERNICUS, Nicolaus, 120, 126, 198, 199, 492, 556, 631, 671
CORIOLIS, Gaspard, 98, 120, 199
CORNU, Alfred, 50
COSTA DE BEAUREGARD, Olivier, 562, 588, 614, 684
COT, Donatien, 66, 67, 76, 233, 683, 688
COULOMB, Jean, 67, 230, 683
COUPRY, Jean, 64
COURRIER, Robert, 69, 227
COX, Jacques François, 543
CRAMER, Harald, 101
CROWTHER, James Gerald, 673

D
DANJON, André, 67, 202
DARMOIS, Georges, 67, 125, 566, 567, 589, 604–606, 612
DARRIEUS, Georges, 66, 76, 233, 542, 688
DARWIN, Charles, 667, 668
DARWIN, Charles Galton, 96, 184
DAVY, Humphry, 667
DECOMBE, Louis, 542
DEJEAN de FONROQUE, Numa, 50, 185, 209
DELAUNEY, Julien Félix, 542
DEMOCRITUS, 669
DEMOZAY, Lucien, 542, 543
DESCARTES, René, 506, 543
DESTOUCHES, Jean-Louis, 638
DETOEUF, Auguste, 673
DEVAUX, Pierre, 67
DIAMILLA-MULLER, Demetrio Emilio, 164
DIRAC, Paul, 523
DIVE, Pierre, 575
DUDLEY, Horace Chester, 8
DUGAS, René, 64, 65, 217, 220, 430, 506, 518, 550, 564, 605, 683
DUHEM, Pierre, 35, 37, 631
DURAND, Émile, 520, 521, 539, 628

E
EDDINGTON, Arthur, 517, 545, 565–568, 589, 594, 601, 605, 612
EINSTEIN, Albert, 37, 57, 125, 383, 430, 502, 503, 506, 507, 521, 523, 526, 534, 537, 538, 543, 550, 560, 562, 563, 565, 566, 568–571, 575–579, 583–585, 587, 589, 590, 594–597, 599, 601, 602, 604, 606, 609, 611–616, 630–632, 638, 640–643, 672, 676, 677, 679, 693
ESCLANGON, Ernest, 4, 38, 41, 56, 57, 73, 76, 86, 157, 214, 374–381, 392, 410, 427–431, 447, 450–453, 469–471, 473, 479–481, 483–487, 489–491, 494–497, 501, 504, 548, 586, 588, 592, 604, 606, 607, 618, 624, 625, 634, 640, 646, 648, 649, 651–654, 657, 658, 681, 682, 690

F
FARADAY, Michael, 7, 8, 506, 542, 543, 575, 667, 673
FENELON, 59
FER, Francis, 521, 522, 539
FERMAT, Pierre de, 667, 678
FEVRIER, Paulette, 539
FEYNMAN, Richard, 532, 533
FITZGERALD, George Francis, 556, 563, 574
FIZEAU, Hippolyte, 528, 590, 601–603, 607, 630
FOCH, Adrien, 126, 200, 578, 628
FOUCAULT, Léon, 44, 53, 54, 93–95, 98, 119, 120, 126, 127, 137, 139, 140, 164, 165, 167, 168, 172–175, 177, 178, 180–183, 185, 189, 193, 196, 199, 200, 204, 207, 209, 210, 213, 234, 243, 260, 271, 281, 320, 321, 374, 430, 492, 545, 552, 587, 620, 653
FRANCOIS, Henri, 67
FRESNEL, Augustin, 44, 506, 507, 520, 528, 531, 533, 590, 601–603, 615, 630
FRIANG, Claude, 684

G
GALABERT, 53, 66
GALE, Henry G., 545, 553, 587
GALILEI, Galileo, 102, 198, 199, 215, 553, 556, 565, 568, 578, 615, 667, 669, 670, 686
GALOIS, Evariste, 674
GAUSSIN, L., 542
GEOFFROY SAINT-HILAIRE, Etienne, 668
GILBERT, Philippe, 121
GOGUEL, Jean, 148, 229, 588

GRAVITY RESEARCH FOUNDATION, 53, 66, 233
GUERIN, Jean, 66, 222
GUINIER, Georges, 521

**H**
HAAG, Jules, 126, 323
HARVEY, William, 665
HEAVISIDE, Oliver, 631
HEGEL, Georg Wilhelm, 203
HEISENBERG, Werner, 523, 526, 616
HELY, Jean, 227, 228, 503, 511, 512, 514–516, 575, 599
HERACLITUS, 669
HOLTON, G., 570

**I**
ILLINGSWORTH, Kenneth Knight, 588
INFELD, Leopold, 506, 534, 583
INSTITUT DE RECHERCHE DE LA SIDERURGIE (I.R.S.I.D), 38, 42, 45, 52, 53, 61, 64, 66, 70, 71, 86, 92, 131, 142, 143, 148, 160, 171, 173, 177, 192, 197, 213, 216, 217, 219–225, 232, 241, 314, 332–334, 339, 341, 345, 348, 355, 359, 364, 366, 368, 371, 447–449, 471, 484, 486, 492, 496, 500, 504, 619, 620, 622, 623, 642, 646–648, 656, 683, 686, 687
INSTITUT GEOGRAPHIQUE NATIONAL (I.G.N.), 38, 64, 138, 332–334, 339, 343, 346–348, 351–354, 356–359, 361–368, 371, 372, 447–449, 471, 484, 486, 496, 497, 500, 504, 586, 622, 642, 647, 648, 656, 690
IVES, Herbert E., 603

**J**
JAMES, William, 71
JANSON, Georges, 686
JEVONS, Stanley, 36, 37, 664
JONES, H. Spencer, 202
JOOS, Georg, 415, 588
JORDAN, Pascual, 616

**K**
KAHAN, Théo, 570, 571
KAMPE DE FERIET, Joseph, 51, 65–67, 76, 113, 220, 233, 683, 688
KASSER, Michel, 333, 371, 647, 656
KASTLER, Alfred, 502
KELVIN, William THOMSON Lord, 105
KENNEDY, Roy J., 415, 588
KEPLER, Johannes, 205, 630, 669–671, 695
KHAIKIN, Semen Emmanuilovich, 522
KOESTLER, Arthur, 536, 669, 670
KOLMOGOROV, Andrey Nikolaevich, 528, 529
KOYRE, Alexandre, 671
KRYLOV, Nikolay Mitrofanovich, 522
KUERTI, Gustav, 412

**L**
LA ROCHEFOUCAULD, Gilles de, 684
LACOSTE, Georges, 81, 83
LAGRANGE, Joseph Louis, 64, 121, 125, 191, 215, 631
LAMARCK, Jean-Baptiste, 668
LANGEVIN, Paul, 614
LAPLACE, Pierre-Simon, 510–512, 523, 524, 535
LATHAM, Robert, 39, 68
LATIL, Pierre de, 67
LAVOISIER, Antoine Laurent de, 678
LE VERRIER, Urbain, 200
LEAUTE, André, 66, 67, 76, 216, 233, 683, 688
LEIBNIZ, Gottfried Wilhelm von, 678
LENNUIER, Robert, 603
LEONE, Fred C., 412
LERAY, Jean, 192, 216, 234, 339, 340, 686
LEVEUGLE, Jules, 557, 570
LEVY, Maurice, 605, 630
LICHNEROWICZ, André, 565, 566, 597
LION, Moïse, 164
LODGE, Oliver, 37, 382, 550, 556
LONGDEN, Aladine Cummings, 174, 175
LORENTZ, Hendrik Antoon, 37, 38, 383, 422, 430, 523, 538, 550, 557, 559–564, 568–577, 582, 584, 585, 589, 594, 595, 599, 603, 615
LORENZ, Ludwig, 503, 511, 512, 515, 516
LUMIERE, Auguste, 8, 236, 665–668, 673

**M**
MACAIRE, Marcel, 69, 684
MACH, Ernst, 502
MARCHAL, Christian, 570
MASCART, Eleuthère, 37, 382, 550
MASON, Stephen F., 671
MAUPAS, Charles, 684
MAXWELL, James Clerk, 37, 209, 506, 531, 533, 537, 543, 571, 572, 575, 597, 599, 600, 612
McCUSKEY, Sidney Wilcox, 412
MENDEL, Gregor, 667, 678
MICHELMORE, Peter, 550

MICHELSON, Albert A., 59, 228, 240, 382, 383, 385, 392, 412–416, 422, 423, 426, 430, 431, 455, 491, 538, 539, 545, 548, 552, 553, 556, 557, 562, 569–571, 573, 575–578, 581–585, 587, 589, 592, 593, 597, 603, 606, 615, 631, 632, 636, 641, 652, 653
MILLER, Dayton C., 4, 38, 41, 56, 57, 59, 68, 73, 76, 77, 86, 90, 157, 185, 214, 226, 228, 234, 240, 332, 374, 375, 378, 382–399, 401–403, 405–421, 423–431, 452–456, 458–470, 474–476, 479, 480, 482–491, 494–497, 501, 502, 504, 539, 541, 548, 552, 576–578, 581, 582, 584–589, 592, 593, 598, 603, 604, 606, 607, 613, 615, 617, 618, 624, 625, 631, 633–637, 639–641, 646, 648, 649, 651–653, 655, 658, 681, 682, 684, 690, 691
MINKOWSKI, Hermann, 564–566, 569
MINORSKY, Nikolai Fyodorovich, 522
MONTAIGNE, Paul-Ernest de, 68, 684
MONTEL, Paul, 638
MORLEY, Edward W., 228, 382, 385, 392, 412–416, 428, 430, 431, 455, 539, 552, 569–571, 573, 581, 584, 593, 641, 652, 653, 655

**N**

NATIONAL AERONATICS AND SPACE ADMINISTRATION (N.A.S.A.), 42, 232, 233
NEUMANN, John von, 521
NEWCOMB, Simon, 604
NEWTON, Isaac, 69, 118, 125, 185, 186, 200–203, 205, 238, 506, 507, 510, 518, 531, 533, 536, 542, 550, 553, 566, 567, 579, 597, 604, 605, 662, 679
NICOLLE, Charles, 666
NORDMANN, Charles, 164

**O**

OLLIVE, F., 542
ONO, Yoshimasa A., 570, 571

**P**

PAINLEVE, Paul, 200, 383, 539, 554, 556, 584, 602
PALVADEAU, Claude, 346–349, 354, 355, 359, 365, 647, 648
PANISETTI, 175
PARETO, Vilfredo, 7, 8, 203, 237, 238, 507, 527, 546, 637, 640, 643, 678
PASCAL, Blaise, 666
PASTEUR, Félix, 175
PASTEUR, Louis, 665, 667, 668, 673, 678
PAULI, Wolfgang, 523
PEARSON, Fred, 415
PEASE, Francis G., 415
PERARD, Albert, 66, 67, 76, 233, 683, 688
PERES, Joseph, 67, 230, 232
PERRIN, René, 67, 683
PICARD, Emile, 589
PICART, Luc, 670
PICCARD, Auguste, 415, 416, 584
PLANCK, Max, 63, 206, 498, 520, 525–527, 533, 542, 547, 548, 643, 660, 675, 677, 693
PLATO, 540, 631
POINCARE, Henri, 7, 8, 38, 63, 79, 80, 422, 423, 429, 430, 492, 522, 523, 538, 539, 546, 550, 552–554, 556–562, 564, 569–576, 589, 597, 600, 611, 615, 616, 638, 643, 688
POISSON, Siméon Denis, 510–512
POL, Balthasar van der, 522
POPPER, Karl, 526
POUCHET, 666
PRUNIER, Fernand, 542, 543
PTOLEMY, 492

**R**

RABELAIS, 72
RESAL, Henri, 126
RICARD, Pierre, 45, 64, 65, 217, 220, 222, 223, 683, 692
RICCI-CURBASTRO, Gregorio, 514
RICHET, Charles, 665
RIEMANN, Bernhard, 514, 568, 630, 631
ROCARD, Yves, 522
ROLLAND, Annie, 45, 53, 64, 217, 221, 314, 683
ROUCH, Jules, 164
ROUGIER, Louis, 65, 71, 234, 339, 639, 667, 674
ROUSSEAU, Pierre, 665, 672, 695
ROY, Maurice, 66, 67, 76, 233, 683, 688
RUSSELL, Bertrand, 331, 332, 526, 609, 616, 659–661, 670
RYDBERG, Johannes, 542

**S**

SAIGEY, Jacques Frédéric, 662, 663
SAXL, Erwin J., 169
SCHRÖDINGER, Erwin, 520, 521, 523, 525, 526, 533, 534, 539
SCHUREMAN, Paul, 115, 135, 185, 187, 272, 273, 480

SCHUSTER, Arthur, 51, 52, 55, 65, 80, 103, 113, 116, 117, 154–156, 172, 218
SCHWARZSCHILD, Karl, 125, 566, 567, 604
SESMAT, Augustin, 37, 382, 385, 430, 550, 560, 584, 601, 606, 611, 641
SHANKLAND, Robert S., 412, 413, 570, 571, 576, 585
SMART, William Marshall, 203, 567, 670
SOCIETE FRANCAISE D'ASTRONAUTIQUE, 53, 66, 233
SOCIETE FRANCAISE DE PHYSIQUE, 174
STACHEL, John, 570
STAHEL, Emile, 415, 416, 584
STILWELL, G. R., 603
STOYKO, Nicolas, 135, 200
SUDRE, René, 67
SYNGE, John Lighton, 538, 539

T

TAILLE, Renaud de la, 570
TARDI, Pierre, 66, 67, 76, 142, 220, 233, 332, 339, 346, 365, 683, 686, 688
TAYLOR, Geoffrey Ingram, 533
TERMIER, Pierre, 693
THIBAUD, Jean, 533
THIRY, René, 66, 67, 76, 216, 233, 566, 683, 688
TISSERAND, Félix, 125, 203
TOCQUEVILLE, Alexis de, 77
TONNELAT, Marie-Antoinette, 506, 589, 605, 606

U

UGAROV, Vladimir Aleksandrovich, 538, 578, 599

V

VALLERY-RADOT, René, 673
VARCOLLIER, Henri, 542, 575
VARLAN, G., 191
VIGIER, Jean-Pierre, 533, 539, 543
VIGNAL, Jean, 332
VILAR, Albert, 542
VILLAT, Henri, 54, 148, 191, 192, 215
VILLEMUR, Alain and Anne-Marie, 75
VITT, Aleksandr Adol'fovich, 522
VOLET, 203
VON LAUE, Max, 565, 589, 599

W

WADE, Nicolas, 671
WALRAS, Léon, 678
WEBER, Wilhelm Eduard, 630
WEGENER, Alfred, 667
WEYL, Hermann, 565, 589, 614
WHEWELL, William, 662
WHITTAKER, Edmund, 37, 38, 506, 511, 536, 550, 556, 558, 564, 631
WIENER, Norbert, 667
WILLIGEN, Volkert van der, 53
WOLD, Herman, 530
WOLF, Rudolf, 130

Y

YOUNG, Thomas, 532

Z

ZEEMAN, Peter, 550
ZHOU, S. W., 68

## Structures éditoriales du groupe L'Harmattan

**L'Harmattan Italie**
Via degli Artisti, 15
10124 Torino
harmattan.italia@gmail.com

**L'Harmattan Hongrie**
Kossuth l. u. 14-16.
1053 Budapest
harmattan@harmattan.hu

---

**L'Harmattan Sénégal**
10 VDN en face Mermoz
BP 45034 Dakar-Fann
senharmattan@gmail.com

**L'Harmattan Cameroun**
TSINGA/FECAFOOT
BP 11486 Yaoundé
inkoukam@gmail.com

**L'Harmattan Burkina Faso**
Achille Somé – tengnule@hotmail.fr

**L'Harmattan Guinée**
Almamya, rue KA 028 OKB Agency
BP 3470 Conakry
harmattanguinee@yahoo.fr

**L'Harmattan RDC**
185, avenue Nyangwe
Commune de Lingwala – Kinshasa
matangilamusadila@yahoo.fr

**L'Harmattan Congo**
67, boulevard Denis-Sassou-N'Guesso
BP 2874 Brazzaville
harmattan.congo@yahoo.fr

**L'Harmattan Mali**
Sirakoro-Meguetana V31
Bamako
syllaka@yahoo.fr

**L'Harmattan Togo**
Djidjole – Lomé
Maison Amela
face EPP BATOME
ddamela@aol.com

**L'Harmattan Côte d'Ivoire**
Résidence Karl – Cité des Arts
Abidjan-Cocody
03 BP 1588 Abidjan
espace_harmattan.ci@hotmail.fr

**L'Harmattan Algérie**
22, rue Moulay-Mohamed
31000 Oran
info2@harmattan-algerie.com

**L'Harmattan Maroc**
5, rue Ferrane-Kouicha, Talaâ-Elkbira
Chrableyine, Fès-Médine
30000 Fès
harmattan.maroc@gmail.com

---

## Nos librairies en France

**Librairie internationale**
16, rue des Écoles – 75005 Paris
librairie.internationale@harmattan.fr
01 40 46 79 11
www.librairieharmattan.com

**Librairie l'Espace Harmattan**
21 bis, rue des Écoles – 75005 Paris
librairie.espace@harmattan.fr
01 43 29 49 42

**Lib. sciences humaines & histoire**
21, rue des Écoles – 75005 Paris
librairie.sh@harmattan.fr
01 46 34 13 71
www.librairieharmattansh.com

**Lib. Méditerranée & Moyen-Orient**
7, rue des Carmes – 75005 Paris
librairie.mediterranee@harmattan.fr
01 43 29 71 15

**Librairie Le Lucernaire**
53, rue Notre-Dame-des-Champs – 75006 Paris
librairie@lucernaire.fr
01 42 22 67 13

www.ingramcontent.com/pod-product-compliance
Lightning Source LLC
LaVergne TN
LVHW050050061026
849532LV00001BA/9
* 9 7 8 2 3 4 3 1 6 8 7 0 8 *